Disaster Resilience and Green Growth

Over the years, the relationship between environment and disasters has received significant attention. This is largely due to the emerging recognition that environmental changes - climate change, land-use and natural resource degradation make communities more vulnerable to disaster impacts. There is a need to break this nexus through environment based and sustainability inclusive interventions. Science – technology and economic measures for disaster risk management, hence, need to adapt more integrated approaches for infrastructure and social resilience. Environmental and anthropogenic factors are key contributors to hazard, risk, and vulnerability and, therefore, should be an important part of determining risk-management solutions.

Green growth approaches have been developed by emphasizing sustainability inclusion and utilizing the benefits of science-technology interventions along policy-practice linkages with circular economy and resource efficiency. Such approaches recognize the perils of traditional material-oriented economy growth models that tend to exploit natural resources, contribute to climate change, and exacerbate disaster vulnerabilities, Green growth integrated approaches are rapidly becoming as preferred investment avenue for mitigating climate change and disaster risks and for enhancing resilience. This includes ecosystem-based and nature-based solutions with potential to contribute to the resilience of infrastructure, urban, rural and peri-urban systems, livelihoods, water, and health. They can lead to food security and can further promote people-centric approaches.

Some of the synergistic outcomes of green growth approaches include disaster risk reduction, climate change mitigation and adaptation, resilient livelihoods, cities, businesses and industry. The disaster risk reduction and resilience outcome of green growth approaches deserve special attention, both for the academic and policy communities. Scholars and professionals across the domains of DRR, CCA, and green growth are in need of publications that fulfill their knowledge needs concerning the disaster resilience outcomes of green growth approaches. Keeping the above background in view, the book series offers comprehensive coverage combining the domains of environment, natural resources, engineering, management and policy studies for addressing disaster risk and resilience in the green growth context in an integrated and holistic manner. The book series covers a range of themes that highlight the synergistic outcomes of green growth approaches.

The book series aims to bring out the latest research, approaches, and perspectives for disaster risk reduction along with highlighting the outcomes of green growth approaches and including Science-technology-research-policy-practice interface, from both developed and developing parts of the world under one umbrella. The series aims to involve renowned experts and academicians as volume-editors and authors from all the regions of the world. It is curated and developed by authoritative institutions and experts to serve global readership on this theme.

Mukhtar Ahmed • Shakeel Ahmad
Editors

Disaster Risk Reduction in Agriculture

Springer

Editors
Mukhtar Ahmed
Department of Agronomy
Pir Mehr Ali Shah Arid Agriculture
University
Rawalpindi, Pakistan

Shakeel Ahmad
Department of Agronomy
Faculty of Agricultural Sciences and
Technology Bahauddin Zakariya University
Multan, Pakistan

ISSN 2662-4885 ISSN 2662-4893 (electronic)
Disaster Resilience and Green Growth
ISBN 978-981-99-1762-4 ISBN 978-981-99-1763-1 (eBook)
https://doi.org/10.1007/978-981-99-1763-1

This Springer imprint is published by the registered company Springer Nature Singapore Pte Ltd.
The registered company address is: 152 Beach Road, #21-01/04 Gateway East, Singapore 189721, Singapore

Paper in this product is recyclable.

Preface

The economic losses from natural disasters (e.g. drought, heavy rain, flood, heat waves) mainly driven by climate change are increasing day by day. In 2022 alone, they caused economic losses of $313 billion. According to the Intergovernmental Panel on Climate Change (IPCC), sixth assessment (AR6) report, it has been documented that the adverse impact of human-induced climate change will continue to intensify and there will be more adverse impacts on biodiversity and ecosystem, agricultural crop production and water availability. Future generations will experience more frequent and severe climate extremes, if urgent action is not taken. Since agriculture is open to the vagaries of nature, it will be severely affected by future climate extremes. The impacts of climate-related extremes disrupt food production, water supply and damage infrastructure. Between 2006 and 2016, 26% of the economic impact caused by climate-related disasters was recorded in the agriculture sector, mainly in developing countries. In the case of droughts, up to 83% of the damage and losses are in agriculture. Developing countries like Pakistan are more vulnerable to climate change, and it is visible in the form of drought, heat waves and flood. Devastating floods of 2022 affected ~ 33 million people across the country and damaged major cash crops. Thus, it is necessary to evaluate the impacts of disasters to avoid their cascading effects and develop disaster risk reduction strategies. Disaster risk reduction is defined as actions: *aimed at preventing new and reducing existing disaster risk and managing residual risk, all of which contribute to strengthening resilience and therefore to the achievement of sustainable development.* Disaster risk reduction (DRR) in agriculture is of paramount importance, especially in the face of a changing climate. Climate change is leading to increased frequency and severity of weather-related disasters such as droughts, floods, storms and heatwaves. These events can have devastating impacts on agriculture, which is not only a critical source of livelihood for many people but also a key component of global food security. The strategies that can be used to have DRR include developing and improving early warning systems for weather-related disasters, promotion of crop diversification to reduce vulnerability to changing climate conditions, developing and promoting the use of crop varieties that are more resilient to climate change, including drought-resistant and heat-tolerant crops, efficient water management that includes the development of irrigation systems and water storage facilities to mitigate the impact of droughts and erratic rainfall patterns, implementation of soil conservation techniques to reduce soil erosion, building of infrastructure that can withstand extreme weather events, encouraging the use of agricultural insurance and risk transfer mechanisms to help farmers recover from disasters, providing training and capacity-building programmes for farmers to help them adopt more resilient agricultural practices, promotion of community-based approaches to DRR, investment in research to better understand the specific climate risks in a given region and monitor changes over time, implementation of policies that support DRR in agriculture and promotion of international cooperation and partnerships to address climate change and its impact on agriculture. Disaster risk reduction in agriculture under a changing climate is an ongoing and dynamic process. It requires a multi-faceted approach that combines scientific knowledge, community engagement, government policies and international cooperation to build resilience and adapt to the challenges posed by climate change. This book *Disaster Risk Reduction in Agriculture* presents the views of agricultural experts. The 26 chapters—contributed by internationally recognized scientists across the globe—have been written under the theme of DRR in agriculture. The book covers a wide range of subject areas, from climate change and farming system to water management strategies for agricultural disasters, from agricultural producers' behavioural adaptation to climate change disaster to impacts of disasters on soil and their management, from conservation agriculture—a sustainable approach for disaster risk reduction—to the role of horticulture in disaster risk management and from use of artificial intelligence (AI) for DRR in agriculture to key challenges and financial needs to promote climate smart agriculture (CSA). As far as possible, the language of the chapters has been kept simple so that educated non-expert readers may enjoy reading and may benefit from the information provided herein. This book will serve as an educational tool for budding scientists, will provide a comprehensive overview for advanced researchers and will lay guidelines for important policy decisions.

Rawalpindi, Pakistan — Mukhtar Ahmed
Multan, Pakistan — Shakeel Ahmad

Acknowledgements

This book is the outcome of the dedication and efforts of editors Dr. Mukhtar Ahmed and Prof. Dr Shakeel Ahmad. We would like to sincerely thank all our valuable national and international reviewers. Many eminent researchers and academicians have contributed to the preparation of the proposal of this book. We highly appreciate the kind support and encouragement from various national and global researchers. In the end, we are highly thankful to the International Publisher (Springer Nature), its administration, and all the respected members who have provided significant support throughout the entire process from proposal to the publication of this book.

Contents

Editors and Contributors

About the Editors

Mukhtar Ahmed's research focuses on the impact of climate change on crop ecology, crop physiology, cropping system, and rain-fed ecosystem management. He has specialization in the field of environmental sciences, precision agriculture, agroecosystem modeling, climate change, and climate variability with teaching and research experience. He has research experience from Sydney University, Australia, where he worked on the application of APSIM as a decision support tool and rainfall forecasting using generalized additive models.

Shakeel Ahmad is Professor of Agronomy at Bahauddin Zakariya University, Multan, Pakistan since 2016. In 2006, he received PhD from the University of Agriculture, Faisalabad. Later, he completed postdoctoral research from the University of Georgia, USA. He has been devoting himself to teaching and research regarding field crops focusing on crop modeling, climate change, and adaptation strategies. He earned Research Productivity Award for 5 years from Pakistan Council for Science and Technology through the Ministry of Science and Technology, Government of Pakistan, Islamabad.

Contributors

Syed Mohsin Abbas Department of Horticulture, Faculty of Agricultural Sciences, University of the Punjab, Lahore, Pakistan

Muhammad Abid Department of Soil Science, Faculty of Agricultural Sciences and Technology, Bahauddin Zakariya University, Multan, Pakistan

Zoha Adil Department of Plant Pathology, University of Agriculture, Faisalabad, Pakistan

Sohaib Afzaal Citrus Research Institute, Sargodha, Pakistan

Sohaib Afzal Department of Agronomy, Faculty of Agricultural Sciences and Technology, Bahauddin Zakariya University, Multan, Pakistan

Md Shamim Ahamed Department of Biological and Agricultural Engineering, University of California, Davis, CA, USA

Irfan Ahmad Department of Forestry and Range Management, University of Agriculture, Faisalabad, Pakistan

Mobin ud Din Ahmad Commonwealth Scientific and Industrial Research Organisation, Canberra, ACT, Australia

Sajid Rashid Ahmad College of Earth and Environmental Sciences, Punjab University Lahore, Lahore, Pakistan

Shakeel Ahmad Department of Agronomy, Faculty of Agricultural Sciences and Technology, Bahauddin Zakariya University, Multan, Pakistan

Mukhtar Ahmed Department of Agronomy, Pir Mehr Ali Shah Arid Agriculture University, Rawalpindi, Pakistan

Niaz Ahmed Department of Soil Science, Faculty of Agricultural Sciences and Technology, Bahauddin Zakariya University, Multan, Pakistan

Salman Ahmed Department of Plant Pathology, University of Sargodha, Sargodha, Pakistan

Rogers Akatwijuka Department of Environmental Sciences, Kabale University, Faculty of Agriculture and Environmental Sciences, Kabale University, Kabale, Uganda

Muhammad Arslan Akbar Department of Breeding and Genetics, Cholistan University of Veterinary and Animal Sciences, Bahawalpur, Pakistan

Mazhar Ali Department of Environmental Sciences, COMSATS University Islamabad, Islamabad, Pakistan

Muhammad Arif Ali Department of Soil Science, Bahauddin Zakariya University, Multan, Punjab, Pakistan

Muhammad Usman Ali Department of Civil Engineering, Bahauddin Zakariya University, Multan, Pakistan

Sajid Ali Department of Agronomy, Faculty of Agricultural Sciences, University of the Punjab, Lahore, Pakistan

Alishpa Anum Faculty of Agricultural Sciences, University of the Punjab, Lahore, Pakistan

Muhammad Aon Department of Soil Science, Bahauddin Zakariya University, Multan, Punjab, Pakistan

Muhammad Arshad Institute of Environmental Science and Engineering, NUST, Islamabad, Pakistan

Muhammad Ashfaq Institute of Agricultural and Resource Economics, University of Agriculture, Faisalabad, Pakistan

Muhammad Ashraf Department of Agricultural Engineering, KFUEIT, Rahim Yar Khan, Pakistan

Muhammad Asif Department of Forestry and Range Management, University of Agriculture, Faisalabad, Pakistan

Department of Agricultural Engineering, Bahauddin Zakariya University, Multan, Pakistan

Muhammad Asim Citrus Research Institute, Sargodha, Pakistan

Mehwish Aslam Institute of Soil and Environmental Sciences, PMAS-Arid Agriculture University, Rawalpindi, Pakistan

Atique-ur-Rehman Department of Agronomy, Faculty of Agricultural Sciences and Technology, Bahauddin Zakariya University, Multan, Pakistan

Masood Iqbal Awan Department of Agronomy, Sub-Campus Depalpur, Okara, University of Agriculture, Faisalabad, Pakistan

Tahir Awan Punjab Agriculture Research Board, Lahore, Pakistan

Natal Ayiga Department of Social Work and Social Administration, Faculty of Arts and Social Sciences, Kabale University, Kabale, Uganda

M. Ahsan Azhar Department of Soil Science, Faculty of Agricultural Sciences and Technology, Bahauddin Zakariya University, Multan, Pakistan

Memoona Aziz Department of Agronomy, Faculty of Agricultural Sciences, University of the Punjab, Lahore, Pakistan

Tauseef Khan Babar Department of Entomology, Faculty of Agricultural Sciences and Technology, Bahauddin Zakariya University, Multan, Pakistan

Muhammad Azhar Inam Baig Department of Agricultural Engineering, Bahauddin Zakariya University, Multan, Pakistan

Abhishek Banerjee State Key Laboratory of Cryospheric Sciences, Northwest Institute of Eco-Environment and Resources, Chinese Academy of Sciences, Lanzhou, China

Fiza Batool Department of Soil Science, Faculty of Agricultural Sciences and Technology, Bahauddin Zakariya University, Multan, Pakistan

Brahim Benzougagh Geophysics and Natural Hazards Laboratory, Department of Geomorphology and Geomatics (D2G), Scientific Institute, Mohammed V University in Rabat, Avenue Ibn Batouta, Agdal, PO, Rabat, Morocco

Muhammad Hamza Tariq Bhatti Faculty of Agricultural Sciences, University of the Punjab, Lahore, Pakistan

Bushra Department of Agronomy, Faculty of Agricultural Sciences, University of the Punjab, Lahore, Pakistan

Sobia Chohan Department of Plant Pathology, Faculty of Agricultural Sciences and Technology, Bahauddin Zakariya University, Multan, Pakistan

Muhammad Dawood Department of Biosciences, Comsat University, Sahiwal, Punjab, Pakistan

Asim Faraz Department of Livestock and Poultry Production, Bahauddin Zakariya University, Multan, Pakistan

Omer Farooq Department of Agronomy, Faculty of Agricultural Sciences and Technology, Bahauddin Zakariya University, Multan, Pakistan

Shahid Farooq Department of Plant Protection, Faculty of Agriculture, Harran University, Sanliurfa, Turkey

Ammara Fatima Department of Environmental Science, Lahore College for Women University, Lahore, Pakistan

Denis Frolov Lomonosov Moscow State University, Geographical Faculty, Moscow, Russia

Mohsin Hafeez International Water Management Institute, Lahore, Pakistan

Hamza Haider Faculty of Agricultural Sciences, University of the Punjab, Lahore, Pakistan

Idrees Haider Department of Soil Science, Bahauddin Zakariya University, Multan, Punjab, Pakistan

Muhammad Hammad Department of Agricultural Engineering, Bahauddin Zakariya University, Multan, Pakistan

Amna Hanif Department of Environmental Sciences, COMSATS University Islamabad, Islamabad, Pakistan

Shakeel Hanif Fodder Research Institute, Sargodha, Pakistan

Ehsan Ul Haque Citrus Research Institute, Sargodha, Pakistan

Amna Hasnain Assistant Professor, Institute of Management Sciences, Bahauddin Zakariya University, Multan, Pakistan

Akbar Hayat Citrus Research Institute, Sargodha, Pakistan

Rifat Hayat Institute of Soil and Environmental Sciences, PMAS-Arid Agriculture University, Rawalpindi, Pakistan

Abrar Hussain Department of Agronomy, Faculty of Agricultural Sciences and Technology, Bahauddin Zakariya University, Multan, Pakistan

Mubshar Hussain Department of Agronomy, Bahauddin Zakariya University, Multan, Pakistan

School of Veterinary and Life Sciences, Murdoch University, Murdoch, WA, Australia

Sajjad Hussain Department of Horticulture, Bahauddin Zakariya University, Multan, Punjab, Pakistan

Department of Environmental Sciences, COMSATS University Islamabad, Islamabad, Pakistan

Syeda Maryam Hussain Department of Livestock Production and Management, Pir Mehr Ali Shah Arid Agriculture University, Rawalpindi, Pakistan

M. Zafar ul Hye Department of Soil Science, Faculty of Agricultural Sciences and Technology, Bahauddin Zakariya University, Multan, Pakistan

Fariha Ilyas Department of Soil Science, Bahauddin Zakariya University, Multan, Punjab, Pakistan

Muhammad Imran Department of Environmental Sciences, COMSAT University Islamabad, Islamabad, Pakistan

Hafiz Muhammad Ishaq Department of Livestock and Poultry Production, Bahauddin Zakariya University, Multan, Pakistan

Nimra Ishfaq Department of Soil Science, Faculty of Agricultural Sciences and Technology, Bahauddin Zakariya University, Multan, Pakistan

Amir Ismail Institute of Food Science and Nutrition, Bahauddin Zakariya University, Multan, Pakistan

Abdul Jabbar Department of Agronomy, University of Agriculture, Faisalabad, Pakistan

Mansoor Javed Department of Agronomy, University of Sargodha, Sargodha, Pakistan

Hafiz Muhammad Rashad Javeed Department of Environmental Sciences, COMSATS University Islamabad, Vehari Campus, Vehari, Pakistan

Sarvet Jehan Institute of Soil and Environmental Sciences, PMAS-Arid Agriculture University, Rawalpindi, Pakistan

Mustafa Kan Kırşehir Ahi Evran University, Agricultural Faculty, Department of Agricultural Economics, Kirsehir, Turkey

Muhammad U. Khan Department of Energy Systems Engineering, Faculty of Agricultural Engineering and Technology, University of Agriculture, Faisalabad, Pakistan

Narmeen Khan Faculty of Agricultural Sciences, University of the Punjab, Lahore, Pakistan

Rana Muhammad Rafique Khan College of Earth and Environmental Sciences, Punjab University Lahore, Lahore, Pakistan

Zahid M. Khan Department of Agricultural Engineering, Bahauddin Zakariya University, Multan, Pakistan

Fatima Latif Department of Soil Science, Faculty of Agricultural Sciences and Technology, Bahauddin Zakariya University, Multan, Pakistan

Hamza Maqsood Department of Agronomy, Faculty of Agricultural Sciences, University of the Punjab, Lahore, Pakistan

Sajid Masood Department of Soil Science, Faculty of Agricultural Sciences and Technology, Bahauddin Zakariya University, Multan, Pakistan

Yasir Mehmood Department of Plant Pathology, Faculty of Agricultural Sciences and Technology, Bahauddin Zakariya University,, Multan, Pakistan

Khuram Mubeen Department of Agronomy, Muhammad Nawaz Sharif University of Agriculture, Multan, Pakistan

Muhammad Mubeen Department of Environmental Sciences, COMSATS University Islamabad, Islamabad, Pakistan

Ghulam Mustafa Department of Agriculture (Extension and Adoptive Research) Punjab, Lahore, Pakistan

Syed Atif Hasan Naqvi Department of Plant Pathology, Faculty of Agricultural Sciences and Technology, Bahauddin Zakariya University, Multan, Pakistan

Bukhtawar Nasir Centre of Agricultural Biochemistry and Biotechnology, University of Agriculture, Faisalabad, Pakistan

Javaria Nasir Institute of Agricultural and Resource Economics, University of Agriculture, Faisalabad, Pakistan

Wajid Nazir Department of Plant Pathology, Faculty of Agricultural Sciences and Technology, Bahauddin Zakariya University, Multan, Pakistan

Yasir Niaz Department of Agricultural Engineering, KFUEIT, Rahim Yar Khan, Pakistan

Ikramullah Qadri State Bank of Pakistan, Karachi, Pakistan

Rafi Qamar Department of Agronomy, College of Agriculture, University of Sargodha, Sargodha, Pakistan

M. Farooq Qayyum Department of Soil Science, Faculty of Agricultural Sciences and Technology, Bahauddin Zakariya University, Multan, Pakistan

Khalid Rasheed Soil and Water Testing Laboratory, Multan, Pakistan

Haroon Ur Rashid Department of Forestry and Range Management, University of Agriculture, Faisalabad, Pakistan

Mujahid Rasool Faculty of Education University College of the North (UCN), Manitoba, Canada

Muhammad Arham Raza Faculty of Agricultural Sciences, University of the Punjab, Lahore, Pakistan

Ateeq ur Rehman Department of Plant Pathology, Faculty of Agricultural Sciences and Technology, Bahauddin Zakariya University, Multan, Pakistan

Atique ur Rehman Department of Agronomy, Faculty of Agricultural Sciences and Technology, Bahauddin Zakariya University, Multan, Pakistan

Muhammad Rizwan Veterinary Sciences, Bahauddin Zakariya University, Multan, Pakistan

Tabukeli Musigi Ruhiiga Department of Environmental Sciences, Kabale University, Faculty of Agriculture and Environmental Sciences, Kabale University, Kabale, Uganda

Muhammad Saeed Department of Horticulture, Faculty of Agricultural Sciences, University of the Punjab, Lahore, Pakistan

Hamza Salahudin Department of Agricultural Engineering, Bahauddin Zakariya University, Multan, Pakistan

Aashir Sameen Department of Agronomy, Pir Mehr Ali Shah Arid Agriculture University Rawalpindi, Rawalpindi, Pakistan

Adnan Sami Faculty of Agricultural Sciences, University of the Punjab, Lahore, Pakistan

Naeem Sarwar Department of Agronomy, Faculty of Agricultural Sciences and Technology, Bahauddin Zakariya University, Multan, Pakistan

Yusuf Kadir Şener Turkish Republic of Ministry of Agriculture and Forestry, Kaman District Directorate of Agriculture and Forestry, Kirsehir, Turkey

Khadija Shabbir Department of Environmental Sciences, COMSATS University Islamabad, Islamabad, Pakistan

Saad Shafaat Department of Agronomy, College of Agriculture, University of Sargodha, Sargodha, Pakistan

Muhammad Shafiq Department of Horticulture, Faculty of Agricultural Sciences, University of the Punjab, Lahore, Pakistan

Azeem Ali Shah International Water Management Institute, Lahore, Pakistan

Zainab Shahbaz Department of Forestry and Range Management, University of Agriculture, Faisalabad, Pakistan

Nuhammad Adnan Shahid Horticultural Science Department, University of Florida, Institute of Food and Agricultural Sciences, North Florida Research and Education Center, Quincy, FL, USA

Muhammad Shahzad Department of Agronomy, University of Poonch, Rawalakot, Azad Jammu and Kashmir, Pakistan

Redmond R. Shamshiri Agromechatronics, Leibniz Institute for Agricultural Engineering and Bio-economy, Potsdam, Germany

Aamir Shezhad Department of Agronomy, Faculty of Agricultural Sciences, University of the Punjab, Lahore, Pakistan

Muhammad Shoaib Department of Agricultural Engineering, Bahauddin Zakariya University, Multan, Pakistan

Muhammad Sultan Department of Agricultural Engineering, Bahauddin Zakariya University, Multan, Pakistan

Nasir Ali Tauqir Department of Animal Nutrition, The Islamia University of Bahawalpur, Bahawalpur, Pakistan

Wycliffe Tumwesigye Department of Economics and Environmental Management, Bishop Stuart University, Mbarara, Uganda

Benson Turyasingura Institute of Tourism and Hospitality, Kabale University, Kabale, Uganda

Sami Ul-Allah College of Agriculture, BZU Bahadur Sub-Campus, Layyah, Pakistan

Ummad ud din Umar Department of Plant Pathology, Faculty of Agricultural Sciences and Technology, Bahauddin Zakariya University, Multan, Pakistan

Abdul Waheed Department of Livestock and Poultry Production, Bahauddin Zakariya University, Multan, Pakistan

Allah Wasaya Department of Agronomy, University of Layyah, Layyah, Pakistan

Tauqeer Ahmad Yasir Department of Agronomy, University of Layyah, Layyah, Pakistan

Saqaina Younas Department of Agronomy, Faculty of Agricultural Sciences, University of the Punjab, Lahore, Pakistan

Adnan Zahid Department of Agronomy, Faculty of Agricultural Sciences, University of the Punjab, Lahore, Pakistan

Uzair Zahid State Bank of Pakistan, Karachi, Pakistan

Climate Change and Farming System: A Review of Status, Potentials, and Further Work Needs for Disaster Risk Reduction

Mukhtar Ahmed

Abstract

Management of farming and food system under changing climate and increased population pressure is critical for the future of biodiversity. High yield farming results in higher emissions of greenhouse gases which are the main driver of climate change, thus damaging the farming system across the globe. The farming system and climate change have strong interactions among each other, and different agroclimatic conditions determine the types of the farming system. Therefore, to have sustainable and resilient farming systems, it is important to understand key drivers that determine farmers' adaptive capacity under changing climate. The main drivers includes environmental pressure, water availability, crop characteristics, and socioeconomic conditions of farmers. However, farmers give less importance to climate change which leads to clear gaps in the implementation of outreach activities. Farmers with small land holdings are more vulnerable to climate change as compared to larger-scale farmers as they have lower resources to adopt new technologies. For designing of effective adaptation policies, it is important to understand what adjustments farmers make to cope with climate change. A previous study about the list of adjustments has shown flaws as more importance was given to climate change as compared to other socioeconomic drivers that alter farmers behaviours. Thus, non-climatic drivers should be part of analysis to design effective climate adaptation strategies and remove potential flaws. Hence after assessing all drivers, the following climate adaptation measures are recommended for farm-level adjustments to minimze the

M. Ahmed (✉)
Department of Agronomy, Pir Mehr Ali Shah Arid Agriculture University, Rawalpindi, Pakistan
e-mail: ahmadmukhtar@uaar.edu.pk

M. Ahmed, S. Ahmad (eds.), *Disaster Risk Reduction in Agriculture*, Disaster Resilience and Green Growth, https://doi.org/10.1007/978-981-99-1763-1_1

impact of disaster: altering the planting date, revising and updating crop varieties, implementing crop switching, and promoting diversification. A case study of smallholder crop-livestock system in sub-Saharan Africa was used to suggest possible adaptation and mitigation options under changing climate. These include (1) providing cushions to farmers against risks (e.g. insurance scheme, weather forecasting, and early warnings system), (2) improving skills and capacity of farmers and value chain actors, and (3) fostering farm investments (e.g. credit facilities, land tenure security, and value chain development). The reviewed work also emphasized the implementation of effective insurance schemes and weather-index insurance systems (WIIS) that can significantly benefit smallholder farmers. Furthermore, economic incentives, technical support, and collaborative networks can facilitate adoption of sustainable agricultural practices by farmers.

Keywords

Climate change · Farming system · Diversification · Farm-level adjustments · Adaptation and mitigation options · Smallholder crop-livestock system · Insurance · Weather-index insurance systems

1 Introduction

Farming system research, development, and extension work involves the integration of plants, animals, and soil at the paddock and farm level. A sustainable farming system needs to have five important characteristics, i.e. it should be (1) purposeful (have goals and allocate resources to achieve these goals), (2) dynamic (change over time in response to internal or external factors), (3) stochastic (uncertain future behaviour and difficult to predict), (4) open (interact with the environment), and (5) abstract (conceptual rather than purely physical in nature) as proposed by Dillon (1992). However, climate is the main defining factor of different farming systems across the globe. The farming system and climate change have strong interactions among each other, and different agroclimatic conditions determine the types of the farming system (Hutchinson et al. 2005). Climate change is visible in the form of extreme variations in weather, e.g. wind, precipitation, and temperature. Maskrey et al. (2007) reported that poor agricultural communities of the developing world is the most affected by the climate change although the developing world is only contributing 10% to the global greenhouse gas emissions. The impact of climate change on developing countries is more visible as their economy is predominantly agriculture based which is open to the vagaries of nature. Thus, it poses serious challenges to the socioeconomic and ecological systems. The World Bank also highlighted this aspect in their report and stated that the poorest people in the south Asian regions are suffering most due to climate change. Poor people in the region are in a poverty trap due to the frequent occurrence of extreme weather events in recent decades. The number of warm days and nights has increased and in future intensity, frequency, and length of heat waves will increase across the globe. Climate change will alter the rainfall intensity and duration, and the occurrence of heavy perception will be more in future with spatio-temporal variability (Field et al. 2012). This variability will lead to the change in water availability to crops, thus affecting crop yield and income of the farm. Maon et al. (2009) stated that the occurrence of abnormal disasters has increased from 125 per year to 400–500 since 1980, and it is mainly because of climate change. Furthermore, climate change, food security, and poverty have strong interactions among each other. Climate change impacts will be more visible in coming decades particularly on agriculture as it is the climate-sensitive sector. Around 2.5 billion people will be affected due to climate change and variability. Adaptation can be a good option to reduce the adverse impact of climate change on agriculture. It has been elaborated in many studies that the detrimental impact of climate change on agriculture can be offset by different farm-level adaptation measures (Deressa et al. 2009; Smit and Skinner 2002). Gbetibouo (2009) further highlighted the importance of adaptive capacity of farming community as it can change the degree of impact of climate change on the agriculture sector. Different localized adaptation strategies have been suggested by researchers, and they have been shown in Fig. 1 (Deressa et al. 2009; Hussain and Mudasser 2007; Mendelsohn and Davis 2001; Smit and Skinner 2002). Similarly, some other adaptation measures have been adopted to mitigate the issue of climate change. It includes adjusting the ratio of livestock to cropping or crop area, adjustment in the whole farm water-use efficiency, actions to minimize livestock emissions, and on-farm diversification (OFD) (Hayman et al. 2012; van Zonneveld et al. 2020). Willett et al. (2019) described OFD as a promising strategy for farmers to adapt to climate change. Figure 1 shows that how we need to work with farmers to make good decisions about OFD of pasture, cropping, and agroforestry systems. These seven steps are (Fig. 2) useful for all types of farmers particularly for smallholder farmers. Farmers with small land holdings are more vulnerable to climate change as compared to larger-scale farmers as they have lower resources to adopt new technologies. Thus smallholder farmers should be the main target of climate smart interventions also called as climate smart

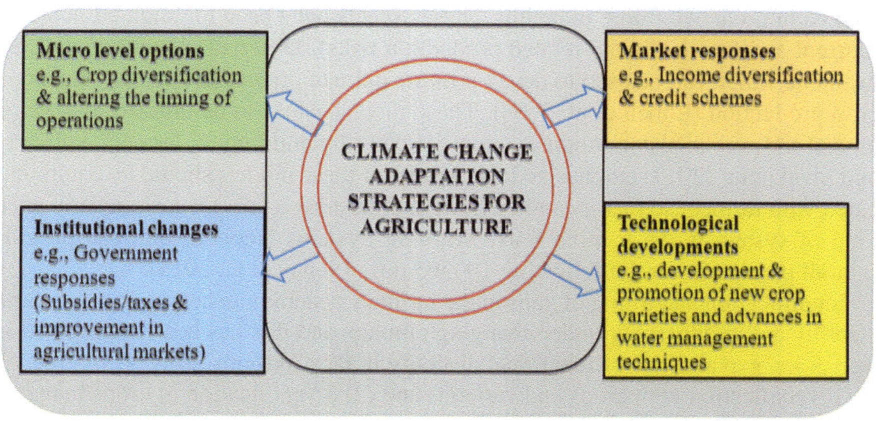

Fig. 1 Possible adaptation strategies for agriculture

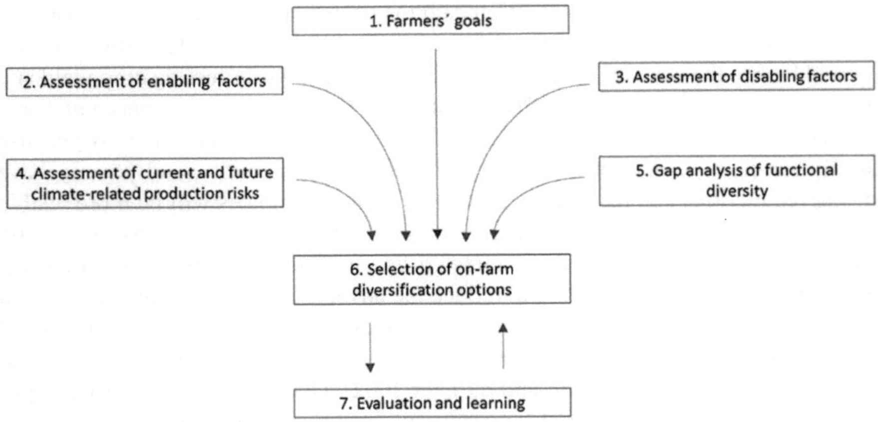

Fig. 2 Steps to develop on-farm diversification strategies for climate change adaptation (Source: van Zonneveld et al. 2020)

agriculture. Lipper et al. (2014) reported that climate smart agriculture can help farmers to adapt to climate change. van Zonneveld et al. (2020) proposed a seven-step decision-making process that can help to connect researchers and practitioners with farmers. Defining farmers goals (Step 1) starts by understanding the goals of different farm households in different farming systems. Schroth and Ruf (2014) stated that farmers do diversification by considering multiple goals as they consider cereals for food security, legumes for nutrition, cash crops for income, and forage and off-season crops for animal production. Step 2 (enabling factors) comprises components of extension, farmer organization, indigenous knowledge, and consideration of underutilized crops, selecting the right variety, insurance, and markets. Disabling factors (Step 3) elaborates successful adoption of OFD, and it depends on farmers' skills and financial status. It includes scale effects, labour constraints, farm size, and land ownership. The identification of climate risks can enable farmers to engage in proactive future planning. These aspects have been highlighted in step 4 (current and future climate-related production risks). Different climate models are available which can be used to develop future climate projections and its impact on crop production (Lobell et al. 2008). These models can also be utilized to develop annual and perennial commodities, aiding in capacity building for farmers. Pulwarty and Sivakumar (2014) emphasized that results of these models should be communicated with farmers so that it can be effective at farming scale. One example of such kind of system is the Famine Early Warning Systems Network that can provide rainfall prediction for the following 10–365 days (Senay et al. 2015).

Identification and filling of gaps in the farming systems can help to increase farm stability and productivity under changing climate, and this has been highlighted in step 5, i.e. gap analysis of functional diversity in farm systems. This is possible via (1) diversification with crops and varieties and (2) diversification of crops/management practices to foster ecological functions. Step 6 considers the selection of OFD options which involve the development of a decision model to select good OFD crops and management practices. Thomas et al. (2007) reported that participatory evaluation is a cost-effective way to evaluate crops, varieties, and management practices, and it is referred to as step 7 (evaluation and learning). The proposed seven steps of OFD plan by van Zonneveld et al. (2020) can help farmers to have diversification plans as an option for climate change adaptation. However, Harvey et al. (2014) reported that OFD is not a good option to reduce the vulnerabilities of climate change, and Hansen et al. (2019) suggested that off-farm diversification could be a better option for farmers to adapt to climate change.

2 On-Farm Diversification: Key Components of Climate Change Adaptation and Mitigation

Diversification is an effective strategy for both adaptation and mitigation to minimize the adverse effects of climate change on farming systems. The benefit of OFD on soil health has been elaborated by Baldwin-Kordick et al. (2022) and concluded that agricultural diversification resulted in higher crop yield, reduced inputs requirements, and decreased environmental footprints. Similarly, Kemboi et al. (2020) reported diversification as an important coping mechanism for climate change. Vernooy (2022) reviewed crop diversification as a climate-resilient strategy and concluded that it can give multiple benefits to the farming communities. The benefits include increased household income and yield, improved nutrition and food security, and reduced poverty. Belay et al. (2017) investigated how smallholder farmers understand climate change and what adaption strategies they used to minimize the impact of climate change. They highlighted the impact of climate change on smallholder farming activities in Ethiopia (Fig. 3). Their results showed that

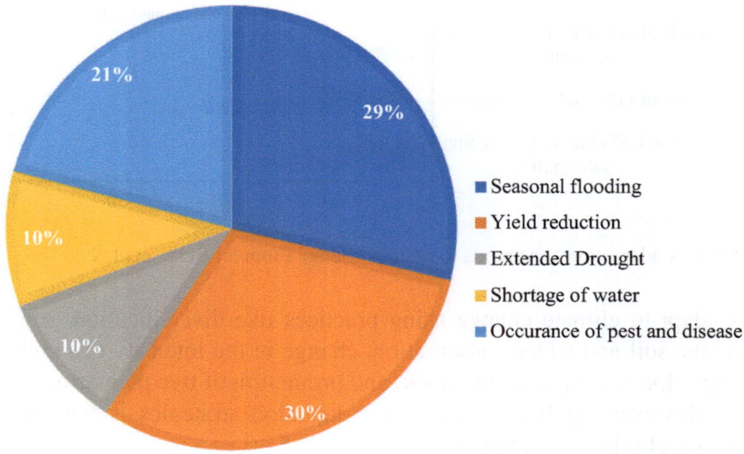

Fig. 3 Climate change impact on smallholder farming activities in Ethiopia (Source: Belay et al. 2017)

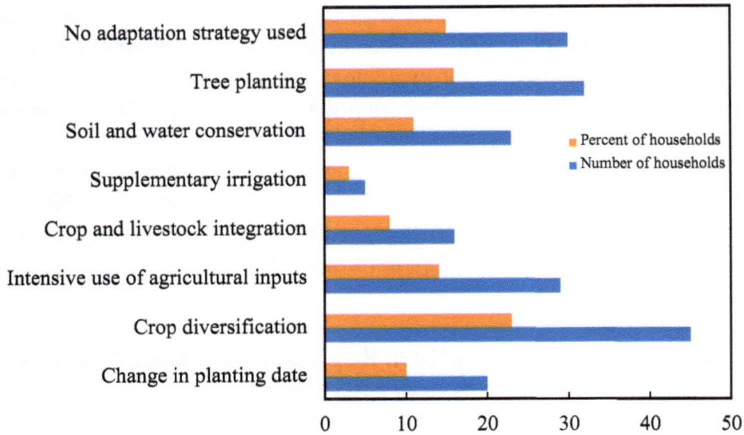

Fig. 4 Possible adaptation strategies to climate change (Source: Belay et al. 2017)

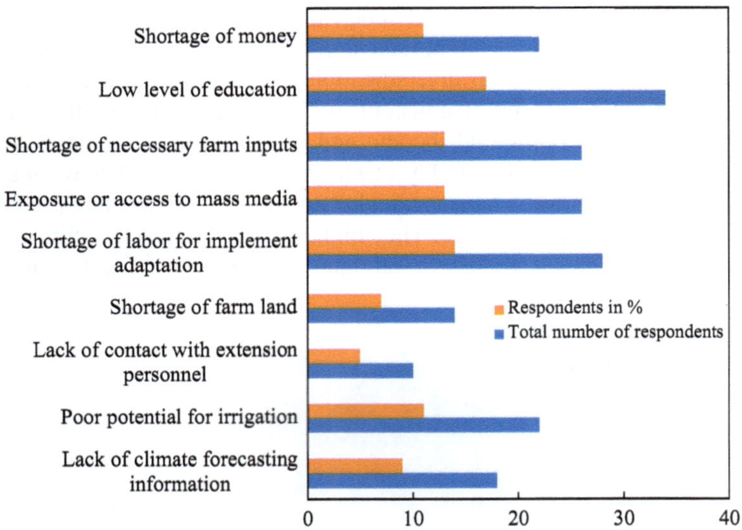

Fig. 5 Primary adaptation constraints to climate change (Source: Belay et al. 2017)

farmers adapt to climate change using practices like diversification, sowing date adjustments, soil and water conservation, change in the intensity of input application, integration of crop with livestock, and promotion of tree plantation with crops (Fig. 4). However, application of these adaptation strategies is not up to mark because of multiple constraints as shown in Fig. 5.

Adaptation strategies like cultivation of a large number of diverse species, integration of livestock with crop production (crop-livestock integration), and use of better adapted crops/varieties have already been recommended by different researchers. The use of legume with cereals could help to improve soil health as well as control pest and insect attacks (Yu et al. 2015). Schlenker and Lobell (2010) suggested the use of crops and crop varieties that can grow well under harsh climatic conditions. These includes sorghum, fonio (*Digitaria* spp), and finger millet for cereals, and cowpea for legumes. Furthermore, varieties with different maturity timings could be suitable for conditions where drastic increase or decrease in temperature or rainfall could disturb the crop yield or survival (Dinar et al. 2012). However, traditional indigenous varieties are also good as they can adapt to climate extremes easily (Vigouroux et al. 2011).

3 Climate-Proof Crops in the Farming Systems

Wild relatives of different staple crops, e.g. wheat, can give 50% higher yield in hot climatic conditions as compared to traditional recommended cultivars as they have less genetic variability. Wheat crop which provides maximum global calories is vulnerable to climate change since it has limited variations. Hence, it is important to consider wild relatives as potential options to develop climate-proof crops. This type of work is going on in CIMMYT so that we can have genetic resources that can have climate resilience (Fig. 6). Satori et al. (2022) reported that climate change will be a big threat for smallholder farming systems. Thus, to combat this threat development of climate-resilient crops is needed using crop wild relatives. Nair (2019) presented the potential use of crop wild relatives to combat global warming and enhance global food security. Similarly, Renzi et al. (2022) reviewed possible adaptive traits that can be used to improve the performance of cultivars in extreme environments.

Fig. 6 Drone shot at the CIMMYT wheat fields, near Sonora, Mexico. Photo credit: CIM

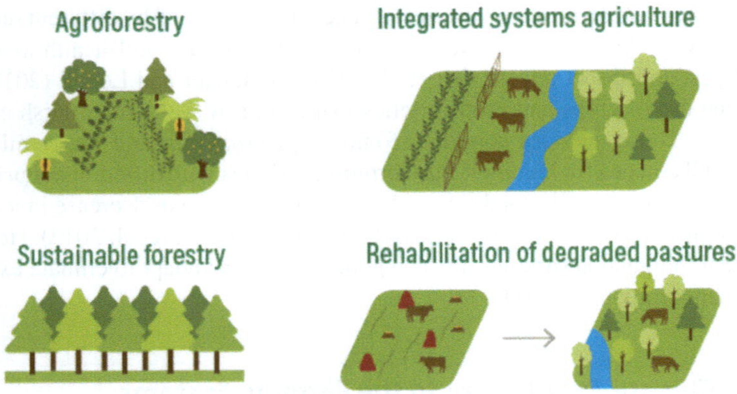

Fig. 7 Four agricultural interventions that can power climate adaptation (Source: WRI)

4 Climate Change and Farm-Level Adaptation

Farm-level adaptation response to climate change is an effective approach to design adaptation strategies at the national and regional scale. A list of farm-level adaptation measures has been recommended by different researchers, and it includes crop diversification, crop switching, changing and updating crop varieties, tree plantation, conservation agriculture, and changing the planting date (Tessema et al. 2019). However, to have accurate farm-level adaptations a new approach proposed by Tessema et al. (2019) should be used. Estimates have shown that climate change could reduce agricultural productivity by 17% by 2050, thus proving to be a big threat for the farming communities. The World Resources Institute (WRI) suggested four agricultural interventions that can power climate adaptation as shown in Fig. 7. The interventions include (1) integrating crop-livestock-forestry systems, (2) rehabilitating degraded pastures, (3) plant agroforestry systems, and (4) pursuing sustainable forestry. Abid et al. (2019) suggested that adaptive measures of smallholder farmers to climate change could be enhanced by providing up-to-date information and training. Similarly, outreach activities should be focussed on interventions that have adaptive and mitigative properties as suggested by Arbuckle Jr. et al. (2015).

5 Smallholder Crop-Livestock System: Case Study in Sub-Saharan Africa

The impact of climate change is seen in sub-Saharan Africa. Temperatures across the African continent have been increased with the projection of drier climate for southern Africa while wet climate for eastern Africa (Descheemaeker et al. 2016). The effect of climate change on a mixed farming system has been shown in Fig. 8. It shows that climate change is affecting the system by altering the individual as well as interactive component. Amejo et al. (2019) stated that majority of the

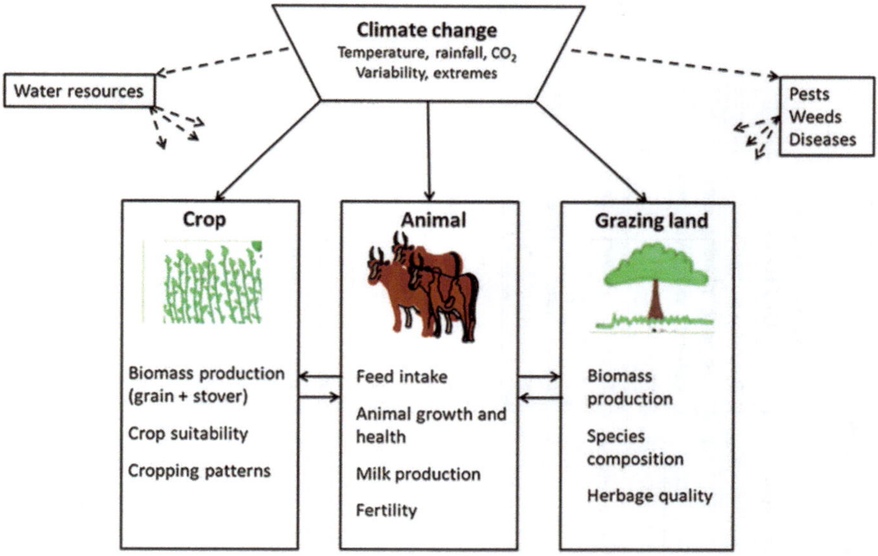

Fig. 8 Effect of climate change on mixed crop-livestock system (Source: Descheemaeker et al. 2016)

smallholder are more vulnerable to climate change due to lack of access to insurance and credit. Furthermore, most of the African smallholders are net food buyers which makes them vulnerable to price shocks under extreme climatic conditions. This shock under climate extremes remains for the longer period, throwing smallholder farmers into poverty traps (Dercon 2004). Crop production in the African continent is not up to mark as 96% of agriculture is rainfed (Cooper et al. 2008). Previous data showed that rainfall intensity and duration have been significantly changed which resulted in the increased occurrence of drought and shortening of growing seasons (Rurinda et al. 2014). A framework was developed to elaborate the vulnerability of smallholder farmers to climate change considering experts and local farmers knowledge (Fig. 9). The framework has three core components of vulnerability, i.e. (1) exposure, (2) sensitivity, and (3) adaptation. Subsystems of this framework include cropping, livestock, and availability of natural resources. Indicators like food self-sufficiency and cattle ownership were used to see the impact of climate change on these subsystems. Climatic features such as frequency of drought, increased rainfall variability, and temperature were identified as main drivers. Adaptation was classified into operational and strategic. However, adoption of these adaptation options depends on the availability of and access to both biophysical and socioeconomic resources as well as support provided by different institutions (Fig. 9).

In mixed smallholder crop-livestock system, livestock's vulnerability depends upon the availability of feed. Generally, crop residues with low nutritive value are used as animal diet and hence it becomes unavailable during dry seasons, thus

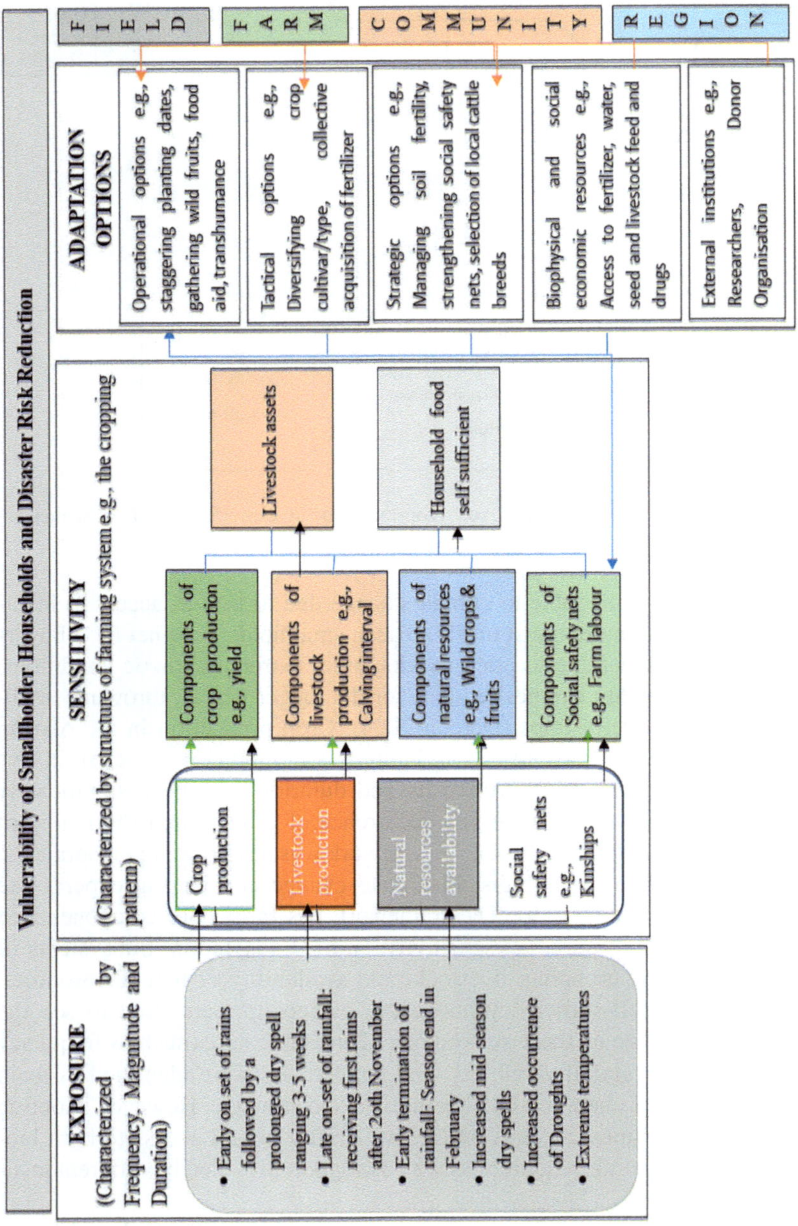

Fig. 9 Vulnerability analysis of smallholders (Source: Rurinda et al. 2014)

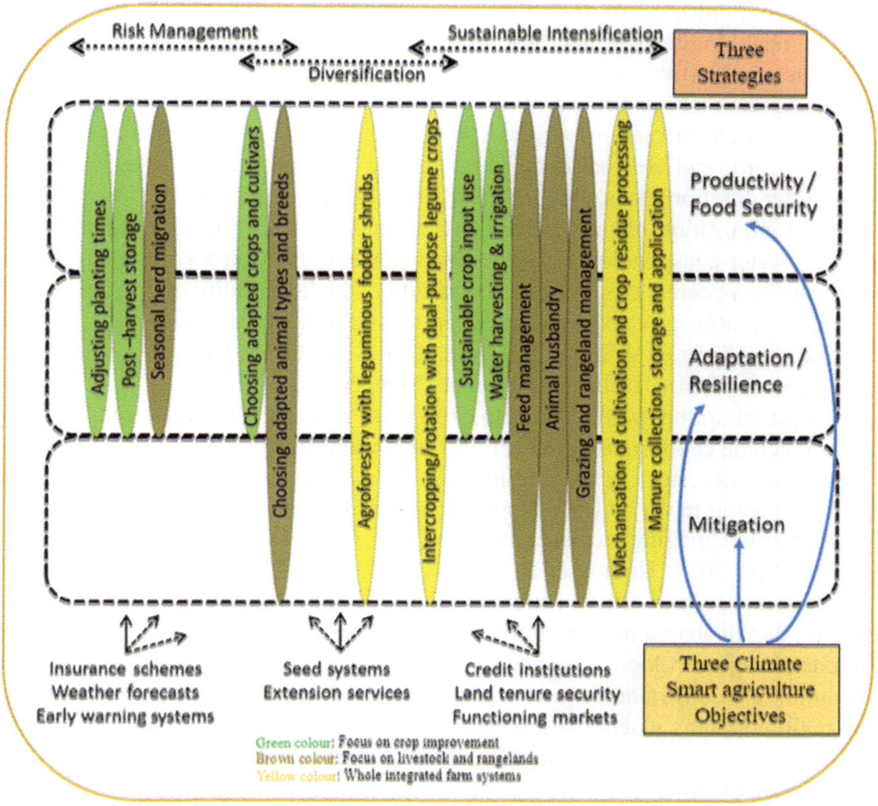

Fig. 10 Farm-level climate smart options (Source: Descheemaeker et al. 2016)

leading to poor animal health (Valbuena et al. 2015). The impact of climate change will be stronger on livestock as climate change will deteriorate crop residues and forage production. Furthermore, in most of sub-Saharan Africa, grazing contributes to the 10–90% animal diet which will be unavailable in future as grazing resources are already under threat due to climate change. Similarly, African livestock systems are contributing more to the emissions of greenhouse gases as compared to other regions. It is mainly due to the usage of fodder sources with low digestibility (Gerber et al. 2013) and emissions from manure during storage, processing, and application (Seebauer 2014). Adaptation and mitigation which are two of three pillars (third pillar: food security) of climate smart agriculture (CSA) are needed to uplift small-holder crop-livestock system. Promising options for smallholder farms have been presented in Fig. 10. which includes (1) risk management, (2) diversification, and (3) sustainable intensification with focus on crops, livestock, and rangeland or on the whole farm system. The suggested objective of risk management is to reduce the variance of an individual or whole system outcome (e.g. fodder or milk yield) while intensification aims to increase the mean of individual or whole system outcome.

However, the objective of diversification is broad as it will shift both the variance and the mean. Figure 10 elaborates that the adaptation or mitigation option is only workable if objectives of increasing food security are fulfilled at first; secondly increasing resilience and capacity building of smallholders is important than mitigation. Suggested adaptation options for the African households include seasonal migration, crop and livestock diversification, tree plantation, use of drought-resistant and shorter duration crops, choice of animal types and breeds, planting dates adjustments, minimization of post-harvest losses, integrated soil fertility management, soil and water conservation, irrigation management, dual-purpose crops as animal diet, improving cereal crop residues palatability and digestibility through chemical/biological treatment or mechanical chopping and grazing management (Campbell et al. 2014; Niang et al. 2017; Milgroom and Giller 2013; Vanlauwe et al. 2015; Oosting et al. 2014; Descheemaeker et al. 2009, 2016).

Mitigation options which can help to improve mixed smallholder crop-livestock system include choosing adapted animal breeds, improved feed through diversification (e.g. agroforestry), better feeding and feed management, improved animal husbandry, keeping fewer better fed animals, improving animal and herd productivity, rangeland and grazing management, and improvement in manure management (Mbow et al. 2014; Hristov et al. 2013; Gerber et al. 2013; Oosting et al. 2014; Thornton and Herrero 2010; Rufino et al. 2006). However, there are challenges in the adoption of adaptation and mitigation options in the African continent, and these include (1) agro-ecological, sociocultural, economic, and institutional dimensions; (2) multiscale constraints (e.g. resource constrained, poor community organization and malfunctioning extension services, poor market infrastructure, inputs cost and unavailability, price uncertainty, and typical communal land tenure system of African rangelands); and (3) farm size, risk, and livestock multi-functionality (e.g. small farm and shrinkage of farm due to population growth) (Ojiem et al. 2006; Jones and Thornton 2009; Cavatassi et al. 2011; Harris and Orr 2014; van Vliet et al. 2015).

Suggested adaptation and mitigation options will improve farm performance by reducing climate vulnerability. However, cultural norms, absence of insurance and credit facilities, and marketing incentives are preventing the suggested options. Similarly, the toughest constraints and barriers to adoption include increased population pressure, small farm size, poor access to inputs, market dysfunction, high investment risk, land tenure insecurity, and less support provided by the institutions. Porter et al. (2014) suggested transformative change as a good option to reduce disaster risks, and it is possible through (1) providing cushions to farmers against risks (e.g. insurance scheme, weather forecasting, and early warnings system), (2) improving skills and capacity of farmers and value chain actors, and (3) fostering farm investments (e.g. credit facilities, land tenure security, and value chain development). Müller (2013) in their work suggested implementation of effective insurance schemes as a risk management strategy which can help to keep large livestock herd. Similarly, Greatrex et al. (2015) emphasized the importance of implementing weather-index insurance systems (WIIS) that can significantly benefit smallholder farmers.

6 Farm Size in Pakistan and Climate Smart Agriculture

Pakistan is an agricultural country as its economy mostly relies on the agricultural sector. However, the agricultural sector is at stake due to the recent extreme climate events. Pakistan is worst affected by climate change despite the fact that Pakistan has been a low producer of carbon dioxide gasses (i.e. 0.2 million metric tons of CO_2) (Smadja et al. 2015). However, environmental disasters are causing a huge economic losses to the country. Pakistan ranks among the top 10 most affected countries in terms of fatalities and mortalities due to long-term climate risks from 2000 to 2019 as reported in the Global Climate Risk Index (GCRI)-2021. Disasters like heatwaves, floods, glacier melting, and droughts are common features of Pakistan climate. Around $30.1 billion loss was caused by the recent 2022 flood and still the economy is struggling to revive as the inflation rate has been seen at its peak. The damage was more severe in Sindh (total loss of $20.4 billion or 68% of the total loss) and Baluchistan province. Since 80% of Pakistan population lives beside Indus basin, they are facing multiple threats due to climate change and climate extremes, also due to poor infrastructure and resource management. Pakistan is also very poor in the three important pillars of disaster risk reduction that includes (i) Preparation (ii) Response and (iii) Rehabilittaion. Furthermore, it has been estimated that it costs $12 billion per annum (4% of the country's GDP) to improve Pakistan's poor water resource management. Pakistan has the largest and longest glaciers where glacial lake outburst flooding (GLOF) is common. A survey was conducted by IPSOS to gather information that how much Pakistanis are familiar with climate change and what are the causes of floods in Pakistan. The survey report shows (Fig. 11) that most of the Pakistanis were not considering climate change as a major problem and consider climate change as a less important factor which contributes to the floods. Hence, it is necessary that people should be informed and guided about recent problems in Pakistan, and it is possible through introduction of climate change as a major subject.

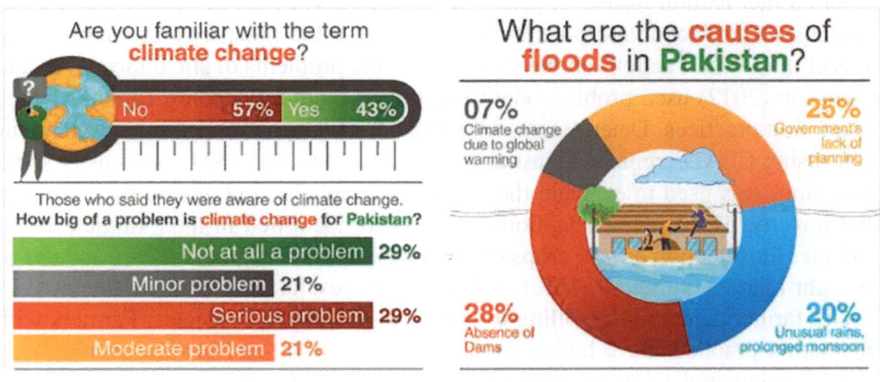

Fig. 11 IPSOS survey report about the terms "Climate change" and "Causes of floods" in Pakistan

Agriculture is still the backbone of Pakistan as it contributes 20% to GDP and gives employment to 43%. Around 2/3rd of Pakistan's population lives in rural areas and directly or indirectly they are connected to agriculture. More than 75% of the total crop output comes from major crops like wheat, rice, sugarcane, maize, cotton, fruits, and vegetables. Hence agriculture is the only sector which can ensure food security and reduce poverty. But the agriculture sector is under threat due to extreme climate variability and climate change acting as a big barrier to achieve foot security and alleviate poverty. Higher temperature has shown impact on the cropping system by increasing evapotranspiration, crop water demands, and heat stress on crops. It has been reported that with 1 °C rise in temperature cereal crops, e.g. wheat yield, will be reduced by 5–7% (Aggarwal and Sivakumar 2011) while it can decline to 7–21% with an increase in temperature of 1.5–3 °C. However, 14–23% increase in wheat yield was reported in Chitral districts of Pakistan due to rise in temperature (Hussain and Mudasser 2007). Ahmad et al. (2013) studied the impact of rise in temperature on rice yield and reported that rice yield will be decreased by 15% from 2012 to 2039, 25% from 2040 to 2060, and 36% from 2070 to 2099. Similarly, decreased rainfall also affects crop production as with 6% decrease in rainfall, net irrigation water requirement could be increased by 29%. Hence almost 1.3 million farms will be negatively affected due to this change. Different adaptation strategies have been adopted to minimize the impact of climate change. It includes sowing date adjustments, shifting to climate-proof crops, usage of stress-tolerant crops, change in the fertilizer, and irrigation usage. However, implementation of these adaptation strategies is difficult due to changes in the farm size.

Farm size in the South Asian countries has been decreasing mainly because of increased population pressures and extreme climate events. Farm size in Pakistan has been decreased from 5.3 ha in 1971 to 2.6 ha in 2010 agricultural census. Hence most of agricultural farming in Pakistan is smallholders. More than 90% farms are smaller than 12 acres out of which 67% are even five acres (2 ha) which will further decrease in future. Some land holdings are so small that they are no longer economically viable. Furthermore, small size is a major limiting factor that hinders in the application of modern tools (Fig. 12). Naseer et al. (2016) and Phambra et al. (2020) reported that natural disasters, pest attacks, access to financial markets, and unfavourable macro-economic policies are threatening the prosperity of small farms; thus climate-smart actions are needed to solve the problems of small farms. Ali and Erenstein (2017) used probit model to identify factors influencing climate change adaptation practices. Determinants of a number of adaptation practices were analysed using CLAD (censored least absolute deviation), and PSM (propensity score matching) was used to evaluate the impact of adaptation options on food security and poverty. Three major adaption options were identified in their studies, and it includes introduction of new crops (25%), change in sowing time (22%), and use of drought-tolerant varieties (15%). Furthermore, results showed that literate and young farmers were more willing to use these adaptation strategies. Farmers with adaptation practices have higher food security (8–13%) and lower poverty levels (3–6%) compared to those who were not opting these practices. Thus, climate change adaptation practices at farm level can reduce weather risks. Climate smart

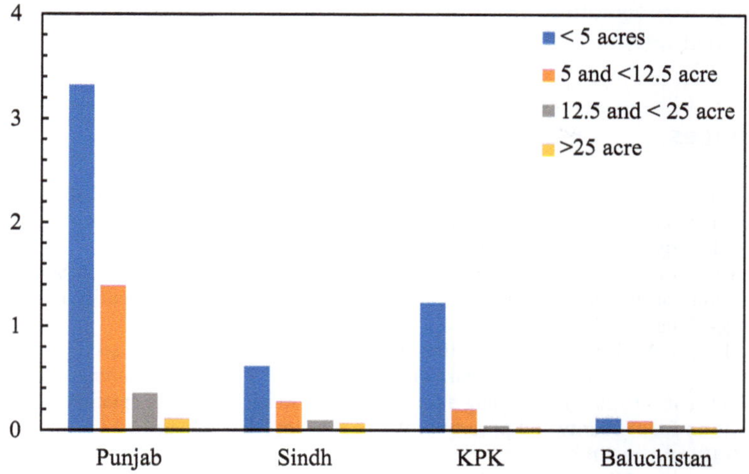

Fig. 12 Province-wise distribution of farm in Pakistan (Source: Pakistan Bureau of Statistics 2017)

agriculture in Pakistan has been further elaborated in the report of International Center for Tropical Agriculture (CIAT) (CIAT 2017).

7 Conclusion

Farming systems are under threat due to extreme climate events, and the impact of climate calamities will be more severe on smallholder farming systems. Hence, farmers, policy makers, and researchers should sit together to develop a framework where they can implement the suggested adaptation and mitigation options to reduce the disaster. This framework should be developed based on real-time data supplemented by modelling studies. Similarly, the range of adaptation and mitigation options should be investigated by considering mixed smallholder crop-livestock system individually as well as whole. Furthermore, changes in the interactions between farm components should also be quantified in response to climate change for designing sustainable adaptation and mitigation strategies. Importance should also be given to the heterogeneity in the biophysical and socioeconomic context as it will also have an effect on the selection of adaptation and mitigation options. This is possible firstly by modelling the effects of climate change, adaptation, and mitigation at the farm level by considering diversity into account. Secondly, conduction of cost and benefit analysis of the proposed adoption strategy will help to make assessments more realistic. Meanwhile a combination of scientific knowledge with indigenous local traditional knowledge via participatory approach will enhance adaptive capacity of local farming communities and can make the system climate smart or climate proof or climate resilient. Finally, long term implementation of policies, early warning system, implementtaion of indigenous techniques/

knowledge, colloborative efforts and climate finance is needed to reach to the end user in a real way.

References

Abid M, Scheffran J, Schneider UA, Elahi E (2019) Farmer perceptions of climate change, observed trends and adaptation of agriculture in Pakistan. Environ Manag 63(1):110–123. https://doi.org/10.1007/s00267-018-1113-7

Aggarwal P, Sivakumar MV (2011) Global climate change and food security in South Asia: an adaptation and mitigation framework. In: Climate change and food security in South Asia. Springer, Cham, pp 253–275

Ahmad M, Iqbal M, Khan M (2013) Climate change, agriculture and food security in Pakistan: adaptation options and strategies. Pakistan Institute of Development Economics, Islamabad

Ali A, Erenstein O (2017) Assessing farmer use of climate change adaptation practices and impacts on food security and poverty in Pakistan. Clim Risk Manag 16:183–194. https://doi.org/10.1016/j.crm.2016.12.001

Amejo AG, Gebere YM, Kassa H, Tana T (2019) Characterization of smallholder mixed crop–livestock systems in integration with spatial information: in case Ethiopia. Cogent Food Agric 5(1):1565299. https://doi.org/10.1080/23311932.2019.1565299

Arbuckle JG Jr, Morton LW, Hobbs J (2015) Understanding farmer perspectives on climate change adaptation and mitigation: the roles of trust in sources of climate information, climate change beliefs, and perceived risk. Environ Behav 47(2):205–234. https://doi.org/10.1177/0013916513503832

Baldwin-Kordick R, De M, Lopez MD, Liebman M, Lauter N, Marino J, McDaniel MD (2022) Comprehensive impacts of diversified cropping on soil health and sustainability. Agroecol Sustain Food Syst 46(3):331–363. https://doi.org/10.1080/21683565.2021.2019167

Belay A, Recha JW, Woldeamanuel T, Morton JF (2017) Smallholder farmers' adaptation to climate change and determinants of their adaptation decisions in the Central Rift Valley of Ethiopia. Agric Food Secur 6(1):24. https://doi.org/10.1186/s40066-017-0100-1

Campbell BM, Thornton P, Zougmoré R, van Asten P, Lipper L (2014) Sustainable intensification: what is its role in climate smart agriculture? Curr Opin Environ Sustain 8:39–43. https://doi.org/10.1016/j.cosust.2014.07.002

Cavatassi R, Lipper L, Narloch U (2011) Modern variety adoption and risk management in drought prone areas: insights from the sorghum farmers of eastern Ethiopia. Agric Econ 42(3):279–292. https://doi.org/10.1111/j.1574-0862.2010.00514.x

CIAT (2017) Climate-smart agriculture in Pakistan. CSA country profiles for Asia series. International Center for Tropical Agriculture, Palmira

Cooper PJM, Dimes J, Rao KPC, Shapiro B, Shiferaw B, Twomlow S (2008) Coping better with current climatic variability in the rain-fed farming systems of sub-Saharan Africa: an essential first step in adapting to future climate change? Agric Ecosyst Environ 126(1–2):24–35. https://doi.org/10.1016/j.agee.2008.01.007

Dercon S (2004) Growth and shocks: evidence from rural Ethiopia. J Dev Econ 74(2):309–329. https://doi.org/10.1016/j.jdeveco.2004.01.001

Deressa TT, Hassan RM, Ringler C, Alemu T, Yesuf M (2009) Determinants of farmers' choice of adaptation methods to climate change in the Nile Basin of Ethiopia. Glob Environ Chang 19(2):248–255

Descheemaeker K, Raes D, Nyssen J, Poesen J, Haile M, Deckers J (2009) Changes in water flows and water productivity upon vegetation regeneration on degraded hillslopes in northern Ethiopia: a water balance modelling exercise. Rangel J 31(2):237–249. https://doi.org/10.1071/RJ09010

Descheemaeker K, Oosting SJ, Homann-Kee Tui S, Masikati P, Falconnier GN, Giller KE (2016) Climate change adaptation and mitigation in smallholder crop–livestock systems in sub-Saharan Africa: a call for integrated impact assessments. Reg Environ Chang 16(8):2331–2343. https://doi.org/10.1007/s10113-016-0957-8

Dillon J (1992) The farm as a purposeful system, Miscellaneous publication no 10. Department of Agricultural Economics and Business Management, University of New England, Armidale

Dinar A, Hassan R, Mendelsohn R, Benhin J (2012) Climate change and agriculture in Africa: impact assessment and adaptation strategies. Routledge, London

Field CB, Barros V, Stocker TF, Dahe Q (2012) Managing the risks of extreme events and disasters to advance climate change adaptation: special report of the intergovernmental panel on climate change. Cambridge University Press, Cambridge

Gbetibouo GA (2009) Understanding farmers' perceptions and adaptations to climate change and variability: the case of the Limpopo Basin, South Africa, vol 849. Intl Food Policy Res Inst, Washington

Gerber PJ, Hristov AN, Henderson B, Makkar H, Oh J, Lee C, Meinen R, Montes F, Ott T, Firkins J, Rotz A, Dell C, Adesogan AT, Yang WZ, Tricarico JM, Kebreab E, Waghorn G, Dijkstra J, Oosting S (2013) Technical options for the mitigation of direct methane and nitrous oxide emissions from livestock: a review. Animal 7:220–234. https://doi.org/10.1017/S1751731113000876

Greatrex H, Hansen J, Garvin S, Diro R, Le Guen M, Blakeley S, Rao K, Osgood D (2015) Scaling up index insurance for smallholder farmers: recent evidence and insights. CCAFS Report

Hansen J, Hellin J, Rosenstock T, Fisher E, Cairns J, Stirling C, Lamanna C, van Etten J, Rose A, Campbell B (2019) Climate risk management and rural poverty reduction. Agric Syst 172:28–46

Harris D, Orr A (2014) Is rainfed agriculture really a pathway from poverty? Agric Syst 123:84–96. https://doi.org/10.1016/j.agsy.2013.09.005

Harvey CA, Rakotobe ZL, Rao NS, Dave R, Razafimahatratra H, Rabarijohn RH, Rajaofara H, MacKinnon JL (2014) Extreme vulnerability of smallholder farmers to agricultural risks and climate change in Madagascar. Philos Trans R Soc B Biol Sci 369(1639):20130089. https://doi.org/10.1098/rstb.2013.0089

Hayman P, Rickards L, Eckard R, Lemerle D (2012) Climate change through the farming systems lens: challenges and opportunities for farming in Australia. Crop Pasture Sci 63(3):203–214. https://doi.org/10.1071/CP11196

Hristov AN, Ott T, Tricarico J, Rotz A, Waghorn G, Adesogan A, Dijkstra J, Montes F, Oh J, Kebreab E, Oosting SJ, Gerber PJ, Henderson B, Makkar HPS, Firkins JL (2013) Special topics—mitigation of methane and nitrous oxide emissions from animal operations: III. A review of animal management mitigation options. J Anim Sci 91(11):5095–5113. https://doi.org/10.2527/jas.2013-6585

Hussain SS, Mudasser M (2007) Prospects for wheat production under changing climate in mountain areas of Pakistan–an econometric analysis. Agric Syst 94(2):494–501

Hutchinson MF, McIntyre S, Hobbs RJ, Stein JL, Garnett S, Kinloch J (2005) Integrating a global agro-climatic classification with bioregional boundaries in Australia. Glob Ecol Biogeogr 14(3):197–212

Jones PG, Thornton PK (2009) Croppers to livestock keepers: livelihood transitions to 2050 in Africa due to climate change. Environ Sci Pol 12(4):427–437. https://doi.org/10.1016/j.envsci.2008.08.006

Kemboi E, Muendo K, Kiprotich C (2020) Crop diversification analysis amongst smallholder farmers in Kenya (empirical evidence from Kamariny ward, Elgeyo Marakwet County). Cogent Food Agric 6(1):1834669. https://doi.org/10.1080/23311932.2020.1834669

Lipper L, Thornton P, Campbell BM, Baedeker T, Braimoh A, Bwalya M, Caron P, Cattaneo A, Garrity D, Henry K (2014) Climate-smart agriculture for food security. Nat Clim Chang 4(12):1068–1072

Lobell DB, Burke MB, Tebaldi C, Mastrandrea MD, Falcon WP, Naylor RL (2008) Prioritizing climate change adaptation needs for food security in 2030. Science 319(5863):607–610. https://doi.org/10.1126/science.1152339

Maon F, Lindgreen A, Vanhamme J (2009) Cross-sector collaboration for disaster relief supply chain enhancement: mingling corporate expertise with humanitarians' Willpower. Working Paper 08/21, Site de Louvain-la-Neuve–Place des Doyens, 1–1348

Maskrey A, Buescher G, Peduzzi P, Schaerpf C (2007) Disaster risk reduction: 2007 global review. Consultation edition prepared for the global platform for disaster risk reduction first session, Geneva, Switzerland, p. 5–7

Mbow C, Van Noordwijk M, Luedeling E, Neufeldt H, Minang PA, Kowero G (2014) Agroforestry solutions to address food security and climate change challenges in Africa. Curr Opin Environ Sustain 6:61–67. https://doi.org/10.1016/j.cosust.2013.10.014

Mendelsohn R, Davis E (2001) Global warming and the American economy. A regional assessment of climate change impacts. Edward Elgar, Cheltenham

Milgroom J, Giller KE (2013) Courting the rain: rethinking seasonality and adaptation to recurrent drought in semi-arid southern Africa. Agric Syst 118:91–104. https://doi.org/10.1016/j.agsy.2013.03.002

Müller C (2013) African lessons on climate change risks for agriculture. Annu Rev Nutr 33(1):395–411. https://doi.org/10.1146/annurev-nutr-071812-161121

Nair KP (2019) Chapter Four - Utilizing crop wild relatives to combat global warming. In: Sparks DL (ed) Advances in agronomy, vol 153. Academic, New York, pp 175–258. https://doi.org/10.1016/bs.agron.2018.09.001

Naseer A, Ashfaq M, Abid M, Razzaq A, Hassan S (2016) Current status and key trends in agricultural land holding and distribution in Punjab, Pakistan: implications for food security. J Agric Stud 4(4):14–27

Niang A, Becker M, Ewert F, Dieng I, Gaiser T, Tanaka A, Senthilkumar K, Rodenburg J, Johnson J-M, Akakpo C (2017) Variability and determinants of yields in rice production systems of West Africa. Field Crop Res 207:1–12

Ojiem JO, de Ridder N, Vanlauwe B, Giller KE (2006) Socio-ecological niche: a conceptual framework for integration of legumes in smallholder farming systems. Int J Agric Sustain 4(1):79–93. https://doi.org/10.1080/14735903.2006.9686011

Oosting SJ, Udo HMJ, Viets TC (2014) Development of livestock production in the tropics: farm and farmers' perspectives. Animal 8(8):1238–1248. https://doi.org/10.1017/S1751731114000548

Pakistan Bureau of Statistics (2017) Pakistan bureau of statistics: agricultural census. In: Government ofPakistan. https://www.pbs.gov.pk/

Phambra AM, Tahir S, Imran M (2020) Current structure of landholdings and importance of small farms in Pakistan. Pak J Econ Stud 3(1):47–64

Porter JR, Xie L, Challinor AJ, Cochrane K, Howden SM, Iqbal MM, Lobell DB, Travasso MI (2014) Food security and food production systems. In: Field CB, Barros VR, Dokken DJ et al (eds) Climate change 2014: impacts, adaptation, and vulnerability. Part A: global and sectoral aspects. Cambridge University Press, Cambridge, pp 485–533

Pulwarty RS, Sivakumar MV (2014) Information systems in a changing climate: early warnings and drought risk management. Weather Clim Extremes 3:14–21

Renzi JP, Coyne CJ, Berger J, von Wettberg E, Nelson M, Ureta S, Hernández F, Smýkal P, Brus J (2022) How could the use of crop wild relatives in breeding increase the adaptation of crops to marginal environments? Front Plant Sci 13:886162. https://doi.org/10.3389/fpls.2022.886162

Rufino MC, Rowe EC, Delve RJ, Giller KE (2006) Nitrogen cycling efficiencies through resource-poor African crop–livestock systems. Agric Ecosyst Environ 112(4):261–282. https://doi.org/10.1016/j.agee.2005.08.028

Rurinda J, Mapfumo P, van Wijk MT, Mtambanengwe F, Rufino MC, Chikowo R, Giller KE (2014) Comparative assessment of maize, finger millet and sorghum for household food security in the face of increasing climatic risk. Eur J Agron 55:29–41. https://doi.org/10.1016/j.eja.2013.12.009

Satori D, Tovar C, Faruk A, Hammond Hunt E, Muller G, Cockel C, Kühn N, Leitch IJ, Lulekal E, Pereira L, Ryan P, Willis KJ, Pironon S (2022) Prioritising crop wild relatives to enhance agricultural resilience in sub-Saharan Africa under climate change. Plants People Planet 4(3):269–282. https://doi.org/10.1002/ppp3.10247

Schlenker W, Lobell DB (2010) Robust negative impacts of climate change on African agriculture. Environ Res Lett 5(1):014010. https://doi.org/10.1088/1748-9326/5/1/014010

Schroth G, Ruf F (2014) Farmer strategies for tree crop diversification in the humid tropics. A review. Agron Sustain Dev 34(1):139–154. https://doi.org/10.1007/s13593-013-0175-4

Seebauer M (2014) Whole farm quantification of GHG emissions within smallholder farms in developing countries. Environ Res Lett 9(3):035006. https://doi.org/10.1088/1748-9326/9/3/035006

Senay G, Velpuri NM, Bohms S, Budde M, Young C, Rowland J, Verdin J (2015) Drought monitoring and assessment: remote sensing and modeling approaches for the famine early warning systems network. In: Hydro-meteorological hazards, risks and disasters. Elsevier, Amsterdam, pp 233–262

Smadja J, Aubriot O, Puschiasis O, Duplan T, Grimaldi J, Hugonnet M, Buchheit P (2015) Climate change and water resources in the Himalayas. Field study in four geographic units of the Koshi basin, Nepal. J Alpine Res 2015:2910

Smit B, Skinner MW (2002) Adaptation options in agriculture to climate change: a typology. Mitig Adapt Strat Glob Change 7(1):85–114

Tessema YA, Joerin J, Patt A (2019) Climate change as a motivating factor for farm-adjustments: rethinking the link. Clim Risk Manag 23:136–145. https://doi.org/10.1016/j.crm.2018.09.003

Thomas DS, Twyman C, Osbahr H, Hewitson B (2007) Adaptation to climate change and variability: farmer responses to intra-seasonal precipitation trends in South Africa. Clim Chang 83(3):301–322

Thornton PK, Herrero M (2010) Potential for reduced methane and carbon dioxide emissions from livestock and pasture management in the tropics. Proc Natl Acad Sci 107(46):19667–19672. https://doi.org/10.1073/pnas.0912890107

Valbuena D, Tui SH-K, Erenstein O, Teufel N, Duncan A, Abdoulaye T, Swain B, Mekonnen K, Germaine I, Gérard B (2015) Identifying determinants, pressures and trade-offs of crop residue use in mixed smallholder farms in Sub-Saharan Africa and South Asia. Agric Syst 134:107–118. https://doi.org/10.1016/j.agsy.2014.05.013

van Vliet JA, Schut AGT, Reidsma P, Descheemaeker K, Slingerland M, van de Ven GWJ, Giller KE (2015) De-mystifying family farming: features, diversity and trends across the globe. Glob Food Sec 5:11–18. https://doi.org/10.1016/j.gfs.2015.03.001

van Zonneveld M, Turmel M-S, Hellin J (2020) Decision-making to diversify farm systems for climate change adaptation. Front Sustainable Food Syst 4:32. https://doi.org/10.3389/fsufs.2020.00032

Vanlauwe B, Descheemaeker K, Giller KE, Huising J, Merckx R, Nziguheba G, Wendt J, Zingore S (2015) Integrated soil fertility management in sub-Saharan Africa: unravelling local adaptation. Soil 1(1):491–508. https://doi.org/10.5194/soil-1-491-2015

Vernooy R (2022) Does crop diversification lead to climate-related resilience? Improving the theory through insights on practice. Agroecol Sustain Food Syst 46(6):877–901. https://doi.org/10.1080/21683565.2022.2076184

Vigouroux Y, Mariac C, De Mita S, Pham J-L, Gérard B, Kapran I, Sagnard F, Deu M, Chantereau J, Ali A, Ndjeunga J, Luong V, Thuillet A-C, Saïdou A-A, Bezançon G (2011) Selection for earlier flowering crop associated with climatic variations in the Sahel. PLoS ONE 6(5):e19563. https://doi.org/10.1371/journal.pone.0019563

Willett W, Rockström J, Loken B, Springmann M, Lang T, Vermeulen S, Garnett T, Tilman D, DeClerck F, Wood A (2019) Food in the Anthropocene: the EAT–Lancet Commission on healthy diets from sustainable food systems. Lancet 393(10170):447–492

Yu Y, Stomph T-J, Makowski D, van der Werf W (2015) Temporal niche differentiation increases the land equivalent ratio of annual intercrops: a meta-analysis. Field Crop Res 184:133–144. https://doi.org/10.1016/j.fcr.2015.09.010

Disaster Risk Reduction in Agriculture

Omer Farooq, Naeem Sarwar, Sohaib Afzal,
Khuram Mubeen, Atique ur Rehman, Mukhtar Ahmed,
and Shakeel Ahmad

Abstract

Disaster risk could be a serious injury or even loss of human life, destruction or damage of property or assets of society due to exposure to some hazardous events. In contrast, disaster risk reduction is the protection of lives and livelihood of the persons of that society who are most susceptible to disaster, caused by either nature or humans or collectively both. Federal Emergency Management Agency of the United States has identified 18 types of natural hazards, including coastal flooding, avalanche, drought, cold wave, earthquake, heat wave, hail, ice storm, hurricane (tropical cyclone), lightning, riverine flooding, landslide, tornado, strong wind, volcanic activity, tsunami, severe winter weather and wildfire. The atmosphere of our globe responds to all the changes occurring in it. Ultimately, interruptions in the total solar irradiance have enhanced the mean global temperature by at least 0.2 °C to 0.85 °C between 1880 and 2012, as claimed in the 5th IPCC assessment report. To date, the rise in temperature has resulted in intense variations to natural and human systems, among these increases are floods, droughts and a few other kinds of risky weather; rise in sea level; and major loss of biodiversity. To counteract the adversaries of disasters,

O. Farooq (✉) · N. Sarwar · S. Afzal · A. u. Rehman · S. Ahmad
Department of Agronomy, Faculty of Agricultural Sciences and Technology, Bahauddin Zakariya University, Multan, Pakistan

K. Mubeen
Department of Agronomy, Muhammad Nawaz Sharif University of Agriculture, Multan, Pakistan

M. Ahmed
Department of Agronomy, Pir Mehr Ali Shah Arid Agriculture University, Rawalpindi, Pakistan

M. Ahmed, S. Ahmad (eds.), *Disaster Risk Reduction in Agriculture*, Disaster Resilience and Green Growth, https://doi.org/10.1007/978-981-99-1763-1_2

risk reduction measures are critical. Structural interventions like the construction of dikes (dykes), dams, storm or hill torrent water control, watersheds, sediments management and groundwater storage, etc., can significantly aid in reducing the natural disaster risk. Disaster hazards can be reduced by the migration of people to safer places. However, forced migration affects the socioeconomics of the people making their livelihoods more vulnerable. Agricultural crops and other vegetation alter with flood disaster and over time run off the volume and intensity determines further land use changes. Migration aids in drought vulnerability by unemployment. Infrastructure health systems and cleanliness, food security and water security will improve socioeconomics, which assists in reducing disasters vulnerability. Moreover, insurance, diversity in agricultural practices, improved drinking water supplies, improved urban planning and access to markets will improve the lifestyle of the disaster-prone areas.

Keywords

Disaster risk · Structural interventions · Migration · Insurance · Diversity in agricultural practices

1 Introduction

Disaster risk could be a serious injury or even loss of human life, destruction or damage of property or assets of a community or society over a specific time duration, probably due to exposure to some hazardous events. It includes the losses that are sometimes critical to calculating or even difficult to measure. However, with the advancement of information on prevailing hazards and the configurations of population and socioeconomic development, disaster risk can be at least judged and assessed. Categorically, disaster risk is divided into acceptable risk and residual risk; the first one is the degree to which disaster risk is considered tolerable or adequate depending on prevailing economic, social, cultural, environmental and political conditions. Engineering uses acceptable risk to consider the measures (structural and non-structural) desired to minimize expected harm to property, people, services and systems up to a certain acceptable level. Residual risk is a type of disaster risk that remains uniform when operative disaster risk reduction actions are in place, whereas emergency responses and retrieval capacities must be sustained. Such risks often imply a continuous mechanism to respond and support in emergencies, response and recovery, and preparedness in connection with socioeconomic structures, especially for risk-transferring mechanisms and safety measures (Dar and Alam 2020; Kelman 2018).

2 Disaster Risk Reduction (DRR)

DRR is the protection of lives, livelihood of communities and the persons who are most susceptible to disaster or crisis whether it is caused by nature and humans or both. DRR restricts its harmful impact on the most affected (Seddiky et al. 2020). Disaster risk reduction, also called disaster risk management (DRM) is a logical way of assessing, identifying and reducing the risks of a disaster. It aims to reduce socioeconomic vulnerabilities to disaster risks and deal with the environmental and other hazards that trigger them. The importance of DRR has gained rapid attention because recent years have observed a significant increase in such incidences and the impact of disasters that are causing huge economic and human losses. Moreover, it is seen that the impact of natural hazards is different in both developed and underdeveloped countries. In underdeveloped countries, property losses are relatively small, but casualties are higher as compared to developed countries, whereas disaster prevention and mitigation measures are well established in developed countries; hence, life damages are relatively less, but economic losses are quite high (Jayaraman et al. 2018). Around 230 million people were affected by floods, droughts and storms between 2000 and 2007 (FAO 2006).

Disaster risk reduction emphasizes on lessening and avoiding the adverse effects of natural hazards through well plan activities and methods for prevention, mitigation and preparedness. Global report suggests that natural disasters have happened more repeatedly and more destructively in the previous decade due to the hazards related to weather and climate shifts in a speedy way that ultimately damage the life of human beings (FAO 2006). Previous history and awareness related to the characteristics and occurrence of climatic irregularities, policies and regaining plans would add great value if one intends to reduce the hazard damage on production level like yield (Hay 2007). Well-organized promotion and field development of disaster-related policies, legislation and regulatory framework play a key role in creating and enabling a productive, sustainable environment for disaster risk reduction efforts (UNFCCC 2008).

3 Natural Disaster

Natural disasters sometimes called natural hazards are natural phenomena that might have a destructive effect on individuals, animals and the environment. The events, natural disasters and natural hazards are not the same but are related. A natural disaster harms the community following the real happening of natural hazards. Broadly speaking, natural disasters can be grouped into two comprehensive categories, biological and the other one is geophysical (Burton et al. 1993). We can understand the difference between a natural hazard and a disaster; for example, an earthquake is considered a hazard but may shift to a disaster, like in the 1906 San Francisco earthquake disaster. Such hazards can usually be triggered by human activities or actions like construction, drainage and changing land use (Gill and Malamud 2017). According to the National Risk Index assembled by the Federal

Emergency Management Agency (FEMA) of the United States, 18 types of natural hazards were counted as coastal flooding, avalanche, drought, cold wave, earthquake, heat wave, hail, ice storm, hurricane (tropical cyclone), lightning, riverine flooding, landslide, tornado, strong wind, volcanic activity, tsunami, severe winter weather and wildfire (Zuzak et al. 2022). Additionally, sometimes there are dust storms and also tornados. Due to prevailing climatic changes, several hazards have a higher risk of occurrence. It has also been observed that natural hazards can often cause additional natural hazards, e.g., ash and lava are secondary hazards of volcanic activity. Whereas National Risk Index only counts primary natural hazards, their resultant products or after-effects are not counted.

4 Global Climate Pattern

The atmosphere of our globe responds to all the changes occurring in it or even the changes occurring beneath it, various types and amounts of gasses entering it. Because the whole globe is in a state of motion constantly, the air is also in a continuous running position making the climate change scenario a purely universal issue (Ahmed 2020). Our earth has a very balanced system due to its relative position to the sun, the ultimate source of energy of our environment and all the gasses entering our system contribute towards the greenhouse effect. This effect is from a fixed proportion of water vapours (about 60%), carbon dioxide (about 25%), ozone (about 8%) and other minute quantities of minor gasses like nitrous oxide and methane (Kiehl and Trenberth 1997). Clouds do have a crucial role in the greenhouse effect because near about 31% (55 petawatts, PW) of the total energy (175 PW) coming from the sun is reflected back by the clouds. At the same time, the remaining (120 PW) energy is emitted back to space as infrared radiation after being absorbed by the atmosphere, land, or ocean (Kiehl and Trenberth 1997). In recent decades, the earth's climatic pattern has experienced exceptional warming (IPCC 2012). Occasional volcanic eruptions during the last century resulted in the release of various gasses and their debris into our atmosphere, which strongly disturbed the energy flows; nonetheless, the resultant cooling effect lasted only for a few years (Houghton et al. 2001). All these changes or interruptions in the total solar irradiance has enhanced the mean global temperature by at least 0.2 °C, whereas, in the last 50 years, human activities have been detected as a major reason influencing climate change pattern (Houghton et al. 2001), while the average global temperature was enhanced by 0.85 °C in between 1880 and 2012, as claimed in 5th IPCC assessment report (IPCC 2013)

To date, the rise in temperature has resulted in intense variations to natural and human systems, among these increases are floods, droughts and a few other kinds of risky weather; rise in sea level; and major loss of biodiversity. Such changes initiate extraordinary hazards to susceptible populations (IPCC 2012, 2014a, b; Mysiak et al. 2016). Interestingly, underdeveloped or developing countries are more affected by these climatic hazards and experience food security, which ultimately raises problems of poverty, malnutrition and migration (IPCC 2012). Drastically the most concerned community is those living in coastal regions, small islands, megacities and high mountainous areas (Albert et al. 2017). Whereas, according to the future prediction by AOGCMs, by 2100, the rise in temperature and associated warming would be between 1.2 °C and 4.1 °C. Conclusively, a huge continual composition temperature with sea level commitment is apparent in the simulations and is slowly recognized over the coming centuries. Furthermore, by the year 3000, the rising temperature range is 1.9–5.6 °C, whereas surface temperatures approach equilibrium relatively fast, sea level continues to rise for many centuries.

Changes in climate directly affect the ecosystem (Shrestha et al. 2012; Xu et al. 2009) that directly trigger the incidence of many dangerous events across the globe (IPCC 2014a, b; Vargas et al. 2017; Sorokin 2017). The consistency of high-temperature events, hot days and nights, has increased, whereas low-temperature events, cold days and nights, have declined across the globe (IPCC 2013). Erratic rainfall patterns, including floods and droughts, have augmented in many corners of the globe (Goswami et al. 2006; IPCC 2013, 2014a), though the number of rainy days has dropped (Rani and Sreekesh 2017). These results show that extremes will probably increase in this century. On the other hand, the number of cold days and nights due to increased in high temperatures are decreasing (IPCC 2013). Due to the increase in temperature, the sea level will rise, which at a later stage, may lead to overall global cooling (Sorokin and Mondello 2017).

5 Measures to Reduce Hazard

To lessen the possible risk, structural interventions like dykes and dams have been utilized in the past. Additionally, underground storage has historically been used to safeguard against drought and flood risks. We will look at how these hazard-reduction strategies may affect the adverse hazard and how they may be affected.

5.1 Dikes or Dykes or Levees

Hill torrent command area, primarily rainfed, contains dike-surrounded fields containing water (Mubeen and Veldman 2022). An embankment build or naturally developed due to sediment deposited alongside of a river by overflowing water to stop water from going in a certain direction is termed a levee. However, its breakage may cause damages. To lessen the risk of flooding, dykes have been constructed along significant stretches of the river systems (Ward et al. 2017) and hill torrents. Flood levees and dikes can affect the danger of drought and how droughts can raise the likelihood that they will fail. Levee and dike failure can make drought hazards worse. For instance, Vicuna et al. (2006) simulated the financial harm that Californian farmers would suffer from probable levee breakage. Additionally, they discovered that this would result in additional expenses for urban water users due to a shortage. Lower infiltration and groundwater recharge can result from the entrainment of rivers within dikes and levees, even in the absence of levee failure. Opperman et al.

(2009) asserted that re-establishing the connection between rivers and their floodplains might boost agricultural output and reduce the need to drain upstream reservoirs, thereby expanding options for water supply, hydropower and recreational activities.

Dikes and levees may occasionally be moistened during dry spells to lessen the likelihood of failing, reducing the amount of water for other uses. The Dutch Water Act establishes a hierarchy of surface water uses during dry seasons (Ministry of Transport, Public Works and Water Management 2010), with safety and the avoidance of irreparable harm, including protecting the stability of dikes and levees, receiving the greatest priority. Building dikes and levees can result in more development in the regions they protect, raising the flood risk (known as the levee effect) (Wiles et al. 2005; Di Baldassarre et al. 2018). The rise in socioeconomic activity and exposure may also stress the existing water resources, raising the chance of drought. Adverse effects of the reverse hazard on measures. At the same time, several instances of dikes and levees are failing due to dry conditions. Van Baars and Van Kempen (2009) claim that between 1134 and 2006, 5% of Dutch dike collapses were caused by drought. The dike failure in Wilnis in 2003, which resulted in 600 homes flooding and the evacuation of 2000 people, is a well-known example (Van Baars 2005). Drought reduced the peat dike's weight relative to the water force, which led to the failure and caused horizontal sliding (Van Baars 2005; Van Baars and Van Kempen 2009).

5.2 Dams

Dams and reservoirs can serve various functions, such as storing water to lessen flood risk and supplying water during probable dry spells. Most dams provide a variety of purposes over a year or even just a season. Thirty per cent of dams now in place serve diverse purposes, while 8.5% are predominantly used for flood management, and 17% are used for further water delivery or lowering drought risks (Lehner et al. 2011). Dams are challenging to manage because they have many distinct functions, and competing interests and goals may have unforeseen consequences. For example, flood protection encourages little water storage in the reservoir, lessening drought preparedness. Contrarily, drought prevention favours high water storage, which increases the risk of dam overflowing or failure in exceptionally heavy rain. When reservoirs guard against floods and droughts, their management can be changed to account for each risk. For instance, the Folsom Reservoir in California must dedicate 40% of its capacity to flood management (United States Congress House Committee on Resources 1997).

When replenishment is sluggish or non-existent, as in 1997, this can raise the chance of a drought. Because reservoirs had been depleted before the 2018 Kerala floods in India, the subsequent drought became worse (Lal et al. 2020). Dams are essential for flood management because they lessen downstream floods and high flows (Sordo-Ward et al. 2012). Compared to dams with fixed-crest spillways, gated spillways have higher levels of water preservation and flood reduction, but they are also more prone to operational failure, which can raise the risk of flooding downstream. As the dams are built to securely withstand floods of a specific scale, dam managers are less concerned about the safety of their structures during floods of smaller intensity. At the same time, dam care becomes a primacy when greater floods are a threat. Due to a false sense of security, this problem is compounded when flash floods happen in between dry seasons (Mediero et al. 2007). Additionally, this has been demonstrated that constructing dams to generate vast reservoirs may cause significant amounts of water to evaporate (Bond et al. 2008), increasing drought hazards. Additionally, dams may create a sense of security, enhancing downstream exposure and susceptibility. Likewise, in an investigation, Di Baldassarre et al. (2018) cite examples that the dams can cause supply/demand cycles and reservoir effects, where a rise in water demand can result in higher extraction rates and a greater sensitivity to drought due to a reliance on the reservoir. The flood hazard may suffer if drought management efforts are prioritized. An illustration would be the 2011 Brisbane floods, which followed a prolonged drought and was recorded by Van den Honert and McAneney (2011). Researchers observed the dam's water discharges, which provide most of Brisbane's water. Their analysis suggests that underestimating future rainfall estimates and adopting “no rainfall” scenarios may have resulted in less-than-ideal dam operations.

Dams and reservoirs can suffer severe effects from floods and droughts even if they are good long-term solutions for minimizing both flood and drought threats. Most reservoirs experience the sedimentation of suspended sediments as the flow velocity in the reservoirs decreases (Vorosmarty et al. 2003). Floods dramatically increase river sedimentation, reducing the amount of water that can be stored to lessen the effects of drought, as river sedimentation relies on its flow velocity. Vargas et al. (2017) reported dangers to levees and dams from high silt and debris flow, which are made worse by drought-related wildfires. The notorious Devil's Gate dam of southern California, which has transformed into a large debris basin as a result of numerous post-fire flood episodes, serves as a prime illustration (Karlamangla 2014).

5.3 Hill Torrent/Storm Water Control and Upstream Management

Urbanization has a variety of effects on the hydrological cycle. For instance, increased imperviousness can intensify runoff formation and speed up runoff response to precipitation, resulting in shorter concentration times and potential effects on flooding downstream, or decreased evapotranspiration, infiltration and groundwater recharge can lead to a failure in river base flow further advanced peak discharges. The hydrological cycle is altered in metropolitan settings by an established system of sealed zones, piped drainages and flow conveyance. SCM (Stormwater Control Measures). Fletcher et al. (2015) are becoming increasingly well-liked in this situation as an addition to or replacement for sub-surface piped systems. At the same time, SCM includes a variety of technologies that aim to alter the water management system of urban areas in order to lessen the risk of flooding or enhance pollution control.

Similarly, upstream solutions aim to keep water in the landscape and lessen the risk of floods and droughts. Room for the river includes dike setback, decreasing flood plains, reconnecting the side channels and removing bank defenses. Since SCM are relatively new, only one of the dangers is frequently considered (UACDC 2010). Following researchers (Ashley et al. 2017; Rauch et al. 2017) though offer theoretic frameworks for thorough evaluations of SCM; actual application-specific experiences are not offered. The improved recreational advantages of areas designed for stormwater storage during dry seasons are also mentioned. Examples include the East Lents Floodplain Project in Portland, Oregon, USA (Hoang et al. 2018), which involved landscaping with retention ponds, and multipurpose areas (Rosenzweig et al. 2019).

There are instances of severe infiltration causing local flooding due to significant groundwater recharge in Perth, Australia (Locatelli et al. 2017), as well as instances where the systems have been shown to manage floods well despite relatively modest measures (Lowe et al. 2017). SCM that rely on plants and green spaces may experience detrimental effects during droughts. For instance, droughts harm plants and mosses on green roofs by preventing the water-holding capacity from working properly. Five succulent species were planted in three different green roof substrates with varying water-holding capacities.

5.4 Underground Storage

In addition to dams, dikes and stormwater control focusing on regulating surface water, DRR methods also utilize the subsurface. For example, Floods can replenish groundwater supplies, which can lessen the effects of drought (e.g. Miguez-Macho and Fan 2012). Groundwater is frequently the most dependable source of water in arid and semi-arid regions, with seasonal flooding in wadi systems serving as the primary process of groundwater recharge. Particularly because abstraction rates outpace recharging in many of these aquifers, catastrophic floods are crucial for groundwater recharge in these places. Water is being stored in the subsurface using methods like managed aquifer recharge (MAR), sand dams, and aquifer storage and recovery (AS&R). However, these methods allow for the subsurface storage and recovery of water that is abundant during the wet season (or wet years) for use during the dry years or seasons. Nonetheless, subsurface storage is mostly used to combat drought, but it can also be used to combat floods, addressing the twin problems of seasonal water scarcity and seasonal floods.

Repercussions of actions taken to tackle the risk posed by the opposing hazard have yet not planned. The use of groundwater instead of surface water may have unforeseen repercussions for flood hazards. When there is a drought, as there was in California from 2008 to 2010, the sinking can be significant, reaching up to 270 mm/year in some regions (Faunt et al. 2016). Flood risk may rise as a result of both the subsidence itself and the decreased storage space. “Underground Taming of

Floods for Irrigation" (UTFI), which was tested in South Asia, is a novel strategy for managing floods and groundwater depletion simultaneously at the river basin scale (IWMI 2017). This involves increasing agricultural output in the area and carefully recharging excess wet season flows in aquifers to safeguard lives and property downstream. Comparing groundwater pumping and capturing flood flows for direct groundwater recharge on private farmlands in the Kings River Basin, California, reveals that the former is more affordable (Bachand et al. 2014). When significant floods occur, areas with substantial inter- or intra-annual rainfall variability might employ MAR to collect and store the water, then pump it to supplement rainwater collected to lessen the impact of drought episodes on agriculture (Rawluk et al. 2013). This method is utilized to catch peak flows in the Chao Phraya River Basin in Thailand, which can greatly lessen flood impacts and create additional income for farmers who can plant crops with high water requirements even during dry years (Pavelic et al. 2012). Additionally, regulated aquifer recharge is regarded as a water resources management strategy that can lessen water crises in the Mediterranean region and the southwest of the United States (Scolobig et al. 2012; Bachand et al. 2014; Maliva and Missimer 2012). It is crucial to keep in mind that MAR systems intended to boost infiltration and water availability during drought should not unintentionally induce flooding in low-lying areas, as happened, for instance, in Mexico when infiltrated wastewater swamped agricultural fields (Jimenez and Chavez 2004). The storage of water in sand riverbeds by means of sand dams improves water supply during dry periods. However, when scoop holes are utilized as an entry point, generating contamination and leading to a shortage of good quality water, the favourable impacts on water safety are negated.

Infiltration into the aquifer can be hampered by flooding because it can harm MAR infrastructure (Lluria 2009). Since overflowing toilets and surface flows can introduce contaminants into the groundwater, more frequent flooding increases the dangers of groundwater pollution. For instance, in 2013, a disastrous flood in Alberta, Canada, resulted in the contamination of drinking water wells along the floodways and flood margins with *E. coli* (Eccles et al. 2013). According to research by Ramachandran et al. (2019) in India, a flood event caused by the polluted Adyar River harmed the area's groundwater quality. When quality criteria are not reached, this groundwater pollution can lead to greater water scarcity.

6 Exposure and Vulnerability Management

6.1 Migration

Because of natural disasters, people frequently relocate from their customary domicile. Floods, in particular, are associated with sudden onset climate-related catastrophes that cause the greatest increases in population displacement (IOM 2019). Most people agree that moving is a significant step toward lowering the risks associated with natural hazards (Bohensky and Leitch 2014; Burrows and Kinney 2016). Migration as a DRR strategy can take the form of a planned relocation by the government, a voluntary action taken by an individual, or a forcible eviction (Bohensky and Leitch 2014; Mortreux et al. 2018). As this section explains, migration can both increase and decrease the danger of natural hazards.

Unplanned, forced migration processes frequently result in migrants with poorer socioeconomic levels and greater vulnerability (Bohensky and Leitch 2014; Wang et al. 2012). Due to their unstable financial situation, scarcity of resources and lack of access to employment prospects and social protection, these migrants are vulnerable. Additionally, due to language issues and mistrust of authorities, migrants may lack knowledge and information about catastrophic events, making them more vulnerable. Wang et al. (2012) observed that they perceived risk substantially less than non-migrants. The features of the areas where migrants settle are another source of risk. Floodplains are frequently a good place for habitation and economic growth but are also vulnerable to flooding. Cities are growing in riskier locations, including mega-deltas or places with inadequate access to amenities and increased urbanization and migration pressure (Kummu et al. 2011).

According to the World Bank, 40% of new migrants who arrived in Dakar, Senegal, between 1998 and 2008 relocated to areas with a high risk of flooding (Foresight 2011), and at the moment, Dakar's peri-urban districts experience significant flooding virtually every wet season. Moving away from floods can also make you more susceptible to drought. For instance, Mozambique saw the greatest flood in the country's history in 2000. Relocating people to new communities was one action performed after the floods. From the severely devastated districts, over 40,000 families were relocated to upland locations that were less susceptible to flooding but more vulnerable to drought (Wiles et al. 2005). These highland areas are very poor for agriculture, have low crop yields, and have farmers who are particularly vulnerable to droughts (Brida et al. 2013). Droughts were viewed as more disastrous than floods by farmers who had relocated to flood-safe places but later experienced water shortages and drought. This frequently forced farmers to relocate to the lowlands, where they were once more susceptible to flooding (Brida et al. 2013).

The growing urbanization of the locations to which migrants move also increases their susceptibility to dangers. More than 80% of the 17 million people who are susceptible to flooding each year live in urban and peri-urban settings (IDMC 2019). Urban sprawl, which has more impermeable surfaces, can worsen erosion and surface water runoff, resulting in more frequent floods. Additionally, excessive urbanization can result in water shortages or drought caused by humans by placing an undue burden on local infrastructure and resources (e.g. De Sherbinin et al. 2007). As a result, people who migrate in pursuit of land, resources, employment and means of subsistence may become more susceptible to recurrent hydrological extremes. Urbanization caused a significant decrease in water infiltration and an increase in runoff and erosion, which over the past century has intensified floods in the city (Lasda et al. 2010).

6.2 Adverse Effects on Measurements of the Opposite Hazard

Additionally, the risks associated with drought or floods themselves may have detrimental effects on migration as a DRR strategy. Because they either lack the means to migrate or do not have the means to travel as far as they would desire, vulnerable people exposed to a natural hazard may encounter considerable migration barriers (Black et al. 2013). For instance, it has been discovered that exposure to drought events decreases migration flows in many situations, such as a decline in Burkinabe migrants abroad (Henry et al. 2004).

6.3 Agricultural Practices and Land Use Changes

Of the culturable area in the hill torrent command system, merely 10% on average is irrigated, indicating the subsistence nature of agriculture. Sorghum and millet dominate up to 40% of the cropped area, followed by wheat (23%), chickpea (22%) and other crops (15%). Limited soil moisture, lack of improved cultivars, non-availability of technological soil-water-crop management, etc., results in very low crop yields (Shafiq 2013).

Extreme droughts and floods significantly influence agriculture, one of the human endeavours that use the most water. Numerous DRR strategies are employed to lessen the risk of agricultural drought and flood due to the close connection between the agricultural sector and the water cycle. These measures include subsurface storage, migration, dams and reservoirs, dikes and levees and dams and reservoirs.

Local communities are involved in constructing consecutive stone dams in Brazil to form micro-basins to preserve soil moisture (Gutierrez et al. 2014). This led to the construction of more than 3000 successive dams between 2001 and 2009, which produced microclimates that enhanced forestation, restored riverine vegetation, restored degraded areas, increased biodiversity and reduced the risk of drought. These little dams might also help with flood mitigation (e.g. Navarathimam et al. 2015). Actions to improve rainfall infiltration in the soil are frequently taken to lessen the likelihood of agricultural drought.

Cross-slope barriers are one example that might cause issues during periods of severe rainfall because the decreased drainage capacity can cause crops to become soggy and produce less (Lluria 2009; Makurira et al. 2009). Additionally, the same result has been shown when conservation agriculture is used (Dile et al. 2013). To decrease runoff, enhance infiltration and lessen the risk of flooding downstream, water-harvesting measures are frequently used in headwater catchments of rural semi-arid and arid regions. These treatments may be utilized to increase the productivity of rainfed agriculture, restore land productivity with insufficient precipitation, and reduce the danger of desertification and drought (Prinz et al. 1996). For example, according to research by Al-Seekh and Mohammad (2009), stone terraces and semi-circular bunds in the West Bank minimize runoff by 65–85% when compared to a control site. The main benefits of water-harvesting treatments include their simplicity, low cost, ability to be replicated, efficiency, and adaptability (Reij et al. 1988). However, poorly planned or upscaled interventions may lead to more topsoil erosion and gully formation, which would then cause more sedimentation and raise the danger of flooding downstream.

Technologies for irrigation that use less water have a strong potential to lower water demand, lowering the likelihood of agricultural drought. Systems of irrigation using a drip or micro-sprinkler are more effective than those with a pivot or flood. Spate irrigation is an old method of irrigation that uses seasonal river and stream flooding to fill irrigation channels. It is particularly popular in desert and semi-arid environments. Applying spate irrigation thereby combines flood and drought mitigation (Gevaert et al. 2020). Antwi-Agyei et al. (2018) analyse the measure's possible drawbacks in light of research conducted in northern Ghana. For instance, the loss of ecosystem services like flood control could result from the conversion of natural forest areas to farmland. The related stream bank farming and deforestation can worsen erosion, which can cause rivers to become more sedimented and increase the likelihood of floods. This hurts food production (Surminski et al. 2016; Saaf et al. 2019). Utilizing several crops and cropping techniques as a drought or flood risk measure might result in complicated relationships.

When choosing crops to cultivate in flood-prone areas, producers consider the flood regime and the crops' vulnerability to floods (Klaus et al. 2016). For instance, farmers in sub-Saharan Africa plant short-lived crops and adjust the planting and harvesting time to avoid periods of heavy rainfall (Sani and Chalchisa 2016) or dry spells (Ochieng et al. 2017). In many agricultural areas, re- or afforestation of degraded land, also known as Eco-DRR, is seen as a workable flood mitigation method for DRR and climate change adaptation (FAO 2019). However, during dry spells, plantations' increased evapotranspiration and decreased groundwater recharge can dramatically lower dry season flow and result in water shortages. When plantation trees were built over degraded grasslands in Fiji, the dry season flows were reduced, which led to limitations in the city's water supply (Waterloo 1994). The installation of eucalypt plantings in Argentina resulted in a 50% reduction in groundwater recharge days and an average loss of 0.38 m in groundwater level (Jobbagy and Jackson 2004). Reforestation's effects on dry season minimum flows are significantly influenced by soil infiltration conditions (Bruijnzeel 2004). In Panama, higher infiltration rates and lower peak flow discharge during the rainy season were the causes of increased baseflow in a forested catchment compared to disturbed catchments. Farmers can modify their farming and harvesting operations when an especially dry or wet season is anticipated, thanks to early warning systems (EWS). Particularly in tropical and subtropical regions, where several seasonal rainfall predictions are currently in operation, seasonal forecasts have proven to greatly benefit (Murphy et al. 2001).

Local mistrust can result from incorrect forecasts, such as farmers being hit by a flood event while preparing for a drought season, which can result in significant losses (Ciurean et al. 2018; Murphy et al. 2001). Local trust in seasonal forecasts and the organizations and governments that provide themes time to develop (Pavelic et al. 2012; Tall et al. 2012). In farming communities without access to or with limited use of scientific projections, traditional forecasting techniques have proved crucial (Recha et al. 2008). However, the use of some indicators by farmers (such as the beginning of rain) has decreased in accuracy and dependability as a result of an increased exposure to irregular, more frequent, severe extreme occurrences, which has negative effects on crop output (Reid et al. 2009).

6.4 Negative Effects of the Opposite Hazard on Measurements

Farmers using rainfed agricultural methods must decide when to plant their crops at the start of the rainy season. Crop failure is more likely if the season gets off to a bad start. Later in the season, seeding could be a useful drought risk mitigation strategy if it stays too dry. However, during the start of the season, excessive rain and flooding might cause nutrients to leak out of the root zone, jeopardizing its effectiveness (Bussmann et al. 2016). Additionally, careful crop selection can lessen the effects of agriculture (Klaus et al. 2016). Sub-Saharan African farmers experiencing decreased precipitation have shifted from high to low water-requirement crops (Sani and Chalchisa 2016). This involves the possibility of decreased yields during periods of higher precipitation and flooding (Patt and Schroter 2008). Drought and heat make re- or afforestation more difficult for trees because they increase the risk of fire and pest infestation (Allen et al. 2015). For instance, fire has been highlighted as the main danger to the success of forestry programmes in the Philippines (Ancog et al. 2016). Forest fires have significantly increased runoff, peak flows and erosion after the fire, which has resulted in destructive floods and debris flows (De Graff 2018). In this regard, droughts may undermine the effectiveness of forestation efforts to lower flood hazards.

6.5 Socioeconomic Factors of Vulnerability and Preparedness

The risk of socioeconomic vulnerability can increase due to successive drought or flood events (Gallina et al. 2016). For instance, flood-induced migration enhances drought susceptibility through social marginalization, and drought-induced unemployment can result in severe financial challenges during floods (Rockstrom 2003). In addition, reduced socioeconomic vulnerability to one type of hazard can affect the risk of another type of hazard (Dilling et al. 2015). However, the majority of scientific articles examine all risks at once. The presumption is that reducing socioeconomic vulnerability benefits all threats or is "no remorse" (i.e., reducing flood vulnerability also benefits drought and vice versa) (e.g. Dilling et al. 2015; White et al. 2001).

Improvements to infrastructure (Kalantari et al. 2019), health care and cleanliness (Few 2007), and food and water security are only a few examples of actions that tend to promote overall socioeconomic development and hence decrease vulnerability to natural hazards (Pelletier et al. 2016), access to markets (Bebbington 1999), urban planning (Houghton 2012), insurance (Surminski et al. 2016), and diversification of agricultural operations or drinking water supplies (Head 2014) are some further examples. Both extremes can benefit from measures to promote readiness, knowledge, education, or information (early warning systems). However, these do not necessarily result in actions that reduce vulnerability, for instance, because the most vulnerable groups in society lack agency (Navarathimam et al. 2015; Sangita 2016). However, some of these actions may also result in maladaptation, unintentionally making people more susceptible to floods and/or droughts.

Repercussions of actions taken to tackle the risk posed by the opposing hazard have yet not planned. The UNDRR defines a society's preparation as its ability to respond to disasters and recover from their effects, and risk perception has an impact on preparedness. Risk perception deals with how individuals and organizations judge the seriousness and propensity of a risk event (Urquijo and De Stefano 2016). Scolobig et al. (2012) cite insufficient awareness as one factor contributing to inadequate preparedness for natural catastrophes. Due to biases in risk information, faith in weather services, people's memories and risk adversity, societies' perceptions of risk may deviate from reality (Loucks 2015). As a result, if one risk is prioritized, another risk may suffer from a lack of preparation, increasing its risk. On the other hand, being ready for one specific hazard might raise general hazard awareness, regardless of the type of hazard, reducing the chance of another hazard (Siegel et al. 2003). The impression of risk is significantly influenced by the media. Following a thorough examination of the daily news for 25 years in Catalonia's most-read newspaper (NE Spain), Llasat et al. (2009) despite the fact that floods are also a significant risk in this area, statistics reveal that most news stories were about droughts and forest fires, which were then followed by floods and heavy rains. This may result in an erroneous perception of low flood risk, which has an impact on the community and individual behaviour.

Early warning systems for flooding are plagued by uncertainty and false alarms, which could incur significant expenditures. For instance, reservoir management might release water based on information from a flood early warning, but if the anticipated flood does not occur or is less severe than anticipated, this could lead to a water shortage (Rogers and Tsirkunov 2010). Disaster relief efforts by donors and NGOs can raise vulnerability by fostering dependency and undercutting local initiatives. Insurance, microcredit programmes, and agricultural diversification have been found to lower incentives for action and impede investment (Loucks 2015; Salim et al. 2019). Another illustration of how such methods to reduce exposure to one risk might make people more vulnerable to the other risk can be seen in Mexico City, where those living in unofficial settlements without access to piping can purchase water from water trucks. This might result in more people living in poverty and less ability to deal with flooding (Eakin et al. 2016).

There are also several instances of water governance and policy. For instance, in Southeast Queensland, Australia, during the Millennium Drought of 2001–2008, the state government launched significant changes in water governance, such as the centralization of power in place of more cooperative methods for water management. Stakeholder disagreement and levels of mistrust increased as a result. During the 2010 flood disaster, the centralized system was unable to stop construction in the floodplain, increasing flood exposure (Head 2014).

6.6 Adverse Effects on Measurements of the Opposite Hazard

Intangible in nature, the majority of vulnerability-reducing strategies are frequently unaffected by an occurrence. The distribution of scarce resources and readiness are just a few examples of indirect impacts. Crisis management during a flood or drought disaster diverts focus, resources and priority away from other water-related problems, potentially raising the probability of the opposite extreme. As illustrated by this comment from an Australian local government representative: "You forget, because of 10 years of drought, that land floods," it is to be expected that flood memory decays more quickly during a prolonged drought (Bohensky and Leitch 2014).

In the hill torrent-affected areas, communication and transport charges affect the local economy as in the rural areas hit by natural disasters of hill torrents like the one this year, 2022 rain floods in Pakistan resulted in a significant reduction of production rate compared with markets. This variation in rate could be attributed to the mode of transport (Shafiq 2013). Therefore, due to consideration should be given now to avoid humanitarian and ecosystem disturbances in the coming years (Mubeen and Veldman 2022).

7 Conclusion

Natural hazards are beyond our control, while human activities drastically change our global's climatic pattern. Furthermore, underdeveloped or developing countries are more susceptible to these climatic changes. So, all human efforts should be diverted to minimize the disturbance in the natural flow of energy and also sustainable ecosystem technologies should be implemented in every run of life where agriculture can play a vital role. Strong legislation in this regard can be a dynamic strategy, and developed countries should play a big role in achieving the target of minimizing disaster risk management. Otherwise, the result shows that extremes will be increasing daily until the end of this century. Whereas, among various strategies, disaster risk can be minimized through interventions like the construction of levee, dikes, dams and aquifer recharge. Sediments and silts should be managed both upstream and downstream as it intensifies the flood disaster. Moreover, disaster risk can be managed through migration, health system improvement, cleanliness, food and water security, insurance, agricultural diversity, etc. It will improve the socioeconomics of the disaster-affected people, and system vulnerability can be reduced.

References

Ahmed M (2020) Introduction to modern climate change. Andrew E. Dessler: Cambridge University Press, 2011, 252 pp, ISBN-10: 0521173159. Sci Total Environ 734:139397. https://doi.org/10.1016/j.scitotenv.2020.139397

Albert S et al (2017) Heading for the hills: climate-driven community relocations in the Solomon Islands and Alaska provide insight for a 1.5°C future. Reg Environ Chang 2017:1–12. https://doi.org/10.1007/s10113-017-1256-8

Allen CD, Breshears DD, McDowell NG (2015) On underestimation of global vulnerability to tree mortality and forest die-off from hotter drought in the Anthropocene. Ecosphere 6:129. https://doi.org/10.1890/ES15-00203.1

Al-Seekh SH, Mohammad AG (2009) The effect of water harvesting techniques on runoff, sedimentation, and soil properties. Environ Manag 44:37–45. https://doi.org/10.1007/s00267-009-9310-z

Ancog RC, Florece LM, Boy Nicopior O (2016) Fire occurrence and fire mitigation strategies in a grassland reforestation area in the Philippines. Forest Policy Econ 64:35–45. https://doi.org/10.1016/j.forpol.2016.01.002

Antwi-Agyei P, Dougill AJ, Stringer LC, Codjoe SNA (2018) Adaptation opportunities and maladaptive outcomes in climate vulnerability hotspots of northern Ghana. Clim Risk Manag 19:83–93. https://doi.org/10.1016/j.crm.2017.11.003

Ashley RM, Digman CJ, Horton B, Gersonius B, Smith B, Shaffer P, Baylis A (2017) Evaluating the longer term benefits of sustainable drainage. Proc Inst Civ Eng 171:57–66. https://doi.org/10.1680/jwama.16.00118

Bachand PAM, Roy SB, Choperena J, Cameron D, Horwath WR (2014) Implications of using on-farm flood flow capture to recharge groundwater and mitigate flood risks along the Kings River, CA. Environ Sci Technol 48:13601–13609. https://doi.org/10.1021/es501115c

Bebbington A (1999) Capitals and capabilities: a framework for analyzing peasant viability, rural livelihoods and poverty. World Dev 27:2021–2044. https://doi.org/10.1016/S0305-750X(99)00104-7

Black R, Arnell NW, Adger N, Thomas D, Geddes A (2013) Migration, immobility and displacement outcomes following extreme events. Environ Sci Pol 27:S32–S43. https://doi.org/10.1016/j.envsci.2012.09.001

Bohensky E, Leitch A (2014) Framing the flood: a media analysis of themes of resilience in the 2011 Brisbane flood. Reg Environ Chang 14:475–488. https://doi.org/10.1007/s10113-013-0438-2

Bond NR, Lake PS, Arthington AH (2008) The impacts of drought on freshwater ecosystems: an Australian perspective. Hydrobiologia 600:3–16. https://doi.org/10.1007/s10750-008-9326-z

Brida A-B, Owiyo T, Sokona Y (2013) Loss and damage from the double blow of flood and drought in Mozambique. Int J Global Warm 5:514–531. https://doi.org/10.1504/IJGW.2013.057291

Bruijnzeel LA (2004) Hydrological functions of tropical forests: not seeing the soil for the trees? Agric Ecosyst Environ 104:185–228. https://doi.org/10.1016/j.agee.2004.01.015

Burrows K, Kinney PL (2016) Exploring the climate change, migration and conflict nexus. Int J Environ Res Public Health 13:443. https://doi.org/10.3390/ijerph13040443

Burton I, Kates RW, White GF (1993) The environment as hazard. Guilford Press, New York

Bussmann A, Ahmed Elagib N, Fayyad M, Ribbe L (2016) Sowing date determinants for Sahelian rainfed agriculture in the context of agricultural policies and water management. Land Use Policy 52:316–328. https://doi.org/10.1016/j.landusepol.2015.12.007

Ciurean R, Gill J, Reeves HJ, O'Grady S, Aldridge T (2018) Review of multi-hazards research and risk assessments. Open report OR/18/057. British Geological Survey, Nottingham

Dar RUN, Alam M (2020) Understanding disaster risk, its components and reduction. In: International conference on building resilient and sustainable societies: emerging social and economic challenges. Jamia Millia Islamia, New Delhi

De Graff JV (2018) A rationale for effective post-fire debris flow mitigation within forested terrain. Geoenviron Disast 5:7. https://doi.org/10.1186/s40677-018-0099-z

De Sherbinin A, Carr D, Cassels S, Jiang L (2007) Population and environment. Annu Rev Environ Resour 32:345–373. https://doi.org/10.1146/annurev.energy.32.041306.100243

Di Baldassarre G, Kreibich H, Vorogushyn S, Aerts J, Arnbjerg-Nielsen K, Barendrecht M, Bates P, Borga M, Botzen W, Bubeck P, De Marchi B, Carmen Llasat M, Mazzoleni M, Molinari D, Mondino E, Mård J, Petrucci O, Scolobig A, Viglione A, Ward PJ (2018) An interdisciplinary research agenda to explore the unintended consequences of structural flood protection. Hydrol Earth Syst Sci 22:5629–5637. https://doi.org/10.5194/hess-22-5629-2018

Dile YT, Karlberg L, Temesgen M, Rokstrom J (2013) The role of water harvesting to achieve sustainable agricultural intensification and resilience against water related shocks in sub-Saharan Africa. Agric Ecosyst Environ 181:69–79. https://doi.org/10.1016/j.agee.2013.09.014

Dilling L, Daly ME, Travis WR, Wilhelmi OV, Klein RA (2015) The dynamics of vulnerability: why adapting to climate variability will not always prepare us for climate change. Wiley Interdiscip Rev Clim Chang 6:413–425. https://doi.org/10.1002/wcc.341

Eakin H, Lerner AM, Manuel-Navarrete D, Aguilar BH, Martínez-Canedo A, Tellman B, Charli-Joseph L, Fernandez Alvarez R, Bojorquez-Tapia L (2016) Adapting to risk and perpetuating poverty: household's strategies for managing flood risk and water scarcity in Mexico City. Environ Sci Pol 66:324–333. https://doi.org/10.1016/j.envsci.2016.06.006

Eccles KM, Checkley S, Sjogren D, Barkema HW, Bertazzon S (2013) Lessons learned from the 2013 Calgary flood: assessing risk of drinking water well contamination. Appl Geogr 80:78–85. https://doi.org/10.1016/j.apgeog.2017.02.005

FAO (2006) The role of local level institutions in reducing vulnerability to natural disasters and in sustainable livelihoods development. Natural Resources Management and Environment Department, SD Dimensions. http://www.fao.org

FAO (2019) Forests for resilience to natural, climate and human-induced disasters and crises. Food and Agricultural Organization of the United Nations, Forestry Department, Rome. https://doi.org/10.4060/ca6920en

Faunt CC, Sneed M, Traum J, Brandt TJ (2016) Water availability and ground subsidence in the Central Valley, California, USA. Hydrogeol J 24:675–684. https://doi.org/10.1007/s10040-015-1339-x

Few R (2007) Health and climatic hazards: framing social research on vulnerability, response and adaptation. Glob Environ Chang 17:281–295. https://doi.org/10.1016/j.gloenvcha.2006.11.001

Fletcher TD, Shuster W, Hunt WF, Ashley R, Butler D, Arthur S, Trowsdale S, Barraud S, Semadeni-Davies A, Bertrand-Krajewski J-L, Mikkelsen PS, Rivard G, Uhl M, Dagenais D, Viklander M (2015) SUDS, LID, BMPs, WSUD and more – the evolution and application of terminology surrounding urban drainage. Urban Water J 12:525–542. https://doi.org/10.1080/1573062X.2014.916314

Foresight (2011) Migration and global environmental change: future challenges and opportunities. Government Office for Science, London. https://assets.publish ing.service.gov.uk/government/uploads/system/uploads/attachment_data/file/287717/11-1116-migration-and-global-environmental-change.pdf

Gallina V, Torresan S, Critto A, Sperotto A, Glade T, Marcomini A (2016) A review of multi-risk methodologies for natural hazards: consequences and challenges for a climate change impact assessment. J Environ Manag 168:123–132. https://doi.org/10.1016/j.jenvman.2015.11.011

Gevaert AI, Van der Meulen RC, Groen J (2020) Towards sustainable groundwater use in African drylands. Report AW_307.2_AG_190984. Acacia Water, Gouda

Gill JC, Malamud BD (2017) Anthropogenic processes, natural hazards, and interactions in a multi-hazard framework. Earth-Sci Rev 166:246–269

Goswami BN, Venugopal V, Sengupta D, Madhusoodanan MS, Xavier PK (2006) Increasing trend of extreme rain events over India in a warming environment. Science 314:1442–1445

Gutierrez APA, Engle NL, De Nys E, Molejon C, Martins ES (2014) Drought preparedness in Brazil. Weather Clim Extremes 3:95–106. https://doi.org/10.1016/j.wace.2013.12.001

Hay A (2007) Extreme weather and climate events, and farming risks. In: Sivakumar MVK, Motha R (eds) Managing weather and climate risks in agriculture. Springer, Berlin, pp 1–19

Head BW (2014) Managing urban water crises: adaptive policy responses to drought and flood in Southeast Queensland, Australia. Ecol Soc 19:33. https://doi.org/10.5751/ES-06414-190233
Henry S, Schoumaker B, Beauchemin C (2004) The impact of rainfall on the first outmigration: a multi-level event-history analysis in Burkina Faso. Popul Environ 25:423–460. https://doi.org/10.1023/B:POEN.0000036928.17696.e8
Hoang L, Fenner RA, Skenderian M (2018) A conceptual approach for evaluating the multiple benefits of urban flood management practices: evaluating the multiple benefits of urban flood management practices. J Flood Risk Manage 11:S943–S959. https://doi.org/10.1111/jfr3.12267
Houghton A (2012) Health impact assessments: a tool for designing climate change resilience into green building and planning projects. J Green Build 6:66–87. https://doi.org/10.3992/jgb.6.2.66
Houghton JT, Ding YDJG, Griggs DJ, Noguer M, van der Linden PJ, Dai X, Maskell K, Johnson CA (eds) (2001) Climate change 2001: the scientific basis: contribution of Working Group I to the third assessment report of the Intergovernmental Panel on Climate Change. Cambridge University Press, Cambridge
IDMC (2019) GRID 2019. Global report on internal displacement. Internal Displacement Monitoring Centre (IDMC) & Norwegian Refugee Council, Geneva. https://www.internal-displacement.org/sites/default/files/publications/documents/2019-IDMC-GRID.pdf
IOM (2019) International migration law-n. 34 Glossary on migration. International Organization for Migration, Geneva
IPCC (2012) Managing the risks of extreme events and disasters to advance climate change adaptation. A special report of working groups I and II of the Intergovernmental panel on climate change. Cambridge University Press, Cambridge
IPCC (2013) Summary for policymakers. Climate change 2013: the physical science basis. In: Stocker TF, Qin D, Plattner GK, Tignor M, Allen SK, Boschung J, Nauels A, Xia Y, Bex V, Midgley PM (eds) Contribution of working Group I to the Fifth Assessment Report of the Intergovernmental Panel on Climate Change. Cambridge University Press, Cambridge, pp 3–29
IPCC (2014a) Climate change 2014. Synthesis report summary for policymakers. IPCC, Geneva
IPCC (2014b) Climate change 2014: impacts, adaptation, and vulnerability. Part A: global and sectoral aspects. In: Contribution of working group II to the fifth assessment report of the intergovernmental panel on climate change. Cambridge University Press, Cambridge
IWMI (2017) Underground taming of floods for irrigation (UTFI). International Water Management Institute. http://utfi.iwmi.org/
Jayaraman TK, Choong CK, Ng CF, Bhatt M (2018) Natural disasters and tourism-led economic growth: a case study of Fiji: 1980–2014. In: Handbook of small states. Routledge, Milton Park, pp 573–590
Jimenez B, Chavez A (2004) Quality assessment of an aquifer recharged with wastewater for its potential use as drinking water source: "El Mezquital Valley" case. Water Sci Technol 50:269–276. https://doi.org/10.2166/wst.2004.0141
Jobbagy EG, Jackson RB (2004) Groundwater use and salinization with grassland afforestation. Glob Chang Biol 10:129911312. https://doi.org/10.1111/j.1365-2486.2004.00806.x
Kalantari Z, Ferreira CSS, Koutsouris AJ, Ahlmer A-K, Cerda A, Destouni G (2019) Assessing flood probability for transportation infrastructure based on catchment characteristics, sediment connectivity and remotely sensed soil moisture. Sci Total Environ 661:393–406. https://doi.org/10.1016/j.scitotenv.2019.01.009
Karlamangla S (2014) L.A. County supervisors OK debris clearance for Devil's Gate Dam. Los Angeles Times. https://www.latimes.com/local/countygovernment/la-me-1113-devils-gate-2-20141113-story.html
Kelman I (2018) Lost for words amongst disaster risk science vocabulary? Int J Disaster Risk Sci 9(3):281–291
Kiehl JT, Trenberth KE (1997) Earth's annual global mean energy budget. Bull Am Meteorol Soc 78(2):197–208
Klaus S, Kreibich H, Merz B, Kuhlmann B, Schroter K (2016) Large-scale, seasonal flood risk analysis for agricultural crops in Germany. Environ Earth Sci 75:1289. https://doi.org/10.1007/s12665-016-6096-1

Kummu M, De Moel H, Ward PJ, Varis O (2011) How close do we live to water? A global analysis of population distance to freshwater bodies. PLoS One 6(6):e20578. https://doi.org/10.1371/journal.pone.0020578

Lal P, Prakash A, Kumar A, Srivastava PK, Saikia P, Pandey AC, Srivastava P, Khan ML (2020) Evaluating the 2018 extreme flood hazard events in Kerala, India. Remote Sens Lett 11:436–445. https://doi.org/10.1080/2150704X.2020.1730468

Lasda O, Dikou A, Papapanagiotou E (2010) Flash flooding in Attika, Greece: climatic change or urbanization? Ambio 39:608–611. https://doi.org/10.1007/s13280-010-0050-3

Lehner B, Liermann CR, Revenga C, Vorosmarty C, Fekete B, Crouzet P, Doll P, Endejan M, Frenken K, Magome J, Nilsson C, Robertson JC, Rodel R, Sindorf N, Wisser D, Nilsson C (2011) High-resolution mapping of the world's reservoirs and dams for sustainable river-flow management. Front Ecol Environ 9:494–502. https://doi.org/10.1890/100125

Llasat MC, Llasat-Botija M, Barnolas M, Lopez L, Altava-Ortiz V (2009) An analysis of the evolution of hydrometeorological extremes in newspapers: the case of Catalonia, 1982–2006. Nat Hazards Earth Syst Sci 94:1201. https://doi.org/10.5194/nhess-9-1201-2009

Lluria MR (2009) Successful application of managed aquifer recharge in the improvement of the water resources management of semi-arid regions: examples from Arizona and the Southwestern USA. Bol Geol Min 120:111–120

Locatelli L, Mark O, Steen Mikkelsen P, Arnbjerg-Nielsen K, Deletic A, Roldin M, John Binning P (2017) Hydrologic impact of urbanization with extensive stormwater infiltration. J Hydrol 544:524–537. https://doi.org/10.1016/j.jhydrol.2016.11.030

Loucks A (2015) Perspectives on socio-hydrology: simulating hydrologic-human interactions. Water Resour Res 51:4789–4794. https://doi.org/10.1002/2015WR017002

Lowe R, Urich C, Domingo NS, Mark O, Deletic A, Arnbjerg-Nielsen K (2017) Assessment of urban pluvial flood risk and efficiency of adaptation options through simulations - a new generation of urban planning tools. J Hydrol 550:355–367. https://doi.org/10.1016/j.jhydrol.2017.05.009

Makurira H, Savenije HHG, Uhlenbrook S, Rockstrom J, Senzanje A (2009) Investigating the water balance of on-farm techniques for improved crop productivity in rainfed systems: a case study of Makanya catchment, Tanzania. Phys Chem Earth 34:93–98. https://doi.org/10.1016/j.pce.2008.04.003

Maliva RG, Missimer TM (2012) Managed aquifer recharge. In: Maliva RG, Missimer TM (eds) Arid lands water evaluation and management. Environmental science and engineering (environmental engineering). Springer, Berlin

Mediero L, Garrote L, Martin-Carrasco F (2007) A probabilistic model to support reservoir operation indecisions during flash floods. Hydrol Sci J 52:523–537. https://doi.org/10.1623/hysj.52.3.523

Miguez-Macho G, Fan Y (2012) The role of groundwater in the Amazon water cycle: 1. Influence on seasonal streamflow, flooding and wetlands. J Geophys Res Atmos 117:017539. https://doi.org/10.1029/2012JD017539

Ministry of Transport, Public Works and Water Management (2010) Water act. Ministry of Transport, Public Works and Water Management, Hague

Mortreux C, Safra de Campos R, Adger WN, Ghosh T, Das S, Adams H, Hazra S (2018) Political economy of planned relocation: a model of action and inaction in government responses. Glob Environ Chang 50:123–132. https://doi.org/10.1016/j.gloenvcha.2018.03.008

Mubeen K, Veldman R (2022) Hill torrent floods impacting the Southern Punjab Pakistan. https://thewaterchannel.tv/thewaterblog/hill-torrent-floods-impacting-the-southern-punjab-pakistan/

Murphy SJ, Washington R, Downing TE, Martin RV, Ziervogel G, Preston A, Todd M, Butterfield R, Briden J (2001) Seasonal forecasting for climate hazards: prospects and responses. Nat Hazards 23:171–196. https://doi.org/10.1023/A:1011160904414

Mysiak J, Surminski S, Thieken A, Mechler R, Aerts J (2016) Brief communication: Sendai framework for disaster risk reduction–success or warning sign for Paris? Nat Hazards Earth Syst Sci 16(10):2189–2193

Navarathimam K, Gusyev MA, Hasegawa A, Magome J, Takeuchi K (2015) Agricultural flood and drought risk reduction by a proposed multi-purpose dam: a case study of the Malwathoya

River Basin, Sri Lanka. In: 21st international congress on modelling and simulation, Gold Coast, Australia. https://pdfs.semanticscholar.org/9a0f/0f56a35b29cc50681361613d22387a215f35.pdf

Ochieng J, Kirimi L, Makau J (2017) Adapting to climate variability and change in rural Kenya: farmer perceptions, strategies and climate trends. Nat Resour Forum 41:195–208. https://doi.org/10.1111/1477-8947.12111

Opperman JJ, Galloway GE, Fargione J, Mount JF, Richter BD, Secchi S (2009) Sustainable floodplains through large-scale reconnection to rivers. Science 326:1487–1488. https://doi.org/10.1126/science.1178256

Patt A, Schroter D (2008) Perceptions of climate risk in Mozambique: implications for the success of adaptation strategies. Glob Environ Chang 18:458–467. https://doi.org/10.1016/j.gloenvcha.2008.04.002

Pavelic P, Kriengsak S, Saraphirom P, Nadee S, Pholkern K, Chusanathas S, Munyou S, Tangsutthinon T, Intarasut T, Smakhtin V (2012) Balancing-out floods and droughts: opportunities to utilize floodwater harvesting and groundwater storage for agricultural development in Thailand. J Hydrol 470–471:55–64. https://doi.org/10.1016/j.jhydrol.2012.08.007

Pelletier B, Hickey GM, Bothi KL, Mude A (2016) Linking rural livelihood resilience and food security: an international challenge. Food Secur 8:469–476. https://doi.org/10.1007/s12571-016-0576-8

Prinz D, Pereria L, Feddes RA, Gilleym JR, Lessaffre B (1996) Water harvesting past and future. In: Proceedings of the NATO advanced research workshop, Vimeiro, Portugal

Ramachandran A, Krishnamurthy RR, Jayaprakash M, Shanmugasundharam A (2019) Environmental impact assessment of surface water and groundwater quality due to flood hazard in Adyar River Bank. Acta Ecol Sin 39:125–132. https://doi.org/10.1016/j.chnaes.2018.08.008

Rani S, Sreekesh S (2017) Variability of temperature and rainfall in the Upper Beas Basin, Western Himalayas. In: Mal S, Singh RB, Huggel C (eds) Climate change, extreme events and disaster risk reduction. Springer, Cham

Rauch W, Urich C, Bach PM, Rogers BC, De Haan FJ, Brown RR, Mair M, McCarthy DT, Kleidorfer M, Sitzenfrei R, Deletic A (2017) Modelling transitions in urban water systems. Water Res 126:501–514. https://doi.org/10.1016/j.watres.2017.09.039

Rawluk A, Curtis A, Sharp E, Kelly BF, Jakeman AJ, Ross A, Arshad M, Brodie R, Pollino CA, Sinclair D, Croke B, Quereshi ME (2013) Managed aquifer recharge in farming landscapes using large floods: an opportunity to improve outcomes for the Murray-Darling Basin? Aust J Environ Manage 20:34–48. https://doi.org/10.1080/14486563.2012.724785

Recha CW, Shisanya CA, Lakokha GL, Kinuthia RN (2008) Perception and use of climate forecast information among smallholder farmers in semi-arid Kenya. Asian J Appl Sci 1:123–135. https://doi.org/10.3923/ajaps.2008.123.135

Reid H, Cannon T, Berger R, Alam M, Milligan A (2009) Community based adaptation to climate change. Participatory learning and action. International Institute for Environment and Development, London. https://pubs.iied.org/pdfs/14573IIED.pdf

Reij C, Mulder P, Begeman L (1988) Water harvesting for plant production. World Bank technical paper 91. World Bank, Washington

Rockstrom J (2003) Resilience building and water demand management for drought mitigation. Phys Chem Earth 28:869–877. https://doi.org/10.1016/j.pce.2003.08.009

Rogers D, Tsirkunov V (2010) Costs and benefits of early warning systems. In: Global assessment report on disaster risk reduction. UNDRR, Geneva

Rosenzweig B, Ruddell BL, McPhillips L, Hobbins R, McPhearson T, Cheng Z, Chang H, Kim Y (2019) Developing knowledge systems for urban resilience to cloudburst rain events. Environ Sci Pol 99:150–159. https://doi.org/10.1016/j.envsci.2019.05.020

Saaf E-J, Figueres C, Waterloo MJ, de Wit G, Nicolin V (2019) Niger-Niamey, Niger River. DRR-Team mission report DRR218NE01, The Hague, The Netherlands

Salim W, Bettinger K, Fisher M (2019) Maladaptation on the waterfront: Jakarta's growth coalition and the Great Garuda. Environ Urban 10:63–80. https://doi.org/10.1177/0975425318821809

Sangita K (2016) Transnational feminism and women's activism: building resilience to climate change impact through women's empowerment in climate smart agriculture. Asian J Women's Stud 22:497–506. https://doi.org/10.1080/12259276.2016.1242946

Sani S, Chalchisa T (2016) Farmers' perception, impact and adaptation strategies to climate change among smallholder farmers in Sub-Saharan Africa: a systematic review. J Resour Dev Manage 26:121. https://doi.org/10.5539/jas.v5n4p121

Scolobig A, De Marchi B, Borga M (2012) The missing link between flood risk awareness and preparedness: findings from case studies in an Alpine Region. Nat Hazards 63:499–520. https://doi.org/10.1007/s11069-012-0161-1

Seddiky MA, Giggins H, Gajendran T (2020) International principles of disaster risk reduction informing NGOs strategies for community based DRR mainstreaming: The Bangladesh context. Int J Disaster Risk Reduct 48:101580

Shafiq CM (2013) Hill torrent management initiatives in southern part of Punjab an overview, impact analysis and way forward. In: 72nd annual session of Pakistan engineering congress, pp 289–314

Shrestha UB, Gautam S, Bawa KS (2012) Widespread climate change in the Himalayas and associated changes in local ecosystems. PLoS ONE 7(5):e36741. https://doi.org/10.1371/journal.pone.0036741

Siegel JM, Shoaf KI, Afifi AA, Bourque LB (2003) Surviving two disasters: does reaction to the first predict response to the second? Environ Behav 35:637–654. https://doi.org/10.1177/0013916503254754

Sordo-Ward A, Garrote L, Martín-Carrasco F, Bejarano MD (2012) Extreme flood abatement in large dams with fixed-crest spillways. J Hydrol 466–467:60–72. https://doi.org/10.1016/j.jhydrol.2012.08.009

Sorokin LV (2017) The experience of disaster risk reduction and economic losses reduction in Malaysia during the water crisis 1998 in the context of the next El Niño strongest on record maximum 2015. In: Mal S, Singh RB, Huggel C (eds) Climate change, extreme events and disaster risk reduction. Springer, Cham

Sorokin LV, Mondello G (2017) Entering the new +2 °C global warming age and a threat of world ocean expansion for sustainable economic development. In: Mal S, Singh RB, Huggel C (eds) Climate change, extreme events and disaster risk reduction. Springer, Cham

Surminski S, Bouwer LM, Linnerooth-Bayer J (2016) How insurance can support climate resilience. Nat Clim Chang 6:333–334. https://doi.org/10.1038/nclimate2979

Tall A, Mason SJ, Van Aalst M, Suarez P, Ait-Chellouche Y, Diallo AD, Braman L (2012) Using seasonal climate forecasts to guide disaster management: the red cross experience during the 2008 West Africa floods. Int J Geophys 2012:986016. https://doi.org/10.1155/2012/986016

UACDC (2010) Low impact development: a design manual for urban areas. Arkansas University Community Design Center, Fayetteville. http://www.bwdh 2o.org/wp-content/uploads/2012/03/Low_Impact_Development_Manual-2010.pdf

UNFCCC (2008) Integrating practices, tools and systems for climate risk assessment and management and strategies for disaster risk reduction into national policies and programmes. Technical Paper. http://unfccc.int/documentation/documents/advanced_search/items/3594

United States Congress House Committee on Resources (1997) Flood control projects and ESA: hearing before the committee on resources, house of representatives. In: One Hundred Fifth Congress, First Session, on H.R. 478, a Bill to Amend the Endangered Species Act of 1973 to Improve the Ability of Individuals and Local, State, and Federal Agencies [sic] to Comply with that Act ... April 10, 1997. U.S. Government Printing Office, Washington, DC

Urquijo I, De Stefano L (2016) Perception of drought and local responses by farmers: a perspective from the Jucar River Basin, Spain. Water Resour Manag 30:577–591. https://doi.org/10.1007/s11269-015-1178-5

Van Baars S (2005) The horizontal failure mechanism of the Wilnis peat dyke. Geotechnique 55:319–323. https://doi.org/10.1680/geot.2005.55.4.319

Van Baars S, Van Kempen IM (2009) The causes and mechanisms of historical dike failures in the Netherlands. E-water report. European Water Association, Hennef

Van den Honert RC, McAneney J (2011) The 2011 Brisbane floods: causes, impacts and implications. Water 3:1149–1173. https://doi.org/10.3390/w3041149

Vargas G, Hernández Y, Pabón JD (2017) La Niña event 2010–2011: hydroclimatic effects and socioeconomic impacts in Colombia. In: Mal S, Singh RB, Huggel C (eds) Climate change, extreme events and disaster risk reduction. Springer, Cham

Vicuna S, Hanemann M, Dale L (2006) Economic impacts of delta levee failure due to climate change: a scenario analysis. In: California Climate Change Center report series number 2006-007. California Climate Center at UC Berkeley, Berkeley

Vorosmarty CJ, Meybeck M, Fekete B, Sharma K, Green P, Syvitski JPM (2003) Anthropogenic sediment retention: major global impact from registered river impoundments. Glob Planet Chang 39:169–190. https://doi.org/10.1016/S0921-8181(03)00023-7

Wang M-Z, Amati M, Thomalla F (2012) Understanding the vulnerability of migrants in Shanghai to typhoons. Nat Hazards 60:1189–1210. https://doi.org/10.1007/s11069-011-9902-9

Ward PJ, Jongman B, Aerts JCJH, Bates PD, Botzen WJW, Diaz Loaiza A, Hallegatte S, Kind JM, Kwadijk J, Scussolini P, Winsemius HC (2017) A global framework for future costs and benefits of river-flood protection in urban areas. Nat Clim Chang 7:642–646. https://doi.org/10.1038/NCLIMATE3350

Waterloo MJ (1994) Water and nutrient dynamics of Pinus Caribaea plantation forests on degraded grassland soils in southwest Viti Levu, Fiji. PhD Dissertation, Vrije Universiteit Amsterdam, The Netherlands

White GF, Kates RW, Burton I (2001) Knowing better and losing even more: the use of knowledge in hazards management. Glob Environ Change Part B Environ Hazards 3:81–92. https://doi.org/10.1016/S1464-2867(01)00021-3

Wiles P, Selvester K, Fidalgo L (2005) Learning lessons from disaster recovery: the case of Mozambique. World Bank, Washington

Xu J, Grumbine RE, Shrestha A, Eriksson M, Yang X, Wang Y, Wilkes A (2009) The melting Himalayas: cascading effects of climate change on water, biodiversity, and livelihoods. Conserv Biol 23(3):520–530. https://doi.org/10.1111/j.1523-1739.2009.01237.x

Zuzak C, Mowrer M, Goodenough E, Burns J, Ranalli N, Rozelle J (2022) The national risk index: establishing a nationwide baseline for natural hazard risk in the US. Natural Hazards, pp 1-25

Agricultural Producers' Behavioral Adaptation to Climate Change Disaster in Turkiye

Yusuf Kadir Şener and Mustafa Kan

Abstract

Studies on behavioral economics have been increasing in recent years, and the factors affecting the behavior of both producers and consumers, especially in times of crisis, have started to be of interest. This study has investigated what factors can be determined to establish the behavioral intentions of the producers about agricultural drought in Nevşehir Province, located in the Central Anatolia Region, which is one of the important regions in terms of agricultural production in Turkiye. In this context, two districts (Hacıbektaş (Dry Agriculture System) and Derinkuyu (Irrigated Agriculture System)) that can represent the province in terms of dry agriculture and irrigated agriculture system selected by Purpose Sampling Method, and questionnaires filled in by interviewing a total of 212 producers in these two districts. In the study, while trying to determine statistically significant differences with basic statistical analyses such as chi-square, T-test, and Likelihood ratio, with artificial neural networks, in determining the "Behavioral Intentions" of producers on agricultural drought, "Farms and Farmers Typology" variables and "Planned Behavior Theory Subjective Norms," "Attitudes," and "Perceived Behavioral Control" variables were considered. As a result of the analysis, it was determined that the behavioral intentions of the producers in taking precautions about agricultural drought were most affected by the

This study was compiled from Yusuf ŞENER's Master's Thesis conducted at Kırşehir Ahi Evran University, Institute of Science and Technology, Department of Agricultural Economics.

Y. K. Şener (✉)
Turkish Republic of Ministry of Agriculture and Forestry, Kaman District Directorate of Agriculture and Forestry, Kirsehir, Turkiye

M. Kan
Kırşehir Ahi Evran University, Agricultural Faculty, Department of Agricultural Economics, Kirsehir, Turkiye
e-mail: mustafa.kan@ahievran.edu.tr

M. Ahmed, S. Ahmad (eds.), *Disaster Risk Reduction in Agriculture*, Disaster Resilience and Green Growth, https://doi.org/10.1007/978-981-99-1763-1_3

variables in the "Subjective Norm" group of the farmers, together with the variables belonging to "Farms and Farmers Typology." As a result, it can be said that it is important to increase the external pressure criteria that constitute the subjective norms in the fight against agricultural drought. In this context, raising awareness, encouraging organized activities, and even creating obligations to direct producers to behave at some points will significantly affect the fight against drought. For this reason, both official institutions, NGOs, and the private sector must encourage and even promote the producer in this direction.

Keywords

Behavioral economics · Agricultural drought · Artificial neural networks · Agriculture · Turkiye

Abbreviations

ADCSAP	Agricultural Drought Combat Strategy and Action Plan
EEA	European Environment Agency
EU	European Union
IPCC	Intergovernmental Panel on Climate Change
MoAF	Ministry of Agriculture and Forestry
MoENR	Ministry of Energy and Natural Resources-Turkiye
UN	United Nations
UNFCCC	United Nations Framework Convention on Climate Change
WMO	World Meteorological Organisation

1 Introduction

The adverse effects of global climate change, one of the most important problems experienced throughout human history, are making itself feel more and more every day (Ahmed, 2020). Although the negative impact on almost all sectors varies from region to region on a global scale, the scenarios created for the coming years do not change the fact that there will be significant losses worldwide. According to scientific findings, the new order established by humanity after the Industrial Revolution has radically changed the global climate in a way that has never been seen before. Greenhouse gases in the atmosphere, including carbon and methane, reached their highest levels compared to the preindustrial era. While the global atmospheric level, which was 280 ppm on average before industrialization and never exceeded 300 ppm, reached 413.2 ppm in 2020, the global annual average temperature in 2021 was approximately 1.11 ± 0.13 °C above the preindustrial average of 1850–1900 (WMO 2022). In the 4th Report of the Intergovernmental Panel on Climate Change

(IPCC), it is stated that climate change is certain (IPCC 2007), and in the 5th Report, the impact of human activities on climate change is clear (IPCC 2013). At the last point we have reached, the cost of global warming continues to increase cumulatively. While focusing on the certainty of the impact of human activities on the formation of climate change in the 6th Report, the latest report of the IPCC, the conclusions and predictions can be summarized as follows (IPCC 2021):

- Global surface temperature will continue to rise under all emissions scenarios considered, at least until mid-century.
- Many anomalies in the climate system are getting bigger and directly related to increasing global warming.
- Continued global warming is expected to further intensify the global water cycle, including international monsoon rains and the severity of rains and drought events.
- In scenarios with increased CO_2 emissions, ocean, and land carbon sinks are predicted to be less effective in slowing the accumulation of CO_2 in the atmosphere.
- Many changes from past and future greenhouse gas emissions, particularly changes in the ocean, ice layers, and global sea level, are irreversible for centuries or even millenniums.

In this context, no country alone has to bear the costs of climate change, and its struggle alone will not be enough to mitigate and adapt to the effects of global climate change. For this reason, global action, partnership, and cooperation are essential in the fight against climate change (Peker et al. 2019). To prevent the most severe impacts of climate change, member states of the United Nations Framework Convention on Climate Change (UNFCCC) have agreed to reduce the global average temperature rise to below 2 °C, which has been ongoing since preindustrial times (EEA 2008). The Paris Agreement, which came into effect in 2016 and is still valid today, is one of the global milestones in the fight against climate change. When it comes to climate change economics, it is understood that the economic losses caused by the negative impact of global climate change on the sectors and the price increases (inflationary pressure) caused by the extra costs to be taken to eliminate these losses are understood. Rather, climate change economics, based on the economic damage that occurs and the cost of preventing them, is based on a standard neoclassical model of rational behavior (based on benefit or profit maximization). In recent years, criticism of this issue has led to the increasing presence of behavioral economy-based models within this structure (Gowdy 2008; van den Bergh et al. 2000; Dubois et al. 2019). Ensuring the changes in human activities that lie at the beginning of the climate change problem, the effects of which can lead to global disasters, is one of the most important elements of the fight against climate change. That's because household consumption accounts for 72% of total greenhouse gas emissions (Hertwich and Peters 2009). According to Cafaro (2011), individuals report that they can save up to 15 billion tons (gigatons) of carbon by 2060 simply by changing their eating habits to avoid meat or by abandoning air travel. When evaluated in this context, behavioral economics shows that it can offer important results in understanding people and directing behaviors in the fight against climate change. The most important argument for combating climate change is to create an aware social structure. Studies on this subject show that there is a strong relationship between quantitative concepts such as awareness, risk perception, social norms, attitude, and struggle with environmental problems (Kollmuss and Agyeman 2002; Doss and Morris 2001; Arbuckle et al. 2015; Dong et al. 2018; Jellason et al. 2019; Yılmaz and Güleç 2021). This situation shows a positive tendency toward effective practices in the fight against climate change by affecting people's behaviors in the right direction.

In the fight against climate change, technological advances, including environmentally friendly innovations and productivity increases in the technology used, have an important place. In addition, numerous behavioral adaptations by private and public actors will increase the benefits of technological innovations, enable the best changes in the climate system, and thus reduce greenhouse gas emissions over time (Seo 2017). One of the most important issues here is understanding the factors in ensuring the spread of innovations in the innovation process, and behavioral economics studies are important in this regard. There are many models and theories for spreading and adopting innovation (Turan 2017). Using these models, it is possible to analyze the issues that determine human behavior, especially in the spread of environmentally sensitive technologies that emerge as innovations.

In this study, it is aimed to make an examination on the behavior of farmers in the agricultural sector, especially in the fight against the problems that arise due to climate change, and to evaluate how farmer behaviors affect the negative effects of climate change, especially in the fight against agricultural drought, with an empirical study on this subject. When we look at the studies on combating climate change in the agricultural sector in Turkiye, there is a need for quantitative studies and studies in which both qualitative and quantitative mixed studies are carried out together with qualitative studies within the scope of the behavioral economy.

2 The Studies on Human Behavioral in the Combat Against Climate Change Disaster in Turkiye

Some gases in the atmosphere create a greenhouse effect and ensure that the earth remains at its current temperature. It causes global warming due to increasing the greenhouse gases in the troposphere layer. People's activities are involved in the increase in the effects of global warming. Examples of these are the increases in the use of fossil fuels and rapid population growth (Akın 2006). It is stated that 197 countries are currently parties to the United Nations Framework Convention on Climate Change (UNFCCC), which entered into force in 1994. Turkiye joined the UNFCCC as 189 parties on May 24, 2004. In addition, our country became a party to the Kyoto Protocol, which was adopted in 1997 and entered into force in 2005, on August 26, 2009 (Ministry of Environment, Urbanization and Climate Change (MoEUCC) 2022). The Paris Agreement entered into force on November 4, 2016, with at least 55 parties comprising 55% of global greenhouse gases ratifying the agreement on October 5, 2016. Turkiye signed the Paris Agreement on April 22, 2016, at the High-Level Signing Ceremony with the representatives of 175 countries (Ministry of Environment, Urbanization and Climate Change (MoEUCC) 2022). There are many studies within the scope of combating climate change in Turkiye. Turkiye's "Climate Change Adaptation Strategy and Action Plan" was established in November 2011, and measures were taken in some areas within this scope. In addition, the work of the researchers is also useful to the literature, and it has been observed that there are literature publications on climate change through different sectors. Some of the studies on other factors, whether directed at human behavior, are listed below.

In his comprehensive research, Çuhadar (2021) conducted a survey study with 162 people in proportion to the number of producers in Afşin, Ekinözü, and Elbistan districts that received the least rainfall according to the rainfall data for many years in the Ceyhan Basin to measure the knowledge levels of drought-affected producers, to determine their thoughts about drought and to determine the factors affecting the degree of drought impact. As a result, it is stated that farmers' knowledge about drought is insufficient, and the factors affecting the degree to which farmers are affected by drought are age, agricultural income share, amount of irrigated land, amount of irrigated land, education, and groundwater use.

In their studies (2017), Mancı and Eren (2017) conducted a face-to-face survey with 140 producers determined by sampling method to determine the risks faced by agricultural enterprises producing under irrigated conditions in the Harran Plain and to determine their attitudes toward risk. It was stated that they were analyzed using statistically appropriate methods. As a result, it was determined that the most common risks were diseases and pests, climatic conditions, and unstable prices. Furthermore, it is stated that the level of education, marital status, the variable of whether to have agricultural insurance before, the status of using credit, and the premium support for agricultural insurance of the state are determined to be effective in the decision of producers to take out agricultural insurance.

Tüzer and Doğan (2021) stated that it is based on cause-and-effect relationship studies ranging from human-induced greenhouse gas emissions to climate change. As a result of the study, it is said that scientists predict that the increase in carbon dioxide emissions will have effects on the climate system. However, these scientific studies should be evaluated by the conditions of their respective periods. As a result, it has been said that global warming and climate change should not be seen as a physical process completely independent of human behavior and decisions.

Şen (2022) Definition of climate change and its variability in general and its effects on different sectors, socioeconomic situation, and especially water resources in Turkiye in particular. Again, recommendations were made by referring to the principles of precaution and adaptation in the fight against climate change in Turkiye.

Şener (2021), "The Effects of Global Climate Change on the Eğirdir Lake Basin and Drought Analysis" aims to determine the effects of climate change in the Eğirdir Lake Basin. It was mentioned that the climatic characteristics of the region were determined, and meteorological and satellite-based drought analyzes were carried out. Within the scope of the study, CanESM5 model outputs were used for precipitation, prediction in temperature data, and drought analysis. Some index methods have been used to perform current drought analyses of the Lake Eğirdir basin and drought predictions for the future. In the drought forecasts for the future, it is stated that extremely dry periods are predicted to be experienced in 2055 and 2072 at the earliest in the analyzes made in the 12-month time scale.

Engindeniz (2010), in his study titled "Analysis of the Attitudes and Behaviors of Tomato Producers in Izmir Regarding Irrigation and Drought," analyzed the effects of the meteorological and agricultural drought experienced in İzmir between 2007 and 2008 on tomato producers. The primary data of the study were obtained from producers producing field tomatoes in Bergama, Ödemiş, and Torbalı districts of İzmir. It was mentioned that the secondary data were obtained from institutions and organizations. In addition, the research data were obtained by survey forms from 86 producers with a proportional sampling method. In the analysis of the data, first, the socioeconomic characteristics of the producers were examined, and then the attitudes and behaviors of the producers regarding irrigation and drought were analyzed. Finally, the attitudes and behaviors of the producers on the issue of agricultural irrigation were evaluated. It is stated that the Likert scale is used. As a result, he mentioned that water resources should be well managed in our country during rainy and dry periods.

Akyüz (2019), in his study titled "Analysis of Farmers' Perceptions and Behaviors for Climate Change Adaptation Policies: The Case of Küçük Menderes Basin," aims to determine the climate change adaptation policies for the Küçük Menderes Basin in line with farmers' perceptions and behaviors and to reveal the willingness of producers to contribute and pay financially to these policies. It is stated that agriculture in the Küçük Menderes Basin, which is at risk in terms of climate change, is carried out intensively in Bayındır, Ödemiş, Tire, and Beydağ districts, and face-to-face surveys were conducted with 171 producers. Clustering analysis was used to determine the factors affecting the perception and behavior of the producers by using factor analysis and to obtain producer groups by using the results of this analysis. In addition, it was stated that the election trial model was used to create a policy for adaptation to climate change in line with the opinions and preferences of the producers and to reveal the willingness of the producers to pay for this policy.

3 Case Study from Nevşehir Province of Turkiye

3.1 Research Area

The study was conducted in two districts of Nevşehir Province in the Central Anatolia Region of Turkiye. Among the regions most affected by the drought in Turkiye is the Central Anatolia Region (Akşan 2021). Nevşehir Province is also one of the provinces most affected by drought in recent years and is seen as one of the Medium-Arid provinces in Turkiye between July 2020 and June 2022 (MGM 2022). In the study titled "Turkiye Agricultural Drought Prevention Strategy and Action Plan (2013–2017)" published by the Ministry of Agriculture and Forestry in 2013, it is estimated that the general temperature increase in the Mediterranean Basin, which includes Turkiye, will reach 1–2 °C. When examined in Turkiye in particular, the expected temperature increase is predicted to be between 2.5 and 4 °C. It is estimated that these temperature increases will reach 4 °C in the Aegean and Eastern Anatolia regions and 5 °C in the inner regions where Nevşehir Province is located (MoAF 2013). In the field of research, the study on the irrigated and dry agricultural systems was carried out separately. Nevşehir Province has 22.131 ha of irrigated agricultural area (NTOİM 2021), and the Hacıbektaş district, which is one of the districts selected as the research area, represents the dry agricultural system of the province (20.76% of the dry agricultural areas) and Derinkuyu district represents the irrigated agricultural system of the province (55.58% of the irrigated agricultural areas).

3.2 Sampling Method

According to the 2021 Farmer Registration System (ÇKS), there are 19,679 farmers in Nevşehir province (Fig. 1). 13.76% (2.707 farmers) of these farmers are in the Hacı Bektaş district, and 8.87% (1.745 farmers) are in the Derinkuyu district. Therefore, the research area covers 22.63% of the total number of farmers in Nevşehir province. Therefore, the research area can represent the region in terms of agricultural systems. A total of 212 farmers were interviewed in the research area, and the number of these farmers constitutes approximately 5% of the number of farmers in the research area (when extreme values are subtracted).

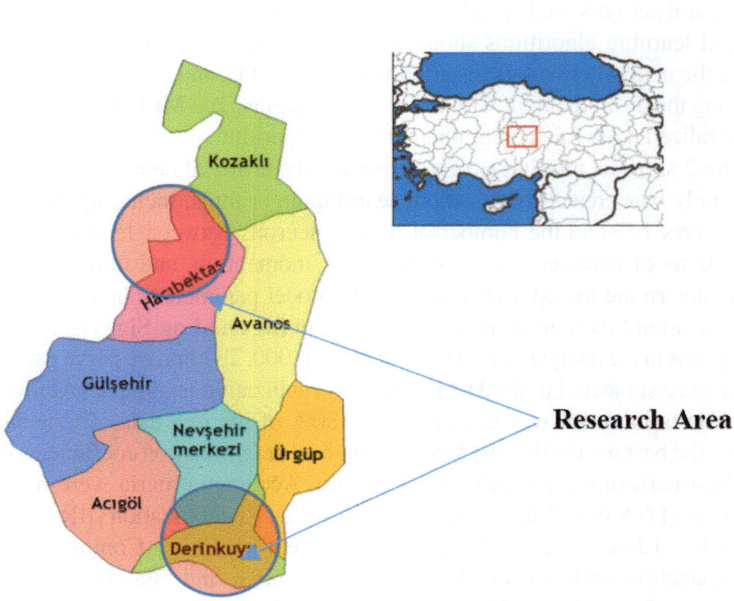

Fig. 1 Research area (Nevşehir province) map in Turkiye

3.3 Statistical Analyses

In the study, artificial neural networks were used to analyze factors influencing the behaviors that farmers intend to show to combat agricultural drought. In this context, four types of variable groups are discussed. The variables used in the study are presented in Table 1. When the table is examined, the variables discussed; include the variables of farmer and farm typology and the variables of attitude, subjective norms, and perceived behavioral controls used in the theory of planned behaviors (Ajzen 2002). In the study, the effects of these variables on the behavioral intention variable developed by producers to combat agricultural drought were tried to be revealed. In addition to 5-point Likert-type scales, percentage, year, and units such as Man Labour Force Unit (MLU) (Erkuş et al. 1995) and Large Animal Unit (LAU) (Erkuş et al. 1995) were used to measure the variables.

Artificial neural networks can be defined as structures consisting of simple interconnected adaptive processing elements that can perform parallel calculations to interpret multivariate data structures (Schalkoff 1997). Today, studies on artificial neural networks, which are widely used in agricultural economics, are increasing day by day, and very successful results are obtained. Artificial neural networks are successfully used as an alternative to traditional methods of predicting and classifying agricultural economy data. This study used a multilayered sensor (MLP) neural network (Fig. 2). MLP networks are one model that performs supervised learning. MLP can consist of an input layer, one or more hidden layers, and an output layer (Liu et al. 2013). The inputs to be used in the analysis and their corresponding outputs are presented to the neural network, and then the training process of the network is carried out. In neural network analysis, the architecture of the network design, training parameters, and learning algorithms should be determined in advance. After the training process, the neural network is restarted with new inputs, and evaluations are made by comparing the desired output with the network outputs. The MLP network works with the generalized delta rule and gradient descent algorithm.

Figure 2 shows a multilayer sensor model. The artificial neural network designed in this study was created with various learning algorithms, including the number of hidden layers 1–3 and the number of hidden neurons between 15 and 50. Different combinations of parameters of learning rates, momentum, and initial weights were tested to determine the optimal values of the model parameters. In addition, two different activation functions were worked with Tan-Sig and Log-Sig. The highest number of epochs in the analysis was determined as 10,000. To improve nerve performance, the data were subjected to the D-Min-Max Normalization technique (Akıllı and Atıl, 2020). The training set was determined by 80% of the raw data. The network that produces the best results (lowest MSE and highest determination coefficient) is treated as the best-performing network in the test set. Accuracy criteria were measured at every stage of NN modeling. In this study, Bayesian Regularization (BR), Levenberg-Marquardt (LM), Scaled Conjugate Gradient (SCG), Conjugate Gradient Backpropagation with Powell-Beale Restarts (CGB), and Brayden Fletcher Gold Farlo Shanno Quasi Newton Backpropagation (BFG) learning algorithms were examined. SCG and LM algorithms are presented in the findings. Information on neural network architecture and training parameters is summarized in Table 2.

Table 1 Questions about variables used in artificial neural networks

Farms and farmers typology	Attitudes		Subjective norms	Behavioral intention
• Production system/province name • Farmer's age (year) • Total manpower unit in the household • Crop production experience (year) • The share of agricultural income (%) • Bovine animal unit • Attitude to risk • Equipment adequacy for crop production • Total agriculture area (Ha) • Rainfed area (Ha) • Agricultural drought awareness status • Drought affect status • Finding agricultural support useful • Finding effectiveness of drought support in dry years	• I think agricultural drought is a natural process; nature is in a constant state of equilibrium and will eventually balance out • I think that the behavior of producers has a very important effect on the formation of agricultural drought • If the Agricultural Drought continues like this, I think it will be a great destruction for the producers engaged in agricultural production in the region • I think that agricultural drought is the most important negative effect of climate change • I think that the most important problem in the region is agricultural drought • I think the agricultural drought is the cause of the most important loss in our income • The public is doing its part in the fight against agricultural drought in the region • The private sector is doing its part in the fight against agricultural drought in the region • People are doing their part in combating agricultural drought in the region • Turkiye is affected more negatively than many countries in the world due to agricultural drought • Nevşehir province is affected by agricultural drought more than the general of Turkiye • Your location is more affected by agricultural drought than Nevşehir province in general • I think there is an opportunity to combat agricultural drought • Agricultural drought is a drought that occurs with a lack of precipitation	• Agricultural drought occurs with the realization of high-temperature values over many years • Wind is an effective factor in the formation of agricultural drought • Agricultural drought causes soil degradation and loss of plant diversity • Drought is the most important factor limiting agriculture • There is no crop production or animal production in places where there is a drought • Agricultural drought is the worst consequence of climate change • Instead of struggling with agricultural drought, it is necessary to adapt to the new situation • I believe that the use of new technology can be a solution to reduce the effects of agricultural drought	• Other members of my family support me in combating agricultural drought • Producers, where I live, are in solidarity to combat agricultural drought • There are sufficient supports to minimize the risks experienced in combating agricultural drought • New information emerging today will facilitate the fight against agricultural drought day by day • There are new technologies/products that I can reach around me in order to adapt to the new situation in agricultural drought situations • I feel personally responsible for combating agricultural drought Perceived behavioral control • I find it difficult to combat agricultural drought • Producers cannot individually combat agricultural drought • I have enough knowledge to combat/adapt to agricultural drought • I have sufficient tool equipment to combat/adapt to agricultural drought • I am doing alternative studies to prevent the loss of income of my family when agricultural drought occurs • The state is dealing with the fight against agricultural drought, and we have nothing to do about it • I can reduce the effects of agricultural drought on my business whenever I want • I know places that will help me to combat/adapt to agricultural drought	• It is necessary to select drought-tolerant/resistant plant species • When choosing a seed variety, it is necessary to consider whether it is drought-tolerant • Agricultural meteorology should be constantly followed, and measures should be taken against the events that may occur • It is necessary to follow the climatic events in the past and take precautions accordingly • Agricultural insurance is required • Drought insurance is required • Alternative income generating works should be done by taking into account the risk of drought in agriculture • There should always be a Plan B in combating/adapting to agricultural drought • It is necessary to constantly search for new technologies/products for agricultural drought control/adaptation • As soon as I feel that there may be an agricultural drought, it is necessary to update my agricultural input (pesticide, fertilizer, etc.) use plan • If I had a problem with agricultural drought in the previous year, different practices should be applied during the production period • It is necessary to use sustainable farming techniques • It is necessary to use certified seed • It is necessary to organize the fight/adaptation to agricultural drought • Communication with public institutions related to agriculture should be strong for combating/adapting to agricultural drought • It is necessary to receive training in agricultural drought control/adaptation

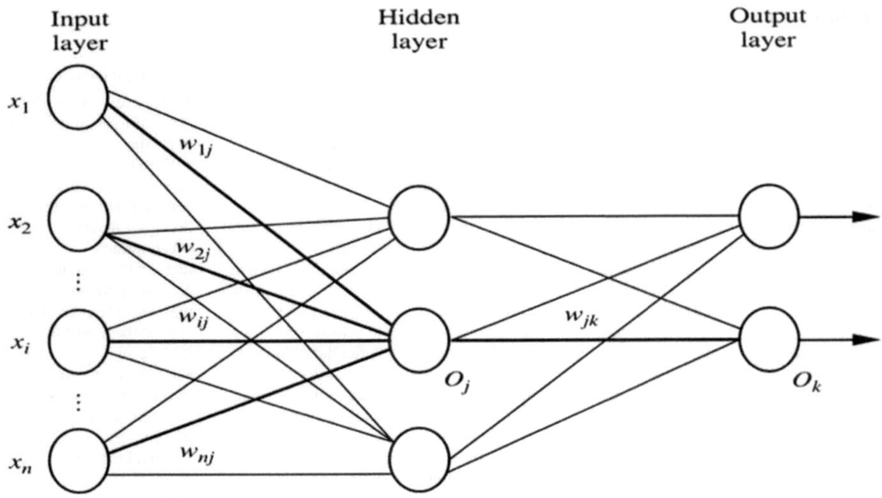

Fig. 2 Multilayer perceptron (Han and Kamber 2006)

Table 2 Neural network parameters

NN structure	Descriptions
Model	Multilayer perceptron
Connections	Feed-forward
Layer	1–3
Input node	
Hidden node	15–75
Output node	1
Activation function	Tan-Sig, Log-Sig
Training parameters	Descriptions
Mode	Supervised
Algorithms	Back-propagation[a]
Weight updates	Each Epoch
Learning rate	0.01–0.03
Momentum coefficient	0.80–0.95

[a] Bayesian Düzenlileştirme (BR), Levenberg-Marquardt (LM), Scaled Conjugate Gradient (SCG), Conjugate Gradient Backpropagation with Powell-Beale Restarts (CGB) ve Brayde Fletcher Gold Farlo Shanno Quasi Newton Backpropagation (BFG)

4 Research Findings

The research area is at the lower middle level in terms of development in Turkiye. Nevşehir Province is among the provinces in the 4th level in the six-level evaluation within the scope of the socioeconomic development index (SEGE) study (STB 2017), and there are eight districts, including the Central district. The province is a province with high tourism potential and is one of the important production areas in terms of agriculture. Especially cereals and tuberous plants constitute the important plant production pattern of the province (NTOİM 2021). Hacıbektaş district is at the 4th level in the 6-level evaluation of the district SEGE-2022, and Derinkuyu is the 5th level district (STB 2022). It presented some statistical data describing the farmers interviewed in the research area in Table 3. There was no statistically significant difference between the variables in terms of age, education, number of households, number of male labor units, and crop production experiences, which are similar in terms of sociodemographic data. As can be seen from Table 3, it is seen that producers over the age of 50 have a level of education between primary and high school schools, and about half of them are included in the scope of social security, which shows that they make agriculture a priority profession. They have an average of 3.50 family members and a 2.52 MLU labor force. These data are similar to the studies on agricultural enterprises in different provinces in the Central Anatolia region of Turkiye (Özdemir and Kan 2020a, b). When the table is examined, the experience of the producers of livestock in the producers in Hacıbektaş district, which

Table 3 Descriptive statistics of the farmers in the research area

Variables		District names				Average	T-test/ chi-square value[a]
		Derinkuyu (irrigated area)		Hacıbektaş (rainfed area)			
		Mean	%	Mean	%		
Age		51.17		50.85		51.04	0.21
Education level	Illiterate		0.00		0.00		5.70[a]
	Literate		0.79		3.49		
	Primary/secondary school		62.70		51.16		
	High school		32.54		36.05		
	University		3.97		9.30		
Proportion of those who practice agriculture as a profession (%)			48.41		41.86		0.88[a]
Household member (head)		3.56		3.41		3.50	0.85
Total manpower unit (MPU)		2.58		2.44		2.52	0.90
Experience in crop production (years)		33.99		31.73		33.08	1.36
Experience in animal production (years)		11.29		17.52		13.82	3.16***
Agricultural income (TL/year %)		135,349	85.04	80,225	75.51	113,458	4.25***
Nonagricultural income (TL/year %)		23,817	14.96	26,023	24.49	24,712	−0.54
Agricultural area (own area Ha)		11.12	67.94	14.63	63.15	12.54	−1.86*
Agricultural area (rented area Ha)		5.25	32.06	7.45	32.15	6.54	−1.48
Agricultural area (sharecropping area Ha)		0.00	0.00	1.09	4.69	0.44	−1.81*
Irrigated area (Ha %)		14.95	91.35	4.46	19.26	10.69	−8.08***
Rainfed area (Ha %)		1.42	8.65	18.70	80.74	8.43	7.28***
Animal number (animal unit)		2.72		10.02		5.69	−4.55***

Statistically significant at *90%, **95%, ***99% confidence levels

represents the dry agriculture system, is higher than the producers in the Derinkuyu district, which represents the irrigated agriculture system. This situation was found to be statistically important. The experience in crop production is similar in both districts. However, with the effect of the irrigated agriculture system, the income from agriculture is higher in the Derinkuyu district, and the difference was found to be statistically significant. When the table is examined, the producers interviewed in both districts are similar to each other in terms of both sociodemographic and crop production experience and business size.

Many factors are effective in the formation of behaviors. Studies on this subject have been carried out on factors such as the farming goals and objectives of the decisions of producers engaged in agricultural production, their behaviors related to traditions, ways of coping with stress and stress, optimism about farming, and satisfaction with their work, risk-taking, and management behaviors, as well as environmental factors, land ownership status, market conditions, policy changes, factors related to financing, age and family width, farm layout, etc. (Akçaöz et al. 2005). Table 4 gives some attitudes that may affect the behavior of producers in the research area. When the table was examined, it was determined that the producers in both districts were similar to each other in terms of risk and did not tend to take more risks. While it was determined that the producers in the dry agriculture system had a more positive attitude in terms of technology, it was seen that the producers in the dry agriculture system tended to act in a more organizational movement, and these situations were found to be statistically important.

Although there are common situations in both production systems about temperature increase, decrease in precipitation, and negative impact of drought, there are producer opinions about the fact that drought affects more in the dry agricultural system more, which is statistically important. When the table is examined, the

Table 4 Some behavioral attitudes of the farmers in the research area

Variables		District names		Chi-square/ likelihood ratio
		Derinkuyu (irrigated area) %	Hacıbektaş (rainfed area) %	
Attitude to risk	Risk lover	30.16	34.52	1.80
	Risk neutral	9.52	4.76	
	Risk averse	60.32	60.71	
Attitude to technology	Technology lover	84.13	88.37	5.79**
	Technology neutral	9.52	11.63	
	Technology averse	6.35	0.00	
Collaborative behavior culture in the region	Yes, there is	3.97	13.95	8.65**
	Limited	15.87	20.93	
	No, there isn't	80.16	65.12	

Statistically significant at *90%, **95%, ***99% confidence levels

results obtained coincide with the expected situation in the research, and it is consistent with the theory that the region has been in the regions affected by drought in recent years and that this is felt more in the dry agricultural system. When the studies conducted on this subject in Turkiye were examined, Engindeniz (2010) stated that 79% of the producers in Izmir thought that the cause of the drought was insufficient rainfall in the study aimed at determining the attitudes and behaviors of tomato producers about drought. Çuhadar (2021) reports in his doctoral study conducted in the Ceyhan Basin that 75.9% of farmers know what agricultural drought is, and about 60% of them see agricultural drought as a complete lack of rainfall and that the degree to which producers are affected by drought decreases as the amount of irrigated land increases. At the beginning of the measures against agricultural drought, he stated that farmers choose drought-resistant products. In their study in the Ceyhan Basin, Çuhadar (2021) report that informative studies should be carried out on the importance of technological methods for the solution of drought, the measures they can take for environmental protection, and the importance of environmental practices on drought and that the behavior of individuals should change first to solve the drought (Table 5).

When the study aims to combat agricultural drought and producer behaviors are examined, the studies conducted are carried out by the producers to choose more drought-resistant products (Engindeniz 2010; Çuhadar 2021; Ashraf et al. 2014; Udmale et al. 2014), changing the irrigation method (Engindeniz 2010; Çuhadar 2021; Ashraf et al. 2014) and reducing excessive water consumption (Engindeniz 2010; Ashraf et al. 2014), economization of spending and consumption habits (Ashraf et al. 2014), making changes to the production dates of products (Udmale et al. 2014), leaving the land empty until the drought passes (Udmale et al. 2014). Also, several authors have pointed out adaptation activities undertaken by farmers after observing climate change and variability (Carlton et al. 2016; Habiba et al. 2012; Maddison 2007; Udmale et al. 2014). In this study, the strategies proposed by the producers to combat drought are presented in Table 6. When Table 6 was examined, it was determined that some strategies were statistically different in the dry and irrigated agricultural systems. While the producers in the dry agriculture system are more willing to take out drought insurance and alternative income areas, it is seen that the producers in the irrigated agriculture system are more willing to choose drought-tolerant crops/varieties and use new technologies in the fight against drought. While the higher risk of drought in the dry agricultural system pushes the producers to find alternative b plans, in the irrigated agriculture system, the producers exhibit a behavior open to innovations such as new varieties, new products, and the use of new technology. Although the proportion of producers who believe in the organized struggle against drought in both production systems is similar, the number of producers who are unstable in the dry agriculture system and who do not believe in organized struggle in the irrigated agricultural system is higher.

The neural network designed within the scope of the study was operated on four different models. The data obtained in the survey study were first subjected to Basic Components Analysis to determine the system inputs. In addition, the correlation values between the variables were also examined. Accordingly, in Model 1 analysis;

Table 5 Farmers' perspectives on agricultural drought in the research area

Variable		District names/production system Derinkuyu (irrigated area) %	Hacıbektaş (rainfed area) %	Chi-square/ likelihood ratio
Agricultural drought awareness status	Yes	82.54	88.37	1.35
	Partially	17.46	11.63	
	No	0.00	0.00	
Agricultural drought definition	Low precipitation levels	53.97	40.70	9.48**
	Realization of moisture loss in the soil	13.49	13.95	
	Decreased irrigation in agricultural fields	18.25	13.95	
	Realization of high-temperature values for long years	14.29	31.40	
Drought situation in the region in the last 20 years	Yes	94.44	93.02	4.87*
	Partially	5.56	3.49	
	No	0.00	3.49	
Temperature change in the last 20 years in the region	Increased	94.44	100.00	4.94*
	Decreased	4.76	0.00	
	No change	0.79	0.00	
Rainfall change in the last 20 years in the region	Increased	2.38	0.00	2.78
	Decreased	96.83	100.00	
	No change	0.79	0.00	
Drought affect status	Very little	0.79	0.00	44.96***
	Little	30.95	11.63	
	Middle	48.41	23.26	
	Extreme	11.90	36.05	
	Too much	7.94	29.07	

Statistically significant at *90%, **95%, ***99% confidence levels

"Farms and Farmers Typology" variables and all questions related to the "Attitude towards Behavior," "Subjective Norms," and "Perceived Behavioral Control" scales were created to explain "Behavioral Intent" views on climate change are included. The neural network findings for Model 1 are shown in Table 7. The results of the analysis show that the neural network configured with the LM algorithm, where the number of hidden neurons in Model 1 is 30, is more successful than the others.

Model 2 includes questions from the variables "Farms and Farmers Typology" to obtain predictions about intent opinions. The results indicate that neural networks configured with the number of 50 hidden neurons, LM algorithm, and Log-Sig functions are more successful than others (Table 8).

Table 6 Farmers behavioral intentions to combat agricultural drought

Behavioral intentions		District names/production strategy: Derinkuyu (irrigated area)	District names/production strategy: Hacıbektaş (rainfed area)	Chi-square
It is necessary to select drought-tolerant/resistant plant species	Strongly disagree	0.00	0.00	7.69*
	Do not agree	4.76	11.63	
	Undecided	7.14	9.30	
	Agree	76.19	59.30	
	Absolutely agree	11.90	19.77	
When choosing a seed variety, it is necessary to consider whether it is drought-tolerant	Strongly disagree	0.00	0.00	1.61
	Do not agree	11.11	8.14	
	Undecided	7.14	10.47	
	Agree	47.62	43.02	
	Absolutely agree	34.13	38.37	
Agricultural meteorology should be constantly followed, and measures should be taken against the events that may occur	Strongly disagree	0.00	0.00	3.79
	Do not agree	2.38	5.81	
	Undecided	5.56	8.14	
	Agree	65.08	53.49	
	Absolutely agree	26.98	32.56	
It is necessary to follow the climatic events in the past and take precautions accordingly	Strongly disagree	0.00	0.00	1.94
	Do not agree	5.56	8.14	
	Undecided	11.90	10.47	
	Agree	58.73	51.16	
	Absolutely agree	23.81	30.23	
Agricultural insurance is required	Strongly disagree	0.79	0.00	2.79
	Do not agree	8.73	4.65	
	Undecided	6.35	4.65	
	Agree	65.87	67.44	
	Absolutely agree	18.25	23.26	
Drought insurance is required	Strongly disagree	3.17	4.65	14.99***
	Do not agree	34.13	16.28	
	Undecided	20.63	18.60	
	Agree	27.78	51.16	
	Absolutely agree	14.29	9.30	
Alternative income generating works should be done by taking into account the risk of drought in agriculture	Strongly disagree	1.59	1.16	13.54***
	Do not agree	30.95	10.47	
	Undecided	7.94	11.63	
	Agree	43.65	61.63	
	Absolutely agree	15.87	15.12	
There should always be a Plan B in combating/adapting to agricultural drought	Strongly disagree	0.00	0.00	0.61
	Do not agree	7.94	8.14	
	Undecided	8.73	11.63	
	Agree	61.11	56.98	
	Absolutely agree	22.22	23.26	

(continued)

Table 6 (continued)

Behavioral intentions		District names/production strategy: Derinkuyu (irrigated area)	District names/production strategy: Hacıbektaş (rainfed area)	Chi-square
It is necessary to constantly search for new technologies/products for agricultural drought control/adaptation	Strongly disagree	0.00	0.00	11.20**
	Do not agree	1.59	8.14	
	Undecided	3.97	3.49	
	Agree	74.60	55.81	
	Absolutely agree	19.84	32.56	
As soon as I feel that there may be an agricultural drought, it is necessary to update my agricultural input (pesticide, fertilizer, etc.) use plan	Strongly disagree	1.59	1.16	1.34
	Do not agree	15.08	16.28	
	Undecided	16.67	18.60	
	Agree	46.83	39.53	
	Absolutely agree	19.84	24.42	
If I had a problem with agricultural drought in the previous year, different practices should be applied during the production period	Strongly disagree	0.00	0.00	1.09
	Do not agree	7.14	5.81	
	Undecided	9.52	6.98	
	Agree	59.52	66.28	
	Absolutely agree	23.81	20.93	
It is necessary to use sustainable farming techniques	Strongly disagree	0.79	0.00	2.18
	Do not agree	0.00	1.16	
	Undecided	8.73	8.14	
	Agree	66.67	67.44	
	Absolutely agree	23.81	23.26	
It is necessary to use certified seed	Strongly disagree	0.00	0.00	4.62
	Do not agree	10.32	3.49	
	Undecided	13.49	12.79	
	Agree	56.35	55.81	
	Absolutely agree	19.84	27.91	
It is necessary to organize for the fight/adaptation to agricultural drought	Strongly disagree	0.00	2.33	16.71***
	Do not agree	22.22	9.30	
	Undecided	2.38	11.63	
	Agree	50.79	44.19	
	Absolutely agree	24.60	32.56	
Communication with public institutions related to agriculture should be strong for combating/adapting to agricultural drought	Strongly disagree	0.00	0.00	0.32
	Do not agree	0.00	0.00	
	Undecided	0.79	1.16	
	Agree	57.14	53.49	
	Absolutely agree	42.06	45.35	
It is necessary to receive training in agricultural drought control/adaptation	Strongly disagree	0.00	0.00	2.13
	Do not agree	0.00	1.16	
	Undecided	0.00	0.00	
	Agree	51.59	45.35	
	Absolutely agree	48.41	53.49	

Statistically significant at *90%, **95%, ***99% confidence levels

Table 7 Model 1 results for determining behavioral intention (farms and farmers typology + attitudes + subjective norms + perceived behavioral control)

YSA model	Hidden neuron	Training algorithm	Error criteria					
			Education		Test			
			Log-Sig	Tan-Sig	Log-Sig		Tan-Sig	
			HKO	HKO	HKO	R^2	HKO	R^2
Model 1	15	LM	0.0118	0.0213	0.0379	0.7290	0.0341	0.4847
	20		0.0083	0.0102	0.0371	0.7258	0.0228	0.7941
	30		0.0085	0.0101	0.0318	0.8039	0.0358	0.8830
	40		0.0111	0.0088	0.0371	0.3211	0.0227	0.8491
	50		0.0076	0.0085	0.0266	0.7238	0.0316	0.7039
	15	SCG	0.0172	0.0141	0.0314	0.5125	0.0252	0.7234
	20		0.0117	0.0085	0.0244	0.5825	0.0318	0.8039
	30		0.0133	0.0102	0.0220	0.7466	0.0285	0.8239
	40		0.0088	0.0112	0.0277	0.7285	0.0289	0.7764
	50		0.0104	0.0138	0.0271	0.6983	0.0187	0.7499

Table 8 Model 2 results for determining behavioral intention (farms and farmers typology)

YSA model	Hidden neuron	Training algorithm	Error criterias					
			Education		Test			
			Log-Sig	Tan-Sig	Log-Sig		Tan-Sig	
			HKO	HKO	HKO	R^2	HKO	R^2
Model 2	15	LM	0.0148	0.0123	0.0442	0.6826	0.0416	0.4939
	20		0.0208	0.0163	0.0271	0.2821	0.0261	0.4468
	30		0.0285	0.0352	0.0453	0.4720	0.0468	0.5752
	40		0.0301	0.0163	0.0428	0.6446	0.0363	0.4586
	50		0.0096	0.0241	0.0389	0.7020	0.0295	0.5430
	15	SCG	0.0183	0.0174	0.0259	0.2552	0.0295	0.2345
	20		0.0256	0.0153	0.0258	0.5671	0.0288	0.3184
	30		0.0175	0.0184	0.0227	0.4889	0.0291	0.4403
	40		0.0158	0.0197	0.0283	0.4387	0.0272	0.4620
	50		0.0198	0.0224	0.0293	0.3713	0.0302	0.4336

In Model 3, there are questions in the "Farms and Farmers Typology" variables and the "Attitude Towards Behavior" scale to obtain predictions about intention views. The results report that neural networks configured with the number of 50 hidden neurons and the SCG algorithm are more successful than the others (Table 9).

Model 4 includes questions on the "Farms and Farmers Typology" variables and the "Subjective Norms" scale to obtain predictions about intentional views. The analysis observed that the neural networks configured with the number of 50 hidden neurons and the SCG algorithm were more successful than the others. The results of the analysis show that the neural network configured with the LM algorithm, where the number of hidden neurons is 30, is more successful than the others (Table 10).

Model 5 includes "Farms and Farmers Typology" variables and questions on the "Perceived Behavioral Control" scale to obtain predictions about intention views. In

Table 9 Model 3 results for determining behavioral intention (farms and farmers typology and + attitudes)

YSA model	Hidden neuron	Training algorithm	Error criterias					
			Education		Test			
			Log-Sig	Tan-Sig	Log-Sig		Tan-Sig	
			HKO	HKO	HKO	R^2	HKO	R^2
Model 3	15	LM	0.0102	0.0219	0.0226	0.5422	0.0289	0.3684
	20		0.0114	0.0129	0.0274	0.6920	0.0351	0.5621
	30		0.0173	0.0100	0.0232	0.4836	0.0257	0.4760
	40		0.0176	0.0159	0.0261	0.3562	0.0252	0.4066
	50		0.0112	0.0162	0.0335	0.5582	0.0216	0.2875
	15	SCG	0.0123	0.0110	0.0189	0.6970	0.0229	0.4532
	20		0.015	0.0149	0.0254	0.5272	0.0253	0.5835
	30		0.0142	0.0129	0.0259	0.5735	0.0236	0.6422
	40		0.0142	0.0099	0.0162	0.5665	0.0252	0.6397
	50		0.0110	0.0224	0.0219	0.7390	0.0336	0.7316

Table 10 Model 4 results for determining behavioral intention (farms and farmers typology and subjective norms)

YSA model	Hidden neuron	Training algorithm	Error criterias					
			Education		Test			
			Log-Sig	Tan-Sig	Log-Sig		Tan-Sig	
			HKO	HKO	HKO	R^2	HKO	R^2
Model 4	15	LM	0.0130	0.0302	0.0309	0.4384	0.0449	0.4127
	20		0.0128	0.0115	0.0251	0.4226	0.0329	0.4383
	30		0.0158	0.0343	0.0296	0.2042	0.0550	0.8662
	40		0.0203	0.0234	0.0341	0.3214	0.0345	0.3276
	50		0.0128	0.0104	0.0292	0.4295	0.0358	0.7424
	15	SCG	0.0208	0.0110	0.0331	0.1533	0.0271	0.3595
	20		0.0161	0.0110	0.0307	0.4392	0.0329	0.3362
	30		0.0137	0.0168	0.0199	0.5854	0.0253	0.3630
	40		0.0165	0.0128	0.0252	0.4232	0.0230	0.4385
	50		0.0187	0.0142	0.0273	0.3960	0.0257	0.4458

the Model 4 analysis, it is seen that the neural network configured with the LM algorithm, where the number of hidden neurons is 40, is more successful than the others (Table 11).

In the general evaluations of the results of the analysis, it is seen that the artificial neural networks method can be used successfully in modeling agricultural economy data. As can be seen in the tables, the performance of neural network architecture designed in different models and data structures varies. Accordingly, Model 1 exhibits a more successful structure in predicting intention views than other models. Model 4 comes in second place in the predictions regarding the intention view.

Table 11 Model 5 results for determining behavioral intention (farms and farmers typology and perceived behavioral control) sonuçlari

YSA model	Hidden neuron	Training algorithm	Error criterias					
			Education		Test			
			Log-Sig	Tan-Sig	Log-Sig		Tan-Sig	
			HKO	HKO	HKO	R^2	HKO	R^2
Model 5	15	LM	0.0215	0.0117	0.0414	0.3100	0.0468	0.7008
	20		0.0221	0.0219	0.0420	0.4601	0.0394	0.1498
	30		0.0189	0.0154	0.0284	0.2261	0.0440	0.4522
	40		0.0268	0.0148	0.0408	0.8578	0.0527	0.7227
	50		0.0170	0.0304	0.0440	0.4218	0.0571	0.7399
	15	SCG	0.0171	0.0206	0.0328	0.2321	0.0272	0.3701
	20		0.0173	0.0124	0.0330	0.3115	0.0482	0.4991
	30		0.0218	0.0133	0.0328	0.4452	0.0302	0.3847
	40		0.0218	0.0133	0.0375	0.4894	0.0386	0.5423
	50		0.0200	0.0127	0.0331	0.5483	0.0413	0.3658

5 Conclusions and Recommendations

Drought is one of the most important negative consequences of climate change. In the scenarios of adaptation to climate change and combating climate change, drought is of particular importance for the agricultural sector. The drought experienced in many countries, especially in Europe, in 2022 is one of the most important reasons behind the food crises. In recent years, humanity has been struggling with pandemic-based problems such as COVID-19 and water and food crises, which are of vital importance such as drought. The inadequacy of water and food, shown as one of the most important crisis causes of the future, reveals that these points should be focused on at the beginning of the measures that mankind should take precautions in this century. It is known that people turn to nonrational behavior in times of crisis instead of rational behavior in economic behavior. For this reason, behavioral economics has become more prominent in the economics literature in recent years.

In this study, where the effective factors in determining the behavioral intentions of the agricultural producers to take precautions against agricultural drought were investigated, it was determined that the variables in the "Subjective Norm" group of the farmers together with the variables belonging to "Farms and Farmers Typology" were most affected by the behavioral intentions of the producers in the fight against agricultural drought. Subjective norms reflect an individual's perception of social pressures to perform or not to perform the behavior (Ajzen 1991). The assumption is that individuals are more likely to undertake the behavior that is regarded as desirable by significant others. Subjective norms refer to the belief that an important person or group of people will approve and support a particular behavior (Ham et al. 2015). As a result, it is important to increase the external pressure criteria that constitute subjective norms in the fight against drought. In this context, creating awareness, encouraging organized activities, and even creating obligations to direct producers to behavior at some points will have an important positive effect in the fight against drought. For this reason, it is important that both official institutions and NGOs, together with the private sector, encourage and even promote the producers in this direction. Especially in both agricultural systems, it is seen that it is important to act horizontally and vertically between both producers and institutions to create an important awareness about working together to combat drought.

The fact that Nevşehir is within the working area of the Konya Plain Project (KOP) Regional Development Administration in the regional development studies being implemented in Turkiye can be seen as an opportunity to ensure regional unity to combat drought, to create financial support and to create an integrated structure in terms of institutional organization. In this context, R&D and extension studies financed by the KOP Regional Development Administration on the fight against drought must be disseminated to create awareness in the region. For this reason, it is recommended to use agricultural extension instruments more actively in terms of institutional cohesion and sharing and dissemination of results.

Turkiye has important strategies and action plans, especially for determining the working principles on many issues. One of them is the Agricultural Drought Prevention Strategy and Action Plan (ADCSAP) prepared by the Ministry of Agriculture and Forestry of the Republic of Turkiye and has been prepared for the period 2023–2027. Within the scope of the plan, Agricultural Drought Provincial Crisis Centers were established with the participation of different institutions and organizations established in the provinces to evaluate the possibility of possible drought in advance and to take the necessary measures by the "Regulation on the Procedures and Principles of the Duties of Agricultural Drought Management" is an important action in the fight against drought.

References

Ahmed M (2020) Introduction to Modern Climate Change. Andrew E. Dessler: Cambridge University Press, 2011, 252 pp, ISBN-10: 0521173159. Sci Total Environ 734, 139397. https://doi.org/10.1016/j.scitotenv.2020.139397

Ajzen I (1991) The theory of planned behavior. Organ Behav Hum Decis Process 50(2):179–211. https://doi.org/10.1016/0749-5978(91)90020-T

Ajzen I (2002) Perceived behavioral control, self-efficacy, locus of control, and the theory of planned behavior. J Appl Soc Psychol 32:665–683

Akçaöz H, Özkan B, Kızılay H (2005) Tarımsal üretimde çiftçilerin tutum ve davranışları: çiftçilik amaçları ölçeği (FOS). Anadolu J AAR 15(2):104–125

Akıllı A, Atıl H (2020) Evaluation of normalization techniques on neural networks for the prediction of 305-day milk yield. Turk J Agric Eng Res 1:354–367

Akın G (2006) Küresel Isınma Nedenleri ve Sonuçları. Ankara Üniv Dil Tarih-Coğrafya Fakültesi Dergisi 46(2):29–43

Akşan GN (2021) Türkiye'de Kuraklık Analizi, Pamukkale Üniversitesi Fen bilimleri Enstitüsü İnşaat Mühendisliği Anabilimdalı, Basılmamış Yüksek Lisans Tezi, Denizli. http://acikerisim.pau.edu.tr:8080/xmlui/handle/11499/38624

Akyüz Y (2019) İklim Değişikliğine Uyum Politikalarına Yönelik Çiftçi Algı Ve Davranışlarının Analizi: Küçük Menderes Havzası Örneği. Ege Üniversitesi, Fen Bilimleri Enstitüsü, Tarım Ekonomisi Anabilim Dalı, Tarım Politikası ve Yayım Doktora Programı. Doktora Tezi. İzmir

Arbuckle JG, Morton LW, Hobbs J (2015) Understanding farmer perspectives on climate change adaptation and mitigation: the roles of trust in sources of climate information, climate change beliefs, and perceived risk. Environ Behav 47(2):205–234

Ashraf M, Routray JK, Saeed M (2014) Determinants of farmers' choice of coping and adaptation measures to the drought hazard in northwest Balochistan, Pakistan. Nat Hazards 73(3):1451–1473. https://doi.org/10.1007/s11069-014-1149-9

Cafaro P (2011) Beyond business as usual: alternative wedges to avoid catastrophic climate change and create sustainable societies. In: Arnold D (ed) The ethics of global climate change. Cambridge University Press, Cambridge, pp 192–215. https://doi.org/10.1017/CBO9780511732294.010

Carlton JS, Mase AS, Knutson CL, Lemos MC, Haigh T, Todey DP, Prokopy LS (2016) The effects of extreme drought on climate change beliefs, risk perceptions, and adaptation attitudes. Clim Chang 135(2):211–226. https://doi.org/10.1007/s10584-015-1561-5

Çuhadar M (2021) Çiftçilerin Tarımsal Kuraklık Hakkında Bilgi Düzeyleri ve Düşünceleri: Ceyhan Havzası Örneği. Türk Tarım Doğa Bilimleri Dergisi 8(4):1151–1159. https://doi.org/10.30910/turkjans.981234

Dong Y, Hu S, Zhu J (2018) From source credibility to risk perception: how and when climate information matters to action. Resour Conserv Recycl 136:410–417

Doss CR, Morris ML (2001) How does gender affect the adoption of agricultural innovations? Agric Econ 25:27–39

Dubois G, Sovacool B, Aall C, Nilsson M, Barbier C, Herrmann A, Bruyère S, Andersson C, Sköld B, Nadaud F, Dorner F, Moberg KR, Ceron JP, Fischer H, Amelung D, Baltruszewicz M, Fischer J, Benevise F, Louis VL, Sauerborn R (2019) It starts at home? Climate policies targeting household consumption and behavioral decisions are key to low-carbon futures. Energy Res Soc Sci 52:144–158

Engindeniz S (2010) İzmir'de domates üreticilerinin sulama ve kuraklıkla ilgili tutum ve davranışlarının analizi. Ege Üniv Ziraat Fakültesi Dergisi 47(3):321–330

Erkuş A, Bülbül M, Kıral T, Açıl AF, Demirci R (1995) Tarım Ekonomisi. Ankara Üniversitesi Ziraat Fakültesi, Eğitim, Araştırma ve Geliştirme Vakfı Yayınları No: 5, Ankara

European Environment Agency (EEA) (2008) Sera gazı emisyonları azaltımı. https://www.eea.europa.eu/tr/themes/climate/intro

Gowdy JM (2008) Behavioral economics and climate change policy. J Econ Behav Organ 68:632–644

Habiba U, Shaw R, Takeuchi Y (2012) Farmer's perception and adaptation practices to cope with drought: perspectives from Northwestern Bangladesh. Int J Disaster Risk Reduct 1:72–84. https://doi.org/10.1016/j.ijdrr.2012.05.004

Ham M, Jeger M, Ivković AF (2015) The role of subjective norms in forming the intention to purchase green food. Econ Res Ekonomska Istraživanja 28(1):738–748. https://doi.org/10.1080/1331677X.2015.1083875

Han J, Kamber M (2006) Data mining: concepts and techniques, 2nd edn. Morgan Kaufmann Publishers, San Francisco

Hertwich EG, Peters GP (2009) Carbon footprint of nations: a global, trade-linked analysis. Environ Sci Technol 43(16):6414–6420

IPCC (2007) Technical summary. In: Climate change 2007: the physical science basis. Contribution of Working Group I to the Fourth Assessment Report of the Intergovernmental Panel on Climate Change. Cambridge University Press, Cambridge. https://www.ipcc.ch/site/assets/uploads/2018/02/ar4-wg1-ts-1.pdf

IPCC (2013) Technical summary. In: Climate change 2013: the physical science basis. Contribution of Working Group I to the Fifth Assessment Report of the Intergovernmental Panel on Climate Change. Cambridge University Press, Cambridge. https://www.ipcc.ch/site/assets/uploads/2018/02/WG1AR5_TS_FINAL.pdf

IPCC (2021) Technical summary. In: Climate change 2021: the physical science basis. Contribution of Working Group I to the Sixth Assessment Report of the Intergovernmental Panel on Climate

Change. Cambridge University Press, Cambridge, pp 33–144. https://www.ipcc.ch/report/ar6/wg1/downloads/report/IPCC_AR6_WGI_TS.pdf

Jellason NP, Baines RN, Conway JS, Ogbaga CC (2019) Climate change perceptions and attitudes to smallholder adaptation in Northwestern Nigerian Drylands. Soc Sci 8(2):31. https://doi.org/10.3390/socsci8020031

Kollmuss A, Agyeman J (2002) Mind the gap: why do people act environmentally and what are the barriers to pro-environmental behavior? Environ Educ Res 8(3):239–260

Liu M, Wang M, Wang J, Li D (2013) Comparison of random forest, support vector machine and back propagation neural network for electronic tongue data classification: application to the recognition of orange beverage and Chinese vinegar. Sensors Actuators B Chem 177:970–980. https://doi.org/10.1016/j.snb.2012.11.071

Maddison D (2007) The perception of and adaptation to climate change in Africa (Policy Research Working Paper 4308). The World Bank, Washington

Mancı A, Eren M (2017) Harran Ovası Tarımsal İşletmelerinde Risk Analizi. Harran Tarım Gıda Bilimleri Dergisi 21(4):456–465. https://doi.org/10.29050/harranziraat.299355

MGM (2022) Kuraklık Analizi. https://www.mgm.gov.tr/veridegerlendirme/kuraklik-analizi.aspx

Ministry of Environment, Urbanization and Climate Change (MoEUCC) (2022) Republic of Turkey, Climate Change Strategy 2010–2023. https://webdosya.csb.gov.tr/db/iklim/editordosya/iklim_degisikligi_stratejisi_EN(2).pdf

MoAF (2013) Türkiye Tarımsal Kuraklıkla Mücadele Stratejisi ve Eylem Planı (2013-2017), Ankara 2013. 59s. https://www.tarimorman.gov.tr/TRGM/Belgeler/Duyurular/2013_2017_Kuraklik_Eylem_Plani.pdf

NTOİM (2021) Nevşehir Tarım ve Orman İl Müdürlüğü Brifing Raporu-2021. https://nevsehir.tarimorman.gov.tr/Belgeler/A%C4%9Fustos%202021%20BR%C4%B0F%C4%B0NG.pdf

Özdemir HÖ, Kan A (2020a) Risk behaviours of agricultural holdings managers on management and decision making process in agricultural production; Kırşehir province case. J Glob Innov Agric Soc Sci 8(1):35–42

Özdemir HÖ, Kan M (2020b) Tarım işletmelerinin yönetiminde kullanılan tarımsal bilgi kaynakları: Kırşehir ili örneği. Türk Tarım Doğa Bilimleri Dergisi 7(2):500–509

Peker K, Kan M, Nadeem M (2019) Corporate governance of climate change adaptation. J Glob Innov Agric Soc Sci 7(1):1–5

Schalkoff RJ (1997) Artificial neural networks. McGraw-Hill, New York, p 448

Şen Z (2022) İklim Değişikliği ve Türkiye. Çevre Şehir İklim Dergisi 1(1):1–19

Şener E (2021) Küresel İklim Değişikliğinin Eğirdir Gölü Havzasına Etkileri ve Kuraklık Analizi. Süleyman Demirel Üniversitesi, Fen Bilimleri Enstitüsü, Jeoloji Mühendisliği Anabilim Dalı, Doktora Tezi

Seo SN (2017) An introduction to the behavioral economics of climate change for provision of global public goods. Academic Press, New York, pp 1–32. https://doi.org/10.1016/B978-0-12-811874-0.00001-5

STB (2017) Türkiye İlleri SEGE-2017 Sonuçları. https://www.sanayi.gov.tr/merkez-birimi/b94224510b7b/sege

STB (2022) Nevşehir İlçeleri SEGE-2022 Sonuçları. https://www.sanayi.gov.tr/merkez-birimi/b94224510b7b/sege

Turan H (2017) Tüketici Yenilikçiliği ve İnternetten Alışveriş: Kayseri'de Bir Uygulama. Yayımlanmamış Yüksek Lisans Tezi, Nuh Naci Yazgan Üniversitesi Sosyal Bilimler, Kayseri

Tüzer M, Doğan S (2021) İklim Değişikliğinin Bilimsel Temelleri. Sosyal Bilimler Araştırma Dergisi 10(3):639–656

Udmale P, Ichikawa Y, Manandhar S, Ishidaira H, Kiem AS (2014) Farmers' perception of drought impacts, local adaptation and administrative mitigation measures in Maharashtra State, India. Int J Disaster Risk Reduct 10:250–269. https://doi.org/10.1016/j.ijdrr.2014.09.011

van den Bergh JCJM, Ferrer-I-Carbonell A, Munda G (2000) Alternative models of individual behaviour and implications for environmental policy. Ecol Econ 32:43–61

WMO (2022) The state of global climate change. Press Release Number: 18052022. https://public.wmo.int/en/media/press-release/four-key-climate-change-indicators-break-records-2021

Yılmaz V, Güleç PAG (2021) Üniversite öğrencilerinin küresel iklim değişikliğine yönelik görüşlerinin araştırılması: bir yapısal eşitlik model önerisi. İzmir İktisat Dergisi 36(1):1–12. https://doi.org/10.24988/ije.202136101

Water Management Strategies for Agricultural Disasters

Muhammad Ashraf, Yasir Niaz, Mohsin Hafeez, Mobin ud Din Ahmad, and Azeem Ali Shah

Abstract

Agriculture is one of the most important sectors to meet food demands of the people around the globe and main source of livelihood of the rural community, especially in developing and under-developing countries. The increased frequency of water-related natural disasters like floods, droughts, and extreme weather events (i.e., temperature and rainfall events) has threatened the food security and the regional economies. These disasters have resulted in partial or complete damage of the crops in different regions of the world. This chapter focuses on the causes, basic principles, assessment methods, and strategies to manage the agricultural disasters due to floods, droughts, and extreme weather events and sufficient details have been presented in this chapter. The water management strategies are comprehensively discussed to address the water-related disasters at different spatial scales, i.e., basin, canal command, and field level. The review on the water-related disaster management strategies at different spatial scales urges the adaption of integrated approach for floods, drought, and extreme weather conditions while addressing these disasters for agriculture sector. The intense and frequent rainfall events have a strong connection with rising temperature which is comprehensively discussed. However, further research-based evidences for the rural areas are required. Flood water storage in dams and subsurface may have significant potential for the sustainable agriculture and the

M. Ashraf · Y. Niaz (✉)
Department of Agricultural Engineering, KFUEIT, Rahim Yar Khan, Pakistan
e-mail: yasir.niaz@kfueit.edu.pk

M. Hafeez · A. A. Shah
International Water Management Institute, Lahore, Pakistan

M. u. D. Ahmad
Commonwealth Scientific and Industrial Research Organisation, Canberra, ACT, Australia

M. Ahmed, S. Ahmad (eds.), *Disaster Risk Reduction in Agriculture*, Disaster Resilience and Green Growth, https://doi.org/10.1007/978-981-99-1763-1_4

economic uplift of the developing countries. The latest technological developments to forecast the floods and droughts are also presented in detail. The economic effect of each disaster and their impact on the agricultural community in different countries presented valuable insights and urge to address these issues. Moreover, secondary, dryland salinity and waterlogging issues and their management at different spatial scale are also discussed in this chapter.

Keywords

Agricultural disasters · Floods · Droughts · Extreme weather events · Salinity

Abbreviations

ASR	Aquifer storage and recovery
CAPE	Convective available potential energy
CPC	Climate prediction center
CRED	Centre for research on the epidemiology of disasters
DRR	Disaster risk reduction
DSS	Decision-support system
EFAS	European Commission launched development of a pan-European Flood Awareness System
EIS	Environmental information systems
EM-DAT	Emergency events database
ENSO	El Niño – Southern oscillation
EVOp	Environmental virtual observatory pilot
EWD	Extreme weather disasters
FDMS	Flood disaster management system
FEWS	Flood early warning system
GFM	Global flood monitoring system
GloFAS	Global flood awareness system
HEIS	High efficiency irrigation system
IFIS	Iowa flood information system
IPCC	Intergovernmental Panel on Climate Change
ISD	Integrated statistical and data-driven
ISDI	Integrated surface drought index
MAR	Managed aquifer recharge
NFFS	National Flood Forecasting System
NWP	Numerical weather prediction
PDO	Pacific decadal oscillation
PDSI	Palmer drought severity index
SPEI	Standardized precipitation evapotranspiration index
SPI	Standardized precipitation index

SST	Sea surface temperatures
TCI	Temperature condition index
VCI	Vegetation condition index
VHI	Vegetation health index

1 Introduction

Agriculture accounts for 4% of GDP globally and in some developing countries, it accounts for more than 25%. The agriculture sector has a significant role in economic development, especially the poor and marginal countries. Moreover, the agricultural sector plays a vital role in providing food to meet the ever-increasing population around the globe. Therefore, no poverty and zero hunger are ranked at the top in the sustainable development goals by the United Nations due to their extreme importance. Poverty and agriculture have a strong link for the community living in developing and poor countries. For example, growth in the agriculture sector is two to four times more effective in raising the income of poor communities as compared to the other factors. According to the World Bank Group, the living of about 65% of the poor working adults is dependent on agriculture (https://www.worldbank.org/en/topic/agriculture/overview). Therefore, agricultural development is the most powerful tool to end poverty and feed the ever-increasing population.

Agriculture sector withdraws 70% of fresh water globally which reaches up to 90 for some countries. This demand will be increased in the future for the agriculture sector to meet the food production for the population around the globe. The reallocation of water for other sectors will also create stress on the agriculture sector. Therefore, water as a precious and important commodity should be used wisely, especially the countries already facing the water stress situation. However, the agriculture sector is threatened by different kinds of disasters like pandemics, floods and droughts, extreme weather conditions, and pest attacks. Extreme weather disasters have a significant role in damaging crops partially or completely at the regional scale. Moreover, natural disasters impact crop productivity significantly by physically damaging the crops and livestock. The damages to infrastructure (i.e., storage, irrigation systems, installations, machinery, and equipment) are indirectly impacted due to floods. Consequently, the impact of disasters may eventually cause local food shortages in the years following a disaster. The developing countries having more dependence on agriculture and related workforce are threatened more as compared to most industrialized countries.

2 Agricultural Disaster Impact on the Economy

The country having a significant dependency on agriculture sector may have greater socioeconomic effects. For instance, a 50% fall in agricultural production caused by a drought would translate into a more than 10% drop in aggregate GDP for countries that rely the most extensively on agriculture (Benson and Clay 2004). Therefore, the countries having a significant economic dependence on agriculture sector should give special focus on the protection of domestic agricultural sector. Hence, in the following sections, different water-related risks to crop production and strategies to counter those risks are comprehensively discussed.

3 Flood Risk

3.1 Flood Losses

Flooding has the most devastating impacts because it is most widely distributed natural disasters globally (Guo et al. 2014). Moreover, it is the most frequent disaster as compared to all other types of disasters across the globe. Floods affect about more than 94 million people worldwide every year through displacement from homes, access to safe drinking water, destruction of infrastructure, injury, and loss of life (Guha-Sapir et al. 2015). The CRED/EM-DAT International Disaster Database 2021 showed that the maximum disasters are related to floods and frequency of flood events has also increased (https://www.emdat.be; Accessed September 16, 2022) (Fig. 1).

The global warming may cause intensify and more frequent flooding globally (Hirabayashi et al. 2013). South Asia and South Central China may be more prone to the increased flooding risk (Lu et al. 2016). Floods may affect the agricultural land by either inundating the adjacent river areas or erosion of agricultural land. Moreover, flash floods also cause severe damages to the agricultural land. Generally, floods are generated in the catchment area; however, their effects are experienced in the plain areas sufficiently downstream (Fig. 2). The economic loss was estimated as 74% in 2021 due to floods only when compared with the other disasters. Therefore, an efficient flood forecasting system and appropriate management at different spatial scales, i.e., regional, catchment, and local, are required.

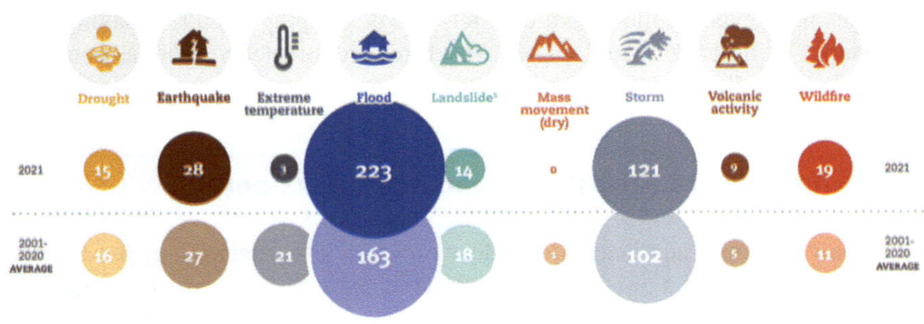

Fig. 1 Occurrence by disaster type 2021 compared to the average of 2001–2020

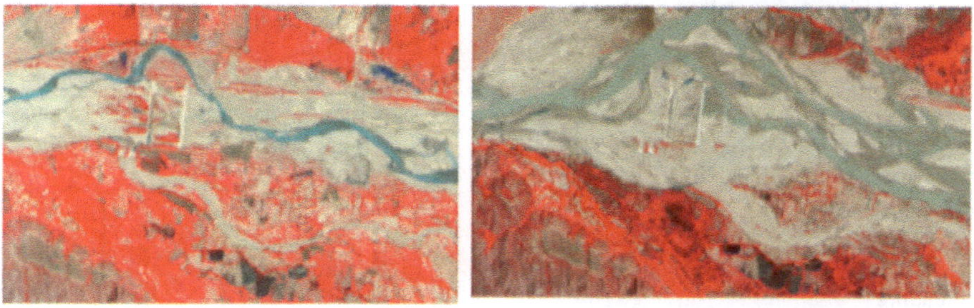

Fig. 2 Breaching of flood bunds and erosion of adjacent agriculture land due to flash flood in hill torrent area of Pakistan. Images were acquired by sentinel satellite on August 11, 2022 (left) and August 31, 2022 and downloaded from sentinel hub-playground

3.2 Flood Analysis and Forecasting Techniques

Many runoff and flood forecasting techniques and methods are used for short- and long-term flood forecasting. Forecast is very important to assess the volume of water, peak discharges, and their timing that can be generated from a particular rainfall storm event for the purpose of planning storages and releases for irrigation to manage floods and seasonal droughts. Recent developments in computational resources, robust and advanced methods, and earth observation technologies have enhanced the capacity to manage the floods and droughts through analysis, modeling, and visualization of disaster-related data. Hydrological modeling systems play a vital role for flood disaster management through simulating the runoff/floods from the catchment areas. Environmental Information Systems (EISs) are getting more attention to scientific community in tackling scientific challenges in disaster monitoring, prediction, and assessment by providing integrated multidisciplinary platforms (Granell et al. 2013). A rapid access to earth observation data through satellites provides opportunities to improve the flood warnings and flood preparation and response measures. The earth observation satellites, airborne and ground-based remote sensing systems have potential to forecast flood more accurately at different spatial scales. Table 1 shows the technical details of quantitative precipitation forecasts used in flood forecasting at a large scale.

Flood disaster management system (FDMS) provides flood reduction capabilities throughout the entire disaster management cycle. FDMS integrates various environmental models and accesses a variety of data sources. These systems have capabilities to model the different phases of FDM and have been widely applied in flood disaster management worldwide. These systems can be classified into three types according to their functional roles:

1. The first type of system integrates atmospheric and hydrological models for simulating floods in the early stage of flood disaster management. Examples of these

Table 1 Technical details of quantitative precipitation forecasts used in large-scale flood forecasting

Product type	Spatial extent	Spatial resolution	Temporal resolution	Forecast range	Uncertainty
Radar nowcasting	~10,000–50,000 km^2	1–4 km	5–60 min	1–6 h	Low
Ensemble radar nowcasting	~10,000–50,000 km^2	1–4 km	5–60 min	1–6 h	↓
Radar-NWP blending	Regional	~2	15–60 min	~6 h	
Limited area NWP	Regional-continental	2–25 km	1–6 h	1–3 days	
Ensemble limited area NWP	Regional-continental	2–25 km	3–6 h	~5–30 days	
Global NWP	Global	~15–100 km	~3–6 h	~5–30 days	
Seasonal forecasts	Global	~15–100 km	~6–24 h	Months	High

Source: Emerton et al. (2016)

systems are Delft-FEWS (Delft Flood Early Warning System), National Flood Forecasting System (NFFS), Global Flood Monitoring System (GFMS), etc.

2. The second type of FDM systems extract the information and assess the flood impacts for emergency response and recovery after occurring of a flood disaster. The NAZCA WebGIS 3D application, Iowa Flood Information System (IFIS), etc. are few examples.
3. The third type of FDM systems manage and share information relevant to the disaster. The Environmental Virtual Observatory pilot (EVOp) statistically evaluates the flood information using web maps (Vitolo et al. 2015), manages several global flood databases, and supports reporting new flood events.

Most of the FDM systems integrate one or several models for providing the flood reduction functions that partially address the one or two phases of disaster management. However, continuous monitoring of the developing disasters is necessary for the robust responses during the entire disaster management phases, because all the four phases of disaster management, i.e., mitigation, preparedness, response, and recovery, are continuous and dependent (Federal Emergency Management Agency 2001). The FDM based on multiple systems can be a less efficient and time-consuming. Therefore, an information system having flexibility for model integration and covering all the phases of disaster management could be the most efficient one.

The floodplain zoning based on hydrodynamic modeling is also a widely adopted technique for reducing flood disasters. These techniques help identifying the flood extents in rivers, the duration of flood recession from floodplain areas, and the areas prone to erosion using 1-D, 2-D hydrodynamic and river bank erosion model.

HEC-RAS, CCHE-2D, and SSIIM are few examples of these models. Moreover, flood frequency analysis and flow duration curves help providing the limits of flood magnitude and discharges when the flows remain more than a particular value. The implementation of legislations based on the results of hydrodynamic modeling coupled with frequency analysis may help reducing damages to agriculture and irrigation infrastructure due to floods.

4 Flash Floods

The intense precipitation events cause to generate flash floods in catchments with dendritic drainage pattern, thus having high hazardous impacts. Generally, these floods are associated with rainfed catchments with steep slopes with minimal snow cover which are nourished by summer precipitation systems. Flash flood is ranked top in the lists of natural hazards for more than 100 countries. Currently, flash flood forecasting is being done at the continental scale because no global flash flood forecasting system exists. These systems generally consider empirical correlations, unit hydrographs, and hydrological modeling-based flood simulations over limited and specific areas. The major limitation of generating a global flash flood forecast is the coarse resolution of Numerical Weather Prediction (NWP) systems which are unable to consider many physical processes at the fine resolution which are responsible for intense precipitation, such as convection. Therefore, there is a potential for better flood forecast by improving the resolution of global NWP. The assessment of the parameters' (like convection) estimation at the subgrid level can also be useful for improving forecast.

5 Droughts

A drought is a relatively extended period when an area experiences shortage of water due to a deficiency of rainfall that results in a significant impact on agriculture and humans. Drought is the least understood disaster as compared with storm events due to its complex temporal-spatial characteristics, whereas storm events have definite beginning and end and can be observed easily.

Drought starts with the temporary changes in climatic factors over the region like increase in temperature, wind speed, and less humidity due to outweigh of weather patterns over climatic conditions. These factors also have direct relationship with the rainfall, e.g., decrease in rainfall intensifies these factors and this drought may be termed meteorological drought. Meteorological drought is a signal for the potential water shortage in the following season if conditions prevail over the region for extended period. The droughts cause significant damages to agricultural crops in the arid regions where agriculture is solely dependent on rainfall. However, these are short-term droughts and have limited effect on agriculture and human. The meteorological drought forces human and livestock for the temporary migration. The prevailing hot weather conditions for longer duration increase the evapotranspiration

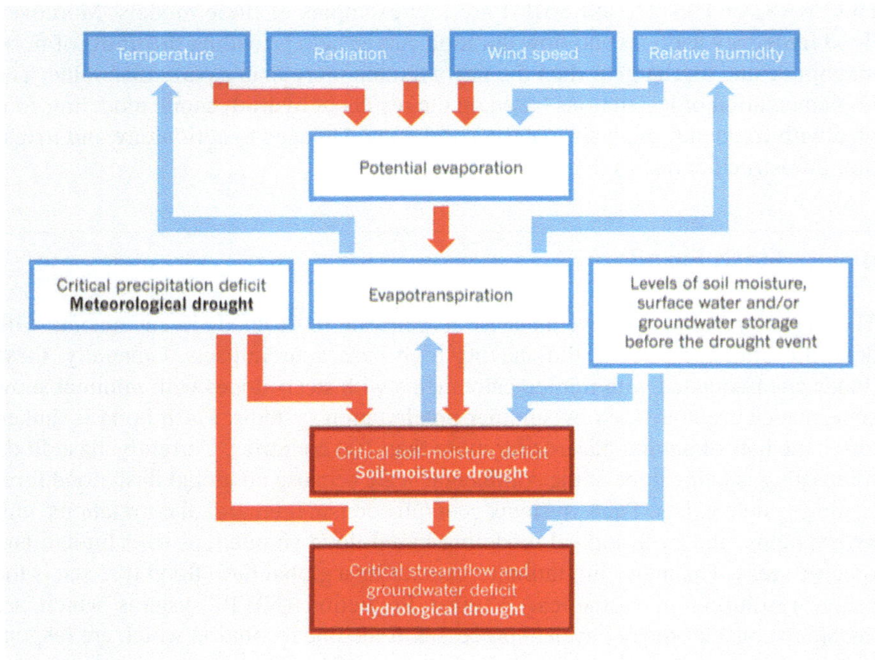

Fig. 3 Interaction between droughts of different types

rates in the region. Therefore, limited irrigation supplies (or available water) cause the soil to become dry which creates the stress in the crops' root zone. The moisture stress causes the reduction in crop yields, which ultimately affects the socioeconomic conditions of the people and the region. This drought is termed agricultural drought. If the water supply to irrigated area is limited due to less precipitation in the source region, then the rivers, irrigation network, and reservoirs will evidently have little water, which ultimately affects the irrigated areas and is termed hydrological drought. Hydrological drought also occurs due to warm weather conditions and less rainfall in the supply zone. Therefore, the impacts of hydrological drought are experienced later and most devastating for the regions solely dependent on the water supplies from the reservoirs or rivers. Hydrological droughts are more severe for the regions where the irrigation supplies are more dependent on rainfall-runoff rather than snow or glacier melt. Figure 3 shows the interaction between droughts of different types.

5.1 Causes of Drought

Droughts are triggered by cyclic weather patterns due to the changes in the amount of moisture and heat in the air, land, and sea. For example, rainfall distribution around the globe is impacted by the air circulation paths through the atmosphere. Generally, the sea surface temperature anomalies cause to alter the air circulation patterns over the globe. These change the precipitation around the globe and cause new weather pattern which in turn affects the water supply and demand. Similarly, the earlier snowmelt causes a reduced water supply for irrigation in summer, if the flows are not regulated through dams. The other causes associated with changes in weather pattern are global warming and climate change (Ahmed et al. 2018). However, the effects of excess water demand, deforestation, and soil degradation are experienced at regional scale. Therefore, assessing the effects of severe drought events on agriculture can help anticipating and adapting the measures to maximize crop yields. Over the past 134 years (1880–2014), the average global temperature has increased by about 0.5 °C (Al-Ghussain 2019).

5.2 Drought Mechanism and Predictability

As discussed in Sect. 5.1, large-scale weather processes which are forced by SST anomaly, air and land surface interactions, and natural and anthropogenic changes generally cause the precipitation occurrence over a particular area (Schubert et al. 2016). These factors help in predicting a meteorological drought to some extent (Wood et al. 2015). The gradual variation of these factors helps in better seasonal forecast (Roundy et al. 2014). The SST anomaly is the major factor among these which has significantly resulted in better forecasting. The teleconnections between weather parameters are a major advancement in drought prediction (Schubert et al. 2004a, b) which combines the ENSO phenomenon (with periods of 2–7 years) for predicting the droughts at seasonal scale. ENSO affects seasonal weather condition over greater spatial extent for the regions including North and South America, Africa, India, Indonesia, Southwest Asia, and Australia (Schubert et al. 2016). ENSO and other data can be found in oceanic and atmospheric data link at NOAA Climate Prediction Center (CPC) website (e.g., https://www.cpc.ncep.noaa.gov/products/monitoring_and_data/oadata.shtml).

Agricultural and hydrological drought predictions are generally made through the hydrological simulations based on estimations of meteorological parameters. The drought occurrence is related with several processes depending on spatio-temporal scales and seasonal variability. However, there are still challenges in explaining drought prediction (Bonsal et al. 2011). The following sections explains the different approaches which can be used for predicting different types of droughts.

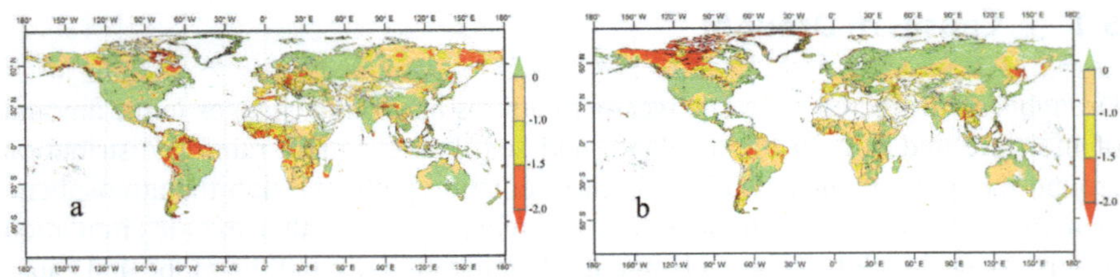

Fig. 4 Spatial distribution map of global drought in 1983 and in 1988: (**a**) shows spatial distribution of global drought in 1983, (**b**) shows spatial distribution of global drought in 1988, green color represents normal, yellow and red color represent drought, less than_2 is severe drought (Source: Wang et al. (2014))

5.3 Drought Monitoring Indices

The common drought monitoring indices include the (1) Palmer Drought Severity Index (PDSI), (2), Standardized Precipitation Index (SPI), (3) Integrated Surface Drought Index (ISDI), (4) Standardized Precipitation Evapotranspiration Index (SPEI), (5) Temperature Condition Index (TCI), (6) Vegetation Condition Index (VCI), and (7) Vegetation Health Index (VHI) (Rhee and Im 2010). The SPEI, SPI, and PDSI are the three most widely used indices for drought prediction. The advantage of SPI is that it can be applicable for water resource application at all temporal scales due to its simplicity and temporal flexibility and it predicted drought better than PDSI for the USA (Hayes et al. 1999). However, the limitation of SPI is that it considers only the precipitation factor, which does not reflect drought due to warming. On a global scale, PDSI can be used for better drought prediction caused by warming. Since PDSI considers the temperature factors, it is unable to evaluate different types of droughts. Therefore, SPEI has advantage over PDSI due to its temporal flexibility determining drought characteristics due to warming (Vicente-Serrano et al. 2010). Moreover, SPEI is better in detecting agricultural drought than SPI. Therefore, it is an effective approach to determine agricultural drought impacts at a global scale (Potop et al. 2012). Figure 4 shows the SPEI drought at a global scale in 1983 and 1988.

5.4 Spatial-Temporal Variation of Drought

The drought analysis is the foremost step of drought disaster mitigation because it provides the characteristics of the drought at a specified spatial scale over the period of time. The studies show that the drought characteristics like intensity and duration over a specified areas are increasing since 1970s and the spatial extent will expand from 1% to 30% in the twenty-first century for the severe drought category (Burke et al. 2006). Moreover, the frequency of severe drought events and their duration are likely to increase (Blunden et al. 2011). However, contradictory findings are reported by different researchers, e.g., see Field et al. (2012) and Dai (2011). In the earlier study, it was concluded that the drought extent has increased on the global scale since 1980s, whereas contrary findings were reported later on. Similarly, for the Pakistan, increasing summer precipitation trend has been reported (Ahmed et al. 2018), whereas a decreasing trend has been reported for the western and southern parts of the country (Hina and Saleem 2019). The complexity of the temporal-spatial characteristics of drought and the conflicting findings of the previous studies urged the reliable framework for the assessment of drought. Therefore, a special report by IPCC fourth assessment concluded that PDSI method over-estimates the frequency of drought events at the global scale. Recently, few researchers have used area method for the temporal-spatial characteristics of drought events on a continent scale.

5.5 Losses Due to Droughts

Losses accounted for drought are much higher as compared with the other natural disasters. The Emergency Event Database (www.em-dat.net) shows that the number of drought events accounts for only 5% of total disasters, but the losses can amount to 30% as compared with all natural disasters. The severe drought can have devastating effects like significant crop yield losses, an increased risk of forest fires, intensified land degradation and desertification, and increased competition for resources and social violence (MacDonald 2007), because of its long-lasting impacts and vast spatial extent. Therefore, the effects of severe drought events on the crop growth process and yields have been the focus of latest research, especially at the regional scale. The severe agricultural drought could considerably reduce agricultural production and drought risk will also increase in the future, which will threaten the world's food security (Schmidhuber and Tubiello 2007).

5.6 Drought Impact on Agriculture

Drought events have severely affected the crop yields and socioeconomic condition of the farming community as well the country and at regional levels. As discussed in Sect. 1, the countries having significant share of GDP from agriculture sector are affected more. In 2002, severe and extreme drought affected the 26 states in the USA, and total losses exceeded $2.7 billion (Wilhite et al. 2007). Wheat and maize resulted 50% yield losses due to the severe drought in the in American southern region during 1930s. The 40% grain production in Ukraine was affected due to the most recent droughts in the Russia (Kogan et al. 2013).

Table 2 Impact of drought on crop yields and socioeconomics

		Effects of drought	
Details	Period of drought	Crops yields	Socioeconomic impact
From 1996 to 2010, south-eastern Australia experienced prolonged dry conditions with rainfall persistently well below average during April to October	Australia's Millennium drought (1995–2009, the most acute period was between 2001 and 2009)	Cotton production decreased by 66% as compared to the normal year	10,000 people directly employed by the cotton industry were impacted by the drought
2011 was the driest year ever for Texas, with an average of only 14.8 inches of rain (Normal year rainfall is 27.25")	Texas Drought (2010–2014)	Crops also suffered, as corn outputs fell by 40% in 2011 and peanut production was down as well	1100 residents depend on tanker trucks to deliver water to the town's storage tank
During the last 50 years, Bangladesh suffered about 20 drought conditions	Bangladesh drought (last 50 years)	Depending on the intensity of drought, the estimated yield reduction of different crops varies from 10% to 70%	Approximately 53% of the Bangladeshi population is affected by droughts
October–December 2020, March–May 2021, October–December 2021, and March–May 2022	Africa Drought 2014–2022	The prices for maize grain are marginally above the 5-year average, and 57% higher than 2018 prices across Southern Africa, and the trend is expected to continue in most areas	18 million people in Ethiopia, Somalia, and Kenya are food insecure. More than 7 million children under age 5 are facing acute malnutrition. Millions of livestock have died

5.7 Drought Impact on Crop Yield

In arid regions, crop yields are significantly impacted by the amount and the frequency of rainfall, for example, 50 to 108% increased yield due to positive response to rainfall for sorghum and corn, respectively. However, the yield of both crops has decreased due to floods. Generally, crop yield is greatly affected by floods during summer seasons. The winter crops especially cereal like wheat and oil seed crops are less affected in irrigated areas due to a shortage of water and, generally, little rainfall is received in the areas where cultivation of these crops is practiced. However, there may be a great difference in crop yields in rainfed areas if compared with the irrigated ones because the crops in rainfed areas are solely dependent on precipitation. Maize crop may have a significant impact on its yield due to its sensitivity to water. Overall, the drought had a greater impact on the crop yield in arid regions than the crops in irrigated areas. Therefore, considering the cultivation of sorghum and other grains having drought-tolerant capabilities may be beneficial for agricultural crop producers in rainfed areas for better yields during drought periods. Table 2 shows the impact of drought events on crop yield in different countries.

6 Flood and Drought Management Strategies

Previously, most researches considered the floods and droughts, which are two extremes of the same hydrological cycle, separately. Therefore, management strategies to address the risk from floods may lead to an increase in the risk from droughts. Di Baldassarre et al. (2018) have comprehensively discussed this issue in the context of reservoir operations. The recent example of this issue is the floods in the hill torrent areas of Pakistan during August 2022 where diversion structures were made to irrigate maximum land without considering the flood impacts in the region. However, the interactions between these closely linked phenomena have been the focus of recent research for better Disaster Risk Reduction (DRR) measures and strategies.

Some structural measures like dikes and levees, dams, flood control measures, subsurface storage, agricultural practices, and vulnerability and preparedness have been a focus for DRR. Ward et al. (2020) have comprehensively discussed the impacts of different interventions like dikes, levees, embankments, and dams on drought. Similarly, strategies designed for DRR like subsurface storage serve as a buffer against both flood and drought hazards. Levee and dikes restraint the river by reducing the floodplain area and thus decrease infiltration and groundwater recharge causing the decline in water table in the nearby areas, if pumping system is used for irrigating the crops. It could be an effective flood hazard strategy. However, a breach in the dikes could result devastating hazard by the loss of agriculture crops due to inundation, loss of land due to erosion and the nearby community, because construction of dikes leads to increase development in the areas protected by dikes. Pakistan experienced the huge loss during 2010 when the dike failure occurred on River Indus and inundated the significant area and damaged the agricultural land and property of the people. Moreover, levees or dikes require to be wetted during the drought to reduce the failure risk.

6.1 Basin Scale Water Management

Water management at basin scale provide the better way to fulfill the agricultural water demand for the downstream stakeholders. The knowledge about the water share of different downstream stakeholders and appropriate management for water releases from the storages may help to manage both extremes, i.e., floods and drought. For example, significant runoff for Tarbela reservoir in Pakistan derived from the snow and glacier melt and reservoir is filled accordingly. Similarly, the runoff for Mangla reservoir is mainly derived from snow melt and rainfall-runoff and reservoir is filled accordingly. The flow releases for agriculture from both the dams are also made accordingly. However, robust data management and analysis system are required to run the complex irrigation system like Indus Basin Irrigation System. It is worth noting that WAA Tool not only helps in efficient operation of reservoirs to achieve equitable and efficient water allocations to different provinces, it also helps reduce shortages (as compared to manual calculations) and the magnitude of extra water could vary from season to season (year to year).

Spate irrigation in the arid regions helps harnessing seasonal floods of rivers and streams to fill irrigation requirement in the cropping area. This technique also addresses the interconnected phenomena, i.e., drought mitigation with flood mitigation. The intensification of agriculture can also contribute to reduce agricultural drought risk. However, planned upscaling of agriculture to the area can have positive impacts, e.g., providing irrigation facilities to the barren areas like desert. The crop zoning based on climate, irrigation facility, and some other factors can be used as a drought or flood risk measure. However, changing precipitation pattern could disrupt agriculture at a regional scale. For example, recent intense precipitation damaged the crops on the area of more than 2 mha in Pakistan. In wake of climate change, farmers in sub-Saharan Africa cultivate short-duration crops and have changed the cropping period to avoid intense rainfall periods or dry spells.

6.2 Interventions in Source Areas

The interventions at catchment level help to reduce the flood and drought impacts on agriculture and intensification of agriculture in the source and downstream areas. Rainwater harvesting at the source region has significant potential to reduce floods, lessen droughts, and agricultual expansion. These interventions are generally made in arid regions to efficiently utilize runoff, especially for irrigation, and increase forestation thus minimizing the drought risk. For example, water and soil conservation methods like water harvesting by constructing small dams and ponds have been widely used in countries like Pakistan and Brazil, where successive small dams/ponds have been built to conserve water and efficiently use it for agriculture by involving local communities. In Potohar region, Pakistan, many small dams and ponds have been constructed that have a significant impact on the expansion of agriculture in the region, improving livelihood and sustainable agriculture (Panhwar et al. 2021). In Brazil, over 3000 successive dams were built during 2001–2009, which resulted in intensive riverine vegetation and decreased drought risk. These small dams can also help flood mitigation by reducing the flood peaks. These water harvesting interventions are simple, economical, adaptable, and efficient. However, interventions in highly erodible catchments may result in increased sedimentation and flood risk downstream.

Dams play a vital role in regulating the river flows by storing a significant runoff for different purposes like irrigation, hydropower, and flood retention. However, dams built for irrigation purposes only consider the water storage for agriculture purpose. These dams provide the water for agriculture during seasonal drought period and thus decrease the chances of adverse effects on crops.

Afforestation in catchment areas is also considered as a viable flood and soil erosion mitigation measure. However, the higher evapotranspiration in dry periods and reduced groundwater recharge can significantly reduce runoff and cause water shortages for downstream users. For example, afforestation reduced the flows from the catchment area causing shortages to the urban water supply in Fiji. Similarly, establishment of Eucalypt plantations caused decline of groundwater level by 0.38 m and resulted 50% decrease in groundwater recharge days in Argentina. The establishing plantation increases the soil infiltration rates that play an important role for generating subsurface flow and causes lower peaks in the wet season. However, persistent high temperatures in forested area during drought period can amplify tree mortality through increased fire hazards. Forest fires caused significant increase in runoff, peak flows, and erosion leading to damaging infrastructure and debris flows, e.g., recent events in Australia, Pakistan, and Thailand.

6.3 Interventions in Plain Areas

The exceptionally high flood event in catchment area may cause severe effect by increasing the chances for dam breaching and extreme flood events downstream of the dam, if it is already at its highest reservoir level. Moreover, the dam and the related irrigation structure lead to extensive agriculture in the downstream areas and even during high water demands over-extraction of groundwater leads to lowering of water table thus increasing the cost of agriculture produce, energy input, and high commodity prices and decreasing the groundwater storage. Cultivating crops in the floodplains also increases the flood exposures and subsequent damages. Legislation for over-extraction of groundwater, floodplain zoning, and robust early warning systems could decrease the floods and drought vulnerability in irrigated and floodplain areas.

Besides dikes and dams to manage surface water, the subsurface storage option is also used for implementing flood management measures. The groundwater recharge during intense precipitation events and diverting flood water to degraded land can help mitigating drought and reducing the flood/intense rainfall events' impact on agricultural lands. In arid and semi-arid catchments, groundwater is often the most reliable source of water for agriculture that can be developed using recharge from the runoff during rainfall events. Heavy and extended rainfall events that cause extreme floods are of significant source for groundwater recharge in these areas. The subsurface is increasingly used for storage of excess runoff and rainfall at both catchment and irrigated areas. The techniques such as sand dams and Managed Aquifer Recharge (MAR) are being used at catchment and irrigated areas, respectively. Subsurface can be used for flood and drought mitigation thus addressing the dual challenges of (seasonal) floods and (seasonal) water scarcity. For example, MAR has significantly reduced flood impacts and encouraged the farmers to grow high-water-demanding crops even in dry years for Chao Phraya River Basin in

Thailand. The MAR has also been proved a successful technique for water resources management in the Mediterranean region and the south-west of the USA.

6.4 Flood and Drought Management at Field Scale

At the field level, different management techniques have been used to increase water productivity like cropping on raised bed can save 35–40% water for the same crop as compared to conventional irrigation method. Moreover, this technique also helps to sustain the high crop yields due to the collection of rain water in the furrows to avoid continuous standing water in the field. Similarly, irrigation to alternate furrows also helps increase crop yields. Significant areas in irrigated agriculture utilize surface irrigation method. Therefore, slight water saving at field level could have a major effect of water utilization at basin scale thus having a huge potential of water saving and addressing both issues. The water saving achieved through implementing the technologies could be used to provide during drought periods. Different tools and technologies have been developed for improving the irrigation efficiency at the field scale, e.g., WinSRFR surface hydraulic modeling tool. The term precision surface irrigation is applicable to these kinds of techniques. Moreover, recharge of excessive water during rainy days into the ground can be used for the groundwater development and addressing the impact of over-abstraction. On-farm water storage also serves as the cushion to both extremes. Moreover, on-farm water storage coupled with HEIS has further potential to increase the application of stored water at the field level. These methods have been proved successful where irrigation supplies are limited, areas are undulated with more hydraulic conductivity, and groundwater quality is not suitable.

In hilly areas, the development of small ponds coupled with the High Efficiency Irrigation System (HEIS) serves both purposes, i.e., reducing flood intensity and recharging groundwater thus increasing agricultural productivity in the region. Moreover, different conventional techniques, like strip farming in the mountainous region, are also effective methods to reduce flood impacts downstream by conserving soil moisture and increasing crop yield. The HEISs have a high great potential to reduce water demand thus eliminating the risk of agricultural drought. Drip irrigation systems are more efficient than pivot irrigation technologies. However, the success of particular efficient irrigation technology depends upon the different factors like topography of the area, soil type, selection of suitable crop, etc. Similarly, providing measures to enhance soil infiltration for precipitation can also help reduce agricultural drought risk.

7 Extreme Weather Disasters

Extreme weather conditions due to climate change have triggered the hydrological extremes globally, especially after the industrial revolution since the middle of twentieth century. Extreme temperatures can cause significant damages to crop thus threatening local and global food security. The climate-related disasters have resulted about one-quarter of all damage and losses in the agricultural sector in developing countries and are expected to become more frequent in the future. Therefore, policy interventions required robust scientific basis to develop effective disaster risk management and adaptive measures (e.g., infrastructure, technology, management, and insurance) to protect the most vulnerable populations and to ensure global food security. A disaster depends not only on the severity of the intensity of weather event but also on the vulnerability and exposure of the natural habitat to it.

Earlier research related to extreme weather conditions has addressed agricultural effects based on the degree days above some threshold. This approach underestimates the effects of EWDs on agriculture; it can affect the exposed system differently depending on the vulnerability of the exposed system. The photosynthesis rates of sweet corn decrease by 30% when the temperature exceeds 30 °C. Knox et al. (2012) estimated that by 2050, with an increase of 1.6 °C temperature, the yield of crops, i.e., wheat, maize, sorghum, and millet in Africa and South Asia, will decrease by 8% compared to current yields. A study forecasted a decrease of maize yield by 4.6% and wheat by 3.8% for each 1 °C.

7.1 Effect of Extreme Weather Conditions at Catchment Scale

The persisted high temperature for few days results in heavy snowmelt in the snow-fed catchments and decrease in the crop yield in the irrigated and rainfed areas. The increased snowmelt also causes the floods in the rivers. Similarly, the intense rainfalls in catchment areas cause devastating impacts downstream by eroding agricultural land and inundating the crops in vast areas. The consequences caused by floods are discussed in Sect. 4.

7.2 Impact on Crop Yields in Irrigated Areas

However, the increase in extreme temperature for few days (heat wave) causes significant loss in the crop yield. The grain crops are expected to reduce their yields at maximum due to extreme temperatures. Similarly, heavy rains also cause significant impact on crop yields due to retention of water in the fields for longer duration, lodging or increase in the pest attacks depending upon the crop type. For example, lodging of crops can reduce the grain crop yield by 20–30% depending upon the crop stage; pest attack on a cotton crop due to rainfall may reduce the crop yield from 20 to 40%. The retention of rainfall water in the field can cause significant loss in the crop yield.

Recent studies have reported that there is a close relationship between the warming of the area and rainfall intensity. The increase in temperature of the area causes more intense rainfall. These studies have focused on the urban areas. Moreover, the studies on the topic have reported the increase in rainfall intensity over urban areas as compared with rural areas. The increase in rainfall due to rise in temperature is more likely to happen in the humid regions. For example, a study in India found that urban areas were 30–40% more likely to receive rainfall than neighboring rural areas (Kishtawal et al. 2010), and comparable patterns have been reported in subtropical North America. However, similar studies should be carried out for the agricultural areas experiencing the warming impact of climate.

7.3 Water Management Strategies for Extreme Weather Conditions

The improvement in early warning systems (EWS), increase in storages especially at the catchment level by constructing large dams, and subsurface storage and small storages at the field level can also be the effective measures to address extreme weather conditions. These interventions not only address the extreme weather conditions issues but also the other hydrological extremes, i.e., drought and floods. The detail on the importance of these interventions is as follows:

- Dams for flood storage and flood reduction in catchment areas will act as buffer during drought and floods as discussed in Sect. 6. Increase in the storages for irrigation can play a vital role for addressing all the climate-related agricultural hazards; otherwise, the countries dependent on agriculture will be affected more.
- On-farm water storages, recharge to groundwater, and application of resource conservation technologies are comprehensively discussed in previous sections (see Sect. 6).
- Early warning systems (EWS) helps farmers to select cropping pattern and harvesting practices as per expected forecast. Short-term forecast (7–15 days) help farmers in planning irrigation scheduling and avoid over-irrigating their fields thus increasing crop yields by providing optimum water to their fields. Seasonal forecasts have significant potential for sustaining agriculture, especially in tropical and subtropical regions. It takes time to build trust of farming community in seasonal forecasts. However, the impacts of wrong forecasts can lead to significant losses and mistrust. Late sowing in the season might be a good drought risk reduction measure if it is too dry in the early days. However, heavy rainfall and flooding during the start of cropping season can leach down the nutrients from the root zone. Furthermore, conscious crop selection can reduce extreme weather impacts. For example, farmers in areas facing shortage of water due to reduced precipitation in sub-Saharan Africa have switched from high- to low-water-requirement crops. Similarly, the farmers in the hill torrents areas have adapted the high-water-requirement crops like rice during the wet season due to abundant water availability.

8 Salinity and Waterlogging

Salinity and waterlogging are generally considered twin problems that arise together and have a strong linkage. Generally, when the waterlogging issue creates in an area it also brings the underground salts to the surface, or during the reclamation of water-logged soils, the salts are left in the root zone. This is also known as secondary salinity which is a major issue in the areas with newly developed irrigation system. Moreover, salinity in dryland soils develops through a rising water table and the subsequent evaporation of the water from the surface. The rise of water table may be due to unequal recharge to groundwater from irrigation system or flooding to the area with underlying impermeable layers. During rising of water table, the groundwater dissolves salts in underlying geological formation and eventually reaching to the soil surface where it evaporates and causes salinity. Dryland salinity can also occur in un-irrigated areas. In Australia, dryland salinity is a major problem that costs Australia over $250 million a year through impacting agriculture, water quality, and the natural environment.

There can be many other causes of salinity in a soil like irrigation with saline water, sea water intrusion into coastal lands as well as into the aquifer due to over-extraction and overuse of fresh water, overuse of fertilizers, use of soil amendments (like gypsum), use of sewage effluent, dumping of industrial brine onto the soil, etc. Figure 5 illustrates the various aspects of the development of soil salinity.

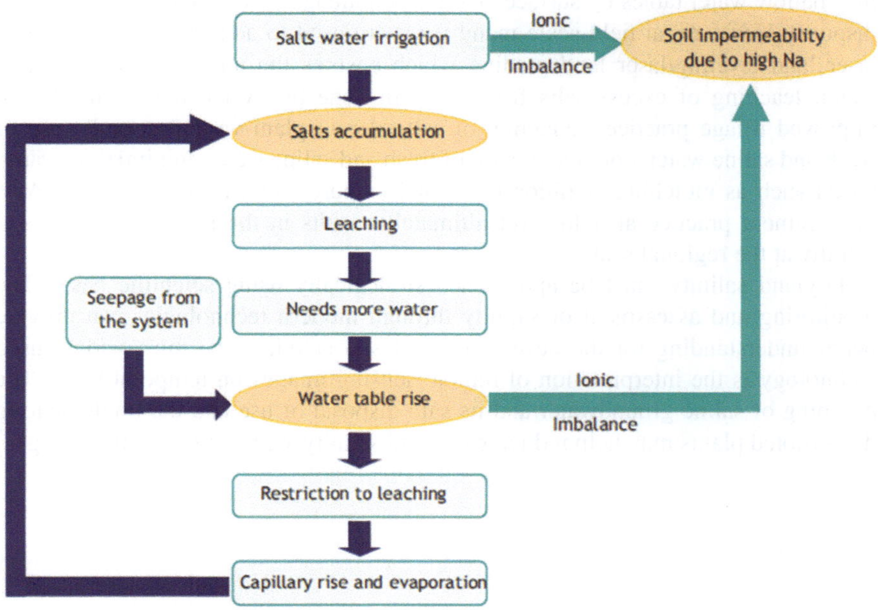

Fig. 5 A hypothetical soil salinization cycle

The excessive sodium concentration in a soil or irrigation supplies causes surface sealing. The soil water sodicity damages the soil structure through a breakdown of soil aggregates during wetting phase of soil. The leaching of sodium ions causes to seal the voids and thus reduces or completely blocks the leaching of water in the soil. However, in desert areas where sandy soils are dominant, surface sealing reduces the infiltration rate thus increasing the irrigation efficiency. However, salinity negatively affects the crops by reducing the yields due to poor growth and patchy crop conditions.

Salinization can reduce the crop yields by 10–25% for most of the crops and may prevent cropping altogether or when it is severe can even lead to desertification. The stage of plant growth has a direct effect on salt tolerance. Generally, the plant tolerates the salinity in the later or developed stages more. The fodder and vegetable crops are more sensitive to soil salinity than row crops. The Electrical Conductivity (EC) of soil is measured to quantify the level of soil salinity. EC of the soil saturation extract or different ratios of soil and water suspension is used for salinity assessment.

8.1 Waterlogging and Salinity Management

Soil salinity is addressed through soil amendments, water and crop management practices for better crop production and to avoid desertification. In irrigated areas, at spatial scale like canal command level, secondary salinity is managed by lowering shallow water tables by surface or vertical drainage system and through its safe disposal. However, at field scale many ways are used to address soil salinity like laser land leveling/laser land grading which ensures the uniform distribution of water, leaching of excess salts from the root zone by over-irrigating the fields, improved tillage practices, adaption of salt-tolerant plant varieties, cyclic use of fresh and saline water, conjunctive use of fresh and saline water, minimizing evaporation such as mulching, addition of animal manure and crop residues, etc. Any, management practice at field level ultimately results in the improvement of soil salinity at the regional scale.

Dryland salinity must be approached strategically using scientific basis. The monitoring and assessment of salinity through modern technologies can provide better understanding for the development of soil salinity. The most widely used technology is the interpretation of remote sensing images on temporal basis. The pumping of saline groundwater and its safe disposal or use and the implementing deep-rooted plants may help reducing dryland salinity, called biological drainage.

9 Conclusions

The comprehensive information on water management strategies for floods, drought, extreme weather disasters, and salinity is discussed in the chapter for sustainable agriculture in future. Most of these disasters are linked with a climate change; therefore, a concept to live with climate change is also applicable to floods, drought, and extreme weather conditions. The following conclusions can be drawn from the discussion made on agricultural disasters.

Significant losses to agricultural land and crops due to floods can be checked by adapting structural (like dams, groynes, etc.) and nonstructural (floodplain zoning, management aquifer recharge, etc.) measures. The dams in catchment areas help regulating flows thus reducing the inundation and erosion of agricultural land downstream. Moreover, groynes also help to check the erosion of land along the rivers. Storage of flood water and capturing the peaks in flash flood stream could help reducing significant land loss and agricultural crops. The diversion and subsurface storage of flood flows also have significant potential in reducing the flood losses thus socioeconomic uplift of the farmer's community. Managed Aquifer Recharge (MAR) has a great potential to act as a cushion against both hydrological extremes at catchment as well as in the plain areas. The interventions for flood management could have significant positive impact on the availability of water for irrigating crops during the drought period. Therefore, the recent research has emphasized to consider both hydrological extremes simultaneously for better flood and drought management.

Climate change studies reveal the overall increased global temperatures in future thus resulting frequent droughts on one side and the intense and high precipitation rates on the other side due to increased evaporation. Therefore, integrated approach for flood and drought management is more effective rather than to consider any hydrological event separately. The researches reveal the strong correlation between rainfall intensity and the increased temperatures in urban areas. However, this aspect lacks for the irrigated agricultural areas and catchment scale for providing better understanding of the impacts of climate change on storm events.

Generally, salinity and waterlogging issue arise in plain areas where natural slope is very low where irrigation system is developed for agricultural activities. Surface and vertical drainage system helps to address the waterlogging issue. However, intensive agriculture in the saline groundwater areas urges the farmer to use it for irrigation which creates the salinity issues in the area. However, many resource conservation technologies help to address the salinity and waterlogging issues and enhancing crop yield at the field level. No subsequent technological development has been made to address the saline water treatment for irrigation. However, salt-tolerant varieties have made it possible to use the saline groundwater. The implementation of policies regarding cropping pattern, cropping intensities, groundwater use, and strengthening monitoring system could help in reducing both issues. The use of renewable energy resources like solar pumping system has pose potential threat to exacerbate both issues, especially the regions where unrestricted groundwater abstraction is made.

References

Ahmed K, Shahid S, Nawaz N (2018) Impacts of climate variability and change on seasonal drought characteristics of Pakistan. Atmos Res 214:364–374

Al-Ghussain L (2019) Global warming: review on driving forces and mitigation. Environ Prog Sustain Energy 38(1):13–21

Benson A, Clay A (2004) Understanding the economic and financial impacts of natural disasters. World Bank, Washington

Blunden J, Arndt DS, Baringer MO (2011) State of the climate in 2010. Bull Am Meteorol Soc 92(6):S1–S236

Bonsal BR, Wheaton EE, Chipanshi AC, Lin C, Sauchyn DJ, Lei W, Stewart RE (2011) Drought research in Canada: a review. Atmosphere-Ocean 49:303–319. https://doi.org/10.1080/07055900.2011.555103

Burke EJ, Brown SJ, Christidis N (2006) Modeling the recent evolution of global drought and projections for the twenty-first century with the Hadley Centre climate model. J Hydrometeorol 7(5):1113–1125

Dai A (2011) Characteristics and trends in various forms of the Palmer Drought Severity Index during 1900–2008. J Geophys Res Atmos 116:D12

Di Baldassarre G, Wanders N, AghaKouchak A, Kuil L, Rangecroft S, Veldkamp TI, Garcia M, van Oel PR, Breinl K, Van Loon AF (2018) Water shortages worsened by reservoir effects. Nat Sustainability 1(11):617–622

Emerton RE, Stephens EM, Pappenberger F, Pagano TC, Weerts AH, Wood AW, Salamon P, Brown JD, Hjerdt N, Donnelly C, Baugh CA, Cloke HL (2016) Continental and global scale flood forecasting systems. WIREs Water 3(3):391–418. https://doi.org/10.1002/wat2.1137

Federal Emergency Management Agency (2001) Information technology architecture, version 2.0: the road to e-FEMA (volume 1). http://www.fema.gov/pdf/library/it_vol1.pdf. Accessed 16 April 2009

Field CB, Barros V, Stocker TF, Dahe Q (eds) (2012) Managing the risks of extreme events and disasters to advance climate change adaptation: special report of the intergovernmental panel on climate change. Cambridge University Press, Cambridge

Granell C, Díaz L, Schade S, Ostlander N, Huerta J (2013) Enhancing integrated environmental modelling by designing resource-oriented interfaces. Environ Model Softw 39:229–246

Guha-Sapir G, Below R, Hoyois P (2015) The CRED/EMDAT International Disaster Database 2015. Universite Catholique de Louvain, Brussels. Available www.emdat.be. Accessed January 27, 2015

Guo E, Zhang J, Ren X, Zhang Q, Sun Z (2014) Integrated risk assessment of flood disaster based on improved set pair analysis and the variable fuzzy set theory in central Liaoning Province, China. Nat Hazards 74(2):947–965

Hayes MJ, Svoboda MD, Wiihite DA, Vanyarkho OV (1999) Monitoring the 1996 drought using the standardized precipitation index. Bull Am Meteorol Soc 80(3):429–438

Hina S, Saleem F (2019) Historical analysis (1981-2017) of drought severity and magnitude over a predominantly arid region of Pakistan. Clim Res 78(3):189–204

Hirabayashi Y, Mahendran R, Koirala S, Konoshima L, Yamazaki D, Watanabe S, Kim H, Kanae S (2013) Global flood risk under climate change. Nat Clim Chang 3:816–821

Kishtawal CM, Niyogi D, Tewari M, Pielke RA Sr, Shepherd JM (2010) Urbanization signature in the observed heavy rainfall climatology over India. Int J Climatol 30(13):1908–1916

Knox J, Hess T, Daccache A, Wheeler T (2012) Climate change impacts on crop productivity in Africa and South Asia. Environ Res Lett 7(3):034032

Kogan F, Adamenko T, Guo W (2013) Global and regional drought dynamics in the climate warming era. Remote Sens Lett 4(4):364–372

Lu Y, Qin XS, Xie YJ (2016) An integrated statistical and data-driven framework for supporting flood risk analysis under climate change. J Hydrol 533:28–39

MacDonald R (2007) Exchange rate economics: theories and evidence. Routledge, Milton Park

Panhwar V, Zaidi A, Ullah A, Edgar TN (2021) Impact of water sector interventions on economy, equity, and environment in the rainfed region of Punjab, Pakistan. Environ Dev Sustain 23(2):2190–2203

Potop V, Možný M, Soukup J (2012) Drought evolution at various time scales in the lowland regions and their impact on vegetable crops in the Czech Republic. Agric For Meteorol 156:121–133

Rhee J, Im J (2010) Monitoring agricultural drought for arid and humid regions using multi-sensor remote sensing data. Remote Sens Environ 114(12):2875–2887. https://doi.org/10.1016/j.rse.2010.07.005

Roundy JK, Ferguson CR, Wood EF (2014) Impact of land-atmospheric coupling in CFSv2 on drought prediction. Clim Dyn 43(1-2):421–434. https://doi.org/10.1007/s00382-013-1982-7

Schmidhuber J, Tubiello FN (2007) Global food security under climate change. Proc Natl Acad Sci 104(50):19703–19708

Schubert SD, Suarez MJ, Pegion PJ, Koster RD, Bacmeister JT (2004a) On the cause of the 1930s Dust Bowl. Science 303(5665):1855–1859. https://doi.org/10.1126/science.1095048

Schubert SD, Suarez MJ, Pegion PJ, Koster RD, Bacmeister JT (2004b) Causes of long-term drought in the US Great Plains. J Clim 17(3):485–503. https://doi.org/10.1175/1520-0442(2004)017%3C0485:COLDIT%3E2.0.CO;2

Schubert SD, Stewart RE, Wang H, Barlow M, Berbery EH, Cai W, Lyon B (2016) Global meteorological drought: a synthesis of current understanding with a focus on SST drivers of precipitation deficits. J Clim 29(11):3989–4019. https://doi.org/10.1175/JCLI-D-15-0452.1

Vicente-Serrano SM, Beguería S, López-Moreno JI, Angulo M, El Kenawy A (2010) A new global 0.5 gridded dataset (1901–2006) of a multiscalar drought index: comparison with current drought index datasets based on the Palmer Drought Severity Index. J Hydrometeorol 11(4):1033–1043

Vitolo C, Elkhatib Y, Reusser D, Macleod CJA, Buytaert W (2015) Web technologies for environmental big data. Environ Model Softw 63:185–198

Wang Q, Wu J, Lei T, He B, Wu Z, Liu M, Mo X, Geng G, Li X, Zhou H, Liu D (2014) Temporal-spatial characteristics of severe drought events and their impact on agriculture on a global scale. Quat Int 349:10–21

Ward PJ, de Ruiter MC, Mård J, Schröter K, Van Loon A, Veldkamp T, von Uexkull N, Wanders N, AghaKouchak A, Arnbjerg-Nielsen K, Capewell L, Carmen Llasat M, Day R, Dewals B, Di Baldassarre G, Huning LS, Kreibich H, Mazzoleni M, Savelli E, Teutschbein C, van den Berg H, van der Heijden A, Vincken JMR, Waterloo MJ, Wens M (2020) The need to integrate flood and drought disaster risk reduction strategies. Water Security 11:100070. https://doi.org/10.1016/j.wasec.2020.100070

Wilhite DA, Svoboda MD, Hayes MJ (2007) Understanding the complex impacts of drought: a key to enhancing drought mitigation and preparedness. Water Resour Manag 21(5):763–774

Wood EF, Schubert SD, Wood AW, Peters-Lidard CD, Mo KC, Mariotti A, Pulwarty RS (2015) Prospects for advancing drought understanding, monitoring and prediction. J Hydrometeorol 16(4):1636–1657. https://doi.org/10.1175/JHM-D-14-0164.1

Disaster Impacts on Soils and Their Management

Fariha Ilyas, Idrees Haider, Muhammad Aon, Niaz Ahmed, Muhammad Arshad, Sajjad Hussain, and Muhammad Arif Ali

Abstract

A natural disaster is a serious problem which occurs over a long or short time period that causes widespread economic, environmental, material or human loss. Any deterioration in soil biological, physical and chemical properties alters the quality of land and reduces the capacity to fully functional. The severity of the hazard caused by natural disasters relies on the community effected and the available infrastructure. Climate change could be considered as the main factor in the aggravation of disaster events because disaster events are increasing with time than in the past. Principle factors involved naturally in land deterioration are floods, wildfires, erosion and drought. This chapter discusses the most common impact of these natural disasters on land and their management strategies.

Keywords

Natural disaster · Quality of land · Climate change · Floods · Wildfires · Erosion · Drought

F. Ilyas · I. Haider · M. Aon · M. A. Ali (✉)
Department of Soil Science, Bahauddin Zakariya University, Multan, Punjab, Pakistan
e-mail: arif1056@bzu.edu.pk

N. Ahmed
Department of Soil Science, Faculty of Agricultural Sciences and Technology, Bahauddin Zakariya University, Multan, Pakistan

M. Arshad
Institute of Environmental Science and Engineering, NUST, Islamabad, Pakistan

S. Hussain
Department of Horticulture, Bahauddin Zakariya University, Multan, Punjab, Pakistan

M. Ahmed, S. Ahmad (eds.), *Disaster Risk Reduction in Agriculture*, Disaster Resilience and Green Growth, https://doi.org/10.1007/978-981-99-1763-1_5

1 Introduction

A natural disaster is a consequential issue that can prolong for a short or long time and can also cause widespread material, human, economic or environmental losses, which usually interfere with the normal life cycles and negatively affect the society and community (Ababa 2021). Any natural problem becomes a natural disaster when the effected community is not appropriately organized to overcome the impacts of natural problem due to limited resources, poverty, exclusion, or being socially unprivileged somehow (Mizutori 2020). Disaster not only affects the nation or society's development but also leads to failure or prior development plans (Kamat 2015). Disasters may be either natural or human-instigated. Natural disaster are normally initiated by natural problems or hazards, while human initiated disasters are due to anthropogenic activities. However, in recent times, it is difficult to differentiate between natural and human-instigated (Gould et al. 2016). According to an estimate, about 520 billion dollars in economic losses annually (UNDRR 2020).

The agriculture sector relies on soil and water, so floods, drought, erosion and wildfires are different forms of disasters that negatively impact agriculture (Kato and Huang 2021). Extreme weather and climatic events influence the economic status of any state, with droughts and floods having the highest adverse effects. Floods and droughts are ranked among the most devastating hazards occurred naturally, and countries around the sea and rivers are most susceptible to their hazards. Floods are caused by overflowing of water-storing bodies like rivers, dams and channels due to the exceeding discharge capacity or due to heavy rainfall/precipitation. Apart from the effect of flood disasters on nation's economy, the land is severely prone to be eroded by their leftover effects (Kamat 2015). Thus, for reduction in the disaster effects, soil erosion prevention and water retention in soil are of significant importance.

In the natural climatic cycle, drought is a dry spell for a longer period of time, which may also have devastating effects on agriculture. It is a slow-onset disaster characterized by a shortage of precipitation, resulting in a water shortage. Soil management may help overcome the risk of disasters, that is, floods, erosion and drought (Kato and Huang 2021). At the same time, the water shortage induces drought and reduces agriculture production, water availability for human consumption and hydroelectricity production. The possible ways to alleviate the disaster risks through proper soil management are discussed in this chapter. For this purpose, selective forms of disaster, i.e., flood, drought, erosion, and wildfires, are focused.

2 Floods

2.1 Impact

Flood occur due to natural calamities and cause widespread losses to land, human environment and lives. Floods may occur at varying locations with varying scales throughout the world. Floods usually initiate from natural sources, but flood occurrence and negative impacts can be further intensified by human actions. Among the most devastating disaster, floods are at the top of the list, which can cause widespread destruction to the environment, human and economic loss (Glago 2021; Vousdoukas et al. 2020). Worldwide, 44% of disasters are associated with floods, of which 24 and 14% contribute to riverine and general floods, respectively (WMO 2021). There could be different types of floods like coastal floods, flash floods and river floods that result from different sources like coastal floods from the rise of sea level, flash floods from extreme rainfall or uneven rains and river/pluvial flood due to the melting of glaciers/ice and a higher level of precipitation in catchment areas. In addition, floods may occur due to cyclones, high tides and sea-level storms. These flooding conditions are further intensified by human activities like encroachments in rivers and canal areas and mismanagement of drainage systems (Kundzewicz 1999).

Flood risk as well as its frequency and impacts are rising due to two main factors(Chan et al. 2022; Yang et al. 2015). First, as sea surface temperatures rise due to climate change, oceans expand thermally, precipitation events become more intense and frequent, and natural land subsidence increases, increasing the risk of flooding (Kaźmierczak and Cavan 2011). Second, due to fast changes in land use, urbanization and economic growth, the assets vulnerable to flood threats, including people and property, are expanding (Yang et al. 2010).

During the international decade of natural disaster reduction (from 1990 to 1999) it was realized that the old paradigm of flood protection was not much effective (Schanze 2006). Because of inherent uncertainties and high costs, absolute protection is both unsustainable as well as unachievable. Risk management has been suggested as being more appropriate in its place, and flood research is increasingly paying greater attention to this paradigm (Schanze 2006). Additionally, regional and environmental policies are also beginning to switch from flood protection to flood risk management in several nations (Osti 2018). Floods risk management needs integration and systematization.

2.2 Vulnerability to Flood Disasters

Numerous physical and societal traits influence an area's vulnerability to flood disasters.

2.2.1 Physical Characteristics

The following physical characteristics contribute to the greatest vulnerability to flood disasters: (1) The marginal hydrological and climatic regime; (2) high rates of sedimentation resulting in decreased reservoir storage; (3) topography and land-use practices that encourage soil erosion; (4) flash flooding conditions and (5) deforestation, which permits increased surface runoff, increased soil erosion and more frequent significant flooding.

2.2.2 Social Characteristics

The following social factors make people more susceptible to flooding disasters: (1) Poverty and low-income levels, which prevent long-term planning and provisioning at the household level; (2) A lack of water control infrastructures; (3) Inadequate maintenance and deterioration of existing infrastructure; (4) A lack of human capital skills for system planning and management; (5) A lack of appropriate and empowered institutions and (6) A lack of appropriate land-use planning it should be emphasized that flooding creates significant risks for the contamination of a region's surface and groundwater resources.

2.3 Risk Management

The goal of flood risk management is to lessen the likelihood and/or consequences of flooding. The management activities can be organized into three tasks with distinct components. Risk analysis (Aerts et al. 2013), risk assessment (Cornwall 2021) and risk reduction are among the primary objectives (Karatzetzou et al. 2022). Risk assessment and reduction are both concerned with activities that can lessen the hazards. Risk analysis gives information on past, present, and future flood risks (Gabriels et al. 2022). Each assignment has a set of goals that must be met in order to succeed. They include everything from danger assessment to the description of post-flood measures (Turay 2022).

2.4 Risk Analysis

Risk analysis is based on analysing the flood hazard and its vulnerability (Turay 2022). Various models have been identified in order to analyse the flood risk such as Source Pathway Receptor Consequence Model (SPRC model) (Hallegatte et al. 2013; Taramelli et al. 2015).

2.4.1 SPRC Model

It displays a straightforward causal chain starting from:

- The "Source" of a flood is the beginning of the danger, which is typically an extreme weather occurrence (e.g. heavy rainfall) that causes floods.
- A hazard travels along the "Pathway" to reach the receptors. Therefore, a pathway is necessary for the realization of a hazard.
- The entities that could be harmed by the hazard are referred to as the "Receptor" (e.g., people, property or the environment).

The influence that a flood may have, whether it be economic, social or environmental, is known as a "consequence." It can be described or expressed numerically (for example, as monetary value), categorically (for example, high, medium, or low), or both:

$$Flood\ risk\ analysis = f\left[p,m,w,t\right)\times\left(i,ac\right)\times\left(s,r\right)\times\left(h,d\right)]$$

In risk analysis, the source of flood and flood pathway represent the flood hazard. Whereas the **source** is determined by the flood probability (p) and other features (m). Two main risk-reduction factors are early warning (w) and source area retention capacity (t). **Pathway** can be defined by the inland discharge of water (i), with different flood interventions (c) and attributes (a). The receptor showed susceptibility (s) with resistance and resilience interventions (r). Consequences provide an idea about the damage or harm (h) with interventions to minimize the damage (d).

In reality, each element at risk and each flood danger are where the causal chain from the SPRC model starts. Additionally, there are intricate relationships between passageways, flood control measures and the exposure of weak points. Multiple feedbacks may make up the interrelations in specific circumstances. A "flood risk system" is referred to here as a system that is presumptively made up of all relevant components and procedures. It speaks of river catchments for inland floods and coastal cells for coastal floods as hydraulically related locations.

The aggregate of the risks associated with each individual component makes up the overall risk associated with a flood risk system. Meteorological, hydrological, hydraulic, economic, social science and ecological methodologies ought to be included based on the SPRC model in concept. As a result, gathering and combining the knowledge from all of these domains becomes difficult (Schanze 2006; Narayan et al. 2011).

However, understanding of flood risk systems and the effects of flood risk-reduction actions will necessarily remain lacking despite the necessary efforts to build more integrated approaches. Therefore, taking uncertainty into account is crucial for flood risk estimation (Stephens and Bledsoe 2022). Aleatory and epistemic uncertainty are two different types of uncertainty. Quantities that are inevitably changing over time, place or populations of people or objects are referred to as aleatory uncertainties. Conversely, incomplete knowledge leads to epistemic uncertainty, correlated with our incapacity to comprehend, quantify and describe the system under study.

This is made obvious by the fact that risk analysis often only takes into account (easily or properly) quantitative elements. Thus, the limited heuristic construction inescapably results in a constrained understanding of the actual flood risk system. In theory, the key justification for seeing flood risk management as a continuous process is aleatory and epistemic uncertainty. The inherent limitations of scientific information on the flood risk system necessitate frequent repeats of the risk analysis in order to account for systemic changes and investigate the presumed impacts of risk-reduction measures. But uncertainty should not just be viewed as a problem that cannot be avoided.

There are techniques for quantifying it, such as Monte Carlo Simulation (Hall et al. 2003). These techniques enable the level of uncertainty to be taken into account, which can be considered as significant supplementary information for flood risk management. Printed or digital risk maps for long-term and real-time flood forecasting and warning systems as operational procedures are the main outcomes of flood risk analysis.

Risk maps, which condense the flood risk system into a static, two-dimensional format, offer details on the likelihood of an impending flood as well as water level, flow rate, sediment transport and other factors. Numerous strategies are used (Veh et al. 2020). Their dependability is based on the accuracy and description of the risk analysis' remaining ambiguity. Risk map systems that are more adaptable, interactive and web-based are currently being developed. In addition, information on current flood events is provided through real-time flood forecasting and warning systems.

They aim to increase the lead time for preparatory activities, including reservoir control and evacuation, among others. They mostly involve hydrological and meteorological modelling (Sebestyen et al. 2021). Additionally, there are currently no full-risk-based systems that would provide accurate information about potential harm. Normal thinking about vulnerability involves crucial peak discharges rather than anticipated damage. However, the value of the current flood (event) management systems is widely recognized.

2.5 Risk Assessment

Depending on individual and societal perspectives as well as the assessment of the tolerance of particular hazards, the findings of scientific evaluations of flood risks can be viewed by society in a variety of ways. Applying this distinction does not imply favouring an objective risk analysis over a subjective estimate, which is less reliable (De Moel et al. 2015). Instead, it recognizes that there is a complementary requirement for information that describes the flood risk system using the theories and methodologies that are now in use, on the one hand, and the reality of individual and group perceptions and weightage of risk as social behaviour, on the other. The following describes a two-step risk assessment method that considers both risk perception and risk weightage.

Risk perception is the idea that how people and groups participating in flood danger management perceive risk generally depends on their individual and collective histories. For instance, people who have already been through an extreme flood event are likely to perceive flood threats differently than others. There are sociological, psychological, and cultural explanations of risk perception (Apel et al. 2004). The psychological-cognitive approaches concentrate on variables affecting (irrational) awareness and decision-making. They are particularly concerned with the importance of both individual and group beliefs, emotions, experiences, and viewpoints when assessing "actual" dangers. These techniques align with the idea of "objective" information from scientific risk analysis as a basis for societal risk perception because they contrast a rational and presumably correct flood danger with subjective awareness and decisions. The potential for accurate interpretation of risk perception elements (De Moel et al. 2015) is still being hotly debated, though.

2.6 Risk Reduction

If hazards are deemed to be intolerable, risk-reduction procedures and tools are implemented. Instruments are defined as "interventions based on mechanisms which lead to measures indirectly or influence human behaviour," while measures are defined as "interventions based on direct physical acts" (Ward et al. 2020). Up until now, the phrases "structural measures" for flood defence interventions and "non-structural measures" for all other initiatives have predominated (Kundzewicz 1999; Ward et al. 2020). The traditional engineering approach to flood risk management may be to blame for it. However, the somewhat general phrase "non-structural measures" requires more clarification in light of the social perspective, which appears to be essential for analysing the interventions' motivating factors. Instead, the word "instrument" is already widely used to describe actions in a number of financial and legal policy sectors. Furthermore, there is a difference between permanent and temporary measures for risk reduction.

2.7 Permanent and Temporary Measures

Permanent solutions are overt physical interventions that affect the flood risk system's physical parameters over time. According to Dadson et al. (2017), permanent solutions include both constructive efforts to increase the resilience and flood resistance of infrastructure and structures as well as engineering work for flood control. Furthermore, in order to decrease peak discharge and matter losses from landscapes (such as suspended and bed load, debris) in upstream catchments and on the floodplains, they also involve actions to boost retention capacity (Schanze 2006).

Temporary measures are direct physical interventions to reduce the risk during ongoing flood events (Schanze 2006). Temporary methods include moving people, animals, or products off-site or elsewhere (evacuation), securing items and goods, and demountable flood protection (such as sandbags and mobile flood barriers) (e.g. locking gas valves). In addition, flood damage recovery (such as removing harmful sediment buildup) is time-limited and thus only temporary.

The structure of flood risk management tools can vary depending on the type of mechanism used, including regulatory, financial and communication tools. In the policy areas of water management (e.g. flood protection acts), spatial planning (e.g. legally binding land-use plans), and environmental protection and nature conservation (e.g. strategic environmental assessments), regulatory instruments are interventions that use legal or other normative mechanisms (Schanze 2006). Financial tools either encourage or impose restrictions on actions related to flood risk, such as incentives for flood-proofing-built structures (e.g. zoning of insurance). The use of informal cooperation, such as the collaboration of relevant players, as well as information transmission through the use of media, literature and brochures as well as education, planning, forewarning and instruction constitutes the foundation of communication tools (e.g. flood warning systems). Programming, regulating, implementing and perhaps even supporting the fulfilment of measures are some of the ways that instruments are utilized.

Following measurements and tools match specific SPRC model coefficients: For instance, retention is focused on the factor "*t*" of sources, flood management on the factor "*c*" of pathways, improvement of resilient and constructive resistance on the factor "*r*" of receptors, and compensation on the element "*d*" of outcomes. They can be categorized as pre-flood, flood event, and post-flood interventions in accordance with the different approaches to managing flood risk (Kundzewicz 1999).

2.8 Pre-flood Prevention

Reduction in the intensity of floods and the vulnerability of existing susceptible elements in flood-prone locations, protection through structural vulnerability protection and readiness through behavioural readiness for likely flood events. Examples include the retention of water and other materials in river floodplains' source areas, risk-aware building design and zoning in flood-prone areas, the design of flood control measures through the construction of reservoirs, polders, and dykes (protection), and the preparation of vulnerable populations, i.e., preparedness (Liu et al. 2018).

2.9 Flood Event Management

The management of flood events includes flood control through operational management of the discharge and water level, flood defense based on flood protection structures, and emergency response as mitigation of damages and harm through evacuation and rescue. In addition, forecasting and warning of an ongoing event is the basis for flood defense and the provision of information to people at risk. For instance, flood control by operational management of reservoirs and polders (defense), flood forecasting and warning through government and unofficial flood warning systems, and risk reduction through evacuation and rescue are examples of emergency response (Liu et al. 2018).

2.10 Post-flood Interventions

Recovery as well as relief and reconstruction are included in post-flood operations. One illustration is how insurance covers flood damage. Measures and instruments can have various effects, efficacy and efficiency. Existing consequences depend on the intended and unwanted/unintended (side-effect) effects of the flood risk system. The achievement of intended outcomes is expressed as effectiveness. The ratio of efficacy to necessary efforts or resources is the efficiency of measures and instruments. For instance, the ratio of the reduced damage potential to the expenses of the measures is what is meant by the cost-efficiency of flood protection measures. In this situation, it requires details on capital expenditures or opportunity costs. Effectiveness often depends on how well measures or instrument mechanisms work, how they affect the flood risk system, and how well financial and non-financial initiatives are implemented. They vary based on the type of intervention and the implementation and site-specific variables (Saeed 2019).

It is beneficial to think about the creation of strategic alternatives in order to reflect the interdependencies between parallel or temporally consecutive measures and instruments in a flood risk system. They discuss alternate intervention concepts built on combinations of distinct alternatives from different policy disciplines. The combination of sociological and natural patterns may offer a broader perspective. Natural patterns like the continuing climate change (Moatty 2017) and societal trends like population growth and GDP growth are only a few examples. Both are qualitative–quantitative scenarios here (Murray et al. 2018). You might consider these scenarios as imagined futures for the flood risk system.

2.11 Management Practices for Flood Control

Climate change is making flood events less predictable. Flood danger is particularly high in metropolitan areas, and the fast growth of cities, especially those located along rivers and coasts, increases the exposure of people and assets to floods (Moatty 2017). Heavy rainfall that is greater than what the earth can absorb and what rivers can carry is one of the things that create floods. Deforestation, inadequate drainage and changes in land use that promote more urbanization and agricultural activity all contribute to flood disasters' increasing intensity and severity. Runoff rates and flood peaks of flooding peaks and runoff rates and flood crests during a higher flood peaks and runoff rates and flood. The capacity of communities to manage drainage infrastructure is tested when urban growth threatens natural drainage and storage areas, increases the amount of impermeable cover, and lowers soil penetration rates. The control of soil and water resources may reduce the likelihood of flooding. The following list of runoff control options for woodlands, rivers, floodplains and estuaries is for this purpose.

2.11.1 Woodland Management

It includes management of catchment hilltops, wood cover installation on slopes, installation of riparian woodland around the coast and formation of slope interceptors (Burgess-Gamble et al. 2018).

2.11.2 River and Floodplain Management

This includes the restoration of wetlands by raising beds and re-meandering, the formation of storage wetlands setup and barriers of wood debris and lodges (Williamson et al. 2015).

2.11.3 Estuarine Management

It includes managed realignment mudflats, saltmarshes and washlands (Burgess-Gamble et al. 2018)

2.11.4 Runoff Management

This could be performed by the formation of headwater drainage tracks, blocking of gullies, sustainable cultivation practices and runoff pathway management by setting water traps (Burgess-Gamble et al. 2018; Quinn et al. 2013).

3 Erosion: A Slow-Onset Disaster

3.1 Impact

It is a kind of land degradation by physical means and refers as the soil particles displacement by the action of some external forces, water and winds. Farming activities like conventional tillage are eventually a major cause of land erosion in agricultural lands. Erosion of soil is highly accelerated by anthropogenic activities and is the principal cause of soil degradation worldwide. Researchers considered that agricultural practices are the main culprit behind accelerated erosion. Unchecked expansion of soil erosion eventually loses the top soil layer, or in severe cases, the original structure of the soil is lost, the land becomes unavailable for agricultural use and often land becomes unable to support any kind of infrastructure—other off-site consequences such as unwanted sedimentation (El-Swaify et al. 1985). Off-site soil sedimentation has some benefits and consequences, like sedimentation in watercourses, channels, rivers and lakes. Intensive agricultural practices and the need for increased cultivation imparted pressure on soil and resulted in the loosening of soil, which can be easily taken away by an external force (Balasubramanian 2017). Socioeconomic conditions of any state can be adversely affected by erosion because of lowering land productivity resulting in lowering fuel, food and feed production. Erosion is a factor involved in silting up water reservoirs, e.g. dams, and origin flooding conditions. Erosion is divided into two ground categories based on its effects: accelerated erosion and geological erosion (Chaudhry 1996).

- Accelerated erosion is the detachment of soil particles by anthropogenic activities. Erosion effects are clearly visible on land in such erosion type because of the high erosion rate.
- Geological erosion occurred to natural effects and had no significant effect on soil productivity and land surface. However, this kind of erosion imparted off-site invisible or slightly visible effects after long-term erosion.

There are two main factors which govern the erosions viz. Water and soil, and erosion caused from these factors are collectively termed as water erosion and wind erosion. Here is a flow chart representing the different erosion types (Fig. 1).

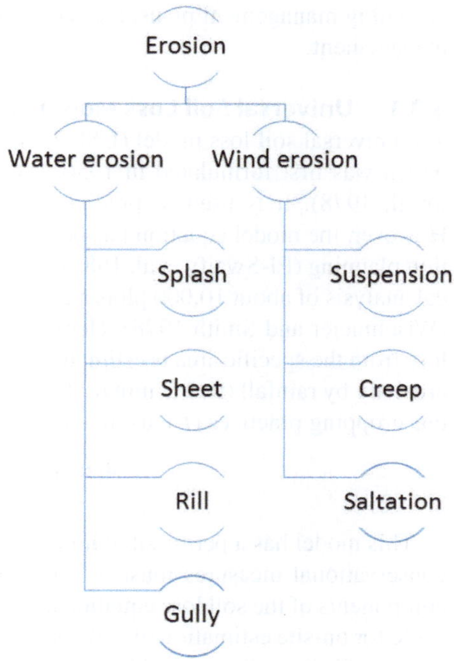

Fig. 1 Classification of land erosion

3.2 Vulnerability to Erosion

The vulnerability to erosion is a function of a number of physical features and social characteristics.

3.2.1 Physical Features

The physical features associated with maximum vulnerability of flood disasters include: (1) Land slope, (2) high-intensity rainfall, (3) loose soil and (4) minimum vegetation favours land erosion.

3.2.2 Social Characteristics

The social characteristics that increase the land erosivity include inadequate planning and management of land and lack of public awareness.

3.3 Erosion Assessment

Models have been used in the planning of conservation, having two major categories, e.g., empirical and process based. These models provide assistance in designing management plans, the scope of the problem and conservation strategies management.

3.3.1 Universal Soil Loss Model (USLE) Model

The universal soil loss model (USLE) is the prime example of an empirical model, which was first formulated in 1950–1960s in the United States (Wischmeier and Smith 1978). It is used to predict soil losses caused by rill and sheet erosion. However, the model equation has been modified and expended for use in conservation planning (El-Swaify et al. 1985). The USLE model was developed after statistical analysis of about 10,000 plots annually from 49 eroded sites in the United States (Wischmeier and Smith 1978). Here we only discussed USLE model. Annual soil loss from the specific area is estimated by slope steepness (*S*), slope-length ratio (*L*), erosivity by rainfall (R: MJ mm h^{-1} ha^{-1} a^{-1}), erodibility (*K*), cropping (*C*) and previous cropping practices (*P*), using the following formula:

$$A(t/\text{ha a}) = \text{SLRKCP}$$

This model has a permissible limit of soil loss, about 12.5, and above this range, conservational measures must be taken. Except rainfall erosivity factor, all other components of the soil loss equation are unit less. However, the model is only applicable for on-site estimation of soil loss and especially when the input data is limited, not for off-site sediment yield estimation. In addition, this model does not estimate the soil loss due to storms (Govers 1996).

For off-site interest, the RUSLE model provides an idea about sediment yield and profile loss. For a comprehensive analysis of off-site soil loss, the WEPP model would be used (El-Swaify et al. 1985). Borrelli et al. (2017) surveyed about 84% of earth lands, used the RUSLE model to determine the erosion extent, and pointed out that soil erosion is increasing at about 2.5%. Human's activities are the prime cause of soil erosion. According to a survey of 202 countries, about 35.8 Pg $year^{-1}$ soil was lost by erosion in 2012. There is a potential to increase the eroded soil, especially in Southeast Asia, Sub-Saharan Africa and South America. South American soils had a higher rate of erosion at about 3.53 Mg ha^{-1} $year^{-1}$, followed by Africa and Asia, 3.51 Mg ha^{-1} $year^{-1}$ and 3.4 Mg ha^{-1} $year^{-1}$ (Borrelli et al. 2017).

3.4 Measures for Erosion Reduction

Erosion control practices are planned according to the three basic themes: mechanical practices, agrostology and agronomical practices. These basic themes are further sub-divided into 11 practices, including tillage management, mulching, organic matter maintenance in soil, intercropping, cover crop, grassland management, contour farming, contour terrace, geo-textile, bunds and micro basin tillage (Ahmad et al. 2020). However, management practices of arable and non-arable land vary especially in case of water erosion.

3.5 Arable Land Conservation

Due to increasing food demand, cultivation area is increasing at a rapid rate; therefore, it is imperative to consider the management of cultivated lands in such a way that their productivity would be sustained. Following practices must be adopted to minimize the water erosion (Chaudhry 1996).

- During rainy seasons, crop cover must be maintained to minimize the direct contact of rain splashes and land, especially in slope lands.
- Field boundaries of about 10–15 cm are necessary to avoid runoff and maximum water absorption in soil.
- No-tillage or minimum tillage practices must be adopted to reduce the erosion rate.
- Mulching is also an effective strategy for land covering.
- Crop rotation and cropping practices must be regulated, which also discourages erosion.
- Grassy outlets, loose stone outlets and brick base outlets are essential for the proper disposal of water from the fields.
- Reduction of slope length is achieved by terrace farming.

3.6 Non-arable Land Management

Land that does not support cultivation for several reasons is termed non-arable. In such areas, the following measure reduced the erosion rate:

- Cultivation of vegetative cover.
- Channel and eyebrow terraces for retaining the water.
- Cultivation of plants in gullied pits.
- Building of check dams to check the water flow.

3.7 Wind Erosion Measures

Wooden poles, field ridges and planks installation are practiced to check the wind flow. Planting windbreaks and shelter belts also assists in reducing wind velocity and erosion rate (Sterk 2003).

4 Drought

4.1 Impact

Drought is characterized by low precipitation and moisture, and its intensity is further determined by climatic variations. Development of drought occurs in varying spatial and temporal scales, breaks out in the substantially vulnerable area and then expands in the surrounding area. Climate change impacts the country's economy significantly, and such meteorological hazards cause annual economic loss. Economic loss caused by drought was about 23.12 billion US dollars from 2010 to 2017 (Su et al. 2018). Drought is becoming the most common meteorological hazard, severely impacting the agricultural economy (IPCC 2012). Drought events have increased with time due to drastic climate change, but in the past, prolonged drought has also been observed. For instance, in 1930, five states of the United States suffered from prolonged drought, letting 2.5 million people homeless. The Republic of China faced severe drought events from 1950s to 1960s. Drought impacts differentially on nature than other natural disaster events because it is low onset and prolonged. A dry week is always set in before monsoon, which is sometimes extended fortnight which often becomes a precursor of drought. The transition of drought is usually unpredictable during the initial stage of drought until it starts hurting the community by water shortage. Duration of drought may last from few days to few months. In comparison with other natural hazards impact, the structural impact of drought is unpredictable (Chandrasekar et al. 2018).

4.1.1 Vulnerability to Drought

The following reasons can be enlisted as the major cause of drought development.

4.1.2 Physical Characteristics

Inconsistency in atmospheric conditions provides a route to anomaly in spatiotemporal precipitation distribution.

4.1.3 Social Characteristics

Anthropogenic activities impact water exchange and alter the energy and momentum, which might greatly impact regional-scale drought events (Huang et al. 2017). Land overexploitation, less land cover grasses and excessive use of groundwater resulted in land deterioration. In semi-arid and arid regions, such degradation of land forms a strong connection with drought and also determines its intensity (Huang et al. 2016).

4.2 Risk Assessment

Drought risk could be assessed in terms of drought probability analysis and risk factors. There are three most used methods of drought risk assessment, e.g. based on the risk factor, probability of hazard loss and based on risk mechanism. It is a qualitative analysis, and attention is paid to selection, optimization and analyse the weighing factor included in calculations of risk (Chen et al. 2020).

4.3 Probability of Hazard Loss

Risk models are formulated considering the hazard loss, and obtained data are analysed statistically. Keeping in view the past drought events, future risk of drought events could be predicted (Ward et al. 2020).

4.3.1 Model-Based Assessment

Risk and loss prediction, stimulation and modelling of loss-causing process (e.g. hydrological cycle and cropping pattern) are used in this method for drought risk assessment of small regions. In this method, risk and consequences are analysed quantitatively (Sun et al. 2015).

4.4 Flash Drought

A short-term drought develops when the temperature suddenly rises and heat waves evolve. This kind of drought has high-intensity and intensity developed rapidly. The present scenario of climate change favours the development of such flash droughts. Although studies focused less on flash drought consensus, some defined the flash drought as lowering the moisture contents below the water available limit, disturbed or abnormal plough layer (Otkin et al. 2018; Sun et al. 2015). Due to its rapid spreading intensity, it is categorized differently than that of normal drought. It has very intense effects on the economy and can be handled with difficulty. If this is not predicted and monitored earlier, it can impart severe effects on the country's economy, agriculture and ecosystem. Low moisture availability, no precipitation, high mean temperature and some anthropogenic activities are involved in flash drought development. Unlike that of normal drought, which develops after low precipitation, flash drought conditions happen after low precipitation along with abnormal rise in mean temperature (hot winds, waves and high radiation rate). These factors increase the evapotranspiration rate, and water can easily escape from the land spaces. Another difference between normal and flash drought is that normal drought prevailed for a longer period, but flash drought only developed during the warm season (NIDIS 2022).

4.4.1 Flash Drought Assessment

High abnormal temperatures and heat waves are the most prominent evidence of flash drought. Further identification is performed by analysing the evapotranspiration rate of soil and the alteration in moisture contents (Chen et al. 2020). Scientists found that flash drought symptoms appeared about two months before complete development, which metrics can easily detect (Christian et al. 2019).

5 Wildfires

5.1 Impact

It is a complex event of climate, land use, weather and urban sprawl. Recently, the intensity of wildfires has been worse than that of nineties fires. Fires are becoming hotter and faster burn than in the past and have more socioeconomic effects. Wildfires are considered one of the most hazardous disasters having biological, physical and ecological consequences on the environment (Alcañiz et al. 2018). Natural disasters contribute about 2% in wildfires while 98% are caused by anthropogenic activities. Massive economic losses have been observed in the past year with wildfires, and a major reason behind this is climatic changes (Raj and Jhariya 2014). Anyhow, the drastic impact of wildfires varies with the fire intensity, regional temperature, fuel availability, vegetational structure and fire return interval. In Algeria, 122,000 acres were burned in 77 fires, Bulgaria faced 31,000 acres burned in 72 fires, 162,000 acres in France in 284 fires, 55,000 acres with 51 fires in Greece, 106,000 acres burned in Kazakhstan, 78,000 acres burned in Morocco, 239,000 acres burned in Portugal, 20,000 acres burned in Russia, 724,000 acres burned in Spain and 34,000 acres burned in Turkey (EFFIS 2022).

Wildfires not only destroy natural ecosystems but also accelerate land erosion due to a reduction in soil cohesion properties. Fires encouraged the survival of "Pyrophytes (fire-resistant plants)" and discouraged the growth of fire-sensitive vegetation. However, fires encouraged the rapid mineralization of organic matter, enriching the soil with essential nutrients for microbes and plant survival (Jhariya and Raj 2014). As a result of wildfires, soil structure, bulk density, and organic matter are severely lessened, and pH value is raised, which can easily destroy the natural ecosystem (Aref et al. 2011; Jhariya et al. 2012). Land water repellency increased with fires because of high temperature and vapour phenomenon, thus lowering the water infiltration rate (Letey 2001). Sand and silt percentages are increased from unburned to burned soil, and a contrary response is found in clay (Nardoto and Bustamante 2003). This may lead to an increase in bulk density (Raj and Jhariya 2014).

5.2 Risk Assessment

Remote sensing and geographical information system are typically used for fire hazard assessment. These are further classified into multi-criteria analysis combined with statistical, machine learning and physics-based techniques.

In multi-criteria analysis, statistical approaches are being used such as probability of fire effected region (Eugenio et al. 2016; Jaafari et al. 2019; Nami et al. 2018; Oliveira et al. 2014). In physics-based methods, fluid mechanics, basic mathematical techniques, combusted biomass, and heat transfer methods are used (Kanga 2018; Herráez et al. 2017). In machine learning techniques, complex algorithms data are used for modelling the nonlinear hidden relationship between land use and topography for estimating wildfires (Oliveira et al. 2012; Tonini et al. 2020). Machine learning includes artificial neural networks (Rodrigues and De la Riva 2014), random forests (Oliveira et al. 2012), logistic reasoning (Guo et al. 2016) and support vector machines (Pourghasemi 2016). The selection of a specific matter depends upon the suitability of a specific model or technique according to the situation. For instance, the artificial neural network had higher accuracy of prediction capability followed by random forest technique than logistic reasoning and support vector machine (Goldarag et al. 2016; Leuenberger et al. 2018). For tropical forest fires, risk assessment is analysed by using logistic reasoning and support vector machine (Bui et al. 2016).

5.3 Pre-fire Management Practices

Fuel management is a key practice in wildfire suppression because fuel plays a key role in fire propagation. Vegetation supplements the fire, and to avoid the propagation of fire, low-inflammable plants should be cultivated (Cui et al. 2019). In many developed countries, the implementation of green fire breaks is used to suppress the propagation of fires. In 1936, green firebreaks were established in the United States, and crested wheatgrass is cultivated around the highway as a green firebreak. Since then, this technique has been adopted in many countries such as Europe (Rocca et al. 2014), China (Cui et al. 2019), Australia (Murray et al. 2018), Africa (Swaine 1992) and South America (Batista et al. 2012). Species used in green firebreaks must have some specific features such as ecologically suitable, silviculture adopted and economically effective species. *Schima superba*, *Schima wallichii*, *Acacia confusa*, conifer species, *Amomum villosum* and crescent wheat grass are some common plant species (Cui et al. 2019).

Vegetation usually fuelled the fires. Plants are divided into categories based on their flammability. Highly flammable plants are termed as active pyrophytes, which enhanced the fires. Active pyrophytes trees (Eucalyptus of Australia) cultivation must be discouraged in bulky forests because such trees are fire-resistant and also contain inflammable oil contents, which encourage fires (Murray et al. 2018).

Trees species composition and timing of thinning had a strong influence on fire events. Therefore, tnning of dry trees should be practiced properly (Cansler et al. 2022).

5.4 Post-fire Management Practices

Post-fire planting weakly impacted the fire events. It is based on three strategies: "restoration, rehabilitation and replacement" (Moreira et al. 2012).

- Restoration means restoring the damaged or burned ecosystem.
- Rehabilitation focuses on restoring the new natural ecosystem, not necessarily the previous ecosystem.
- Replacement or reallocation of new productive ecosystem.

Active and Passive Restoration

In active restoration, direct seeding of replantation of tree technique is adopted. On the other hand, in passive restoration, the natural regeneration of plants is promoted. Unfortunately, managers often ignore natural regeneration techniques, but it is a low-cost technique that promotes the ecosystem's natural buildup.

References

Ababa A (2021) Disasters and emergencies. In: Disasters and emergencies. WHO, Geneva

Aerts JCJH, Lin N, Botzen W, Emanuel K, de Moel H (2013) Low-probability flood risk modeling for New York City. Risk Anal 33(5):772–788

Ahmad NSB, Mustafa FB, Yusoff SM, Didams G (2020) A systematic review of soil erosion control practices on the agricultural land in Asia. In: International soil and water conservation research, vol 8. International Research and Training Center on Erosion and Sedimentation and China Water and Power Press, Beijing, pp 103–115

Alcañiz M, Outeiro L, Francos M, Úbeda X (2018) Effects of prescribed fires on soil properties: a review. Sci Total Environ 613–614:944–957

Apel H, Thieken AH, Merz B, Blöschl G (2004) Flood risk assessment and associated uncertainty. Nat Hazards Earth Syst Sci 4(2):295–308

Aref IM, El Atta HA, Al Ghamde ARM (2011) Effect of forest fires on tree diversity and some soil properties. Int J Agric Biol 13(5):659–664

Balasubramanian A (2017) Soil erosion–causes and effects. Researchgate.Net

Batista AC, Biondi D, França A, De Assunção R, Tres A, Costa R (2012) Evaluation of the flammability of trees and shrubs used in the implementation of green barriers in Southern Brazil. In: Proceedings of the fourth international symposium on fire economics, planning, and policy: climate change and wildfires, PSW-GTR-24, pp 256–264

Borrelli P, Robinson DA, Fleischer LR, Lugato E, Ballabio C, Alewell C, Meusburger K, Modugno S, Schütt B, Ferro V, Bagarello V, Van Oost K, Montanarella L, Panagos P (2017) An assessment of the global impact of 21st century land use change on soil erosion. Nat Commun 8(1):1–13

Bui DT, Le KTT, Nguyen VC, Le HD, Revhaug I (2016) Tropical forest fire susceptibility mapping at the Cat Ba National Park area, Hai Phong City, Vietnam, using GIS-based Kernel logistic regression. Remote Sens 8(4):347

Burgess-Gamble L, Ngai R, Wilkinson M, Nisbet T, Pontee N, Harvey R, Kipling K, Addy S, Rose S, Maslen S (2018) Working with natural processes-evidence directory. Environment Agency

Cansler CA, Kane VR, Hessburg PF, Kane JT, Jeronimo SMA, Lutz JA, Povak NA, Churchill DJ, Larson AJ (2022) Previous wildfires and management treatments moderate subsequent fire severity. For Ecol Manag 504:119764

Chan FKS, Gu X, Qi Y, Thadani D, Chen YD, Lu X, Li L, Griffiths J, Zhu F, Li J, Chen WY (2022) Lessons learnt from Typhoons Fitow and In-Fa: implications for improving urban flood resilience in Asian Coastal Cities. Nat Hazards 110(3):2397–2404

Chandrasekar K, Murthy CS, Seshai M, Roy PS (2018) Agricultural drought assessment and monitoring using geospatial information

Chaudhry MA (1996) Soil erosion and conservation. In: Soil science. Blackwell Science, Malden

Chen LG, Hartman A, Pugh B, Gottschalck J, Miskus D (2020) Real-time prediction of areas susceptible to flash drought development. Atmos 11(10):1114

Christian JI, Basara JB, Otkin JA, Hunt ED (2019) Regional characteristics of flash droughts across the United States. Environ Res Commun 1(12):125004

Cornwall W (2021) Europe's deadly floods leave scientists stunned. Science 373(6553):372–373

Cui X, Alam MA, Perry GL, Paterson AM, Wyse SV, Curran TJ (2019) Green firebreaks as a management tool for wildfires: lessons from China. J Environ Manag 233:329–336

Dadson SJ, Hall JW, Murgatroyd A, Acreman M, Bates P, Beven K, Heathwaite L, Holden J, Holman IP, Lane SN, O'Connell E, Penning-Rowsell E, Reynard N, Sear D, Thorne C, Wilby R (2017) A restatement of the natural science evidence concerning catchment-based "natural" flood management in the UK. Proc R Soc A 473:2199

De Moel H, Jongman B, Kreibich H, Merz B, Penning-Rowsell E, Ward PJ (2015) Flood risk assessments at different spatial scales. Mitig Adapt Strateg Glob Chang 20(6):865–890

EFFIS (2022). https://effis.jrc.ec.europa.eu/apps/effis.statistics/estimates. Accessed 23 October, 2022

El-Swaify SA, Moldenhauer WC, Lo A (1985) Soil erosion and conservation. Wiley, Hoboken

Eugenio FC, dos Santos AR, Fiedler NC, Ribeiro GA, da Silva AG, dos Santos ÁB, Paneto GG, Schettino VR (2016) Applying GIS to develop a model for forest fire risk: a case study in Espírito Santo, Brazil. J Environ Manag 173:65–71

Gabriels K, Willems P, Van Orshoven J (2022) A comparative flood damage and risk impact assessment of land use changes. Nat Hazards Earth Syst Sci 22(2):395–410

Glago JF (2021) Flood disaster hazards; causes, impacts and management: a state-of-the-art review. In: Natural hazards - impacts, adjustments and resilience. IntechOpen, London

Goldarag YJ, Mohammadzadeh A, Ardakani AS (2016) Fire risk assessment using neural network and logistic regression. J Indian Soc Remote Sens 44(6):885–894

Gould KA, Magdalena Garcia M, Remes JAC (2016) Beyond "natural-disasters-are-not-natural": the work of state and nature after the 2010 earthquake in Chile. J Polit Ecol 23(1):93–114

Govers G (1996) Soil erosion process research: a state of the art. Mededelingen van de Koninklijke Academie Voor Wetenschappen, Letteren En Schone Kunsten van België. Klasse Der Wetenschappen, pp 1–53

Guo F, Zhang L, Jin S, Tigabu M, Su Z, Wang W (2016) Modeling anthropogenic fire occurrence in the boreal forest of China using logistic regression and random forests. Forests 7(11):250

Hall JW, Meadowcroft IC, Sayers PB, Bramley ME (2003) Integrated flood risk management in England and Wales. Nat Hazards Rev 4(3):126–135

Hallegatte S, Green C, Nicholls RJ, Corfee-Morlot J (2013) Future flood losses in major coastal cities. Nat Clim Chang 3(9):802–806

Herráez DP, Asensio Sevilla MI, Ferragut Canals L, Cascón Barbero JM, Morillo Rodríguez A (2017) A GIS-based fire spread simulator integrating a simplified physical wildland fire model and a wind field model. Int J Geogr Inf Sci 31(11):2142–2163

Huang J, Ji M, Xie Y, Wang S, He Y, Ran J (2016) Global semi-arid climate change over last 60 years. Clim Dyn 46(3–4):1131–1150

Huang J, Li Y, Fu C, Chen F, Fu Q, Dai A, Shinoda M, Ma Z, Guo W, Li Z, Zhang L, Liu Y, Yu H, He Y, Xie Y, Guan X, Ji M, Lin L, Wang S, Wang G (2017) Dryland climate change: recent progress and challenges. Rev Geophys 55(3):719–778

IPCC (2012) Managing the risks of extreme events and disasters to advance climate change adaptation. Cambridge University Press, Cambridge, p 582

Jaafari A, Mafi-Gholami D, Pham BT, Bui DT (2019) Wildfire probability mapping: Bivariate vs. multivariate statistics. Remote Sens 11:6

Jhariya MK, Raj A (2014) Effects of wildfires on flora, fauna and physico-chemical properties of soil-An overview. J Appl Nat Sci 6(2):887–897

Jhariya MK, Bargali SS, Swamy SL, Kittur B (2012) Vegetational structure, diversity and fuel load in fire affected areas of tropical dry deciduous forests in Chhattisgarh. Vegetos 25(1):210–224

Kamat R (2015) Planning and managing earthquake and flood prone towns. Stoch Env Res Risk A 29(2):527–545

Kanga S (2018) Forest fire simulation modeling using remote sensing & GIS forest fire simulation modeling using remote sensing & GIS. Researchgate 8(June):326–332

Karatzetzou A, Stefanidis S, Stefanidou S, Tsinidis G, Pitilakis D (2022) Unified hazard models for risk assessment of transportation networks in a multi-hazard environment. Int J Disaster Risk Reduct 75:102960

Kato S, Huang W (2021) Land use management recommendations for reducing the risk of downstream flooding based on a land use change analysis and the concept of ecosystem-based disaster risk reduction. J Environ Manag 287:112341
Kaźmierczak A, Cavan G (2011) Surface water flooding risk to urban communities: analysis of vulnerability, hazard and exposure. Landsc Urban Plan 103(2):185–197
Kundzewicz ZW (1999) Flood protection—sustainability issues. Hydrol Sci J 44(4):559–571
Letey J (2001) Causes and consequences of fire-induced soil water repellency. Hydrol Process 15(15):2867–2875. https://doi.org/10.1002/hyp.378
Leuenberger M, Parente J, Tonini M, Pereira MG, Kanevski M (2018) Wildfire susceptibility mapping: deterministic vs. stochastic approaches. Environ Model Softw 101:194–203
Liu CC, Shieh MC, Ke MS, Wang KH (2018) Flood prevention and emergency response system powered by Google Earth Engine. Remote Sens 10(8):1283
Mizutori M (2020) Foreword for the Journal of the International Consortium on Landslides. Landslides 17(4):753–753
Moatty A (2017) Post-flood recovery: an opportunity for disaster risk reduction? In: Floods. Elsevier, Amsterdam, pp 349–363
Moreira F, Arianoutsou M, Vallejo VR, de Heras J, Corona P, Xanthopoulos G, Fernandes P, Papageorgiou K (2012) Setting the scene for post-fire management. Elsevier, Amsterdam, pp 1–19
Murray BR, Martin LJ, Brown C, Krix DW, Phillips ML (2018) Selecting low-flammability plants as green firebreaks within sustainable urban garden design. Fire 1(1):1–4
Nami MH, Jaafari A, Fallah M, Nabiuni S (2018) Spatial prediction of wildfire probability in the Hyrcanian ecoregion using evidential belief function model and GIS. Int J Environ Sci Technol 15(2):373–384
Narayan S, Hanson S, Nicholls RJ, Clarke D (2011) Use of the source-pathway-receptor-consequence model in coastal flood risk assessment
Nardoto GB, Bustamante MMDC (2003) Effects of fire on soil nitrogen dynamics and microbial biomass in savannas of Central Brazil. Pesqui Agropecu Bras 38(8):955–962
NIDIS (2022). https://www.drought.gov/what-is-drought/flash-drought#related
Oliveira S, Oehler F, San-Miguel-Ayanz J, Camia A, Pereira JMC (2012) Modeling spatial patterns of fire occurrence in Mediterranean Europe using multiple regression and random forest. For Ecol Manag 275:117–129
Oliveira S, Pereira JMC, San-Miguel-Ayanz J, Lourenço L (2014) Exploring the spatial patterns of fire density in Southern Europe using geographically weighted regression. Appl Geogr 51:143–157
Osti RP (2018) Integrating flood and environmental risk management: principles and practices. ADB East Asia Working Paper Series
Otkin JA, Svoboda M, Hunt ED, Ford TW, Anderson MC, Hain C, Basara JB (2018) Flash droughts: a review and assessment of the challenges imposed by rapid-onset droughts in the United States. Bull Am Meteorol Soc 99(5):911–919
Pourghasemi HR (2016) GIS-based forest fire susceptibility mapping in Iran: a comparison between evidential belief function and binary logistic regression models. Scand J For Res 31(1):80–98. https://doi.org/10.1080/02827581.2015.1052750
Quinn P, O'Donnell G, Nicholson A, Wilkinson M, Owen G, Jonczyk J, Barber N, Hardwick M, Davies G (2013) Potential use of runoff attenuation features in small rural catchments for flood mitigation. NFM RAF Report. http://www.ydrt.org.uk/
Raj A, Jhariya M (2014) Impact of forest fire on the ecosystem and environment. Reader Shelf
Rocca D, Danti R, Raddi P, Moya B, Moya J (2014) Projet CypFire. Implementation of the "cypress system" as a green firewall. Forêt Méditerr 35(3):275–280. https://hal.archives-ouvertes.fr/hal-03556546/document
Rodrigues M, De la Riva J (2014) An insight into machine-learning algorithms to model human-caused wildfire occurrence. Environ Model Softw 57:192–201
Saeed K (2019) Analysing local perceptions of post-conflict and post-floods livelihood interventions in Swat, Pakistan. Dev Policy Rev 37:274–292

Schanze J (2006) Flood risk management–a basic framework. In: Flood risk management: hazards, vulnerability and mitigation measures. Springer, Dordrecht, pp 1–20

Sebestyen SD, Lany NK, Roman DT, Burdick JM, Kyllander RL, Verry ES, Kolka RK (2021) Hydrological and meteorological data from research catchments at the Marcell Experimental Forest, Minnesota, USA. Hydrol Process 35(3):e14092

Stephens T, Bledsoe B (2022) Simplified uncertainty bounding: an approach for estimating flood hazard uncertainty. Water 14(10):1618. https://doi.org/10.3390/w14101618

Sterk G (2003) Causes, consequences and control of wind erosion in Sahelian Africa: a review. Land Degrad Dev 14(1):95–108

Su B, Huang J, Fischer T, Wang Y, Kundzewicz ZW, Zhai J, Sun H, Wang A, Zeng X, Wang G, Tao H, Gemmer M, Li X, Jiang T (2018) Drought losses in China might double between the 1.5 °C and 2.0 °C warming. Proc Natl Acad Sci U S A 115(42):10600–10605

Sun Y, Fu R, Dickinson R, Joiner J, Frankenberg C, Gu L, Xia Y, Fernando N (2015) Drought onset mechanisms revealed by satellite solar-induced chlorophyll fluorescence: insights from two contrasting extreme events. J Geophys Res Biogeo 120(11):2427–2440

Swaine MD (1992) Characteristics of dry forest in West Africa and the influence of fire. J Veg Sci 3(3):365–374

Taramelli A, Valentini E, Sterlacchini S (2015) A GIS-based approach for hurricane hazard and vulnerability assessment in the Cayman Islands. Ocean Coast Manag 108:116–130

Tonini M, D'andrea M, Biondi G, Esposti SD, Trucchia A, Fiorucci P (2020) A machine learning-based approach for wildfire susceptibility mapping. The case study of the Liguria region in Italy. Geosciences 10(3):105

Turay B (2022) Flood hazard management in a multiple hazard context: a systematic review of flood hazard management during the COVID-19 pandemic in Africa. Discover Water 2(1):6

UNDRR (2020) Status report on target E implementation. https://www.undrr.org/publication/status-report-target-eimplementation-2020

Veh G, Korup O, Walz A (2020) Hazard from Himalayan glacier lake outburst floods. Proc Natl Acad Sci 117(2):907–912

Vousdoukas MI, Mentaschi L, Hinkel J, Ward PJ, Mongelli I, Ciscar JC, Feyen L (2020) Economic motivation for raising coastal flood defenses in Europe. Nat Commun 11(1):2119

Ward PJ, de Ruiter MC, Mård J, Schröter K, Van Loon A, Veldkamp T, Wens M (2020) The need to integrate flood and drought disaster risk reduction strategies. Water Secur 11:100070

Williamson P, Ogunyoye F, Dennis I, Douglas J, Hardwick M, Sayers P, Fisher K, Thorne C, Holmes N (2015) Delivering benefits through evidence: channel management handbook. Environment Agency, Bristol

Wischmeier W, Smith D (1978) Predicting rainfall erosion losses: a guide to conservation planning. USDA, Washington

WMO (2021) State of the global climate 2021: WMO provisional report. https://library.wmo.int/index.php?lvl=notice_display&id=21982#.Y8YuNRdBz3g. Accessed on 14 January 2023

Yang L, Lv Y, Zheng H (2010) Review on research of urban land carrying capacity. Prog Geogr 29(5):593–600

Yang L, Scheffran J, Qin H, You Q (2015) Climate-related flood risks and urban responses in the Pearl River Delta, China. Reg Environ Chang 15(2):379–391

Role of Soil Science in Mitigating Natural and Anthropogenic Disasters

Fatima Latif, Nimra Ishfaq, M. Ahsan Azhar, Sajid Masood, Fiza Batool, M. Zafar ul Hye, Muhammad Abid, Niaz Ahmed, Shakeel Ahmad, M. Farooq Qayyum, Sarvet Jehan, and Khalid Rasheed

Abstract

Disaster is any natural and man-made (anthropogenic) adverse event that results in mass destruction and the ecosystem as a whole. Disasters like catastrophes, earthquakes, avalanches, cyclones, droughts, landslides, floods, hailstorms and fires occur worldwide due to deforestation, climate change, mining activity, soil erosion and tectonic movements. Moreover, these disasters not only limit crop production but also decline the quality of the soil. In this regard, numerous management measures have been adopted to regulate or control all forms of disasters by maintaining soil properties, i.e., soil texture, soil water and soil aggregation. Further, soil quality is improved and exhibits resilience to disasters when soil properties are suitable. Therefore, assessment of soil quality is crucial to improve crop production and conservation of natural resources and can help as a tool for agricultural executives and policymakers to gain a better understanding of how the soil properties reduce or minimize the risk of disasters. This chapter summarizes the accessible information regarding the role of soil science, particularly

F. Latif · N. Ishfaq · M. A. Azhar · S. Masood (✉) · F. Batool · M. Z. u. Hye · M. Abid · N. Ahmed · M. F. Qayyum
Department of Soil Science, Faculty of Agricultural Sciences and Technology, Bahauddin Zakariya University, Multan, Pakistan
e-mail: sajidmasood@bzu.edu.pk

S. Ahmad
Department of Agronomy, Faculty of Agricultural Sciences and Technology, Bahauddin Zakariya University, Multan, Pakistan

S. Jehan
Institute of Soil and Environmental Sciences, PMAS-Arid Agriculture University, Rawalpindi, Pakistan

K. Rasheed
Soil and Water Testing Laboratory, Multan, Pakistan

M. Ahmed, S. Ahmad (eds.), *Disaster Risk Reduction in Agriculture*, Disaster Resilience and Green Growth, https://doi.org/10.1007/978-981-99-1763-1_6

soil properties, in mitigating all kinds of disasters. In addition, it also describes the conservation of soil resources to control or minimize the chances of disasters in future.

Keywords
Anthropogenic · Climate change · Conservation · Natural disasters · Soil quality · Tectonic movement

1 Introduction

Disaster is a natural calamity or a major disturbance that causes poses serious threats to human life, socioeconomic or environmental components (Lalehzari and Kerachian 2020; Zhu et al. 2021). Disasters are of two types, i.e. natural and man-made, of which some could happen once, while others might occur (Sawalha 2018). Man-made or anthropogenic disasters are catastrophic actions resulting from human activities, while natural disasters are catastrophic actions resulting from natural causes. In addition, man-made disasters are classified as sociotechnical and warfare disasters like plant or factory failures and production and transport failures (Moore 2008). In contrast, natural disasters are further categorized into five groups, i.e., geophysical, climatological, hydrological, meteorological, and biological disasters (Rautela 2006).

Agriculture is the backbone of any economy as it feeds 75% of the world's population directly or indirectly. However, in recent years, many risks and pressures have been challenging because of natural and man-made disasters (Cassidy et al. 2019). Pakistan is extremely prone to natural disasters affected by catastrophes such as earthquakes, avalanches, cyclones, drought, landslides, floods, hailstorms and fires. Consequently, agriculture production is negatively influenced by these disasters, in addition to the decline in soil quality (Basile et al. 2020).

It needs to progress some management strategies, plans and policies that alleviate the risks and threats of disasters to human beings and the ecosystem. Subsequently, all disasters are the consequence of human failure to implement suitable disaster management approaches (Li et al. 2021). Although numerous management measures have been used to regulate and control all forms of disasters, assessment of soil quality is crucial for improving the management and land-use systems (Rahmanipour et al. 2014). Because soil quality not only regulates the productivity of plants and the quality of the environment over time (Wander et al. 2002) but also improves the production and conservation of natural resources (Shahab et al. 2013). This chapter summarizes the types and causes of disasters, their effects on ecosystems and management strategies, especially the role of soil properties in mitigating disasters.

2 What Are Disasters?

As described above, a disaster is any sudden and unpleasant incident that damages human life and the ecosystem. It has two types which are as follows.

2.1 Natural Disasters

These types of disasters occur naturally, either by quick or slow onset of actions and have direct and indirect effects on human life, resulting in widespread deaths and environmental disruptions (Sun and Khayatnezhad 2021). These disasters are further classified as:

- Geophysical (earthquakes, landslides, tsunamis and volcanic activity)
- Climatological (extreme temperatures, drought and wildfires)
- Hydrological (avalanches and floods)
- Meteorological (cyclones and storms/wave surges)
- Biological (disease epidemics and insect/animal plagues)

2.2 Man-Made Disasters/Anthropogenic Disasters

The type of disasters that occur in response to humans' environmental and technological activities (Shaluf et al. 2003). This may include:

- Pollution
- Environmental degradation
- Accidents (industrial, technological and transport generally relating to the production, use or transport of hazardous materials)

2.3 Causes of Natural and Man-Made Disasters

The major causes of both kinds of disasters have been discussed below.

2.3.1 Soil Erosion

Soil erosion is the movement of the upper layer of the soil from one place to another place. According to Apollo et al. (2018), soil erosion occurs either through the quick action of erosive agents like snow, water, ice (glaciers), wind (air), plants and animals, including humans or under the slow actions of relatively unseen agents. Moreover, both fast and slow-acting processes not only result in loss of upper soil but also affect surface water quality, crop production and drainage. Also, climate change, intensive agriculture, acid rains, roads, and deforestation are among the most significant human actions concerning their pronounced effects on soil erosion (Julien Pierre 2010).

2.3.2 Deforestation

Deforestation is the clearing away of the prevailing natural vegetation cover, particularly where the vegetation cover is mainly forests (Mawalagedara and Oglesby 2012; Abere and Opara 2012). Deforestation is one of the serious problems of climate change in Pakistan. In general, forests are spread on 4.6 M ha of Pakistan, whereas fast disruption of forests at 1.5% every year poses serious threats to the ecosystem (Ali et al. 2006). Several studies reported that deforestation is a major environmental problem worldwide after food security, land degradation, biodiversity decline, and environmental sustainability (Olson et al. 2004; Foley et al. 2005). Finally, deforestation significantly affects soil erosion and land degradation (Lal 2003). Because the rate of deforestation has a direct relationship with land degradation and soil erosion, such conditions are recognized as albedo, where soil erosion and land degradation are prevalent; therefore, vegetation cover, including forests will reduce these risks (Tariq et al. 2014).

2.3.3 Climate Change

It is defined as the gradual and continuous change in average weather conditions in an area over a certain period of time. In recent years, this is becoming a global hazard that markedly affects agricultural productivity (Leal Filho et al. 2021; Feliciano et al. 2022) and the ecosystem by various means (Fig. 1).

It is well known that climate change is influenced by a wide-range of precipitation and temperature fluctuations. Additionally, it is influenced by abrupt changes in weather conditions, ice and glaciers melting and rise in sea level (Michel et al. 2021; Murshed and Dao 2020). Other factors like industrialization, volcanic eruption, wildfires, and seismic actions also contribute to climate change through greenhouse gases (CH_4, CO_2, N_2O and H_2O) into the environment (Sovacool et al. 2021; Murshed 2022).

2.3.4 Mining

It is well known that soils play an important role in sustaining the environment and crop production in terrestrial ecosystems (Sarker et al. 2020). However, these soils have been degraded worldwide because of metal pollution (Beiyuan et al. 2021; Keshavarzi et al. 2021). Therefore, environmental contamination with mining and metallurgy has been evident in the form of soil and water pollution and the presence of potentially toxic materials or compounds. Consequently, these activities have deteriorated soil quality and led to an undesirable change for living organisms over the twentieth century (Reyes et al. 2020; Izydorczyk et al. 2021). This situation is specifically pertinent when mechanical weathering such as deflation, erosion and thermal pressure work together to active chemical weathering, hydration, or hydrolysis of the silicates and carbonates (Rapant et al. 2006; Wu et al. 2020), thereby releasing contaminants into the environment.

2.3.5 Tectonic Movement

Tectonic movements is a largely accepted scientific theory that reflects the earth's lithosphere covers several tectonic plates, which have been gradually moving since

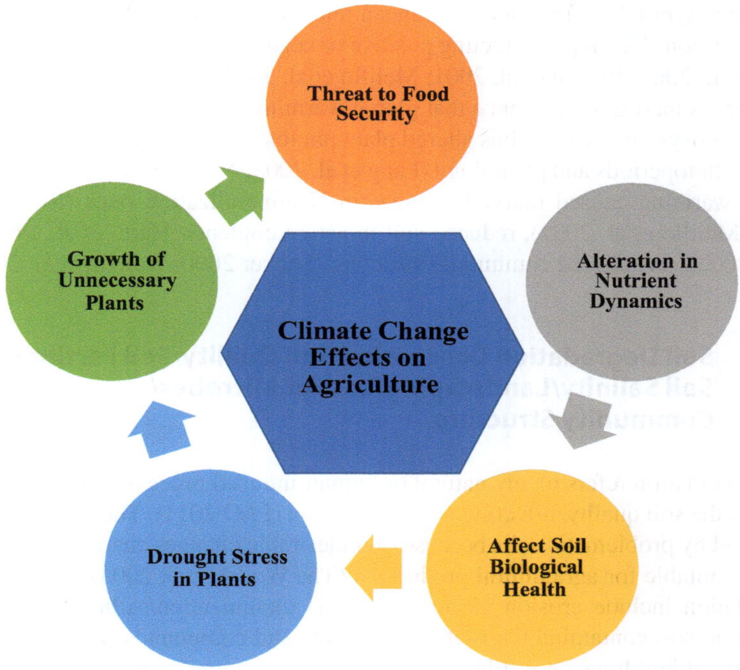

Fig. 1 Processes that affect crop productivity under changing climate scenario and results in food security disaster

about 3.4 billion years ago (Little et al. 1990). As the lithosphere is the outermost shell of the earth's crust, it forms major and minor plates upon movement. According to Read and Watson (1975), these plates move with a relative speed of 0–10 cm yearly, resulting in convergent, divergent and/or transformed structures. Further, these movements evolved in the form of volcanic eruptions, earthquakes, mountain-building, and oceanic channel formation (faults). In addition, the movement of tectonic plates towards subduction sites occurs to disruption of the asthenosphere, which is weakened by the dissipation of heat from earth's mantle (Stern and Robert 2002).

3 Effects of Disasters on the Ecosystem

3.1 Global Warming

According to Houghton et al. (2001), global earth's temperature has been raised to 1.4–5.8 °C in the twenty-first century compared to the twentieth century, with a 0.6 °C increase in the mean annual temperature. This has resulted due to global warming effects of increased atmospheric CO_2 levels, changes in land-use patterns and fossil fuel ignition. A rise in earth's mean temperature can markedly influence the global carbon (C) budget, directing positive or negative responses to change climate (Luo et al. 2001; Rustad et al. 2001; Melillo et al. 2002).

There is increasing evidence that global warming influenced the C fluxes in terrestrial ecosystems, which thus altered plant photosynthesis and respiration (Shaver 2000), photoperiods and phenology (Fang et al. 2003; Norby et al. 2003). Moreover, global warming caused marked changes in N mineralization rates (Rustad et al. 2001; Melillo et al. 2002), reduced soil moisture contents (Harte et al. 1995; Wan et al. 2002), and altered community structure (Shaver 2000; Weltzin et al. 2003).

3.2 Soil Degradation Concerning Soil Quality/Soil Fertility/Soil Salinity/Landscaping/Loss of Microbes/Community Structure

Soil degradation refers to any natural or human-induced process that strongly deteriorates the soil quality, affecting crop production (FAO 2011). These soils are characterized by problematic soils because of undesirable changes characteristics which are not suitable for agricultural production (The World Bank 2008). Causes of soil degradation include erosion (wind or water), alkalinization/salinization, nutrient depletion, soil contamination with toxic metals and compounds (Fig. 2), and construction of buildings on fertile lands. These reasons significantly contribute to the decline in the soil fertility status of topsoil, inhibition of soil microbial community, increase in soil pH, and metals toxicity in soils (Edrisi and Abhilash 2016; Wang 2017). According to Srivastava et al. (2019), anthropogenic activities in the last few

Fig. 2 Causes of soil degradation contributing to food security issues and other types of disasters

years either severely or slightly affected the soils by 25% and 50%, respectively. It is unequivocally clear that these kinds of edaphic problems further inhibit net vegetation cover and soil restoration.

3.3 Glaciers Melting and Flooding

Floods originating from natural glacier lakes are one of the major disasters (Carrivick and Tweed 2016; Veh et al. 2020). Glacier lakes outburst occurs due to seepage or erosion, excessive ice melting, loss of ice-cored dams, and prolonged rainfalls (Westoby et al. 2014; Sattar et al. 2021). Moreover, these outbursts contribute to high magnitudes of water discharge into the receiving rivers (Cenderelli and Wohl 2001).

Recently, it has been observed that climate change has reduced the glaciers' size or volume worldwide and raised serious concerns about hazard or disaster management (IPCC 2019). Consequently, glaciers outburst raised the downstream level and influenced the population diversity and socioeconomic development (Zhang et al. 2015). Although dispersion and locations of glacial lakes have been studied (Shugar et al. 2020), it is necessary to estimate the glaciers' outburst for effective disaster management.

3.4 Food Security Challenges

Generally, food security is defined as the availability of raw food materials, access to food by the people, its utilization and finally, the stability of food materials (Clapp et al. 2021). It is alarming that most of the food policies rely on agriculture and food production (Lang and Barling 2012) though health, trading, and environmental problems have been merged together in recent food policies (Lang and Heasman 2015).

Due to the occurrence and intensity of climate-change-assisted disasters, their negative consequences are probably worse (Islam and Hasan 2016). Small-scale farmers are more significantly impacted by natural disasters. Hence farmers are dependent on agriculture for their livelihood, whereas disasters directly influence their food security (Raihan 2018). The fatalities are socioeconomically unstable, and their livelihoods are demolished, forcing them to change their living patterns or become unemployed. After a natural disaster, many people migrate from their current location to the inner city in search of food and income (Mallick et al. 2017).

In general, the capability of food-insecure persons is least to any natural hazards (Habiba et al. 2016). Therefore, natural disasters not only influence food security but also limit physical and economic access to food. Consequently, health, learning and productivity of the society are compromised (Haen and Hemrich 2007). Therefore, sustainable policy focusing on agricultural production and rural development ensures food quality and accessibility (Kapucu et al. 2013).

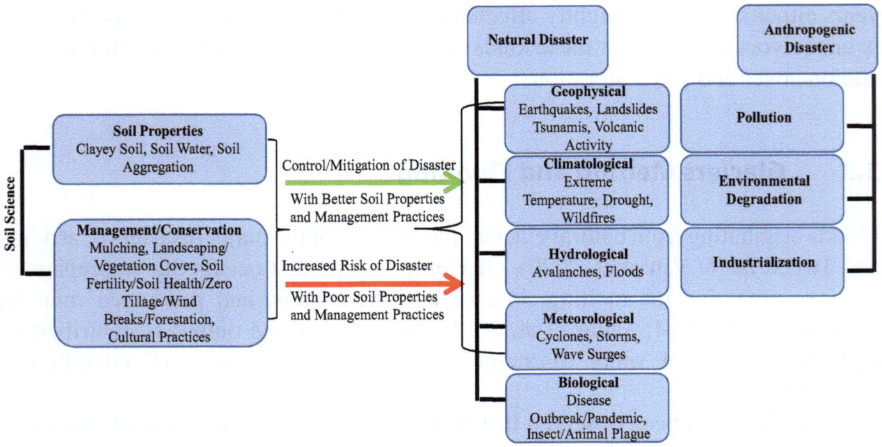

Fig. 3 Schematic diagram exhibiting the role of soil science in the control or mitigation of natural disasters except biological. Green arrow indicates the control or minimization of disaster under better soil properties and management approaches, whereas the red arrow depicts the pronounced effects of poor soil properties and improper soil management

4 Disaster Management

Disaster management is a set of approaches or processes to be taken for an effective preparedness plan to control or mitigate any disaster. Here, we discuss the possible role of soil science, including soil properties, soil management and/or conservation approaches to address natural and man-made disasters effectively (Fig. 3). Soil science plays a crucial role in the mitigation or control of natural disasters either directly or indirectly. Anthropogenic disasters might be compensated by soil science via indirect means. However, there is no direct role of soil science in the control of anthropogenic types of disasters. The roles of soil properties and management approaches have been further discussed below:

4.1 Soil Properties

Soil water balance for hydrological disaster alleviation is considered as crucial for assessing land water accessibility in a specific region, particularly to trace out when and how much excess water occurs. Further, weather fluctuations impact variations in hydrological cycles like drought and floods as hydrological disasters (Mahyuddin et al. 2013). That is why accretion of water resource management exertions is instantly needed. This is highly dependent on soil texture as sandy soils contain macro-pores, whereas clay fraction is dominated by meso- and micro-pores. Therefore, soil water balance can appraise the changes in soil water and water use quantitatively by vegetation (Jonard et al. 2021) and screen the water stress in plants (Kalédjé et al. 2020). In addition, it assesses the incorporation of agricultural irrigation systems in different climatic conditions (Zhang et al. 2019) as well as to analyse the accessibility of water in a specific area (Latha et al. 2010). It will lessen the fatalities and be predicted with the steps facing hazards, and investors could have taken the perception that makes the community more organized to face disaster (Cassidy et al. 2019).

Soil quality is determined by aggregate stability, whereas soil aggregates are basic units of soil structures (Lynch and Bragg 1985). Apart from this, soil aggregate stability is measured by the average weight diameter of soil aggregates (Choudhury et al. 2014). The aggregate stability index is influenced by naturally occurring binding agents like iron, aluminium oxide, organic materials, carbonate fractions, and metal cations (Six et al. 2004). Among the binding agents, organic materials have been proven effective in terms of aggregate stability as compared to inorganic materials (Tisdall and Oades 1982). It is generally believed that these organic binding agents often show resistance to soil erosion and other hazards (Annabi et al. 2017). Aggregate stability is better under clay fractions of soil and contributes to disaster management.

4.2 Mulching

Mulching is a practice used to cover the soil surface with organic residues or plastic materials to conserve soil moisture. It plays an important role in the control or mitigation of disasters and their impacts on crop productivity (Wang et al. 2009; Li et al. 2018). In this regard, the efficiency of mulching material like polythene has been reported in the form of ecosystem conservation because of its stable structure and tensile strength (Sintim and Flury 2017; Borrowman et al. 2020). Some organic mulching materials have been documented to lower soil water permeability and microbial activities and increase soil fertility and structural quality (Shogren 2000; Chen et al. 2019).

Organic mulching is a common and simple method practiced in forestry, landscaping and agriculture. Organic mulch is defined as organic material from residues of crops, paper, wood and animal waste that can be used to fully cover the surface of the soil. Former studies have revealed that mulch covering the soil surface as compared to soil mixed with mulch is more beneficial for water conservation (Singh et al. 1991). Many experiments on organic mulch have established its positive impacts on soil fertility (Liu et al. 2011; Ghosh et al. 2015), reduced wind and soil erosion (Saeed et al. 2019), reduction in evaporation losses (Kader et al. 2019), and enhancement of crop growth and yield (Mupangwa et al. 2007). Additionally, organic mulch material helps to moderate temperature and provides shelter to the soil and roots from exciting temperatures (Einert et al. 1975; Bussière and Cellier 1994), avert pests, controls weeds, and positively impact the nearby microclimate (Thankamani et al. 2016). Studies have been carried out to investigate the impacts of organic mulching on soil and water conservation by employing different mulch materials (Ranjan 2015) under different soil depths (Zhou et al. 2021), and different time scales (Liu 2004). Mulches application to the soil changes its hydraulic functioning as well as its runoff characteristics, thereby reducing runoff and playing role in floods control.

4.3 Landscaping and Vegetation Cover

Landscaping is relatively one of the lethal geological hazards as compared to other hazards. This is formed under gravity as strong forces of inner rock masses deteriorate slope stability (Corominas et al. 2014). Factors influencing landscaping commonly include earthquakes, volcanic activities, land use or land cover, high rainfalls, crystal growth, climate change (Segoni et al. 2018; Mondini et al. 2021), and landslides. Among these factors, landslides cause fatalities in human beings. In this regard, Brabb (1993) proposed that fatalities caused by landslides may be controlled or minimized by 90% through keen attention towards high-risk areas. So far, only 1% of continent's earth has been explored, of which 17% of the area is designated for landslides (Guzzetti et al. 2012; Ma et al. 2020). Landslide susceptibility mapping (LSM) techniques through geographic information systems (GIS) and satellite remote sensing (SRS) to detect landslide-prone areas (Chen et al. 2020a, b; Lee et al. 2017; Brenning 2005). These mapping techniques are separated into two groups, i.e., qualitative and quantitative (Fell et al. 2008). Quantitative and qualitative mapping techniques are used to assess landslide occurrences (Fell et al. 2008). Qualitative techniques mostly depend on the level of a specialist's knowledge, while quantitative approaches rely on arithmetical theories or models (Xue et al. 2013) and are based on probabilistic/arithmetical theories or models (Xue et al. 2013). Both methods have their limits, as poor expertise may result in bad accuracy and precision of qualitative mapping, whereas the use of estimated or not précised data may result in imprecise quantitative mapping (Tsangaratos and Ilia 2017). However, it is clear that these techniques will be helpful in detecting or locating such events in an area under the interactions of geological, geophysical and environmental conditions (Kavzoglu et al. 2019) to take certain measures against such hazards (Choi and Cheung 2013; Iovine and Cohen 2014).

4.4 Soil Fertility/Soil Health Improvement

Land degradation often results from decreased soil fertility, poor soil management and conservation techniques, and climate change. In this regard, regenerative agriculture has received much attention to mitigate the negative impacts of intensive agriculture on soil health (Francis et al. 1986; Giller et al. 2021). Therefore, regenerative agriculture is helpful in the renewal of soil health and the central concept (Schreefel et al. 2020). Moreover, climate change contributes to global warming and changing rainfall subtleties worldwide have augmented soil loss in areas where soils are not generally exposed to drought and flooding (IPCC 2019).

Apart from soil degradation by extensive cultivation, the invasion of high-yielding crop varieties to meet the food security challenges of the ever-increasing population has resulted in losses of nitrogen, phosphorus and carbon. Consequently, nutritious food consumption and access to people have been limited (FAO 2019).

4.5 Zero Tillage

Zero tillage or no tillage refers to the cultivation of crops without disturbing the soils. In some cases, it is attained by minimum disturbance of the soils. This is important in recent times because many soil management practices involve diverse systems of tillage that expose soils to erosion under excessive cultivation of crops and weed control. In zero tillage, much of the soil is always covered by crop residues, whereas minimum tillage relies on the idea of stubble mulching. Zero tillage allows the farmers to sow seed in the soils at the required depth with minimum disturbance in soil structure. For this purpose, special machinery is designed to reduce the need for ploughing. Direct rice seeding with a drill machine is an example of zero tillage.

This method of tillage or cultivation of crops is common in many parts of the world, including Canada, the USA, Australia and North America. The adoption of such innovative techniques not only mitigates soil losses, but also helps farmers to manage large area with minimum labour, energy, and tillage implements (Triplett and Dick 2008). In addition, such techniques are helpful to improve soil health or soil fertility and contribute to the mitigation of soil degradation (Trigo et al. 2009).

4.6 Cultural Practices

Cultural practices are the measures to be taken to improve soil health and increased agricultural productivity under climate change scenario. Further, these practices address global climate change patterns as well as food security challenges (Harvey 2014). These approaches are also called as sustainable because of their role in improving soil physical, chemical and biological properties. Among these, crop rotation, zero or minimum tillage and organic mulching improve soil nutrient dynamics, thereby improving soil structure and water-holding capacity (Rivera-Zayas et al. 2017). Sustainable cultural practices also help in maintaining better landscaping (Deyn and Kooistra 2021), which in turn minimize the chances of natural hazards.

4.7 Wind Breaks/Forestation

A windbreak is commonly defined as any structure that minimizes wind velocity (Wang et al. 2010; Chendev et al. 2015) and is generally related to a native vegetative barrier against the wind. It can be only an element or a system of elements that, over its existence in the airflow lessens the impact of wind velocity not only at the system itself but also at a certain windward and leeward distance. The purpose of windbreaks is achieved by placing the structures or walls at right angles to erosive winds. Moreover, wind velocity is decreased by windbreak at a distance of 2–5 times of upward wind, whereas it is up to 10-30 times the elevation of windbreak for downward wind (Cornelis and Gabriels 2004).

5 Conclusions

In summary, soil properties like soil texture, soil water/soil moisture and soil aggregation play a crucial role in the alleviation of natural disasters as well as minimize the chance of future risks. The underlying phenomenon is to maintain soil quality through increased soil aggregate stability, soil water ratios and efficient landscaping. Additionally, soil management and conservation practices contribute to alleviating natural types of disasters. There is no significant role of soil science in controlling anthropogenic disasters.

References

Abere SA, Opara JA (2012) Deforestation and sustainable development in the tropics: causes and effects. J Educ Soc Res 2:105–109

Ali T, Shahbaz B, Suleri A (2006) Analysis of myths and realities of deforestation in north west Pakistan: implications for forestry. Int J Agric Biol 8:23–67

Annabi M, Raclot D, Bahri H, Bailly JS, Gomez C, Bissonnais YL (2017) Spatial variability of soil aggregate stability at the scale of an agricultural region in Tunisia. Catena 153:157–167

Apollo M, Andreychouk V, Bhattarai SS (2018) Short-term impacts of livestock grazing on vegetation and track formation in a high mountain environment. A case study from the Himalayan Miyar Valley (India). Sustain For 10:951–989

Basile A, Albrizio R, Autovino D (2020) A modeling approach to discriminate soil hydrological properties and slope gradient to water stress in Mediterranean vineyards. Agric Water Manag 241:106–338

Beiyuan J, Fang L, Chen H, Li M, Liu D, Wang Y (2021) Nitrogen of EDDS enhanced removal of potentially toxic elements and attenuated their oxidative stress in a phytoextraction process. Environ Pollut 268:115–719

Borrowman CK, Johnston P, Adhikari R, Saito K, Patti AF (2020) Environmental degradation and efficacy of a sprayable, biodegradable polymeric mulch. Polym Degrad Stab 175:109–126

Brabb E (1993) Proposal for worldwide landslide hazard maps. In: Proceedings of the seventh international conference and field workshop on landslides in Czech and Slovak Republics, Czech Republic and Slovakia, Rotterdam, Brookfield, VT, USA, vol 44, pp 15–27

Brenning A (2005) Spatial prediction models for landslide hazards: review, comparison and evaluation. Nat Hazards Earth Syst Sci 5:853–862

Bussière F, Cellier P (1994) Modification of the soil temperature and water content regimes by a crop residue mulch: experiment and modelling. Agric For Meteorol 68:1–28

Carrivick JL, Tweed FS (2016) A global assessment of the societal impacts of glacier outburst floods. Glob Planet Chang 144:1–16

Cassidy R, Thomas IA, Higgins A (2019) A carrying capacity framework for soil phosphorus and hydrological sensitivity from farm to catchment scales. Sci Total Environ 687:277–286

Cenderelli DA, Wohl EE (2001) Peak discharge estimates of glacial-lake outburst floods and normal climatic floods in the Mount Everest region. Geomorphology 40:57–90

Chen N, Li X, Šimunek J, Shi H, Ding Z, Peng Z (2019) Evaluating the effects of biodegradable film mulching on soil water dynamics in a drip-irrigated field. Agric Water Manag 226:105–788
Chen L, Long C, Wang D, Yang J (2020a) Phytoremediation of cadmium (Cd) and uranium (U) contaminated soils by Brassica juncea L. enhanced with exogenous application of plant growth regulators. Chemosphere 242:125–112
Chen S, Miao Z, Wu L, He Y (2020b) Application of an incomplete landslide inventory and one class classifier to earthquake induced landslide susceptibility mapping. IEEE J Sel Top Appl Earth Obs Remote Sens 13:1649–1660
Chendev YG, Sauer TJ, Ramirez GH, Burras CL (2015) History of east European chernozem soil degradation; protection and restoration by tree windbreaks in the Russian steppe. Sustain For 7:705–724
Choi KY, Cheung RWM (2013) Landslide disaster prevention and mitigation through works in Hong Kong. J Rock Mech Geotech Eng 5:354–365
Choudhury SG, Srivastava S, Singh R, Chaudhari SK, Sharma DK, Singh SK, Sarkar D (2014) Tillage and residue management effects on soil aggregation, organic carbon dynamics and yield attribute in rice-wheat cropping system under reclaimed sodic soil. Soil Tillage Res 136:76–83
Clapp J, Moseley WG, Burlingame B, Termine P (2021) The case for a six-dimensional food security framework. Food Policy 106:102–164
Cornelis WM, Gabriels D (2004) A simple model for the prediction of the deflation threshold shear velocity of dry loose particles. Sedimentology 51:1–13
Corominas J, van Westen C, Frattini P, Cascini L, Malet JP, Fotopoulou S, Catani F, Van Den Eeckhaut M, Mavrouli O, Agliardi F (2014) Recommendations for the quantitative analysis of landslide risk. Bull Eng Geol Environ 73:209–263
Deyn GB, Kooistra L (2021) The role of soils in habitat creation, maintenance and restoration. Philos Trans R Soc B 376:445–622
Edrisi SA, Abhilash PC (2016) Exploring marginal and degraded lands for biomass and bioenergy production: an Indian scenario. Renew Sust Energ Rev 54:1537–1551
Einert AE, Guidry R, Huneycutt H (1975) Mulches for landscape plantings of dwarf crape myrtles
Fang J, Piao S, Field C, Pan B, Gao Y, Zhou QL, Peng C, Tao S (2003) Increasing net primary production in China from 1982 to 1999. Front Ecol Environ 1:293–297
FAO (2011) The state of the world's land and water resources for Food and Agriculture (SOLAW) managing systems at risk. Earthscan, Rome
FAO (2019) Climate-smart agriculture and the sustainable development goals: mapping interlinkages, synergies and trade-offs and guidelines for integrated implementation. FAO, Rome
Feliciano D, Recha J, Ambaw G, MacSween K, Solomon D, Wollenberg E (2022) Assessment of agricultural emissions, climate change mitigation and adaptation practices in Ethiopia. Clim Pol 33:1–18
Fell R, Corominas J, Bonnard C, Cascini L, Leroi E, Savage WZ (2008) Guidelines for landslide susceptibility, hazard and risk zoning for land use planning. Eng Geol 102:85–98
Foley JA, DeFries R, Asner GP, Barford C, Bonan G, Carpenter SR, Chapin FS, Coe MT, Daily GC, Gibbs HK (2005) Global consequences of land use. Science 309:570–574
Francis CA, Harwood RR, Parr JF (1986) The potential for regenerative agriculture in the developing world. Am J Altern Agric 1:65–74
Ghosh BN, Dogra P, Sharma NK, Bhattacharyya R, Mishra PK (2015) Conservation agriculture impact for soil conservation in maize–wheat cropping system in the Indian sub-Himalayas. Int Soil Water Conserv Res 3:112–118
Giller KE, Hijbeek R, Andersson JA, Sumberg J (2021) Regenerative agriculture: an agronomic perspective. Outlook Agric 50:13–25
Guzzetti F, Mondini AC, Cardinali M, Fiorucci F, Santangelo M, Chang KT (2012) Landslide inventory maps: new tools for an old problem. Earth-Sci Rev 112:42–66
Habiba U, Abedin MA, Shaw R (2016) Chapter 6: food security, climate change adaptation, and disaster risk. In: Uitto JI, Shaw R (eds) Sustainable development and disaster risk reduction. Springer, New York

Haen HD, Hemrich G (2007) The economics of natural disasters: implications and challenges for food security. Agric Econ 37:31–45
Harte J, Torn MS, Chang FR, Feifarek B, Kinzig A, Shaw PR, Shen K (1995) Global warming and soil microclimate: results from a meadow-warming experiment. Ecol Appl 5:132–150
Harvey CA (2014) Climate-smart landscapes: opportunities and challenges for integrating adaptation and mitigation in tropical agriculture. Conserv Lett 7:77–90
Houghton JT, Ding Y, Griggs DJ, Van Noguer M, Linden D, Dai PJ, Maskell XK, Johnson CA (2001) Climate change: the scientific basis. Cambridge University Press, New York
Iovine G, Cohen D (2014) Advanced methods in landslide modelling. Nat Hazards 73:1–4
IPCC (2019) Special report on the ocean and cryosphere in a changing climate
Islam MR, Hasan M (2016) Climate-induced human displacement: a case study of Cyclone Aila in the south-west coastal region of Bangladesh. Nat Hazards 2:1051–1071
Izydorczyk G, Mikula K, Skrzypczak D, Moustakas K, Witek-Krowiak A, Chojnacka K (2021) Potential environmental pollution from copper metallurgy and methods of management. Environ Res 197:111–150
Jonard F, De Cannière S, Brüggemann N (2021) Value of sun-induced chlorophyll fluorescence for quantifying hydrological states and fluxes: current status and challenges. Agric For Meteorol 291:38–99
Julien Pierre Y (2010) Erosion and sedimentation. Cambridge University, Cambridge
Kader MA, Nakamura K, Senge M, Mojid MA, Kawashima S (2019) Soil hydro-thermal regimes and water use efficiency of rain-fed soybean (Glycine max) as affected by organic mulches. Agric Water Manag 56:25–76
Kalédjé PSK, Ndam Ngoupayou JR, Rakotondrabe F (2020) Quantitative assessment of water resources by the method of the hydrological balance in the Kadey catchment area (East-Cameroon). Groundw Sustain Dev 10:55–96
Kapucu N, Hawkins CV, Rivera FI (2013) Disaster preparedness and resilience for rural communities. Risk Hazards Crisis Publ Policy 4:215–233
Kavzoglu T, Colkesen I, Sahin EK (2019) Machine learning techniques in landslide susceptibility mapping: a survey and a case study. In: Pradhan SP, Vishal V, Singh TN (eds) Landslides: theory, practice and modelling; advances in natural and technological hazards research, vol 24. Springer, Berlin, pp 283–301
Keshavarzi A, Kumar V, Ertunc G, Brevik EC (2021) Ecological risk assessment and source apportionment of heavy metals contamination: an appraisal based on the Tellus soil survey. Environ Geochem Health 43:2121–2142
Lal R (2003) Soil erosion and the global carbon budget. Environ Int 29:437–450
Lalehzari R, Kerachian R (2020) Developing a framework for daily common pool groundwater allocation to demands in agricultural regions. Agric Water Manag 241:106–278
Lang T, Barling D (2012) Food security and food sustainability: reformulating the debate. Geogr J 178:313–326
Lang T, Heasman M (2015) Food wars: the global battle for mouths, minds and markets. Routledge, Milton Park
Latha C, Saravanan S, Palanichamy K (2010) A semi-distributed water balance model for the Amaravati River basin using remote sensing and GIS. Int J Geometrics Geosci 1:252–263
Leal Filho W, Azeiteiro UM, Balogun AL, Setti AFF, Mucova SA, Ayal D, Oguge NO (2021) The influence of ecosystems services depletion to climate change adaptation efforts in Africa. Sci Total Environ 45:146–414
Lee S, Hong SM, Jung HS (2017) A support vector machine for landslide susceptibility mapping in Gangwon Province, Korea. Sustain For 9:48–76
Li Q, Li H, Zhang L, Zhang S, Chen Y (2018) Mulching improves yield and water-use efficiency of potato cropping in China: a meta-analysis. Field Crop Res 2:50–60
Li A, Mu X, Zhao X, Xu J, Khayatnezhad M, Lalehzari R (2021) Developing the non-dimensional framework for water distribution formulation to evaluate sprinkler irrigation. Irrig Drain 55:36–67

Little W, Fowler HW, Coulson J (1990) The shorter Oxford English dictionary: on historical principles, vol II, 3rd edn. Clarendon Press, London

Liu LJ (2004) Experimental study on effect of mulch cover on soil water content under different rainfall. Collection of extent abstracts of 2004 CIGR International Conference, vol 2, pp 225–556

Liu XJ, Li Y, Yang QS (2011) Effect of yard wastes organic mulches on soil nutrient (International Symposium on Water Resource & Environmental Protection)

Luo Y, Wan S, Hui D, Wallace LL (2001) Acclimatization of soil respiration to warming in tall grass prairie. Nature 413:622–625

Lynch JM, Bragg E (1985) Microorganisms and soil aggregate stability. Adv Soil Sci 17:133–171

Ma Z, Mei G, Piccialli F (2020) Machine learning for landslides prevention: a survey. Neural Comput Applic 33:10881–10907

Mahyuddin M, Sugianto S, Alvisyahrin T (2013) Analysis of land cover on Krueng Aceh watershed pre and post tsunami. Faculty of Agriculture, Universitas Syiah Kuala, Banda Aceh. J Land Res Manag 2:297–303

Mallick B, Ahmed B, Vogt J (2017) Living with the risks of cyclone disasters in the south-western coastal region of Bangladesh. Environment 4:13–45

Mawalagedara R, Oglesby RJ (2012) The climatic effects of deforestation in South and Southeast Asia. Deforestation Around World 23:978–995

Melillo JM, Steudler PA, Aber J, Newkirk D, Lux KH, Bowles F, Catricala PC, Magill A, Ahrens T, Morrisseau S (2002) Soil warming and carbon-cycle feedbacks to the climate. Science 298:2173–2176

Michel D, Eriksson M, Klimes M (2021) Climate change and (in) security in transboundary river basins. Handb Secur Environ 66:119–179

Mondini AC, Guzzetti F, Chang KT, Monserrat O, Martha TR, Manconi A (2021) Landslide failures detection and mapping using synthetic aperture radar: past, present and future. Earth Sci 216:103–574

Moore T (2008) Disaster and emergency management systems. British Standards Institution, London

Mupangwa W, Twomlow S, Walker S, Hove L (2007) Effect of minimum tillage and mulching on maize (Zea mays L.) yield and water content of clayey and sandy soils. Phys Chem Earth 32:15–59

Murshed M (2022) Pathways to clean cooking fuel transition in low and middle income Sub-Saharan African countries: the relevance of improving energy use efficiency. Sustainable Prod Consumption 30:396–412

Murshed M, Dao NTT (2020) Revisiting the CO2 emission induced EKC hypothesis in South Asia: the role of export quality improvement. Geo J 34:166–198

Norby R, Hartz-Rubin J, Verbrugge MJ (2003) Phenological responses in maple to experimental atmospheric warming and CO2 enrichment. Glob Chang Biol 9:1792–1801

Olson JM, Misana S, Campbell DJ, Mbonile M, Mugisha SA (2004) Research framework to identify the root causes of land use change leading to land degradation and changing biodiversity. In: LUCID working paper. International Livestock Research Institute, Nairobi

Rahmanipour F, Arzaioli R, Bahrami HA, Fereydoun Z (2014) Assessment of soil quality in agricultural lands of Qazvin Province, Iran. J Ecol Indicator 40:19–26

Raihan MJ (2018) Effect of seasons on household food insecurity in Bangladesh. Food Energy Secure 7:1–15

Ranjan B (2015) Soil degradation in India: challenges and potential solutions. Sustain For 7:3528–3570

Rapant S, Dietzová Z, Cicmanová S (2006) Environmental and health risk assessment in abandoned mining area, Zlata Idka, Slovakia. Environ Geol 51:387–397

Rautela P (2006) Redefining disaster: need for managing accidents as disasters. Disaster Prev Manag 15:799–809

Read HH, Watson J (1975) Introduction to geology, vol 44. Halsted, New York, pp 13–15

Reyes A, Thiombane M, Panico A, Daniele L, Lima A, Di Bonito M, De Vivo B (2020) Source patterns of potentially toxic elements (PTEs) and mining activity contamination level in soils of Taltal city (northern Chile). Environ Geochem Health 42:2573–2594

Rivera-Zayas J, Sotomayor-Ramírez D, Barnes R (2017) Nitrogen fertilizer rate for improving inbred maize (Zea mays L.) yield on the semi-arid southern coast of Puerto Rico. J Agric Univ 101:185–202

Rustad LE, Campbell JL, Marion GM, Norby RJ, Mitchell MJ, Hartley A, Cornelissen H, Gurevitch CJ (2001) A meta-analysis of the response of soil respiration, net nitrogen mineralization, and aboveground plant growth to experimental ecosystem warming. Ecología 126:543–562

Saeed S, Ali HA, Hamid S, Mohammad J, Seyed FA (2019) Optimization of parameters affecting organic mulch test to control erosion. J Environ Manag 2:249–287

Sarker A, Deepo DM, Nandi R, Rana J, Islam S, Rahman S, Hossain MN, Islam MS, Baroi A, Kim JE (2020) A review of microplastics pollution in the soil and terrestrial ecosystems: a global and Bangladesh perspective. Sci Total Environ 733:139–296

Sattar A, Haritashya UK, Kargel JS, Leonard GJ, Shugar DH, Chase DV (2021) Modeling lake outburst and downstream hazard assessment of the Lower Barun Glacial Lake, Nepal Himalaya. J Hydrol 598:126–208

Sawalha I (2018) Behavioural response patterns: an investigation of the early stages of major incidents. Foresight 20:337–352

Schreefel L, Schulte RPO, De Boer IJM, Schrijver AP, Van Zanten HHE (2020) Regenerative agriculture – the soil is the base. Glob Food Secur 26:100–404

Segoni S, Tofani V, Rosi A, Catani F, Casagli N (2018) Combination of rainfall thresholds and susceptibility maps for dynamic landslide hazard assessment at regional scale. Front Earth Sci 6:85–127

Shahab H, Emami H, Haghnia GH, Karimi A (2013) Pore size distribution as a soil physical quality index for agriculture and pasture soil in Northeastern Iran. Pedosphere 23:312–320

Shaluf I, Ahmadun F, Mustapha S (2003) Technological disaster's criteria and models. Disaster Prev Manag 12:305–311

Shaver GR (2000) Global warming and terrestrial ecosystems: a conceptual framework for analysis. Bioscience 50:871–882

Shogren RL (2000) Biodegradable mulches from renewable resources. J Sustain Agric 16:3–47

Shugar DH, Burr A, Haritashya UK, Kargel JS, Watson CS, Kennedy MC, Strattman K (2020) Rapid worldwide growth of glacial lakes since 1990. Nat Clim Chang 10:939–945

Singh SB, Pramod K, Prasad KG, Kumar P (1991) Response of Eucalyptus to organic manure mulch and fertilizer sources of nitrogen and phosphorus. Van Vigyan 29:200–207

Sintim HY, Flury M (2017) Is biodegradable plastic mulch the solution to agriculture's plastic problem? Environ Sci Technol 51:1068–1069

Six J, Bossuyt H, Degryze S, Denef K (2004) A history of research on the link between (micro) aggregates, soil biota, and soil organic matter dynamics. Soil Tillage Res 79:7–31

Sovacool BK, Grifths S, Kim J, Bazilian M (2021) Climate change and industrial F-gases: a critical and systematic review of developments, sociotechnical systems and policy options for reducing synthetic greenhouse gas emissions. Renew Sustain Energy 141:110–759

Srivastava P, Giri N, Mandal D (2019) 137 Cs technology for soil erosion and soil carbon redistribution. Curr Sci 11:688–889

Stern A, Robert J (2002) Subduction zones. Rev Geophys 40:10–12

Sun X, Khayatnezhad M (2021) Fuzzy-probabilistic modeling the flood characteristics using bivariate frequency analysis and α-cut decomposition. Water Supply 89:446–558

Tariq M, Rashid M, Rashid W (2014) Causes of deforestation and climatic changes in Dir Kohistan. J Pharm Alternat Med 3:28–37

Thankamani CK, Kandiannan K, Hamza S, Saji KV (2016) Effect of mulches on weed suppression and yield of ginger. Sci Hortic 207:125–130

The World Bank (2008) Agriculture for development. World development report 2008. The World Bank, Washington

Tisdall JM, Oades JM (1982) Organic matter and water-stable aggregates in soils. J Soil Sci 33:141–163

Trigo E, Cap E, Malach V, Villarreal F (2009) The case of zero-tillage technology in Argentina. International Food Policy Research Institute, Washington

Triplett GB, Dick WA (2008) No-tillage crop production: a revolution in agriculture. Agron J 100:153–156

Tsangaratos P, Ilia I (2017) Chapter 24: applying machine learning algorithms in landslide susceptibility assessments. In: Samui P, Sekhar S, Balas VE (eds) Handbook of neural computation. Academic Press, Cambridge, pp 433–457

Veh G, Korup O, Walz A (2020) Hazard from Himalayan glacier lake outburst floods. Proc Natl Acad Sci 117:907–912

Wan S, Luo Y, Wallace L (2002) Changes in microclimate induced by experimental warming and clipping in tallgrass prairie. Glob Chang Biol 8:754–768

Wander MM, Walter GL, Nissen TM, Bollero GA, Andrews SS, Cavanaugh-Grant DA (2002) Soil quality: science and process. Agron J 94:23–38

Wang F (2017) Occurrence of arbuscular mycorrhizal fungi in mining-impacted sites and their contribution to ecological restoration: mechanisms and applications. Crit Rev Environ Sci Technol 47:1901–1957

Wang FX, Feng SY, Hou XY, Kang SZ, Han JJ (2009) Potato growth with and without plastic mulch in two typical regions of Northern China. Field Crop Res 110:123–129

Wang X, Zhang C, Hasi E, Dong Z (2010) Has the three norths forest shelterbelt program solved the desertification and dust storm problems in arid and semiarid China? J Arid Environ 74:13–22

Weltzin JF, Bridgham SD, Pastor J, Chen J, Harth C (2003) Potential effect of warming and drying on peatland plant community composition. Glob Chang Biol 9:141–151

Westoby MJ, Glasser NF, Brasington J, Hambrey MJ, Quincey DJ, Reynolds JM (2014) Modelling outburst floods from moraine-dammed glacial lakes. Earth Sci Rev 134:137–159

Wu W, Qu S, Nel W, Ji J (2020) The impact of natural weathering and mining on heavy metal accumulation in the karst areas of the Pearl River Basin, China. Sci Total Environ 734:139–480

Xue L, Xiaoli L, Jinggang L, Qiuliang W, Wulin L, Lifen Z (2013) Factor analysis of earthquake-induced geological disasters of the M7. Lushan earthquake in China. Geod Geodyn 4:22–29

Zhang G, Yao T, Xie H, Wang W, Yang W (2015) An inventory of glacial lakes in the third pole region and their changes in response to global warming. Glob Planet Chang 131:148–157

Zhang B, AghaKouchak A, Yang Y (2019) A water-energy balance approach for multi-category drought assessment across globally diverse hydrological basins. Agric For Meteorol 264:247–265

Zhou J, Obrist D, Dastoor A, Jiskra M, Ryjkov A (2021) Vegetation uptake of mercury and impacts on global cycling. Nat Rev Earth Environ 2:269–284

Zhu P, Saadati H, Khayatnezhad M (2021) Application of probability decision system and particle swarm optimization for improving soil moisture content. Water Supply 28:67–97

Role of Environmental Science for Disaster Risk Reduction in Agriculture

Muhammad Mubeen, Khadija Shabbir, Amna Hanif, Mazhar Ali, Sajjad Hussain, and Shakeel Ahmad

Abstract

In this chapter, the role of various disciplines of environmental sciences in disaster risk reduction in agriculture is discussed. First, some natural disasters like drought, floods, land degradation, and pest outbreaks are highlighted because agriculture faces a significant portion of these disasters. Then, some management practices for mitigation of the effects of these hazards on agriculture are discussed. For example, management of carbon emissions especially through green finance, management of water through smart irrigation and precision farming, the role of beneficial microorganisms in soil quality and plant health, and improving the ecosystem can reduce food security issues arising through natural disasters.

Keywords

Disaster risk reduction · Drought · Floods · Land degradation · Pest outbreaks · Mitigation · Carbon emissions · Green finance · Smart irrigation · Precision farming · Food security

M. Mubeen (✉) · K. Shabbir · A. Hanif · M. Ali · S. Hussain
Department of Environmental Sciences, COMSATS University Islamabad, Islamabad, Pakistan
e-mail: muhammadmubeen@cuivehari.edu.pk

S. Ahmad
Department of Agronomy, Faculty of Agricultural Sciences and Technology, Bahauddin Zakariya University, Multan, Pakistan

M. Ahmed, S. Ahmad (eds.), *Disaster Risk Reduction in Agriculture*, Disaster Resilience and Green Growth, https://doi.org/10.1007/978-981-99-1763-1_7

1 Importance of the Agriculture Sector

Agriculture is an important source of food for both humans and animals. It also improves the gross revenue of the country. In this modern scientific and technological age, most countries still rely on agriculture to provide and enhance the national economy (Sunkad 2020; Akram et al. 2022a, b).

Economic growth and development are significantly influenced by agriculture. First, it is essential to human existence as a source of nourishment. It contributes significantly to economic activity as a supplier of industrial raw materials (Blandford 2011).

2 Natural Disasters and Food Security Issues

The agriculture industry is highly susceptible to natural calamities and risks. Given the considerable dependence of agricultural systems on climate, the obvious rise in the number of weather incidents in the past caused remarkable confront to the agriculture industry (Akram et al. 2019; Naz et al. 2022). Calamities can have a negative impact on fisheries and aquaculture productivity, crop growth, livestock health, and the forestry sector (FAO 2018).

2.1 Climate-Related Risks

The influence of climate change on agriculture is depicted in varied ways in the latest report by the Intergovernmental Panel on Climate Change (Sabagh et al. 2020). In low-latitude areas, agricultural yields are anticipated to decrease due to medium global temperature rises (1–3 °C), higher carbon dioxide concentrations, and accompanying changes in rainfall. However, mid- to high-latitude areas may see a slight positive influence. Less output is predicted to occur in all locations if the worldwide mean temperature rises by more than 3 °C. Several modeling studies' findings indicate that global cereal prices may rise tremendously if temperatures rise above 3 °C. Change in climate may cause the danger of starvation in some areas to rise remarkably, especially in sub-Saharan Africa. More climatic upsets including heat stress, droughts, and flooding will probably occur due to global warming, and the danger of fires, pest and pathogen outbursts, and heat stress will likely increase as well (Mubeen et al. 2022). These will probably make agricultural productivity more inconsistent in several areas and perhaps even in the entire world (Blandford 2011).

2.1.1 Droughts and Extreme Heats

Crops are subjected to intense hyperthermia, and water shortages during droughts can reduce agricultural output if they happen at specific points in the plant's life cycle (Aslam et al. 2013; Akram et al. 2018).

Compared to other zones, the agricultural industry is more severely impacted by drought. Agriculture suffered 83% of all drought-related losses and damages. This steady-outset threat mostly affects the crop and animal sectors (FAO 2018).

Increased temperatures have a variety of negative effects on crops, including reduced photosynthesis, leaf senescence, reduced pollen production and viability, seed abortion, and subsequently lower grain quantity and grain mass (Siebert and Ewert 2014; Amin et al. 2018; Ali et al. 2019; Mubeen et al. 2021).

2.1.2 Flooding

In the present global environment, floods are primarily to a greater extent common and damaging than many hazards (Niles and Mueller 2016; Ali et al. 2017; Teo et al. 2018) because they are strongly linked to societal risks, financial deprivation, and human mortality (Roco et al. 2017; Miah et al. 2018; Mahmood et al. 2019; Hossain et al. 2020; Ahmad and Afzal 2021). Due to its locality in hazard-susceptible areas and the frequent flooding, it experiences (Abbas et al. 2017; Ahmad and Afzal 2021), Pakistan is categorized as the fifth greatest damaged country in the world by natural hazards. Glacier melting, erratic rains, and expanding cycles of monsoon rain are the main causes of recurrent floods in interfiled rivers connecting upriver or downriver (Teo et al. 2018; Ahmad and Afzal 2020). Floods, mainly caused by climate-related calamities, have severely harmed Pakistan's agricultural industry (Hussain et al. 2020; Ahmad and Afzal 2022).

2.2 Land Degradation

One of nature's risks, land degradation, happens when ground cover is detached, exposing sensitive soil and organic materials to being splashed or carried aside. Related outcomes, such as decreased land production, might result from soil salinization. The signs are numerous and concerning: shifting and growing sand dunes, decreased rainfall, the disappearance of pastureland, wind and rain erosion, and loss of biological integrity and biomass (Hussain et al. 2022a, b).

The earth's cultivable areas and meadows, which are necessary for the production of food, water, and clean air, are under pressure as a result of these socioeconomic and environmental methods. Land deterioration and desertification can have complicated effects on person's well-being. Food production and water supplies desiccate, and communities are pushed to relocate to more sociable places as a result of land deterioration and the extension of deserts in some areas.

Currently, 25% of all areas on the planet are damaged. Whereas 3.2 billion persons are impacted at once by land deterioration, particularly small-scale cultivators and countryside societies, lots of further are negatively impacted by malnutrition, rising edibles costs, global climate change, ecological dangers, and the damage of habitat and bionetwork facilities. 24 billion tons of productive land is damaged annually, according to scientists, mostly because of unmanageable cultivation procedures. Because of poor administration of resources, overgrazing, and overcrowding, 95% of the area of the world will eventually become an everlasting desert if this inclination prolongs (Giannini et al. 2003).

2.3 Pest Outbreak

The greatest important threat to cultivators worldwide may come from an outburst of a disease or pest, but it is also usually believed to be one of the most manageable hazards. As stated by the FAO, pests pose a severe threat to cultivation, with yearly yield deficiency ranging from 20% to 40% worldwide. Such damages could exceed 80% if pest repellents are not used for conservation (Savary et al. 2019).

Additionally, an alarming rise in the frequency of transnational outbursts of plant and animal illnesses and pests constantly threatens social stratification. Wheat yield around the globe is at threat due to phytopathology like wheat rust, which can reduce outputs by up to 80%.

It has been documented that locust outbreaks significantly reduce crop production (FAO 2018). When winds are favorable, adult flocks of locusts from barren lands can move up to 150 km each diurnal in quest of nourishment, covering great space and, in the awful situation, scattering from one continent to further. One square kilometer of a minor flock could eat the equivalent of 35,000 people's daily food intake. Feeding areas, on which shepherds rely, are not resistant. The Greater Horn of Africa, both sides of the Red Sea, the Islamic Republic of Iran, India, and Pakistan have all been affected by the outburst that occurred in 2020–2021. Surprisingly, nations that are not frequently afflicted by the pests, such as Uganda and the United Republic of Tanzania, have been impacted, and undernourished populations in the Sahel that are already dealing with different problems encountered the danger of one more pest invasion.

2.4 Damage for Ecosystems

The natural resources and environments particularly supporting cultivation, are also negatively impacted by disasters in both intended and unintended ways. The abovementioned could consist of:

1. Degradation and pollution of surface and groundwater.
2. Enhanced soil erosion.
3. Harm to local wetlands, mangroves, and forests.
4. Soil salinization.
5. Harm to coral reefs.
6. Loss of biodiversity.

Moreover, the relocation of impacted individuals in the wake of catastrophes may inadvertently result in greater demand for natural resources (such as the misuse of water and forest reserves) in the regions close to relocation sites. Drought has devastating effects, particularly in nations with less ability to withstand tremors. It is a key contributor to water scarcity and land degradation. If precautionary actions are not taken, extended or recurrent droughts in dry and desiccated regions can cause dry, barren lands, which are irrecoverable. Water pollution and increased soil deterioration methods are recurrently related to floods. Following a flood, residual contaminants are present in the land (Tariq et al. 2018; Zahoor et al. 2019). Soils deteriorate due to sediments and polluted water, especially in farmed regions. By harming ecosystems and natural resources that are crucial for agricultural output, tropical storms, too, have a significant negative impact on the environment across vast land. Tropical Cyclone Pam which struck Vanuatu in 2015, severely harmed tropical forests as well as marine and coastal habitats like coral reefs and mangroves (FAO 2018).

3 Role of Environmental Sciences in Disaster Risk Reduction for Agriculture

The discipline of environmental sciences is a wide field that integrates with innovative technology in order to mitigate major environmental hazards that cause severe impacts on the well-being of mankind, mainly economic stability. In order to enhance the management of disaster risks, environmental sciences have unleashed various prospects. Owing to the capabilities of this science to visualize, analyze, and forecast disaster risks, it can shape human activities and disaster risk management considerations in agricultural contexts across the globe. It has been noted that disasters adversely affect soil structure and physiology, which in turn affects the agricultural productivity of the specified region. Hence, there is a need to reduce disaster risk and manage it with the help of innovative environmental sciences.

Various efforts and technologies to reduce environmental disaster risk in agricultural sectors include the use of early warning systems (Glade and Nadim 2014; Barnes et al. 2019), improved scale application of disaster risk registers (Marković et al. 2016), disaster monitoring based on Geographic Information System (GIS) and improvement of communication via satellite network in case of emergency situations (Miao and Ding 2015; Hussain et al. 2022c).

3.1 Management of Carbon Emissions

Climate change is a major concern globally as it adversely affects crop growth and yield. Climate change mitigation could be achieved by reducing total carbon emissions from the environment, and this could be made possible by reducing carbon emissions from agricultural production.

For this purpose, eco-friendly and green technology should be acquired for future agricultural development. Green agriculture would become trendy and help maintain good environmental health and ensure low carbon.

It is suggested that government should consider and promote sustainable agricultural development and ensure green agricultural production, which would aid in maintaining low carbon emissions by the agriculture production sector. Reinforce top-level designs of government, improve production technology by replacing it with the most innovative and eco-friendly gadgets and techniques, promoting the replacement of synthetic fertilizers with biofertilizers aimed at carrying agricultural fertilizer loss action to acquire negative increase in usage of synthetic fertilizers if supervised strictly would help governments to aware and warn the farmers about carbon emission and its tax imposition. Moreover, announcing different penalties and incentives would help to encourage the farmer to participate in green production. In conclusion, various tools and techniques, which include strengthening collaboration among international social groups, learning from other countries' technical experiences, and introducing technologies with zero carbon emissions or energy-saving capabilities, would aid in achieving the goal of carbon neutrality at early stages.

3.2 The Role of Beneficial Microorganisms in Soil Quality and Plant Health

When it comes to the quality of soil and health of plant, microorganisms come forward as a savior and play an essential role in maintaining both. Microbes should be included in soil affected by natural hazards or disasters; these could help the soil to regain its physiochemical properties. They add to the fertility of existing soil, increase organic content, and replenish nutrients.

3.2.1 Biofertilizers

Biofertilizers are regarded as sustainable and feasible substitutes for increasing crop growth and yield, improving and restoring soil fertility, and reducing production costs and environmental risks of chemical fertilizers (Zambrano-Mendoza et al. 2021). Biofertilizers have been applied to various crops such as Sudan grass, sugarcane, wheat, cucumber, fan cypress, eggplant, black pepper, sorghum, peanut, corn, sunflower, coffee, rice, potato, coconut, alfalfa, alder, tea, strawberries, beans, neem, flax, oats, tobacco, green soybeans, pepper, tomato, cotton, carrots, pine, beet, lettuce, and many more.

Bacterial groups which aid in nutrient fixation in soil for the betterment of plant growth are termed as bacterial biofertilizers (Fasusi et al. 2021). These could help in the fixation of nitrogen, solubilization of micronutrients such as zinc, phosphorous, and potassium and aid in the secretion of organic compounds, which helps suppress harmful bacteria or provide growth-promoting substances. The most famous applied bacterial biofertilizers include *Rhizobium*, *Azospirillum*, *Bacillus*, and *Azotobacter*. For legume crops, the suitable bacterial biofertilizer is *Rhizobium*, and for nonlegume crops, *Azospirillum* and *Azotobacter* are considered suitable. For sugarcane, specifically, *Acetobacter* is used. Using these bacterial biofertilizers helps enhance sustainable agriculture by promoting crop growth and yield, soil

Table 1 Useful bacteria genera for various crop plants

Phylum	Genus	Host	Benefit
Actinobacteria	*Arthrobacter, Cellulommonas, Rhodococcus*	Maize, pea, rice, wheat	Increase plant vigor, tolerance to biotic and abiotic stress
Firmicutes	*Bacillus, Pontibacillus, Alicyclobacillus*	Apple, barley, pepper, oat	Solubilization of phosphorous and zinc, production of indole acetic acid, hydrogen cyanide
Proteobacteria	*Allidiomarina, Marinobacter, Aquisalimonas*	Barley, bean, cotton, mustard	Increase plant vigor, nitrogen fixation, solubilization of nutrients

fertility, and protecting against phytopathogen attacks. In addition, sustainable agriculture provides environment-friendly substitutes for synthetic fertilizers and pesticides. Some of the useful bacteria for different crop plants are given in Table 1.

3.2.2 Biopesticides

For crop protection, biopesticides are combined with various other compounds like chemical pesticides and play a role in eco-friendly pest management. Many pests or phytopathogens attack the crops, causing great economic losses. Among the microorganisms employed as biofertilizers and biopesticides, one of the most used is the *Bacillus* genus. There are different ways of action of *Bacillus,* which include direct and indirect mechanisms. These mechanisms act concurrently during plant growth. The ability of *Bacillus* to get the supply of nutrients, including phosphorous, nitrogen, potassium, minerals, or modulate plant hormone levels, is considered to be their direct mechanism. In indirect mechanisms, the secretion of substances that restrict pathogen attack or induce resistance to them is included. One of the most used biopesticides worldwide is *Bacillus thurigiensis*.

3.2.3 *Bacillus* spp. Beneficial for Plants

Bacillus genus has multi-strains and various species, which have been extensively utilized as biopesticides, biofertilizers, and as significant biotechnological tools. *Bacillus* genus could restrict different harmful bacteria present in the soil while enhancing plants' growth by various mechanisms (direct and indirect), which can act concurrently with the growth of the plant (Saxena et al. 2020). Thus, these species are considered very efficient on plants' tissues in controlling the colonization of harmful pathogens with the help of antibiosis and by inducing resistance in the plants (on which) pathogens reside.

3.3 Different Techniques to Monitor Different Ecosystems

Ecosystem is usually defined as a group of living organisms that live and grow together and exhibit some interaction among themselves in a specified environment. Different ecosystems require different techniques and information to reduce

Table 2 Surveillance techniques for monitoring disaster risks in different ecosystems

Ecosystems	Techniques/information required
Mangrove	GIS mapping
Swamps	Laws and regulations
Coastal forests	Evaluating the worth of every ecosystem from the economic perspective
Beach	Information on fisheries management

disaster risks. For monitoring disaster risks in mangrove ecosystems, proper surveillance is required with the help of GIS mapping; similarly, various other ecosystems are mentioned in Table 2.

3.4 Importance of Geospatial Data

Over-population posed a severe burden on the agricultural sector, which led farmers to intensive farming and increased load on the environment, for example, the use of more agrochemicals like pesticides, fertilizers, etc. A possible solution to reduce this load is precision farming, achieved by crop management on the basis of area analysis, which requires the use of geospatial data obtained by advanced geospatial technology. Techniques that provide necessary agricultural information collection:

- Farm machinery telemetry provides information about the machines being operated in the agricultural fields by the application MapLogAgri.
- For the purpose of storage, analysis, and publication of sensor data, agrometeorological observations are carried out with the help of a wireless network system and SensLog.
- Remote sensing is applied for monitoring crop conditions, which integrates data from the imagery of various high-resolution satellites. Some tests were conducted, and their results demonstrated that agricultural techniques might be transferred to crisis/emergency management domains.

3.5 Management of Water Through Smart Irrigation and Precision Farming

The domains requiring more disaster risk reduction (DRR) include water and land, as they face more pollution. It is considered that most of the freshwater is used in agricultural practices globally; every country uses almost 70% of its fresh water in agricultural practices, while the excessive water is recharged to the ground or surface water. Precision farming, as discussed earlier, helps to better crop management; it requires geospatial data usually available by Farm Management Information System (FMIS). It is suggested to consider FMIS as a type of GIS (Geographic Information System) because it gives answers to basic queries like what occurred and where it occurred. Smart irrigation implies watering the field to the required extent, and this could help prevent water wastage and increase crop yield. This depends upon site-specific management and takes into account many variables with the help of geospatial data., It helps to reduce inputs and enhances output hence ensuring sustainable agriculture. This helps apply fertilizers and pesticides where necessary, reducing their usage and saving the environment from the harmful risk of these agrochemicals. It helps lead to the reduction of environmental risks. The main component in precision farming is the kind of sensor used because they are expensive and have different capabilities, like repeated small-scale measurements relative to space and time. The site-specific optimization for cultural activities could be one of the ways of reducing the environmental burden produced by agriculture. The authors claim that precision agriculture cannot altogether compensate for these effects, but it can potentially lessen some of these. The basic prerequisite for profitability by precision agriculture techniques is that small-scale in-field differences exist in factors relevant to plant development, for example, variations in soil, water, and fertilizer availability.

3.5.1 Drought Monitoring

Evaluating existing conditions of droughts and predicting future trends could help to reduce disaster risks. Remote Sensing and Geographic Information Systems played a significant part in assessing specific areas' drought vulnerability and helped draft mitigation plans. For this purpose, various sensors and satellites are used to obtain data, including Landsat OLI sensor and Advanced Very High-Resolution Radiometer (AVHRR). Drought indices are generated with the help of remote sensing and GIS techniques. Drought indices further include meteorological and agricultural indices (NDVI).

Land management of drought affected area depends highly on the magnitude of the drought hazard (dH) and future trends of this hazard (Dao and Peduzzi 2003; Hayes et al. 2004; Cardona et al. 2012; Hussain et al. 2021). Before implementing any drought hazard (dH) management plan, it is suggested to always assess simulated dH conditions (Wilhite et al. 2007). Most of the regions which are mostly affected by the drought hazard presently were agricultural lands such as parts of the United States or India and central Europe (Ramankutty et al. 2008). Global food production might depend upon these areas in the future as these were pieces of fertile agricultural land (Foley et al. 2011; Din et al. 2022). There is a prediction of an increase in drought hazards in the amazon region. It is considered as a tip for the climate management department involving large potential feedback to the global carbon cycle (Lenton et al. 2008; Phillips et al. 2009; Lewis et al. 2011). It is necessary to assess the likely range of droughts and their uncertainty in any specific region, so that people in nondrought areas should also be aware of the water shortage problems (Orlowsky and Seneviratne 2013). Governments need to plan mitigation practices and preventive measures for the dH-affected areas and regions with the suspicion of hazard emergence. It should aim to reduce climate change's effects on the environment and human health.

3.5.2 Forest Fire Surveillance

The need to monitor forest fires is fulfilled by satellite imagery. Moderate Resolution Imagery Spectroradiometry (MODIS) instrument shares forest fire products that are based on satellite imagery. It was first launched in December 1999. This covers all the circumference of the earth within a time span of 1–2 days—images obtained from MODIS help to monitor forest fires through a specified region. T. J. Lynham et al. worked to determine the prerequisites for observation in detecting and managing fire, mapping, monitoring, smoke management, and recovery of burned areas. Space-based observations are highly dependent on weather conditions. Poor weather (clouds) would reduce the accuracy of forest fire detection and lags by providing low-resolution images and taking more time in the scanning area.

3.5.3 Floods Forecasting

Floods can damage agricultural lands to a great extent. Flood forecasting is possible with the help of many models, including hydrological (rainfall), hydrodynamic (flood routing), and meteorological models. These help hydrologists determine whether a flood is expected or not. Hydrological models show the representation of the water cycle and are used in forecasting streamflow. These use rainfall data as input which is obtained from RADAR, rain gauge weather, or simulated precipitation from numerical models. Hydrodynamic models help to study the motion of water within a water body.

Warning systems for floods and different forecasting techniques are employed worldwide. Most of the systems are based on utilizing high-resolution altimetric information as LiDAR (Light Detection and Ranging). The Delft-FEWS provides a platform known to be operational in forecasting, which helps to bring various code models to an operational domain. Then the models could be connected with information obtained from external databases and various other formats of files. A little gap between the rainy period and flood becomes a hurdle in adopting an early warning system and mitigation or protective measures. The best way is to predict the probability of a flood before its occurrence and take in mitigatory/defensive measures. To forecast advanced fluvial floods in the Agueda river basin, the Flood Forecast and Alert System (FFAS) was initiated, which worked on the basis of meteorological forecasting. This system provided various benefits, among which the best way to combine the 2D hydrodynamic and real-time hydrologic modeling, which worked on Numerical Weather Prediction (NWP). NWP is a digital terrain surface model with high resolution and it helps forecast flood dissemination by web GIS (which is updated every 6 h).

3.5.4 Storm Prediction

Another major disaster that affects agricultural land and is needed to be reduced or mitigated is Sand and Dust Storms (SDS). The technologies that aid in the surveillance of SDS are Space-based monitoring which includes the use of satellites and various remote sensing techniques, and ground-based monitoring includes the use of handheld cameras, drone cameras, etc., for data collection through images. Space-based monitoring helps to get large-scale monitoring of the ground within few days, and ground-based techniques provide small-scale monitoring of the ground. Recently, AI (artificial intelligence) has been applied in SDS surveillance and prediction to provide accurate and precise results. However, this integration is yet in its initial stages. Hybrid systems characterized as AI-enabled are proposed to be the future trend in SDS monitoring.

3.5.5 Pest Monitoring Through Remote Sensing

Vegetal stress and growth reaction of crops could be projected by the remote sensing techniques. Different research demonstrated various types of plant health indicators, plant diseases, pathogenic infections, and infections symptoms that could not be obtained by the naked eye at the early stage by aerial hyperspectral imaging. Investigating plants by remote sensing data requires a basic concept of the firm and the role played by the foliage and reflectance characteristics. For the calculation of incident light reflection at various wavebands, the sensors have been upgraded and linked to plant-cover and -association. Remote sensing is considered an economically feasible and efficient science forest monitoring (Masood et al. 2022).

Dong et al. (2020) conducted a study in which pest habitat and attack's sensitive indexes were used along with the early forecasting, and then the forecasting model's parameters were optimized to enhance its applicability at the national level. Second, he developed a self-operated system based on a web GIS platform for efficiently identifying pest habitats and early forecasting. In the end, thematic maps of pest and disease forecasting were issued. The experimental objects were the national disease of China the "wheat yellow rust" (*Puccinia striiformis*), and national pest oriental migratory locust (*Locusta migratoria manilensis* (Meyen)).

3.6 Green Finance

It is suggested that green finance would be able to reduce the use of synthetic fertilizers to a negligible extent and alleviate carbon emissions within the time period of one decade. It helped support the goals of reducing carbon emissions and became a pillar of national strategic goals.

Funding to help promote green policies, investments in green projects, and establishment of a green financial structure are the three main aspects of green finance. It helps in the improvement of environmental quality with the collective efforts for the reduction of carbon footprints. It could promote the trend of eco-friendly projects development, minimize resource depletion by increasing utilization efficiency, and helps the public to be aware of the establishment of green consumption concepts.

Previously, environmental concerns were not considered during decision-making processes regarding any developmental projects. Only cost–benefit analyses were carried out to review all economic aspects, which led to the pose of disastrous effects on the environment. By considering environmental aspects, a development team began to consider green development, provide more funds, and guarantees the farmers to provide them with subsidies, green credit, and insurance to encourage farmers to shift toward green practices and reduce the usage of fertilizers.

4 Conclusion

This chapter includes the efforts which could be integrated with environmental sciences and utilized for reducing disaster risks in agriculture and their proper management. These strategies include incorporating green finance in basic agricultural businesses to ensure reduced carbon emissions, which could help reduce the risk or disaster caused by increased carbon proportion in the atmosphere, which is more likely to cause global warming, topsoil erosion, and various other adverse effects. In addition, it discusses the use of technology with the ideas of environmental sciences, such as monitoring pests before their arrival to a specific region or monitoring their movement, location, density, and direction. This helps to deal with natural hazards, accurately dealing with and reducing their potential impacts on agriculture.

References

Abbas G, Ahmad S, Ahmad A, Nasim W et al (2017) Quantification the impacts of climate change and crop management on phenology of maize-based cropping system in Punjab, Pakistan. Agric For Meteorol 247:42–55

Ahmad D, Afzal M (2020) Climate change adaptation impact on cash crop productivity and income in Punjab province of Pakistan. Environ Sci Pollut Res 27:30767–30777

Ahmad D, Afzal M (2021) Impact of climate change on pastoralists' resilience and sustainable mitigation in Punjab, Pakistan. Environ Dev Sustain 23:11406–11426

Ahmad D, Afzal M (2022) Flood hazards and agricultural production risks management practices in flood-prone areas of Punjab, Pakistan. Environ Sci Pollut Res 29(14):20768–20783. https://doi.org/10.1007/s11356-021-17182-2

Akram R, Turan V, Hammad HM, Ahmad S, Hussain S, Hasnain A, Maqbool MM, Rehmani MIA, Rasool A, Masood N, Mahmood F (2018) Fate of organic and inorganic pollutants in paddy soils. In: Environmental pollution of paddy soils. Springer, Cham, pp 197–214. https://doi.org/10.1007/978-3-319-93671-0_13

Akram R, Natasha, Fahad S, Mubeen M et al (2019) Trends of electronic waste pollution and its impact on the global environment and ecosystem. Environ Sci Pollut Res 26:16923–16238. https://doi.org/10.1007/s11356-019-04998-2

Akram R et al (2022a) Research on climate change issues. In: Jatoi WN, Mubeen M, Ahmad A, Cheema MA, Lin Z, Hashmi MZ (eds) Building climate resilience in agriculture. Springer, Cham, pp 255–268. https://doi.org/10.1007/978-3-030-79408-8_17

Akram R, Amanet K, Iqbal J, Fatima M, Mubeen M, Hussain S, Fahad S (2022b) Climate change, insects and global food production. In: Climate change and ecosystems. CRC, Boca Raton, FL, pp 47–60

Ali K, Bajracharyar RM, Raut N (2017) Advances and challenges in flash flood risk assessment: a review. Journal of Geography & Natural Disasters 7(2):1–6

Ali M, Mubeen M, Hussain N, Wajid A, Farid HU, Awais M, Hussain S, Akram W, Amin A, Akram R, Imran M (2019) Role of ICT in crop management. In: Agronomic crops. Springer, Singapore, pp 637–652. https://doi.org/10.1007/978-981-32-9783-8_28

Amin A, Nasim W, Mubeen M (2018) Regional climate assessment of precipitation and temperature in southern Punjab (Pakistan) using SimCLIM climate model for different temporal scales. Theor Appl Climatol 131(1–2):121–131. https://doi.org/10.1007/s00704-016-1960-1

Aslam M, Zamir MSI, Afzal I, Yaseen M, Mubeen M, Shoaib A (2013) Drought stress, its effect on maize production and development of drought tolerance through potassium application. Cercetări Agronomiceîn Moldova 46(2):99–114

Barnes B, Dunn S, Wilkinson S (2019) Natural hazards, disaster management and simulation: a bibliometric analysis of keyword searches. Nat Hazards 97:813–840

Blandford D (2011) The contribution of agriculture to green growth. Report to the OECD, pp 1–36

Cardona O, van Aalst M, Birkmann J, Fordham M, McGregor G, Perez R, Pulwarty R, Schipper E, Sinh B (2012) Determinants of risk: exposure and vulnerability. In: Field C, Barros V, Stocker T, Qin D, Dokken D, Ebi K, Mastrandrea M, Mach K, Plattner GK, Allen S, Tignor M, Midgley P (eds) Managing the risks of extreme events and disasters to advance climate change adaptation. Cambridge University Press, Cambridge, pp 65–108

Dao H, Peduzzi P (2003) Global risk and vulnerability index trends per year (GRAVITY). In: Technical annex and multiple risk integration phase IV. UNDP/BCPR, Geneva

Din MSU, Mubeen M, Hussain S, Ahmad A, Hussain N, Ali MA, Nasim W (2022) World nations priorities on climate change and food security. In: Building climate resilience in agriculture. Springer, Cham, pp 365–384. https://doi.org/10.1007/978-3-030-79408-8_22

Dong Y, Xu F, Liu L, Xiaoping D, Ren B, Guo A, Geng Y, Ruan C, Ye H, Huang W, Zhu Y (2020) Automatic system for crop pest and disease dynamic monitoring and early forecasting. IEEE J Sel Top Appl Earth Obs Remote Sens 13:4410–4418. https://doi.org/10.1109/JSTARS.2020.3013340

FAO (2018) Food and agriculture organization of the United Nations, Rome. http://faostat.fao.org

Fasusi OA, Cruz C, Babalola OO (2021) Agricultural sustainability: microbial biofertilizers in rhizosphere management. Agriculture 11:163

Foley JA, Ramankutty N, Brauman KA, Cassidy ES, Gerber JS, Johnston M, Mueller ND, O'Connell C, Ray DK, West PC, Balzer C, Bennett EM, Carpenter SR, Hill J, Monfreda C, Polasky S, Rockstrom J, Sheehan J, Siebert S, Tilman D, Zaks DPM (2011) Solutions for a cultivated planet. Nature 478:337–342. https://doi.org/10.1038/nature10452

Giannini A, Saravanan R, Chang P (2003) Oceanic forcing of Sahel rainfall on interannual to interdecadal time scales. Science 302(5647):1027–1030

Glade T, Nadim F (2014) Early warning systems for natural hazards and risks. Nat Hazards 70:1669–1671

Hayes M, Wilhelmi O, Knutson C (2004) Reducing drought risk: bridging theory and practice. Nat Hazards Rev 5:106–113

Hossain MS, Arshad M, Qian L, Kächele H, Khan I, Islam MDI, Mahboob MG (2020) Climate change impacts on farmland value in Bangladesh. Ecol Indic 112:106181

Hussain S, Ahmad A, Wajid A, Khaliq T, Hussain N, Mubeen M, Farid HU, Imran M, Hammad HM, Awais M, Ali A, Aslam M, Amin A, Akram R, Amanet K, Nasim W (2020) Irrigation scheduling for cotton cultivation. In: Cotton production and uses. Springer, Singapore, pp 59–80. https://doi.org/10.1007/978-981-15-1472-2_5

Hussain S, Mubeen M, Ahmad A, Masood N, Hammad HM, Amjad M, Waleed M (2021) Satellite-based evaluation of temporal change in cultivated land in southern Punjab (Multan region) through dynamics of vegetation and land surface temperature. Open Geosci 13(1):1561–1577. https://doi.org/10.1515/geo-2020-0298

Hussain S, Amin A, Mubeen M, Khaliq T, Shahid M, Hammad HM, Nasim W (2022a) Climate Smart Agriculture (CSA) Technologies. In: Building climate resilience in agriculture. Springer, Cham, pp 319–338. https://doi.org/10.1007/978-3-030-79408-8_20

Hussain S et al (2022b) Managing greenhouse gas emission. In: Sarwar N, Atique-ur-Rehman, Ahmad S, Hasanuzzaman M (eds) Modern techniques of rice crop production. Springer, Singapore. https://doi.org/10.1007/978-981-16-4955-4_27

Hussain S, Qin S, Nasim W, Bukhari MA, Mubeen M, Fahad S, Aslam M et al (2022c) Monitoring the dynamic changes in vegetation cover using spatio-temporal remote sensing data from 1984 to 2020. Atmos 13(10):1609. https://doi.org/10.3390/atmos13101609

Lenton TM, Held H, Kriegler E, Hall JW, Lucht W, Rahmstorf S, Schellnhuber HJ (2008) Tipping elements in the earth's climate system. Proc Natl Acad Sci 105:1786–1793. https://doi.org/10.1073/pnas.0705414105

Lewis SL, Brando PM, Phillips OL, van der Heijden GMF, Nepstad D (2011) The 2010 Amazon drought. Science 331:554

Mahmood N, Arshad M, Kächele H, Ma H, Ullah A, Müller K (2019) Wheat yield response to input and socioeconomic factors under changing climate: evidence from rainfed environments of Pakistan. Sci Total Environ 688:1275–1285

Marković V, Nagy I, Sik A, Perge K, Laszlo P, Papathoma-Köhle M et al (2016) Assessing drought and drought-related wildfire risk in Kanjiza, Serbia: the SEERISK methodology. Nat Hazards 80(2):709–772

Masood N, Akram R, Fatima M, Mubeen M, Hussain S, Shakeel M, Nasim W (2022) Insect pest management under climate change. In: Building climate resilience in agriculture. Springer, Cham, pp 225–237. https://doi.org/10.1007/978-3-030-79408-8_15

Miah JH, Griffiths A, McNeill R et al (2018) Environmental management of confectionery products: life cycle impacts and improvement strategies. J Clean Prod177:732–751

Miao C, Ding M (2015) Social vulnerability assessment of geological hazards based on entropy method in Lushan earthquake-stricken area. Arab J Geosci 8(12):10241–10253

Mubeen M, Bano A, Ali B, Islam ZU, Ahmad A, Hussain S, Fahad S, Nasim W (2021) Effect of plant growth promoting bacteria and drought on spring maize (Zea mays L.). Pak J Bot 53(2):731–739. https://doi.org/10.30848/PJB2021-2(38)

Mubeen M, Rasul F, Ahmad A et al (2022) Climate change-induced irrigation water problems and resolution strategies: a case study. In: Jatoi WN, Mubeen M, Ahmad A, Cheema MA, Lin Z, Hashmi MZ (eds) Building climate resilience in agriculture. Springer, Cham. https://doi.org/10.1007/978-3-030-79408-8_12

Naz S, Fatima Z, Iqbal P, Khan A, Zakir I, Ullah H, Ahmad S (2022) An introduction to climate change phenomenon. In: Building climate resilience in agriculture. Springer, Cham, pp 3–16. https://doi.org/10.1007/978-3-030-79408-8_1

Niles MT, Mueller ND (2016) Farmer perceptions of climate change: associations with observed temperature and precipitation trends, irrigation, and climate beliefs. Glob Environ Chang 39:133–142

Orlowsky B, Seneviratne SI (2013) Elusive drought: uncertainty in observed trends and short- and long-term CMIP5 projections. Hydrol Earth Syst Sci 17:1765–1781. https://doi.org/10.5194/hess-17-1765-2013

Phillips OL, Aragão LEOC, Lewis SL, Fisher JB, Lloyd J, López-González G, Malhi Y, Monteagudo A, Peacock J, Quesada CA, van der Heijden G, Almeida S, Amaral I, Arroyo L, Aymard G, Baker TR, Bánki O, Blanc L, Bonal D, Brando P, Chave J, de Oliveira ACA, Cardozo ND, Czimczik CI, Feldpausch TR, Freitas MA, Gloor E, Higuchi N, Jiménez E, Lloyd G, Meir P, Mendoza C, Morel A, Neill DA, Nepstad D, Patiño S, Peñuela MC, Prieto A, Ramírez F, Schwarz M, Silva J, Silveira M, Thomas AS, Steege H, Stropp J, Vásquez R, Zelazowski P, Dávila EA, Andelman S, Andrade A, Chao KJ, Erwin T, Fiore AD, Honorio E, Keeling H, Killeen T, Laurance WF, Peña Cruz A, Pitman NCA, Núñez Vargas P, Ramírez-Angulo H, Rudas A, Salamão R, Silva N, Terborgh J, Torres-Lezama A (2009) Drought sensitivity of the Amazon rainforest. Science 323:1344–1347

Ramankutty N, Evan AT, Monfreda C, Foley JA (2008) Farming the planet: 1. Geographic distribution of global agricultural lands in the year 2000. Global Biogeochem Cycles 22:GB1003. https://doi.org/10.1029/2007GB002952

Roco L, Bravo-Ureta B, Engler A, Jara-Rojas R (2017) The impact of climatic change adaptation on agricultural productivity in Central Chile: a stochastic production frontier approach. Sustainability 9(9):1648

Sabagh AE, Hossain A, Islam MS, Iqbal MA, Fahad S, Ratnasekera D, Llanes A (2020) Consequences and mitigation strategies of heat stress for sustainability of soybean (Glycine max L. Merr.) production under the changing climate. In: Plant stress physiology. IntechOpen, London. https://doi.org/10.5772/intechopen.92098

Savary S, Willocquet L, Pethybridge SJ, Esker P, McRoberts N, Nelson A (2019) The global burden of pathogens and pests on major food crops. Nature ecology & evolution 3(3):430–439

Saxena AK, Kumar M, Chakdar H, Anuroopa N, Bagyaraj DJ (2020) Bacillus species in soil as a natural resource for plant health and nutrition. J Appl Microbiol 128:1583–1594

Siebert S, Ewert F (2014) Future crop production threatened by extreme heat. Environ Res Lett 9(4):041001

Sunkad G (2020) The importance of agriculture in present world. https://doi.org/10.13140/RG.2.2.14008.78080

Tariq M, Ahmad S, Fahad S (2018) The impact of climate warming and crop management on phenology of sunflower-based cropping systems in Punjab, Pakistan. Agric For Meteorol 256–257:270–282. https://doi.org/10.1016/j.agrformet.2018.03.015

Teo M, Goonetilleke A, Ahankoob A, Deilami K, Lawie M (2018) Disaster awareness and information seeking behaviour among residents from low socio-economic backgrounds. Int J Disaster Risk Reduct 31:1121–1131

Wilhite DA, Svoboda MD, Hayes MJ (2007) Understanding the complex impacts of drought: a key to enhancing drought mitigation and preparedness. Water Resour Manag 21:763–774

Zahoor SA, Ahmad S, Ahmad A, Wajid A, Khaliq T, Mubeen M, Hussain S, Din MSU, Amin A, Awais M, Nasim W (2019) Improving water use efficiency in agronomic crop production. In: Agronomic crops. Springer, Singapore, pp 13–29. https://doi.org/10.1007/978-981-32-9783-8_2

Zambrano-Mendoza JL, Sangoquiza-Caiza CA, Campaña-Cruz DF, Yánez-Guzmán CF (2021) Use of biofertilizers in agricultural production. In: Technology in agriculture. IntechOpen, London. https://www.intechopen.com/onlinefirst/76918. Accessed 27 Apr 2022

Sustainable Development in Agriculture Beyond the Notion of Minimizing Environmental Impacts

Syed Atif Hasan Naqvi, Ateeq ur Rehman, Sobia Chohan, Ummad ud din Umar, Yasir Mehmood, Ghulam Mustafa, Wajid Nazir, and Amna Hasnain

Abstract

The total economic cost of all-natural disasters has risen 14-fold since the 1950s, and they considerably influence agricultural growth. Geophysical and hydro-meteorological disasters are two categories of natural disasters. Definitions for the different hydro-meteorological disaster categories were presented, including heat waves, cyclones, torrential rains, droughts, floods, and forest fires. Between 1950 and 2005, evidence obtained from various parts of the world showed a rise in the rate of natural disasters. There was extensive information on how natural disasters affected forests, rangeland, and agriculture. Because it increases the risk of natural disasters, environmental damage is one of the main causes, making forestry, rangeland, and agriculture more vulnerable. Traditional concepts of sustainable development focused on finding a balance between environmental concerns and agricultural productivity. The concept of sustainable development must now go beyond the idea of reducing environmental impact; it must also consider managing susceptibility and improving the ability to adapt and respond to natural disasters. Therefore, disaster risk management and mitigation should be a part of a sustainable agricultural growth matrix. The impacts of hydro-meteorological disasters must be significantly minimized via greater usage of

S. A. H. Naqvi (✉) · A. u. Rehman · S. Chohan · U. u. d. Umar · Y. Mehmood · W. Nazir
Department of Plant Pathology, Faculty of Agricultural Sciences and Technology, Bahauddin Zakariya University, Multan, Pakistan
e-mail: atifnaqvi@bzu.edu.pk

G. Mustafa
Department of Agriculture (Extension and Adoptive Research) Punjab, Lahore, Pakistan

A. Hasnain
Assistant Professor, Institute of Management Sciences, Bahauddin Zakariya University, Multan, Pakistan
e-mail: amna.hasnain@bzu.edu.pk

M. Ahmed, S. Ahmad (eds.), *Disaster Risk Reduction in Agriculture*, Disaster Resilience and Green Growth, https://doi.org/10.1007/978-981-99-1763-1_8

weather and climate data and predictions, systems of early warning, suitable land management, and managing natural reserve techniques.

Keywords

Geophysical and hydro-meteorological disasters · Sustainable development · Disaster risk management and mitigation · Systems of early warning · Suitable land management

1 Introduction

Agriculture and climate are intimately related, so climate and weather are the primary factor of agricultural production. Future predictions of climate change due to high carbon dioxide levels indicate temperature increases and rainfall variability, which will significantly impact the agricultural industry. An analysis of food security in Southern Vietnam is presented in this study as a result of climate change throughout Southeast Asia. This interdisciplinary study integrates regional climate models, agricultural science crop models, and risk assessments to create local/regional information products with direct societal applications. The Association of South East Asian Nations region will directly benefit from this study's prospects for multidisciplinary collaboration, which will help it establish effective adaptive methods for food security, planning, risk management, and diversification.

The modern food chain's complex, globalized, and extended network has not kept up with the food supply chain resilience to disturbances. The food supply chain is very much interlinked holistically with its component dynamics, risks, and vulnerabilities to interruptions with the natural disasters. Corporations, communities, and nations all over the world face huge economic challenges as a result of natural disasters. One sector that is particularly vulnerable to natural disasters is agriculture especially the growing impacts of worldwide climate change can be observed in everyday life in the form of global warming, water scarcity and water-related hazards (such as floods and droughts) and the rising temperatures disrupt precipitation patterns and the entire water cycle. Specifically, there has been a strong correlation between extreme weather and natural disasters and the yearly variation in crop output and trade in agricultural items that influence the food supply chain. Food Supply Chain is dependent on climate change and can disrupt food availability, reduce access to food and affect the food quality due to the lack of preparedness of a country to cater such disasters generally and exclusively for the developing countries where natural disasters may increase the intensity of disruptions. So, it is imperitive to identify the challenges and gaps in knowledge which may serve as a next step in the process of increasing food supply chain resilience.

A complex system, agriculture is affected by the interaction of political, social, environmental, and economic forces. Agriculture is generally recognized as one of the sectors most susceptible to climate and weather risks since it is naturally sensitive to climatic conditions. As a consequence of the 14-fold rise in the cost of natural disasters since the 1950s, natural disasters are one of the main development challenges in agricultural meteorology that are also connected to humanitarian aid because they significantly impact agricultural production. Data compiled by the United States Agency for International Development Office of Foreign Disaster Assistance (USAID/OFDA) and the Center for Research in the Epidemiology of Disasters (CRED) in the Emergency Events Database (EM-DAT) indicate that weather-related natural disasters have increased significantly over the past 50 years (CRED 2000). In less advanced countries, there have been an average of 100 to 1600 weather-related disasters over the past 50 years, with an increase from 15 million to 4 billion people affected or killed in the 1990s. These tendencies are a result of the method that natural disasters occur changing, as well as demographic factors that put more people in danger. Natural disasters such as floods, wildland fires, storms, tropical cyclones, and droughts have significantly reduced human life span, destroyed social and economic infrastructure, and harmed the ecosystem during the previous four decades. Höppe (2007) presented results on the increase in the number of significant natural disasters since 1950, which is divided into several risks: geophysical disasters (eruptions, volcanic eruptions, tsunamis, and earthquakes), floods, windstorms, and other weather-related events (droughts, forest fires, and heat waves). Since the 1950s, deaths have climbed by 50% every 10 years, but population growth has only increased by 20% (Kreimer and Munasinghe 1991). Guha-Sapir et al. (2004) calculated that two million deaths and 182 million homeless persons occurred between 1974 and 2003. This chapter "Sustainable Development in Agriculture Beyond the Notion of Minimizing Environmental Impacts" delves into the crucial intersection of disaster risk reduction and agriculture, and urging us to reevaluate our approach to sustainable development in this critical sector. This chapter explored the imperative to transcend the traditional notion of solely minimizing environmental impacts and instead embrace a holistic perspective that integrates disaster risk reduction principles. Agriculture plays an irreplaceable role in providing sustenance and livelihoods to billions of people worldwide. However, this vital sector is increasingly vulnerable to the escalating threats posed by climate change, extreme weather events, and natural disasters. The prevalent focus on minimizing environmental impacts, while essential, has often led to overlooking the necessity of building resilience against unpredictable and catastrophic events. By embracing disaster risk reduction in agriculture, we recognize that sustainability goes beyond the realm of reducing ecological footprints. It encompasses the adaptation and fortification of agricultural systems to withstand shocks, ensuring food security and economic stability in adversity. Employing cutting-edge technologies, sound land-use planning, and climate-smart practices can empower farmers to cope with disaster-induced disruptions effectively. Furthermore, fostering community engagement and awareness is instrumental in developing a culture of resilience. Local knowledge, combined with scientific expertise, forms the bedrock of effective disaster risk reduction strategies, tailored to the unique challenges faced by different regions. Hence, this chapter highlights that it is evident that sustainable development in agriculture must be grounded in a multidimensional framework that accounts for the intricate balance between human needs, environmental preservation, and disaster resilience while embracing this comprehensive approach empowers us to harmonize agriculture's productivity with ecological integrity, ensuring the long-term viability of our food systems.

2 Natural Hazard-Induced Disasters (NHID)

Extreme weather, floods, fires, earthquakes, and droughts are examples of natural hazard-induced disasters (NHID) that can have significant, all-encompassing, and hard effects on the agricultural sector by severely damaging farmland, infrastructure, and agricultural assets, as well as by causing crop and livestock losses. NHIDs can also result in the loss of livestock and crops. The effects on animal welfare and social well-being may also result in costs. Natural disasters have mostly damaging effects in developing countries, as they can seriously inhibit the development of sustainable development and negatively influence rural lives, agricultural growth, and food security. From 2008 to 2018, significant disaster losses worldwide were concentrated in lower-middle-income and least developed nations, where agriculture accounts for a substantial part of informal and formal financial activity (FAOSTAT 2021). A yearly threat to agricultural output is presented by natural disasters including earthquakes, floods, tornadoes, storms, and fires. Agriculture is readily affected by natural disasters and occurrences since it depends on the surroundings, the climate, and the availability of water to survive. The most common agricultural effects of natural disasters and events are damaged water supplies, decreased livestock or harvest, greater susceptibility to disease (plant diseases or animal diseases or human diseases), and damage to irrigation systems and other agricultural infrastructure.

Agricultural production can experience enduring impacts, encompassing crop growth, forest development, and the enhancement of arable lands, which necessitate sufficient time to mature. To mitigate and recover from the long-term effects of natural disasters and events on agriculture and the environment, it is essential to gain insights into effective mitigation strategies. Even though natural disasters and events may severely hamper agricultural productivity, it is imperative to uphold regional, state, and federal environmental laws without compromise as there are resources and assistance available to help in planning, preparing for, recovering from, and responding to natural disasters. It is crucial to recognize that while the impact of a single natural hazard occurrence can be substantial, disaster risk, as a whole, is becoming more intricate and interconnected. The effects of weather-related hazards are projected to intensify and occur more frequently due to climate change. Moreover, the interplay of natural disasters, such as strong storms, insect outbreaks, and droughts, has exacerbated the impacts of the COVID-19 epidemic on agriculture in several nations (FAOSTAT 2021). Due to the interrelations between agriculture and other industries, natural disasters may also have an impact on agriculture and interfere with farm operations by affecting or shutting down vital infrastructure, including drainage and irrigation systems as well as transportation networks that link farmers to logic and markets (FAOSTAT 2018). These same relationships also suggest that, even though lower output is the most obvious consequence, effects may spread across the whole agro-food value chain, damaging rural economies and posing a risk to all aspects of food and nutrition security, particularly in the impacted areas. Disaster risk management (DRM) framework that improves farmers' ability to predict, recover from, respond to, absorb, and successfully adapt to and transform in terms of natural disasters is important for improving agricultural resilience to natural hazards. This applies to more generalized systemic hazards affecting the agriculture sector, such as those brought on by the COVID-19 epidemic and natural disasters. Effective DRM requires that both private and public stakeholders are aware of the dangers that natural disasters pose to their assets and operations and that they take ownership of mitigating those risks. When deciding how to manage disaster risk to promote resilience, it also relies on solid governance structures.

3 Natural Disasters, Agriculture, and Losses

Natural disaster losses have significantly increased (Höppe 2007). The frequency of major natural disasters doubled in the second half of the twentieth century, and annual monetary damages increased by a ratio of more than six (Munich Re 2006). All weather-related natural disasters reached \$1 trillion in economic expenses between 1980 and 2003, with affluent and developing countries bearing between 40% and 60% of the cost, respectively (Munich Re 2004). The increase in the relative frequency of weather-related disasters likened to those not connected to the weather (such as earthquakes or volcanic eruptions) is particularly striking, even though all losses have grown in absolute terms (Vellinga et al. 2001). Inflation-adjusted worldwide insured and economic losses have grown exponentially during the 1950s, reaching a record high in 2004 before being exceeded once more by new loss records in 2005. Record losses of around \$178 billion in dollars, or 0.7% of world gross domestic product (GDP), were reported in 1995, the year of the Kobe earthquake in Japan (Munich Re 2002). Hurricane Katrina in 2005 resulted in the greatest loss from a singular event in history, with economic damage of \$60 billion and net economic losses of \$125 billion. Agriculture and the rural sector are significantly linked to practically every other sector of the economy as sources of labor, food, markets, and raw materials. In reality, the proportion of agriculture in a country's GDP, employment levels, and imports/exports increases with the increase in the level of poverty. Natural disasters, such as floods, droughts, hurricanes, and earthquakes, can have devastating impacts on communities, particularly in impoverished regions. When a country experiences a natural disaster, it can exacerbate poverty levels by destroying homes, infrastructure, and agricultural assets, disrupting livelihoods, and leading to food insecurity. When a country experiences higher levels of poverty, it often means that a significant portion of its population is engaged in subsistence farming or small-scale agricultural activities. In such economies, agriculture becomes a critical sector, contributing a larger share to the overall GDP. As industrial and service sectors may not be as developed or accessible to the impoverished population, agriculture becomes a primary source of income and sustenance. In impoverished regions, where job opportunities in other sectors are limited, agriculture tends to be one of the few available sources of employment. As people struggle to find alternative jobs, they may turn to agricultural activities for livelihoods. This leads to an increase in the number of people employed in the agricultural sector. Poverty often leads to a lack of resources and limited industrial development in a country. As a result, the agricultural sector may not be equipped with advanced technologies or infrastructure to produce high-value products. Consequently, these countries may rely heavily on imports of agricultural goods to meet the needs of their population. On the other hand, they might export raw agricultural commodities or low-value products, which may be the most accessible means for generating income and foreign exchange. One of the major factors impacting the danger of natural disasters is rural poverty. It is disturbing that in the least developed countries (LDCs), agricultural production has not kept pace with the population increase. Although agricultural output increased by an average of 2.5% annually from 1990 to 1999, above the ratio of 1.6% in the prior years, there was essentially no increase in output in terms of per capita, and even a slight decrease occurred (IFAD 2001). Only five nations had positive growth between 1990 and 1999, whereas the per capita growth rates of more than 25 LDCs were negative. Around 75% of the extremely poor currently live in rural areas and depend on agriculture and related industries for their livelihood, according to estimates from McCalla and Ayres (1997) and the IFAD (2001). The extreme poor are predicted to be significantly centered in rural areas between 2020 and 2035, even under optimistic estimates for economic growth and rural-to-urban migration. The conclusion is that people living in LDCs are most susceptible to natural disasters due to their high rates of poverty and low agricultural output. Economic activity disruptions and the redirection of government resources to prepare for and respond to natural disasters have an influence on development. According to UNISDR (2003), developing nations are more affected economically by natural disasters than developed nations; between 1985 and 1999, the expenses of natural disasters in the developing nations were around 13% of GDP, compared to only 2% in the developed nations. In 1998, the Honduran Prime Minister provided a stark example of the potential negative consequences of major weather disasters when he said that Hurricane Mitch, which may have killed 20,000 Central Americans, pushed the economics of the country growth back 20 years (IFRC/RCS 2003). Hurricane-related damages reached 50% of the total GDP in Honduras (Hooke 2000).

Because the majority of the economic effects on this economy are the result of relatively "minor" disasters, the agriculture sector has not gotten the attention it needs from policymakers when mitigating the effects of natural disasters. Disasters that make headlines are often observed by the public and decision-makers. Swiss Re (2002) suggested unspectacular climate anomalies, which the general public perceives as "weird" weather conditions rather than "disastrous," might cause losses on the scale frequently associated with natural disasters. Due to both the changing nature of natural disasters and society's growing susceptibility to them, dealing with such climatic anomalies is becoming more expensive (IPCC 2001). The poor farmers bear the costs that are not covered by national governments, international help, or insurance. The most marginal lands are occupied by the poorest rural residents, forcing them to rely on insecure and extremely susceptible means of subsistence in regions subject to natural calamities like floods, droughts, etc. (UNDP 2004). The most susceptible social groups and nations include those with the fewest resources, knowledge, and skills, as well as those with the least access to technology, weak or unstable organizations, and unequal access to resources and empowerment (Smit et al. 2001). Re-evaluating the topic of sustainable agricultural growth is important in light of the above-mentioned issues. Agriculture production and environmental concerns are balanced in traditional definitions of sustainable development. According to Swindale (1988), sustainability refers to a balance between environmental concerns and human needs. While preserving natural resources, sustainable agricultural systems should meet the demands of both the present and future generations (National Research Council 1991). The improvement of careful usage and environmental source quality that agriculture depends on is seen as a requirement for sustaining agricultural productivity (American Society of Agronomy 1989). The concept of sustainable development must now go beyond the idea of reducing environmental impact; it must also take into account controlling susceptibility and improving the ability to adapt to and react to natural disasters. In this sense, disaster risk management and mitigation should be a part of the matrix for sustainable agricultural growth.

4 What Is Naturally Occurring Disaster?

A natural disaster is a natural phenomenon that has disastrous effects on nearby living things. Although there are several definitions of natural disasters, some of which are more heavily weighted toward death rates. The Centre for Research on the Epidemiology of Disasters (CRED) maintains the emergencies database (EM-DAT), which classifies events as disasters if at least "10 persons are killed and/or 100 or more are affected and/or a request for international aid is made and/or a state of emergency is declared" (CRED 2000). It is obvious that the only portion of this definition that is relevant to agricultural purposes is the last one. In a disaster training program from 1992, the United Nations (UN) defined a disaster as "a substantial disturbance of society's functioning, creating extensive human, environmental, or material losses that exceed the capacity of the affected community to manage using just its own resources." Agriculture might be capable of using this definition if certain portions are properly understood. Natural disasters, according to Anderson (1990), are temporary phenomena brought on by natural hazards that exceed local responsiveness and have a significant impact on a region's social and economic development. Disasters, according to Susman et al. (2019), are the result of the interaction between a susceptible human population and a severe physical environment. Such definitions emphasize the fact that political and socioeconomic factors play a role in explaining why communities are sensitive to their surroundings and suffer natural disasters. Geophysical disasters and hydro-meteorological disasters are categorized as natural disasters by the International Federation of Red Cross and Red Crescent Societies (2003). Droughts, famines, landslides, avalanches, windstorms, extremely high temprature, floods, hurricanes, heat waves, forestscrub fires, and others (waves, surges and pest attack) are a few examples of hydro-meteorological disasters. Earthquakes as well as volcanic eruptions are two examples of geophysical disasters. A prolonged period of time, often a season or more, with consistently lower rainfall results in a drought. Drought is sometimes accompanied by additional climatic variables (including low humidity, strong winds, and extreme temperatures) that can increase the intensity of the event. Drought results from the interaction between human demand for water supply and natural water availability. It is not a simple physical occurrence. Political factors make the exact definition of drought complex; however, there are commonly three sorts of situations referred to as drought, including agricultural drought, which occurs when there is not enough rainfall for average crop or range production. Meteorological drought is caused by a prolonged period of below-average precipitation. Due to soil conditions or agricultural activities, this situation can develop even during periods of average precipitation. Similarly, a hydrologic drought occurs when the water supplies in lakes, aquifers, and reservoirs fall below the numerical average and when more water is used than supplies available, even during periods of average (or above average) rainfall, this condition might emerge.

A heat wave is a protracted period of very hot weather that may also include unusually high humidity. Temperatures that individuals from a hotter region find typical may be regarded as heat waves when they are beyond the normal for a cooler area since the term is relevant to the particular weather pattern. The word is used to describe both "ordinary" weather changes and very hot spells that may only occur once every century. Flood refers to a situation that happens when water overflows a stream's or other body of water's boundaries, whether naturally occurring or created artificially, or accumulated by draining over low-lying areas. Several different events, such as waves, lahars, precipitations, river overflows, tsunamis, storm surges, mudflows, the failure of water-retaining constructions, groundwater leakages, and water backup in sewage systems, can result in a flood, which is a brief inundation of generally dry land by water, suspended matter, and/or debris. A forest fire (also known as a bushfire in Australia) is an unfortunate fact of life that starts in vegetation that is taller than 6 ft (1.8 m). These fires can occasionally be started by burning and heat from surface and ground fires, and they often grow to the size of a large conflagration. Typhoons, hurricanes, and tropical cyclones are various regional names for the same phenomenon. In Bengal, parts of the South Pacific, Southwest Indian Ocean, the Arabian Sea, and along the northern coast of Australia, depressions in the tropics that develop into storms are referred to as tropical cyclones. Typhoons are the name given to these storms in the northwest Pacific, whereas hurricanes are the name given to them in the southeast United States, the Caribbean, Southeast United States, and Central America. An earthquake, a seaquake, volcanic activity, a slump, or a meteorite impact in or near the sea can all result in a series of extremely large waves known as a tsunami. The term "tsunami" (Japanese meaning "great wave in port") is sometimes used incorrectly as a tidal wave. Since the tsunami's constant energy is determined by its height and speed, where the wave reaches land, its height increases as soon as its speed decreases. The waves move quickly, mostly unnoticed when crossing deep water, but they can reach heights of meters or more when they get close to land. On beaches and islands, tsunamis may seriously damage the environment.

5 Disasters Damage in Range Land and Forests

Natural disasters can positively or negatively impact forestry, rangeland, and agriculture. Although the impacts are primarily negative and have a large negative impact on human society, there are certain possible benefits or positive effects (Joy 1991). According to Das (2003), natural disasters can directly or indirectly affect forestry, rangeland, and agriculture. Direct impacts result from the extreme hydrometeorological event's physical damage to plants, animals, and crops. The effects might range from short-term, temporary harm at a specific crop phase to complete crop loss. Within hours after their occurrence, natural disasters bring actual harm to agriculture in the form of destruction of crops that are ready for harvest as well as of entire or partial farm buildings, equipment, machinery, installation, and transport and storage systems. Disasters can also result in indirect damage, which refers to lost potential production as a result of disturbed supply flows for services and goods, lost production capacity, and higher production costs. Low incomes, decreased production, environmental damage, and other disaster-related factors gradually cause these indirect effects to emerge (Das 2003). The impacts of natural disasters can be classified as tangible or intangible, according to Anaman (2003). Tangible impacts are those that can be readily measured in monetary terms. Intangible effects like anxiety or fear of upcoming natural disasters, inconvenience and disruption of farm work, stress-related diseases, and intangible impacts are frequently challenging to quantify in monetary terms because they are not bought or sold in clearly defined markets and therefore do not direct market values.

Famines were a frequent occurrence in pre-twentieth century Asia and Europe due to natural disasters such as drought, excessive cold, diseases, and pests that damaged crops and livestock (Devereux 2000). The ability of an economy to produce income is negatively affected over the long term by the loss of permanent crops like forests or banana trees. Agriculture-related assets might experience a permanent or temporary loss (Charvériat 2000). In contrast to hurricanes, which have the power to wash fertile soil or permanently increase its salinity via flash floods and storm surges, floods render a place unusable for agriculture till the waters recede. Examples of indirect effects include the need to evacuate people in the case of a cyclone landfall, disruptions to houses, stress-related diseases, and worry (Handmer and Smith 1992; Anaman 1996). Natural disasters do the most damage in developing countries. Poor people are especially exposed, as Devereux (2000) said, since they frequently live in marginal areas, depend on high-risk, low-return livelihood systems like rain-fed agriculture, and are susceptible to a variety of economic risks, such as a lack of physical infrastructure. According to the UNDP (United Nations

Development Programme), 24 of the 49 least developed countries are in higher danger of natural disasters. Between two and eight major disasters have hit at least six of them annually during the past 15 years, with long-term implications for human progress (UNDP 2001). Although the absolute value of losses from natural disasters is higher in advanced nations, GDP loss rates in developing countries are 20% higher (Funaro-Curtis 1982). Beyond the indirect or direct losses, the economic implications are highly significant because of their impact on the countries' economy (foreign trade, GDP, price indices, and public finances). Due to its significant role in developing a country's capital and population requirements, the agriculture sector is considered extremely vulnerable. For example, agricultural activity contributed 30.9% of Bangladesh's Gross National Product (GNP), compared to 44.6 and 54.3% in Cambodia and Laos. Vos et al. (1999) estimated that during the last El Niño in Ecuador, 12,000 employees temporarily lost their occupations in sugarcane and banana fields in the lowlands. According to the firm, the unemployment rate in Honduras as the instant result of Hurricane Mitch had risen to an assessed 32%.

6 Natural Disasters and Economics

The economic implications also affect foreign trade-related operations, which are now essential due to national debt. It is considered that industries like agriculture, manufacturing, crafts, and tourism produce the foreign currency needed to keep the payments balance in check. Even more substantially, agricultural export items are important. Due to their position on the principal deltas and on the coastal plains, free zones are more likely to be affected by storms and floods. The 1991 cyclone seriously damaged Bangladesh's Chittagong Free Zone (Normand 1991).

7 Can Natural Disasters Impact be Minimized?

The areas with the least access to technology resources and infrastructure development are those that are most susceptible to climatic extremes, natural disasters, and perhaps climate change. Because each nation must be aware of the entire scope and impact of natural disasters, especially those that are geographically smaller countries, it is not reasonable to expect them to manage on their own. Although socioeconomic damages cannot be completely eliminated, they can be reduced with timely and efficient mitigation measures. The need to reduce the effects of droughts and floods was emphasized at the World Summit on Sustainable Development (WSSD), which was held in Johannesburg in 2002. This was done by implementing strategies such as better weather and climate information use, agricultural practices, early systems of warning, prediction, ecosystem conservation, and land and natural resources management. This is because the current trends in land and water resource deterioration are not sustainable, and reversing them will reduce their impact. The World Summit on Sustainable Development recognized the importance of enabling the transfer of technology for early detection methods and mitigation programs to underdeveloped nations affected by natural disasters.

7.1 Improved Use of Climate and Weather Information and Forecasts

It is not fair to expect growers and sensible solutions to be put into action because the interaction between agricultural production and the weather is so complicated (Wenjia and Hao 2012). Every year or season will carry its own distinct set of conditions, farmers must make choices based on each individual circumstance. Therefore, a cooperative strategy involving farmers, agricultural extension organizations, and the National Meteorological and Hydrological Services (NMHSs) is required. Knowing the influence of climate and weather variables on sustainable agriculture output is one fundamental requirement. This awareness is strong in many situations, and many farmers frequently search for intelligent as well as low-risk solutions. Farmers' interest in assessing the products predicted by the NMHSs should be stimulated by this. Significant progress has been achieved in the research and applications of climate pattern forecasting over the last two decades. The main scientific postulate of seasonal forecasting is that considerable predictability of atmospheric processes can result from lower-boundary forcing, which develops more slowly than weather systems. Some of these initial conditions include the surface temperature and albedo, sea ice cover and temperature, sea-surface temperature (SST), soil humidity, and snow cover, however, not all of these are often considered to be of equal value. Climate variations, usually referred to as anomalies, are changes from the climatic system's typical state for that period (averaged over a number of years, frequently during a 30-year period). The effect of persistent SST anomalies on weather systems, which in turn leads to seasonal climatic anomalies, provides the bulk of the strongest evidence for long-term prediction. The main meteorological variables that impact crop forecasts are rainfall, temperature, and the sun's radiation. Because slowly changing ocean variability substantially affects changes in weather statistics, seasonal climate predictions can provide insight into the future evolution of the climate across timescales of seasons and longer, as indicated by Doblas-Reyes et al. (2006). The field of climate forecasting may now offer a complete, integrated multi-scale (in time and space) prediction system that delivers skillful, helpful forecasts of variables of socioeconomic importance. Using mathematical models of the climatic system, seasonal predictions can be generated. Climate forecasting at the national and regional levels uses a variety of forecast methodologies, including both dynamical methods and empirical-statistical techniques. Operational empirical-statistical techniques that are based on statistical relationships between current observations and future weather conditions include the analysis of general circulation patterns, canonical correlation analysis, time series, analog methods, multiple linear, regression, correlation, and the analysis of climatic anomalies, discriminant, and the optimal climate normal linked to El Niño-Southern Oscillation (ENSO) events. The atmospheric General Circulation Models (GCMs) in a two-tiered prediction system or the dynamically connected atmosphere-ocean GCMs are used in dynamical methods, which are model-based. They are mostly used in large international climate prediction centers. These equation-based dynamical prediction models predict the responses of the global climate system to initial atmospheric conditions and boundary forcing from the underlying land and ocean surfaces. They are an expansion of the numerical techniques used to forecast the weather a few days in advance. Forecasts of future precipitation trends made 3 months or more in advance, may be crucial for land management, forestry, and agriculture because they may be able to predict droughts or heat waves. In terms of preparing to assist shortages of food, lessen the effects of droughts, and ensure energy distribution, these outlooks are strategically important to national strategy. Seasonal predictions are presently effectively utilized at the farm level in industrialized nations to modify seasonal crop planning, while there is still work to be done to get such information relevant for farmers in low-input systems (Meinke and Stone 2005; Salinger et al. 2005). The MARS (Manufacturing Architecture for Resilience and Sustainability) program of the European Union, which has been expanded to the African areas, imply that seasonal predictions are already employed in poor nations for output forecasting to guide policy decision-making (Hansen and Indeje 2004; Rojas et al. 2005).

7.2 Early Warning Systems

The effective systems of early warning and risk assessments that provide decision-makers and the population at risk with timely and reliable data are essential to disaster preparation. Natural disasters cannot always be avoided, but many of them may be prevented by combining risk analysis, early warning systems, and mitigation efforts. This indicates that action may be done to significantly minimize the resulting life's loss and socioeconomic damage. Without doubt, a reliable early warning system that can provide people at risk with correct information quickly and dependably is a precondition for disaster preparedness. The significance of early warning systems is becoming more popular and widespread. The Hyogo Framework for Action 2005–2015 (HFA) was accepted by 168 countries during the Second World Conference on Disaster Reduction (Hyogo, Kobe, Japan, and January 2005). A critical component of disaster risk reduction is identifying, evaluating, and monitoring disaster risks and improving early warnings, the second high-priority area. Over 7000 natural disasters occurred between 1980 and 2005, which claimed the lives of approximately two million people and caused more than one trillion dollars in economic losses. The number of people killed decreased over time; however, the frequency of disasters and their economic effects increased. There has been an almost fourfold increase in the disaster number associated with weather, water, and climate-related hazards and an almost fivefold rise in economic losses, but a nearly threefold drop in fatalities. A number of factors contributed to this remarkable success, including the growth of customized end-to-end initial systems of warning (Jarraud 2006). In order to improve early warning capabilities, World Meteorological

Organization (WMO) interacts with its partners at the regional, national, and international levels. Especially those countries with limited resources should be able to make use of these systems. The WMO's scientific initiatives have been important in advancing our understanding of the climate system. By using defined procedures, systematic observations have been conducted throughout the world. These observations have produced information that may be used to analyze, research, and simulate the environment and its changing weather patterns. The World Meteorological Organization (WMO) maintains a global network for collecting and disseminating observational data as part of its World Weather Watch Programme. The system's 10,000 ground stations, 1000 top stations, 7000 ships, and 3000 aircraft, along with a network of 16 environmental, operational, and meteorological satellites, enable the collection of more than 150,000 observations every day. One hundred eighty-seven National Meteorological Centers, 35 Regional Specialized Meteorological Centers, and 3 World Meteorological Centers are all coordinated by WMO. A better understanding of the relationships between climate variability and change as well as the domains of the climate system, has been made possible by specialized observational programs, such as those for chemical components of the atmosphere and characteristics of the oceans and their circulations. New technologies have advanced our climate system knowledge during the past several years. Numerous specially equipped commercial aircraft, ocean buoys, expendable bathythermographs, operated and automatic weather locations on land, and satellites for issues are considered of the oceans and sparsely populated regions of the world, all adding to the volume of data and the knowledge base.

Disaster prevention can only be successful in regard to any kind of threat, such as flash floods, if there is the adequate lead time for the specific measures to be done in order to save lives and reduce the damage. Lead times can be prolonged more than they could with simply the use of radar systems due to the numerical of weather data prediction (NWP) products. Modern state-of-the-art technologies include enhanced ground- and space-based monitoring systems, more accurate models, and the required telecommunications equipment to rapidly convey data to warning centers and forecast. This is especially true for moderate weather prediction, where the development of composite prediction systems (EPS) has enabled one to assess the degree of forecast uncertainty. To prolong the lead time, particularly in developing nations, such systems must be employed extensively and customized to local conditions. WMO's Flood Forecasting Initiative, which aims to produce quicker and more efficient flood prediction products and services, was started with the goal of improving flood forecasting. The National Meteorological Services (NMSs) and the National Hydrological Services (NHSs) are cooperating well on this, and disaster managers and NMSs are working together to improve capacity. Over the past several decades, there have been substantial improvements in science and technology as well as in the timeliness and accuracy of flood and weather warnings. Today's large-scale weather pattern predictions are just as accurate for 7 days out as they were for 2 days out 25 years ago (Sivakumar 2005). It is feasible to provide extensive information on expected weather systems several seasons in advance, and predictions up to 10 days in the future are also highly accurate. For example, improved national planning made possible by early detection of El Niño occurrences has had a positive impact on several sectors of the economy, including the management of water resources, agriculture, tourism, and fisheries (OP 1996). Thanks to developments in the study of El Niño and in the monitoring of sea-surface temperatures in the Pacific Ocean, scientists at the NMHSs were able to foresee the emergence of the 1997–1998 El Niño event more precisely than any previous occurrence in the past. The Internet and other recent advancements in communication technology have made it possible to quickly and effectively spread information about El Niño throughout the world. These have encouraged global collaboration and coordinated efforts to deal with the resulting repercussions, enabling a number of governments to put the required measures into place. Over the past several years, forecasts for the path of tropical cyclones and the timeliness of warnings have gradually improved. A lot of lives have been saved, and significant amounts of property damage have been prevented because of international efforts, particularly those made under the framework of the WMO's Tropical Cyclone Programme. Government sources, for example, attribute a substantial portion of the decline from over 130,000 to 500 in the death toll in Bangladesh due to developments in evacuation and early warning systems in 1991 and 1994, respectively (Obasi 1997). Information on global and regional climate monitoring and prediction has proved to be exchanged more easily because of the developing Internet. However, a lot of users need help choosing, interpreting, and using the right information. The potential human loss from these disasters has been decreased, thanks, to efficient early warning systems and community education for preventive action. Floods as a disaster also make for a good candidate for both legislative and structural preparation measures (urbanization, zoning plans, and land use laws). In these high-risk areas, preparation for life-saving measures and evacuation strategies should be aggressively encouraged.

7.3 More Efficient Management of Land and Resources

Extreme temperatures and minimal precipitation in arid regions during prolonged natural disasters such as droughts result in poor organic matter production and rapid oxidation. Low organic matter prevents soil from aggregation, increasing the risk of wind and water erosion. For example, there is significant wind and water erosion across the majority of Africa. Along with the present desert, which makes up around 46% of the land mass, about 25% of the land is prone to erosion by water, and about 22% is prone to erosion by wind. On the other hand, due to the force of the rainfall, subsurface and surface drainage, and river flooding, intense thunderstorms, such as those that happen during hurricanes, can erode the soil. Rapidly striking raindrops on soil surfaces generate significant amounts of kinetic energy, which may be exploited to shift soil particles. The easily soluble soil material that is made water solvable by the weak acids in the rainwater can also lead to erosion at this microscopic scale. The initial breakdown and smearing of soil particles by raindrops is only the principal step of the process; afterward comes water's additional erosion by washing away soil particles. The greater the soil particles transported away, the more severe the rainfall and consequent surface runoff. It should be clear from the preceding that natural disasters significantly affect the current agricultural making methods and the soils, requiring the adaptation of farm technologies and management practices to keep soil functions for crop production and ensure defensible agricultural output. Agricultural activities can have a major short-term impact on soil functioning in areas that are regularly susceptible to natural disasters like floods and droughts, and agricultural technology and management can be essential in these processes. Poor irrigation techniques and using drainage with a lot of salt, for example, might make soils too salinized to be used for agricultural production. Other examples include overgrazing in semi-arid areas like the Sahel, which can result in desertification and wind erosion for various reasons. Heavy machinery usage on a regular basis mixed with slow growing crops and soil cover causes soil compaction in temperate areas with high input systems, which can enhance runoff, decrease water penetration, and create runoff. There are several chances to lessen the effects of natural disasters offered by both modern agricultural technology and those that have been used for indigenous technologies for many generations. The efficiency of farm technology is becoming ever more important for the efficiency of different systems of agricultural production at a variety of input levels due to the expected climate change (Sivakumar et al. 2005). Farmers can alter crop management, such as adjusting the sowing date in accordance with the expected seasonal weather, in addition to modifying the crops and cultivars that are employed. The seasonal rainfall pattern is one of the most crucial pieces of information for farmers in semi-arid locations that practice rain-fed agriculture, especially for low-input systems in poor nations (onset of rain, distribution during crop-growing period, and duration of rainy season). As a result, they are able to change their crop selection and planting dates (Stigter et al. 2005; Ingram et al. 2002; Mati 2000). According to Matthews et al. (1997), moving from single cropping to double cropping in high latitudes when increasing temperatures enabled longer projected crop-growing seas may substantially impact regional production. Considering that grain production and ripening periods are pushed to less favorable circumstances later in the season, two short ripening cultivars may be chosen over a longer ripening variety. Unmanageable usable water and competition for water supplies in agriculture are the results of the continually increasing water demand and the slowly growing water supply. This trend has major implications for sustainable agricultural growth, particularly in developing nations. In both developed and developing nations, effective water resource management via suitable farming technology plays and will continue to play a significant role in regions with limited supplies for agricultural production. For example, irrigated agriculture, which was initially used in the ancient Mediterranean, has subsequently developed as a result of experience. However, the majority of Mediterranean nations have used the same irrigation methods for generations. In many regions of Spain and Egypt, for example, ineffective flooded irrigation techniques are still in use (El-Gindy et al. 2001; Neira et al. 2005). In certain Mediterranean European countries, like Spain, expensive modern drip irrigation and sprinkler systems have been installed (Agricultura 1999). The consumption of water is drastically decreased by these innovative methods. In spite of Spain's lower overall production over the previous 15 years compared to nations like Egypt, irrigated crops like maize has seen higher yield. Although there may be various reasons for the differences between Spain and Egypt, the new mechanical irrigation substructures built in Spain have significantly impacted the yield improvements observed (Eitzinger et al. 2007). To grow and improve rain-fed agriculture, watershed management must be improved by constructing structures for spreading, collecting, and storing water and using supplemental irrigation methods. In the coming years, methods like "deficit irrigation" should be taken into consideration as an option; otherwise, irrigated agriculture would become costly (Fereres and Soriano 2007). In addition, it is important to take the necessary steps to reduce irrigation water losses during transportation and on-farm use. Massive resources containing a range of water-quality groundwater aquifers should be managed in accordance with established standards. However, these resources are mainly nonrenewable.

8 Recommendation to Follow to Protect Farm During a Disaster

The weather significantly impacts many of the difficulties that farmers face. The land is permanently altered by hurricanes, floods, fires, droughts, and disease outbreaks, affecting the farms, ranches, and people that depend on their success to support our nation. Creating a disaster management strategy can help farms in overcoming Mother Nature's most severe natural disasters and minimize the loss of infrastructure, livestock, or crops. Expect the unexpected and make plans for it, suggests Mac Miller, Senior Vice President of Ag America Lending (02-Dec-2022, https://agamerica.com/blog/tips-to-prepare-farm-for-natural-disaster/).

8.1 Recommendation-1: Identity Higher Ground Ahead of Flooding

According to North Dakota State University, the first step in lowering floodwater losses is to include flood-fighting equipment in your agricultural supplies, such as wire, plywood, plastic sheeting, sandbags, and rope. The capacity of farm operators to react swiftly to risky situations or rising water is improved by having these requirements available. Additionally, farms can reduce risks by (1) moving priceless and potentially dangerous farm equipment to higher ground, such as livestock, equipment, pesticides, fuel, grain, and feed. Include parts and tools as well. (2) Keep vaccines updated, especially for livestock with exposure risk to flood waters; (3) Increase your fuel supply and store it safely above any flood-prone areas. (4) Turn off the electricity in any structures near floodways, (5) secure any earthen manure storage facilities, and (6) take an evaluation of the livestock, machinery, and dangerous chemicals on your farm.

8.2 Recommendation 2: Implement Land and Water Management Strategies During a Drought

The farm will be ready when a period of dry turns into a drought due to year-round efforts to save water resources. The USDA advises "considering how crops, forage, and other resources have responded to drought in the past and incorporating your strategy into a comprehensive conservation project that considers the types and conditions of all of your resources." One of the most important decisions you should make is to: (1) assess the irrigation systems that better suit the demands of the farm. Utilize resources that measure water usage to lessen water loss due to evaporation, precipitation, and runoff; (2) Build a storage system or store water in field ditches for use during irrigation season; (3) Use conservation tillage to maintain soil moisture; and (4) Implement conservation techniques that promote water infiltration.

8.3 Recommendation 3: Prepare for a Hurricane by Stocking Up on Essentials

According to a University of Florida Extension Specialist, farms and ranches should securely store enough essentials to maintain operations under restricted conditions for at least 1–2 weeks after a hurricane. Farmers must be ready for a range of damages, including wind, flood, fire, and even mold, due to the unpredictable nature of a hurricane. The University of Florida suggests doing the following before hurricane season to help protect your business from harm during a storm: (1) Stock up on enough supplies for animals to last up to 2 weeks, such as hay, feed, medicine, and other medical supplies, (2) Run generators frequently to check their functionality, (3) Check the chainsaws' functionality and have extra fuel on hand. (4) Keep an extra supply of lanterns, flashlights, and batteries for every member of the family and every employee. (5) Be careful to maintain a supply in busy areas and in places you will need to go right away after the storm. (6) Keep a supply of fencing materials on hand in case repairs are needed. (7) Create and keep conservation buffers around water sources, such as grassed streams, filter strips, and riparian buffers. (8) In the case of a drought, ensure that cattle and other grazing animals have access to secure alternative feed sources.

8.4 Recommendation 4: Recognize and Relocate Fire Hazards

To best prepare for a wildfire, the fire wise program at Michigan State University Extension suggests having an emergency recovery plan, action plan, and preplan, i.e., (1) be aware of common fire hazards on farms, such as flammable liquids, straw, hay, as well as animals that cannot escape from fenced-in areas. (2) observe forecasts and recognize "fire weather," (3) establish in advance locations for the water supply, fueling, and evacuation of both humans and livestocks; (4) instruct agricultural workers on proper firefighting techniques; (5) Acquire knowledge about how to handle crops and land following exposure to heat. For example, according to Michigan State University Extension, watering crops right away after a wildfire might pH-shock certain plants and have a long-term negative impact on yield.

8.5 Recommendation 5: Utilize Biosecurity Measures to Cater Disease

The chance of a disease spreading widely can be increased by the movement of people, vehicles, and other livestock on and off the farm. Everyone, from the farm owner to a worker or even a traveling agribusiness representative, is responsible for biosecurity. Livestock must be protected by paying close attention to high-risk regions, especially throughout a natural disaster. (1) Monitoring animals and humans who access the farm, as well as inputs like fertilizer and feed. And water is essential when supplies are few or have been polluted by water or debris. (2) Maintaining good cleanliness on the farm to prevent the transmission of disease and pests through contact with feed, water, and packaging. (3) Take active measures to eradicate weeds and wild animals that may contain zoonotic and plant diseases. (4) Educating staff on best procedures and mandating the keeping of records for all animal purchases, sales, and movements in order to track any outbreaks, (5) Maintaining your agricultural enterprise during a natural disaster can be difficult, even with the finest preparation. If you require aid, consult the USDA's Disaster Resources Guide to locate programs that offer financial or material support in response to your operation's needs during a disaster.

9 Damage, Destruction, and Disaster

Agriculture accounts for 23% of all losses and damages from natural disasters in countries like Canada every year. In reality, between 2005 and 2014, agricultural productivity in developing nations lost $116 billion. Who knows how much the entire world would have lost! These disasters, in fact, have the power to devastate vital agricultural resources, including infrastructure, production cycles, trade flows, and employment. Many people throughout the world rely on farming to survive. But if there was no food to harvest, there would be no money (in certain jobs). Additional disruptions and imbalances in value chains, such as scheduled imports and exports, would result from food security risks. Because agriculture significantly contributes to the gross domestic product (GDP) and employment in a society, this might potentially have an impact on the growth of the economy in general. Which kind of natural disaster does the most harm, do you know? The main culprits are the disasters that are brought on by the climate, including droughts and floods! In actuality, 26% of all damages between 2006 and 2016 were due to losses from climate-related disasters. Since they cause around 19% of all agricultural losses, droughts, in particular, the rate among the worst natural disasters, in reality, prolonged periods of drought might result in a loss of up to 20% of the potential agricultural production.

10 All About Drought

Natural disasters are linked to the climate change include droughts. They happen when a region has little to no precipitation for an extended period of time. Precipitation shortages can result in lower agricultural yields, which significantly impact agriculture. Crops could be gathered in the summer because the northern section of Alberta (north of Calgary) managed to get some moisture throughout the year. However, due to the severe heat waves and lack of rainfall this summer, the southern part of Alberta (south of Calgary) could only produce little to no crops. As a result, there were issues with crop production, marketing, and contracts made for agricultural sales! In fact, according to a statement made by Lynn Jacobson, president of the Alberta Federation of Agriculture, "running at the trouble of buying or supplying their contracts out, and there's quite an issue over some of the government costs and interest fees that have been accused by some of the grain companies." That is a significant problem for people. Having said that, predictions have shown that another drought year is coming due to the lack of moisture in the soil needed for crops to produce and be profitable to sell (profit). However, the idea of building more irrigation systems in Alberta has been considered. Even though this is a great concept, everything depends on how much rain falls in 2022.

11 Possible Solution

Then, is there any way to mitigate all these dangers? For nations, particularly those that are vulnerable to such natural disasters, the United Nations is actively developing disaster risk reduction (DRR) programs. In the coming years, droughts and heat waves will be the biggest threats as temperatures continue to increase as a result of heat-trapping gasses. These "heat-trapping gasses" are greenhouse gasses, and climate change is their main cause. The water cycle is changed by the greenhouse gas effect's absorption and storage of heat. Drought is brought on by the soils being drier and hotter as a result. The world has seen and continues to experience droughts, particularly in South Africa. After that, is there a way to stop climate change? Not quite, but there are a lot of methods we can try to manage this issue. First, let us talk about education. Simply put, many people are unaware of this interesting issue. Because of this, raising awareness is essential to addressing this issue. It is imperative for policymakers, stakeholders, and individuals alike to take up the mantle of proactive disaster risk reduction measures in agriculture. By investing in research, education, and innovative solutions, we can pave the way for a future where agriculture thrives resiliently, safeguarding livelihoods, ecosystems, and the very foundation of our societies. In conclusion, let us move forward with a shared commitment to sustainable development in agriculture that extends beyond environmental considerations, embracing disaster risk reduction as a vital pillar for a thriving, resilient, and secure agricultural landscape. By doing so, we not only fortify the foundation of our food systems but also honor our responsibility to the planet and future generations.

12 Conclusion

After all of this, there is obviously hope for resolving this issue. Yes, Mother Nature's actions are beyond our control, but we can do our part to protect and preserve agriculture and crops as much as we can! As Kiara Nirghin revealed, dreams are useless until you act on them. Our planet will become a bit more sustainable if we can all work together to improve the problems of drought-stricken agriculture production and climate change. We only need to take the first step!

References

Agriculture, Fisheries and Food, MD (1999) The National Irrigation Plan. Water Eng 6(1):13–26

American Society of Agronomy (1989) Decision reached on sustainable agriculture. Agron News 15

Anaman KA (1996) An introductory discussion of cost benefit analysis applied to climate change issues. Macquarie University, Graduate School of the Environment, Sydney

Anaman KA (2003) Assessing the economic and social impacts of extreme events on agriculture and the use of meteorological information to reduce adverse impacts. In: Agrometeorology related to extreme events. WMO, Geneva, p 943

Anderson MB (1990) Analyzing the costs and benefits of natural disaster responses in the context of development, vol 29. World Bank Policy Planning and Research Staff, Environment Department, Washington, DC

Charvériat C (2000) Natural disasters in Latin America and the Caribbean: an overview of risk. Inter-American Development Bank, Washington, DC

CRED (2000) EM-DAT: The OFDA/CRED International Natural Disaster Database 1900-1999. Centre on the Epidemiology of Disasters (CRED), Louvain, Belgium

Das HP (2003) Incidence, prediction, monitoring and mitigation measures of tropical cyclones and storm surges. In: Agrometeorology related to extreme events. WMO no. 943. World Meteorological Organization, Geneva

Devereux S (2000) Famine in the twentieth century. Institute of Development Studies, Brighton

Doblas-Reyes FJ, Hagedorn R, Palmer TN (2006) Developments in dynamical seasonal forecasting relevant to agricultural management. Climate Res 33(1):19–26

Eitzinger J, Utset A, Trnka M, Zalud Z, Nikolaev M, Uskov I (2007) Weather and climate and optimization of farm technologies at different input levels. In: Managing weather and climate risks in agriculture. Springer, Berlin, Heidelberg, pp 141–170

El-Gindy AM, Abdel-Mageed HN, El-Adl MA, Mohamed EMK (2001) Management of pressurized irrigated faba bean in sandy soils. Misr J Ag Eng 18(1):29–44

FAOSTAT (2018) FAOSTAT on-line. United Nations Food and Agriculture Organization, Rome

FAOSTAT (2021) FAOSTAT on-line. United Nations Food and Agriculture Organization, Rome

Fereres E, Soriano MA (2007) Deficit irrigation for reducing agricultural water use. J Exp Bot 58(2):147–159

Funaro-Curtis R (1982) Natural disasters and the development process: a discussion of issues. Evaluation Technologies, Montreal, QC

Guha-Sapir D, Hargitt D, Hoyois P (2004) Thirty years of natural disasters 1974–2003: the numbers. Presses Univ. de Louvain, Louvain-la-Neuve

Handmer J, Smith DI (1992) Cost effectiveness of flood warnings. Report prepared for the Australian Bureau of Meteorology by the Centre for Resource and Environment Studies. Australian National University, Canberra

Hansen JW, Indeje M (2004) Linking dynamic seasonal climate forecasts with crop simulation for maize yield prediction in semi-arid Kenya. Agric For Meteorol 125(1–2):143–157

Hooke WH (2000) US participation in international decade for natural disaster reduction. Nat Hazard Rev 1(1):2–9

Höppe P (2007) Scientific and economic rationale for weather risk insurance for agriculture. In: Managing weather and climate risks in agriculture. Springer, Berlin, Heidelberg, pp 367–375

IFAD (2001) IFAD, Rural Poverty Report 2001: The challenge of ending rural poverty, International Fund for Agricultural Development (IFAD), Rome, 2001

IFRC/RCS (2003) World disasters report: 2002. International Federation of Red Cross and Red Crescent Societies, Geneva, p 239

Ingram KT, Roncoli MC, Kirshen PH (2002) Opportunities and constraints for farmers of West Africa to use seasonal precipitation forecasts with Burkina Faso as a case study. Agr Syst 74(3):331–349

IPCC (2001) Climate change 2001. Third Assessment Report of the Intergovernmental Panel on Climate Change. Cambridge University Press, Cambridge, New York

Jarraud M (2006) Opening statement at the symposium on multi-hazard early warning systems for integrated disaster risk management, 23 May 2006. World Meteorological Organization, Geneva

Joy CS (1991, May) The cost of natural disasters in Australia. In: Climate change impacts and adaptation workshop, climatic impacts centre. Macquarie University, Sydney

Kreimer A, Munasinghe M (1991) Managing natural disasters and the environment. The World Bank, Washington, DC

Mati BM (2000) The influence of climate change on maize production in the semi-humid–semi-arid areas of Kenya. J Arid Environ 46(4):333–344

Matthews RB, Kropff MJ, Horie T, Bachelet D (1997) Simulating the impact of climate change on rice production in Asia and evaluating options for adaptation. Agr Syst 54(3):399–425

McCalla AF, Ayres WS (eds) (1997) Rural development: from vision to action, vol 12. World Bank Publications, Washington, DC

Meinke H, Stone RC (2005) Seasonal and inter-annual climate forecasting: the new tool for increasing preparedness to climate variability and change in agricultural planning and operations. Clim Change 70(1):221–253

Munich Re (2002) Topics geo, annual review: natural catastrophes 2002. Munich Reinsurance Group, Munich

Munich Re (2004) Topics geo, annual review: natural catastrophes 2003. Munich Reinsurance Group, Geo Science Research, Munich, p 56

Munich Re (2006) Topics geo, annual review: natural catastrophes 2005. Munich Reinsurance Group, Munich

National Research Council (1991) Toward sustainability: a plan for collaborative research on agriculture and natural resource management. National Academies Press, Washington, DC

Neira XX, Alvarez CJ, Cuesta TS, Cancela JJ (2005) Evaluation of water-use in traditional irrigation: an application to the Lemos Valley irrigation district, northwest of Spain. Agric Water Manag 75(2):137–151

Normand JM (1991) Le Bangladeshà la dérive. Le Monde, Dossier set documents, 22 juin 1991

Obasi GOP (1997) Address at the opening of the second joint session of the WMO. In: ESCAP Panel on Tropical Cyclones and the ESCAP/WMO Typhoon Committee, p 20

OP OG (1996) Climate climate change variability and predictability

Rojas O, Rembold F, Royer A, Negre T (2005) Real-time agrometeorological crop yield monitoring in eastern Africa. Agron Sustain Dev 25(1):63–77

Salinger MJ, Sivakumar MVK, Motha R (2005) Reducing vulnerability of agriculture and forestry to climate variability and change: workshop summary and recommendations. In: Increasing climate variability and change. Springer, Dordrecht, pp 341–362

Sivakumar MV (2005) Impacts of natural disasters in agriculture, rangeland and forestry: an overview. In: Natural disasters and extreme events in agriculture, pp 1–22

Sivakumar MVK, Das HP, Brunini O (2005) Impacts of present and future climate variability and change on agriculture and forestry in the arid and semi-arid tropics. In: Increasing climate variability and change. Springer, Dordrecht, pp 31–72

Smit B, Pilifosova O, Burton I, Challenger B, Huq S, Klein RJT, Yohe G, Adger N, Downing T, Harvey E, Kane S, Parry M, Skinner M, Smith J, Wandel J, Patwardhan A, Soussana JF (2001) Adaptation to climate change in the context of sustainable development and equity. In: Chapter 18 in Climate Change 2001: impacts, vulnerability, and adaptation. Intergovernmental Panel on Climate Change, United Nations and World Meteorological Organization, Geneva

Stigter CJ, Dawei Z, Onyewotu LOZ, Xurong M (2005) Using traditional methods and indigenous technologies for coping with climate variability. In: Increasing climate variability and change. Springer, Dordrecht, pp 255–271

Susman P, O'Keefe P, Wisner B (2019) Global disasters, a radical interpretation. In: Interpretations of calamity. Routledge, London, pp 263–283

Swindale LD (1988) Agricultural development and the environment: a point of view. In: TAC Secretariat (ed) Sustainable agricultural production: implications for international agricultural research. FAO, Rome, pp 7–14

Swiss Re (2002) Opportunities and Risks of Climate Change. Swiss Reinsurance Company, Zurich, Sigma 2/2002

UNDP (2001) Disaster profile of the least developed nations. United Nations Development Programme, New York

UNDP (2004) Reducing disaster risk: a challenge for development. United Nations Development Programme, New York

UNISDR (2003) Linking disaster risk reduction and climate change adaptation. In: Presentation to the workshop in insurance-related actions to address the specific needs and concerns of developing country parties arising from the adverse effects of climate change and from the impact of the implementation of response measures. UNFCC, Bonn. 14-15 May. Inter-Agency Secretariat for the International Strategy for Disaster Reduction (UNISDR)

Vellinga P, Mills E, Bowers L, Berz G, Huq S, Kozak L, Dlugolecki A et al (2001) Insurance and other financial services. Clim Change:417–450

Vos R, Velasco M, Edgar de Labastida R (1999) Economic and social effects of El Niño in Ecuador, 1997–1998. Inter-American Development Bank, Washington, DC

Wenjia WANG, Hao FENG (2012) The progress and problems in the development of foreign crop models. Water Sav Irrig 8:63–68

Drought Stress in Crop Plants and Its Management

Shahid Farooq, Sami Ul-Allah, and Mubshar Hussain

Abstract

The continuous worldwide climatic changes and population increase pose a danger to global food security. Extreme weather events have a significant impact on agricultural productivity, particularly drought stress. Drought stress hinders plant development and reduces agricultural yields. There are several strategies that plants have developed to increase their resilience to drought. On the other hand, cutting-edge methods are being employed to generate drought resistance in agricultural plants. Seed priming is a common agronomic practice to mitigate the negative effects of drought on crops. Before the seeds germinate, they are subjected to a stress (osmotic solution or solution of diverse chemical and natural components) that imprints stress memory and help the plants endure the negative effects of drought stress. Development of drought-tolerant genotypes, use of arbuscular mycorrhizal fungus (AMF), plant growth promoting rhizobacteria (PGPR), nutrient management and organic amendments may be used to reduce the hazardous effects of drought stress on crop plants. This chapter provides a high-level summary of how agricultural plants react to drought stress at various stages of their development and how management strategies can alleviate the drought impacts.

S. Farooq
Department of Plant Protection, Faculty of Agriculture, Harran University, Sanliurfa, Turkey

S. Ul-Allah
College of Agriculture, BZU Bahadur Sub-Campus, Layyah, Pakistan

M. Hussain (✉)
Department of Agronomy, Bahauddin Zakariya University, Multan, Pakistan

School of Veterinary and Life Sciences, Murdoch University, Murdoch, WA, Australia
e-mail: mubashir.hussain@bzu.edu.pk

M. Ahmed, S. Ahmad (eds.), *Disaster Risk Reduction in Agriculture*, Disaster Resilience and Green Growth, https://doi.org/10.1007/978-981-99-1763-1_9

Keywords

Extreme weather events · Drought stress · Drought resistance · Seed priming

1 Introduction

Food security for the world's continually expanding population is seriously threatened by biotic and abiotic stressors. Drought is one of the most detrimental abiotic factors that significantly reduce crop output around the globe. It causes a 50–70% drop in agricultural productivity (Yadav et al. 2021). Around 55 million people globally are affected by drought stress each year. Additionally, it impacts 40% of the world's population, and by 2030, a drought-related population relocation of 700 million people is predicted (Yadav et al. 2021).

Drought stress induces reactive oxygen species (ROS) production, including free radicals (superoxide, alkoxy, and hydroxyl) and non-radicals (singlet oxygen and hydrogen peroxide) (Farooq et al. 2012). Because of their high reactivity and toxicity, these substances disrupt cellular homeostasis by wreaking havoc on proteins, carbohydrates, lipids and DNA. Plant height, canopy, root growth, and leaf area index are all altered by drought stress (Sissons et al. 2018). Several physiological aspects of plants are negatively impacted by drought stress. These aspects include stomatal conductance, carboxylation efficiency, photosynthetic rate, pressure potential, and transpiration rates (Farooq et al. 2009, 2012). Consequently, drought stress constitutes a significant risk to sustainable agriculture since it negatively impacts crop production. However, to avoid the effects of drought, plants have evolved several adaptations at the morphological, physiological, biochemical, and phonological levels (Farooq et al. 2009, 2012). These changes to the plant's transcriptome, proteome and metabolome lead to cellular biosynthesis and breakdown processes. Therefore, plants engage a number of strategies, such as enhanced synthesis phytohormones, ROS signalling, plant hydraulic status and osmotic adjustment, to maintain growth and production during drought stress (Chhaya et al. 2021; Jogawat et al. 2021).

Primary metabolites are synthesized by plants for fundamental processes, including growth and development, whereas secondary metabolites serve more specialized purposes. Plants produce a diverse range of secondary metabolites to thrive in unfavourable environments (Yadav et al. 2021). A broad range of secondary metabolites has been identified in plants. Plants must produce an appropriate amount of primary and secondary metabolites to thrive in dynamic environments. Signalling, defence, interspecies communication, and enzyme regulation are just a few of the many roles that secondary metabolites play in plants (Farooq et al. 2009). Many of a plant's visual and gustatory characteristics come from its secondary metabolites, which are often aromatic and coloured chemicals. The whole impact of secondary metabolites has yet to be understood, despite much research on the topic. However, advances in science and technology have aided in identifying their roles. For instance, the relevance of secondary metabolites in diverse plant species has been made clear by genome sequencing and gene editing methods (Chen and Arora 2013; Dhanya Thomas et al. 2020; Sen and Puthur 2020).

Crop yield has greatly declined due to changes in the global climate and the existing agricultural system, even though plants often respond to drought stress through a variety of tolerance mechanisms. Multiple methods are used to mitigate the adverse effects of drought stress, including the introduction of drought-tolerant varieties via breeding programmes and exogenous application of growth regulators, osmoprotectants, and plant mineral supplements (Hussain et al. 2018). However, the complexities of drought stress did not allow the agricultural community to benefit from these approaches (Marthandan et al. 2020). Seed priming is a practical and inexpensive approach that may promote and improve seed germination and improve drought tolerance in crop plants (Farooq et al. 2009, 2017; Mehboob et al. 2022). Primed seeds may protect themselves from oxidative stress and retain memories of previous adversity via early activation of cellular defence mechanisms, decreased imbibition time, higher germination promoters, and osmotic control (Farooq et al. 2012, 2015). This chapter deals with the impact of drought stress on different crops and their various morphophysiological attributes. Furthermore, the impacts of different seed priming techniques in improving the drought tolerance of plants are discussed.

2 Drought-Induced Yield Losses in Leading Field Crops

Drought stress is a major problem for many cash crops, including rice (*Oryza sativa* L.), wheat (*Triticum aestivum* L.), maize (*Zea mays* L.), and cotton (*Gossypium hirsutum* L.). Multiple drought events have been documented throughout the last few decades, making farming exceedingly difficult in many nations (Jogawat et al. 2021). Drought-induced yield losses are well reported in almost all kinds of arable crops sown in multiple sol and environment conditions (Table 1).

Maize is among the most important crops in terms of the quantity of grains produced all over the globe (Cai et al. 2020; Wang et al. 2020). A decline in maize production of 39.3% was recorded, with a water reduction of 40% (Hussain et al. 2013; Khan et al. 2015). However, the stress caused by drought has a negative influence on the yield and output of maize crops (Iqbal et al. 2018). For example, Li et al. (2019) indicated that severe drought stress could reduce maize yield in the United States by 37%. The physiological, biochemical, and molecular properties of the plant are all profoundly altered as a result of drought stress (Usmani et al. 2020). Photosynthesis, lipid peroxidation, carbon absorption, free radical production, membrane degradation, and leaf development are all negatively affected by drought stress (Zhang et al. 2018; Hussain et al. 2019).

Wheat, the world's second-most significant grain and staple crop, meets around 22% of the population's dietary demands and accounts for 760 million tons of annual production (FAO 2022). About 70% of wheat harvests are earmarked for human consumption, 19% for animal feed, and 11% for industrial uses, such as

Table 1 Effect of drought stress on yield of arable crops

Crop species	Drought stress level	Growth stage	Crop traits	Decrease over control (%)	References
Sunflower (*Helianthus annuus* L.)	Mild drought	Budding	Achene yield	10	Hussain et al. (2008)
	Mild drought	Vegetative	Achene yield	15	Heidari and Karami (2014)
	Mild drought	Whole season	Achene yield	51	Alahdadi and Oraki (2011)
	Severe drought	Reproductive stage	Achene yield	40	Iqbal et al. (2009)
	Severe drought	Vegetative stage	Achene yield	16	
Wheat (*Triticum aestivum* L.)	Severe stress	Booting to maturity	Grain yield	37	Shamsi et al. (2010)
	Mild stress	Heading	Grain yield	57	Balla et al. (2011)
	Severe stress	Heading to maturity	Grain yield	44	Prasad et al. (2011)
	Mild stress	Anthesis	Grain yield	8	Akram (2011)
	Severe stress	Anthesis	Grain yield	11–39	Jatoi et al. (2011)
	Severe stress	Grain filling to maturity	Grain yield	31	Shamsi et al. (2010)
Rice (*Oryza sativa* L.)	Mild stress	Reproductive	Grain yield	54	Lafitte et al. (2007)
	Severe stress	Reproductive	Grain yield	94	Lafitte et al. (2007)
	Short severe stress	Flowering	Grain yield	54	Lanceras et al. (2004)
	Prolonged severe stress	Flowering and grain filling	Grain yield	84	
	Prolonged mild stress	Flowering and grain filling	Grain yield	52	
Soybean (*Glycine max* L.)	–	Pod set	Grain yield	73–83	Wei et al. (2018)
	–	Reproductive phase	Grain yield	46–71	Samarah et al. (2006)
	–	Pod set	Grain yield	45–50	Kobraee et al. (2011)
	–	Grain filling stage	Grain yield	42	Maleki et al. (2013)

(continued)

Table 1 (continued)

Crop species	Drought stress level	Growth stage	Crop traits	Decrease over control (%)	References
Chickpea (*Cicer arietinum* L.)	–	Reproductive phase	Grain yield	45–69	Nayyar et al. (2006)
	–	Ripening stage	Grain yield	49–54	Samarah et al. (2009)
	–	Anthesis stage	Grain yield	27–40	Mafakheri et al. (2010)
	–	Ripening stage	Grain yield	50	Varshney et al. (2014)
Cowpea (*Vigna unguiculate* L.)	–	Reproductive phase	Grain yield	60	Ogbonnaya et al. (2003)
	–	Reproductive phase	Grain yield	34–66	Ahmed and Suliman (2010)
	–	Pod filling stage	Grain yield	29	Kyei-boahen et al. (2017)
Common bean (*Phaseolus vulgaris* L.)	–	Reproductive phase	Grain yield	58–87	Martinez et al. (2007)
	–	Pod filling stage	Grain yield	40	Ghanbari et al. (2013)
	–	Flowering stage	Grain yield	49	Rosales-Serna et al. (2004)
Pigeon pea (*Cajanus cajan* L.)	–	Reproductive phase	Grain yield	40–55	Nam et al. (2001)
Mung bean (*Vigna radiata* L.)	–	Reproductive phase	Grain yield	26	Al et al. (2012)
	–	Flowering stage	Grain yield	31–57	Ahmad et al. (2015)
Faba bean (*Vicia faba* L.)	–	Grain filling stage	Grain yield	68	Ghassemi-Golezani and Hosseinzadeh-Mahootchy (2009)
Lentil (*Lens culinaris* L.)	–	Pod development	Grain yield	70	Shrestha et al. (2006)
	–	Reproductive phase	Grain yield	24	Allahmoradi et al. (2013)

biofuels. To meet rapidly expanding demand, the wheat output must be increased by up to over 60% by 2050. More than 36% of people throughout the globe eat wheat as their main source of sustenance (Hasanuzzaman et al. 2018). Wheat production is reduced by about 29% as a direct result of the adverse impacts of drought stress (Manickavelu et al. 2012). Wheat under drought stress shows a decrease in a variety of physiological and morphological parameters, including chlorophyll content, plant height, plant dry weight, grain size and transpiration rate. Furthermore, drought induces oxidative stress leading to hormonal instability and reduced anti-oxidant enzyme activity (Raheem et al. 2018). Growth, yield, photosynthesis, nutrient absorption, and metabolic activities are also all significantly stunted as a result of drought stress (Hasanuzzaman et al. 2018)). In addition, it has been shown that wheat plants are unable to fully mature and produce as much fruit when exposed to drought stress during the early stages of their reproductive development (Dolferus et al. 2011).

Rice is a cereal crop that is very common and necessary for human nutrition in several geographic regions of the world. It supplies between 52 and 76% of the calories that are eaten by about half of the world's population (Liu et al. 2006). Rice accounted for around 29% of the total cereal consumption, with a global annual production of 756 million tons (FAO 2022). Rice farming is very thirsty for water when compared to other agricultural endeavours. About 30.9% of all rice fields are utilized for farming that relies only on rainwater (Wassmann et al. 2009; Dixit et al. 2014). Drought may affect almost 23 million acres of rice that are grown using rainwater (Serraj et al. 2011). Drought stress has a major effect on yield and biomass, causing an annual 18 million ton decrease in rice output throughout the globe (O'Toole 2004). Biological activities in leaves, such as protein synthesis, redox balance, and energy metabolism, are all affected by drought stress throughout the reproductive phase of a plant (Wang et al. 2017). Drought stress results in premature leaf senescence, reduced photosynthesis, stunted leaf development and low yield. Additionally, it causes the breakdown of cytokinins and indole acetic acid, poor seed germination, uneven distribution and decreased fertility in the spikelets. Drought stress causes significant changes in glucose metabolism, accumulation of osmolytes and organic acids, and a subsequent yield loss (Chen et al. 2018).

Cotton is a significant cash crop and a vital textile fibre in dry and semi-arid regions (Riaz et al. 2013). It is considered a cash crop because of its high market value as a major source of fibre for the textile industry. It is extensively cultivated in >50 nations. India is the world's top cotton producer (FAO 2022). On the other hand, drought stress reduces cotton's growth, productivity and yield by influencing the biochemical and physiological processes that regulate lint formation and fibre quality. Plants' height, root development, and other morphological traits are all negatively affected by drought stress (Farooq et al. 2012). Cotton leaves have been shown to have reductions in their net photosynthetic rate, transpiration rate, stomatal conductance, water potential, and carboxylation efficiency as a result of the water scarcity condition (Kapoor et al. 2020; Kumar et al. 2001). Although several studies have shown that drought negatively affects agricultural plant productivity and performance, little is known about the processes by which plants detect drought stress and initiate defences.

3 Impacts of Drought Stress on Different Growth Phases

3.1 Impact of Drought Stress on Seed Germination

Germination, the process through which seeds develop into plants, is crucial to agricultural production. Drought stress is especially harmful to seedlings during the germination period. Seeds cannot germinate without water, yet even under ideal conditions, seeds cannot absorb water due to drought stress (Farooq et al. 2012). Drought stress has similar effects on germination and seedling vitality because of its effect on water intake (AL-Quraan et al. 2021). Reduced germination due to drought stress during the planting and early development stages of crops results in weaker stands overall. Low rates of germination were seen in rice and pea seeds when the plants were stressed by drought (Liang et al. 2021). In conjunction with other environmental variables, the low soil water content may hinder germination. Seed germination of maize is severely influenced by drought stress (Ahmad et al. 2019). During the germination and seedling development stages, certain agricultural crops are especially susceptible to cold and dry environments (early phases). Every seed needs its own unique combination of soil moisture and temperature to germinate (Gioria and Pyšek 2017; Lei et al. 2021). Many weed species are thought to be better able to withstand the effects of drought. However, research has revealed that drought stress is particularly damaging to the germination of weed seeds across a number of different plant species (Farooq et al. 2019, 2021a, b). Different populations of the same species may have varying germination rates (Eslami 2011). Nonetheless, seed priming is one agronomic approach that has been used to improve seed germination in several crop species that are under drought stress (Farooq et al. 2017; Haider et al. 2020).

3.2 Impact of Drought Stress on Plant Morphology

Internal factors that drive plants to grow taller are significantly impacted by drought (Lei et al. 2021). Inadequate mitosis slowed cell growth, and rapid leaf drop under drought is responsible for shorter plant height under drought stress (Farooq et al. 2009). The quantity of maize leaves drastically dropped as a result of water stress (Ahmad et al. 2019; Wahab et al. 2022). There is conclusive evidence that the leaves of sweet basil (*Ocimum basilicum*) are much more valuable than the plant's stems and roots due to their role in photosynthesis and the presence of photosynthetic pigments (Damalas 2019). Total plant biomass and leaf area are reduced when plants are subjected to drought regimes that prevent the growth of leaves and disrupt photosynthetic activity (Kumar et al. 2021). Previous research found that when subjected to drought stress, numerous cereals, including wheat and rice, significantly decreased the amount of leaf area (Kumar et al. 2021; Naz and Perveen 2021). The leaf's capacity to endure pressure is reduced when water from the top epidermis evaporates, leading to the leaf rolling. The benefits of this occurrence include a decrease in leaf temperature, a rise in light absorption, and a rise in transpiration rate. When maize leaves were treated with drought stress regimens, both leaf area and leaf rolling increased significantly (Cai et al. 2020).

There is a negative correlation between shoot length and fresh weight under drought stress. A significant reduction in dry weight of the shoot of common bean (*Phaseolus vulgaris* L.) (Widuri et al. 2018). The shoot length of maize crop indicates that the crop requires more water and nutrients to recover from the impacts of drought stress (Tůmová et al. 2018). The length of maize seedlings was also shown to be drastically reduced during drought stress. Since maize crops dry in the shade, their final dry weight is significantly influenced by the shortage of water (Bocchini et al. 2018). Drought stress has a significant influence on the harvest's fresh weight when compared to the control. A robust root system that anchors the plant and draws water and nutrients from the soil is essential under these conditions (Hussain et al. 2013; Begum et al. 2019).

3.3 Impact of Drought Stress on Root Traits

Drought stress alters plants' root architecture and morphology, and it is important to understand how plants respond to abiotic stresses. During drought stress, many plants' root biomass increases as their length becomes more prolonged; thus, allowing them to absorb more water and minerals from the soil. This can lead to the development of better tolerance and adaptation against drought stress conditions (Bhattacharya 2021). Polyethylene glycol (PEG)-induced drought stress decreased length, fresh and dry weight, root length and diameter in maize. The PEG-induced drought stress significantly reduced plant growth, which can lead to serious yield loss (Hu and Chen 2020). Moderate drought may not always significantly limit plant growth at the root level (Valliere et al. 2019). Water stress had little effect on maize's root growth. For instance, *Catharanthus roseus* L. and *Helianthus annuus* L. benefit from drought stress as their root growth increased under moisture deficiency (Sharma et al. 2021). Variations in the root–shoot and root-leaf ratios of plant species are the most important indicator of how drought stress affects harvests (Abideen et al. 2020). Plants can adapt to their surroundings by changing root architecture. Under drought or high salinity stress, plants' roots may become longer and branch out into more soil mass, allowing them to absorb more water and minerals. Several agronomic and breeding methods have been employed to improve the root traits of crop plants under drought stress.

3.4 Impact of Drought Stress on Yield

While certain plant species' yields may be influenced differently by water scarcity, agricultural yields are significantly impacted by a lack of water after anthesis, regardless of the intensity or length of the shortage. Drought stress reduces harvests in a number of ways. In both barley (*Hordeum vulgare* L.) and wheat, drought stress decreased the number of spikes, tillers and grains per plant as well as the grain weight. Both crops saw comparable changes in plant architecture because of drought stress, including fewer spikes, tillers and grains per plant. This suggests that the physiological mechanism through which drought stress lowers plant output in barley and wheat harvests is the same (Istanbuli et al. 2020; Nofouzi 2018). Millet (*Pennisetum glaucum* L.) output declined due to drought stress, which lowered silking and lengthened the anthesis-to-silking period (Kalagare et al. 2021). Production of soybean seeds was hampered by drought stress, which also had an impact on crop germplasm physiology and yield (Maleki et al. 2013). The number of ears and kernels per plant was positively correlated with total grain yield (Ullah and Farooq 2022). The relationship between grain production and harvest index has been investigated in earlier studies (Wellstein et al. 2017).

Crops in many parts of the globe are under constant stress from a lack of water due to a shrinking quantity of freshwater. Because of these alterations, agricultural plants become less productive overall (Table 1). However, the world's rapidly expanding population need a great deal more food production than is now being harvested. For this reason, it will need a combination of strategies to increase agricultural plants' resilience to drought. Various seed priming techniques and their significance in enhancing crop output when under drought stress are described in depth, along with a broad discussion of the methods available for increasing agricultural plants' tolerance to water shortages.

4 Management of Drought Stress

4.1 Development of Drought-Tolerant Cultivars

Genetic diversity is the basis of adaptation of crops to the changing environment and different agroecological conditions. Pyramiding of genes associated with drought stress tolerance into a single genotype may lead it to grow well in water scare conditions. It is the most sustainable to cope with drought stress, as there is no need for extra sources except yield. For this purpose, genetic diversity may be exploited to identify the sources of tolerant genes, and then a proper crossing plan is required to make combinations containing a maximum of these genes. Various efforts have been made in developing/identifying drought-tolerant cultivars for various crops.

With the advancement of molecular techniques and genomics, various genomic regions are identified which are linked with drought stress tolerance and these regions are used as molecular markers to select crop genotypes tolerant to drought stress. Drought tolerance is a complex quantitative character, and many genes are involved in it; therefore, in a single crop, many genes/regions associated with drought stress tolerance may be identified.

Biotechnology has opened new avenues in the development of crops tolerant to biotic and abiotic stresses. With genetic engineering, it is possible to transfer genes from unrelated organisms. Various efforts have been made to identify drought-tolerant genes in xerophytes and to transfer these to crop plants. Liu et al. (2021) isolated *ZxABCG11* from a xerophyte plant (*Zygophyllum xanthoxylum*) and transferred it to Arabidopsis. They found transgenic plants more tolerant to drought and salinity stress. In the coming era, there are opportunities for the development of transgenic drought-tolerant crops which can adapt well to drought environments.

4.2 Nutrient Management and Organic Manure

In a drought environment, the transport of nutrients also decreases along with water. There are some nutrients which can enhance drought stress tolerance in plants. For example, nitrogen (N) and potassium (K) are known to improve drought stress tolerance in plants by improving plant homeostasis, protein synthesis, improving photosynthetic capacity, and antioxidant enzymatic activities (Ul-Allah et al. 2020; Table 2). Moreover, the K application also improves water uptake, water relations and stomatal regulation leading to better growth and development in moderate stress conditions.

Silicon (Si) is another nutrient whose role in drought stress is well described. Application of Si improved the capacity of crop plants to uptake macro and micronutrient under water stress conditions contributing to proper growth and development. Moreover, the foliar application of Si reduces the ROS by enhancing the antioxidant enzymatic activities, leading to regulated crop growth and development under water stress conditions (El-Mageed et al. 2021). In a study on wheat crops, it was observed that the application of Si improves root–shoot ratio, production of osmolytes and antioxidant enzymatic activities under drought stress, leading to improved growth and development relative to the control (Sattar et al. 2020).

Organic matter in the soil is known to conserve water for long periods as the water-holding capacity of organic matter is much higher than the soil. Therefore, soil amendment with different organic matter materials like farmyard manure, biochar and humus are known to improve crops' drought stress tolerance. Therefore, in drought-prone areas, efforts must be made to enhance the organic matter of cultivated lands to enhance their water-holding capacity.

4.3 Alternate Irrigation Techniques

With climate change and industrialization, resources of useable freshwater are being reduced and often less water is available for crops which can be managed using alternate irrigation techniques. Studies have proved that skipping one or two irrigations than normal can increase the crop's water use efficiency (WUE), and comparative less production losses are observed (Table 2). For this purpose, there is a need to identify the most critical crop stages that require water. For example, in the case of wheat tillering, booting and milking stage are the most critical stages that require water and skipping of irrigation. Other than these stages can reduce production losses. Chu et al. (2018) reported improvement in WUE and grain yield in drought-tolerant cultivars under alternate wetting drying irrigation (skip irrigation) in rice and saved up to 50% water. They further reported that this improvement was attributed to improved agronomic traits of drought-tolerant cultivars under alternate wetting techniques. Ul-Allah et al. (2014, p. 15) conducted experiments on different fodder crops and reported that skipping alternate irrigation make crops more efficient water user, as evidenced by increased irrigation water use efficiency and there

Table 2 Effect of management strategies on the performance of different crops under drought stress

Management option	Crop species	Improved traits under drought stress	Reference
Use of arbuscular mycorrhizal fungi/PGPR	Soybean (*Glycine max* L.)	Vegetative growth and grain yield	Leggett et al. (2017)
		Grain yield	Ulzen et al. (2016)
		Management of root osmotic potential	Porcel and Ruiz-lozano (2004)
	Cow pea (*Vigna unguiculata* L.)	Stomatal conductance	Stancheva et al. (2017)
	Pea (*Pisum sativum* L.)	Vegetative growth	Saikia et al. (2018)
	Sunflower (*Helianthus annus* L.)	Dry matter yield and oil quality	Gholamhoseini et al. (2013)
Nutrient management	Different crops	Photosynthetic efficiency and water use efficiency	Waraich et al. (2011)
	Sunflower (*Helianthus annus* L.)	Physiological and yield traits	Hussain et al. (2018)
	Wheat (*Triticum aestivum* L.)	Physiological and antioxidant traits	Aurangzaib et al. (2022)
	Maize	Physiological and metabolic traits	Parveen et al. (2019)
		Water use efficiency and yield	
Organic amendment	Maize (*Zea mays* L.)	Water use efficiency and yield	Romdhane et al. (2019)
		Growth, morphological traits and yield	Minhas et al. (2020)
	Maize (*Zea mays* L.), Egyptian clover (*Trifolium alexandrinum* L.), Sudan grass (*Soghum bicolor* var. sudanese) and oat (*Avena sativa* L.)	Water use efficiency and dry matter yield	Ul-Allah et al. (2015)
	Quinoa (*Chenopodium quinoa*)	Physiological traits and yield	Benaffari et al. (2022)
	Wheat (*Triticum aestivum* L.)	Physiological attributes and grain yield	Omara et al. (2022)
	Barley (*Hordeum vulgare* L.)	Water productivity and yield	Khalifa et al. (2022)

(continued)

Table 2 (continued)

Management option	Crop species	Improved traits under drought stress	Reference
Seed priming	Faba bean (*Vicia Faba* L.)	Physiological and growth traits	Farooq et al. (2020)
	Black gram (*Vigna radiata* L.)	Vegetative growth and seed yield	
	Maize (*Zea mays* L.)	Seedling establishment and grain yield	El-Sanatawy et al. (2021)
	Wheat (*Triticum aestivum* L.)	Grain yield and nutritional quality	Hussain et al. (2018)

are relatively less crop losses. Ullah and Datta (2018) reported that alternate wetting and drying technique improves the root system of the crop, which leads to better growth and development of the crop.

4.4 Seed Priming

Seed priming boosts the ability of plants to endure the effects of stress (Dhanya Thomas et al. 2020; Sen and Puthur 2020). Priming seeds entails exposing them to a mix of natural and synthetic elements in order to elicit a moderate stress response in the developing plant (Paparella et al. 2015). Seed treatments bring about a physiological condition referred to as the "primed state" prior to the germination of the seed, which speeds up a variety of cellular processes (Wojtyla et al. 2016). Because of the primed state, plants are ready to respond quickly when subjected to further stress (Farooq et al. 2012, 2017). Early and uniform germination characterize the seedlings that emerge from the primed seeds, and an overall improvement in many growth aspects may be seen throughout the course of their lives (Hussain et al. 2013; Farooq et al. 2017). Seed priming methods, such as hydropriming, osmopriming, halopriming, UV-B priming and chemical priming, all put plants through minor stress. Plants may acquire resistance to drought stress via the upregulation of stress-responsive genes and proteins like late embryogenesis abundant (LEAs) (Chen and Arora 2013; Dhanya Thomas et al. 2020; Sen and Puthur 2020). Studies suggest that plants may be able to imprint epigenetic memory via early exposure to mild stress, which would prepare them for similar or different forms of stress in the future (Ding et al. 2013; Marcos et al. 2018).

The accumulation of osmolytes was correlated with the level of tolerance shown by the plants (Tabassum et al. 2018). The accumulation of osmolytes during different seed priming processes has been linked to increased drought tolerance in plants (Khan et al. 2019). Plants grown from osmoprimed sorghum and barley seeds showed an increase in amino acid synthesis, glycine betaine synthesis, and total soluble sugar production when exposed to a stressful environment (Zhang et al.

2018). Osmotic stress enhanced the levels of proline, total soluble sugars and total free amino acids in rice seedlings emerging from the seeds that had been hydroprimed, haloprimed, or UV-B primed (Sen and Puthur 2020). Primed rice seedlings showed increased proline production, indicating osmotic potential modification. It has the potential to store carbon and nitrogen for later use in the healing process (Thomas and Puthur 2019). Appropriate solutes also protect macromolecules and keep enzymes stable (Wang et al. 2019). Maintaining a higher water content by priming-induced osmolyte augmentation is another possible strategy for facilitating a speedier recovery from the effects of drought stress.

The antioxidant system is activated during seed priming to scavenge reactive oxygen species (ROS) and reduce oxidative damage (Paparella et al. 2015). Seed priming increases antioxidant levels, which protect plants from the oxidative stress brought on by dryness (Khan et al. 2020; Zheng et al. 2016). The UV-B priming increased the antioxidant capacity of rice by increasing the levels of SOD, CAT, APX, ascorbate, and glutathione while lowering the levels of hydrogen peroxides and superoxides (Dhanya Thomas et al. 2020; Sen and Puthur 2020). Higher gene expression for SOD, CAT and APX was noticed in rice that had been primed with UV-B. Increased SOD, POD and CAT activity were also seen in spermidine-primed rice. The SOD and GR activity in *Brassica juncea* were increased by hydropriming when exposed to osmotic stress (Srivastava et al. 2010). Priming seeds to increase stress tolerance provided a similar reaction in the tolerant variety (Sen and Puthur 2020). The efficient functioning of antioxidant machinery, which is connected to various priming tactics, is obviously going to help plants recuperate and resume their usual activity as soon as possible after the stress has been eliminated from their surroundings.

The fundamental repair systems, such as nucleotide excision repair and base excision repair, are activated during the early stages of germination in order to preserve the integrity of the genome (Chen et al. 2012; Macovei et al. 2010). During the process of seed germination in Arabidopsis, the proteins AtLIG6 and AtLIG4 play a crucial role in ligating DNA double-strand breaks, which has an effect on both the viability and the quality of seeds (Waterworth et al. 2010). Following the application of hydropriming to seeds of *Medicago truncatula*, an increase in gene expression associated with DNA damage repair and anti-oxidation machinery was found. The DNA repair enzyme known as formamidopyrimidine DNA glycosylase (FPG) became more active after being exposed to hydropriming for a period of 4 h (Forti et al. 2020). The growth of tubulin subunits in *Arabidopsis* seeds treated with hydropriming and osmopriming provided additional evidence for the importance of seed priming in the re-initiation of the cell cycle (Gallardo et al. 2001).

4.5 Arbuscular Mycorrhizal Fungi and PGPR

Sustainable agriculture that is ecologically friendly offers a lot of promise for plant-fungus interaction. Through a number of processes, the beneficial microbiome associated with roots and plant tissues reduces plant stress. As abiotic stressors on plants increase, the symbiotic interaction between plant roots and arbuscular mycorrhizal fungus (AMF) is crucial for enhancing mineral feeding. Despite the fact that this symbiosis has existed for more than 450 million years, we are only now starting to understand how it is kept alive. In fact, AMF increases the surface area of plant roots to absorb more water and nutrients (Bahadur et al. 2019). Various species of AMF have been reported that enhance plant tolerance to abiotic stresses, especially drought stress (Table 2). In a study, the use of two AMF species, i.e., *Glumus mossea* and *Glumus etanicatum* improved the yield and the nutritional content of sunflower seeds, with *Glumus etanicatum* being more advantageous (Heidari and Karami 2014). To realize their full potential, one must, however, have a thorough grasp of the drought tolerance processes in plants that are mediated by AMF. Signalling crosstalk that occurs during these adaptive processes is a crucial additional consideration.

The PGPR can indirectly boost the plant immune system against abiotic stress such as drought by improving micronutrient absorption and impacting phytohormone balance. PGPR are able to overcome the endodermis barrier and enter the vascular system from the root cortex. They then flourish as endophytes in the stem, leaves, tubers and other organs. Under drought conditions, the interaction between the plant and the PGPR alters not only the plant but also the soil's characteristics. Plant life under drought stress is largely dependent on the mechanisms induced by PGPR, such as inducing new genes and osmotic responses (Vurukonda et al. 2016).

Thus, PGPR and AMF have the potential to improve drought stress tolerance in crop plant and these can be a strategy to manage drought stress, especially in the areas where moderate drought severely affects the crop production. As soil biota varies under different agroecological conditions, AMF and PGPR must be identified and characterized per agroecological region.

5 Conclusions and Recommendations

Drought negatively impacts plant productivity, which also lowers the overall economic viability of agricultural output. To mitigate the consequences of drought, various strategies have been developed, and each has its own set of advantages and disadvantages. The genetic make-up of the plant, the developmental stage it is in when it is first exposed to stress, the quantity of stress, the amount of time it is under stress, and the developmental stage it is in all have an impact on how the plant reacts. The drought stress response serves as more than just a defence mechanism; it also enables long-term growth and safeguards a thriving ecological succession for future generations. In other words, the reaction to drought stress goes beyond simple defence. Various practices can be adopted to manage drought stress, including the use of drought-tolerant cultivars, alternate irrigation scheduling, nutrient management, seed priming with osmolytes and the use of AMF and PGPR.

References

Abideen Z, Koyro HW, Huchzermeyer B, Ansari R, Zulfiqar F, Gul B (2020) Ameliorating effects of biochar on photosynthetic efficiency and antioxidant defence of *Phragmites karka* under drought stress. Plant Biol 22:259–266. https://doi.org/10.1111/plb.13054

Ahmad A, Selim MM, Alderfasi AA, Afzal M (2015) Effect of drought stress on mungbean (*Vigna radiata* L.) under arid climatic conditions of Saudi Arabia. Ecosyst Sustain Dev 192:185–193

Ahmad S, Kamran M, Ding R, Meng X, Wang H, Ahmad I, Fahad S, Han Q (2019) Exogenous melatonin confers drought stress by promoting plant growth, photosynthetic capacity and antioxidant defense system of maize seedlings. PeerJ 7:e7793. https://doi.org/10.7717/peerj.7793

Ahmed FE, Suliman ASH (2010) Effect of water stress applied at different stages of growth on seed yield and water-use efficiency of cowpea. Agric Biol J North Am 1:534–540

Akram M (2011) Growth and yield components of wheat under water stress of different growth stages. Bangladesh J Agril Res 36:455–468

Al R, Dahanayaka N, Ugs A, Wdrj R, Utd R (2012) Effect of water stress on growth and yield of mung bean (*Vigna radiata* L.). Trop Agric Res Ext 14:2–6

Alahdadi I, Oraki H (2011) Effect of water stress on yield and yield components of sunflower hybrids. Afr J Biotechnol 10(34):6504–6509

Allahmoradi P, Mansourifar C, Saiedi M, Jalali Honarmand S (2013) Effect of different water deficiency levels on some antioxidants at different growth stages of lentil (*Lens culinaris* L.). Adv Environ Biol 7:535–543

Al-Quraan NA, Al-Ajlouni ZI, Qawasma NF (2021) Physiological and biochemical characterization of the GABA shunt pathway in pea (*Pisum sativum* L.) seedlings under drought stress. Horticulturae 7:125. https://doi.org/10.3390/horticulturae7060125

Aurangzaib M, Ahmad Z, Jalil MI, Nawaz F, Shaheen MR, Ahmad M, Hussain A, Ejaz MK, Tabassum MA (2022) Foliar spray of silicon confers drought tolerance in wheat (*Triticum aestivum* L.) by enhancing morpho-physiological and antioxidant potential. SILICON 14(9):4793–4807

Bahadur A, Batool A, Nasir F, Jiang S, Mingsen Q, Zhang Q, Pan J, Liu Y, Feng H (2019) Mechanistic insights into arbuscular mycorrhizal fungi-mediated drought stress tolerance in plants. Int J Mol Sci 20(17):4199. https://doi.org/10.3390/ijms20174199

Balla K, Rakszegi M, Li Z, Bekes F, Bencze S, Veisz O (2011) Quality of winter wheat in relation to heat and drought shock after anthesis. Czech J Food Sci 29:117–128

Begum N, Ahanger MA, Su Y, Lei Y, Mustafa NSA, Ahmad P, Zhang L (2019) Improved drought tolerance by AMF inoculation in maize (Zea mays) involves physiological and biochemical implications. Plan Theory 8:579. https://doi.org/10.3390/plants8120579

Benaffari W, Boutasknit A, Anli M, Ait-El-Mokhtar M, Ait-Rahou Y, Ben-Laouane R, Ben Ahmed H, Mitsui T, Baslam M, Meddich A (2022) The native arbuscular mycorrhizal fungi and vermicompost-based organic amendments enhance soil fertility, growth performance, and the drought stress tolerance of quinoa. Plan Theory 11(3):393

Bhattacharya A (2021) Effect of soil water deficit on growth and development of plants: a review. In: Soil water deficit and physiological issues in plants, pp 393–488

Bocchini M, D'Amato R, Ciancaleoni S, Fontanella MC, Palmerini CA, Beone GM, Onofri A, Negri V, Marconi G, Albertini E, Businelli D (2018) Soil selenium (se) biofortification changes the physiological, biochemical and epigenetic responses to water stress in *Zea mays* L. by inducing a higher drought tolerance. Front Plant Sci 9. https://doi.org/10.3389/fpls.2018.00389

Cai F, Zhang Y, Mi N, Ming H, Zhang S, Zhang H, Zhao X (2020) Maize (*Zea mays* L.) physiological responses to drought and rewatering, and the associations with water stress degree. Agric Water Manag 241:106379. https://doi.org/10.1016/j.agwat.2020.106379

Chen K, Arora R (2013) Priming memory invokes seed stress-tolerance. Environ Exp Bot 94:33–45. https://doi.org/10.1016/j.envexpbot.2012.03.005

Chen K, Fessehaie A, Arora R (2012) Dehydrin metabolism is altered during seed osmopriming and subsequent germination under chilling and desiccation in Spinacia oleracea L. cv.

Bloomsdale: possible role in stress tolerance. Plant Sci 183:27–36. https://doi.org/10.1016/j.plantsci.2011.11.002

Chen T, Feng B, Fu W, Zhang C, Tao L, Fu G (2018) Nodes protect against drought stress in rice (*Oryza sativa*) by mediating hydraulic conductance. Environ Exp Bot 155:411–419. https://doi.org/10.1016/j.envexpbot.2018.07.025

Chhaya, Yadav B, Jogawat A, Gnanasekaran P, Kumari P, Lakra N, Lal SK, Pawar J, Narayan OP (2021) An overview of recent advancement in phytohormones-mediated stress management and drought tolerance in crop plants. Plant Gene 25:100264. https://doi.org/10.1016/j.plgene.2020.100264

Chu G, Chen T, Chen S, Xu C, Wang D, Zhang X (2018) Agronomic performance of drought-resistance rice cultivars grown under alternate wetting and drying irrigation management in Southeast China. Crop J 6(5):482–494. https://doi.org/10.1016/j.cj.2018.04.005

Damalas CA (2019) Improving drought tolerance in sweet basil (*Ocimum basilicum*) with salicylic acid. Sci Hortic (Amsterdam) 246:360–365. https://doi.org/10.1016/j.scienta.2018.11.005

Dhanya Thomas TT, Dinakar C, Puthur JT (2020) Effect of UV-B priming on the abiotic stress tolerance of stress-sensitive rice seedlings: priming imprints and cross-tolerance. Plant Physiol Biochem 147:21–30. https://doi.org/10.1016/j.plaphy.2019.12.002

Ding Y, Liu N, Virlouvet L, Riethoven J-J, Fromm M, Avramova Z (2013) Four distinct types of dehydration stress memory genes in *Arabidopsis thaliana*. BMC Plant Biol 13:229. https://doi.org/10.1186/1471-2229-13-229

Dixit S, Singh A, Kumar A (2014) Rice breeding for high grain yield under drought: a strategic solution to a complex problem. Int J Agron 2014:1–15. https://doi.org/10.1155/2014/863683

Dolferus R, Ji X, Richards RA (2011) Abiotic stress and control of grain number in cereals. Plant Sci 181:331–341. https://doi.org/10.1016/j.plantsci.2011.05.015

El-Mageed A, Taia A, Shaaban A, El-Mageed A, Shimaa A, Semida WM, Rady MO (2021) Silicon defensive role in maize (*Zea mays* L.) against drought stress and metals-contaminated irrigation water. SILICON 13(7):2165–2176

El-Sanatawy AM, El-Kholy AS, Ali MM, Awad MF, Mansour E (2021) Maize seedling establishment, grain yield and crop water productivity response to seed priming and irrigation management in a Mediterranean arid environment. Agronomy 11(4):756

Eslami SV (2011) Comparative germination and emergence ecology of two populations of common lambsquarters (*Chenopodium album*) from Iran and Denmark. Weed Sci 59:90–97

FAO (2022) FAO [WWW Document]. www.faostat.fao.org

Farooq M, Wahid A, Kobayashi N, Fujita D, Basra SMA (2009) Plant drought stress: effects, mechanisms and management. Agron Sustain Dev 29:185–212. https://doi.org/10.1051/agro:2008021

Farooq M, Hussain M, Wahid A, Siddique KHM (2012) Drought stress in plants: an overview. In: Plant responses to drought stress. Springer, Berlin. Heidelberg, pp 1–33. https://doi.org/10.1007/978-3-642-32653-0_1

Farooq S, Shahid M, Khan MB, Hussain M, Farooq M (2015) Improving the productivity of bread wheat by good management practices under terminal drought. J Agron Crop Sci 201:173–188. https://doi.org/10.1111/jac.12093

Farooq S, Hussain M, Jabran K, Hassan W, Rizwan MS, Yasir TA (2017) Osmopriming with CaCl2 improves wheat (Triticum aestivum L.) production under water-limited environments. Environ Sci Pollut Res 24:13638. https://doi.org/10.1007/s11356-017-8957-x

Farooq S, Onen H, Ozaslan C, Baskin CC, Gunal H (2019) Seed germination niche for common ragweed (*Ambrosia artemisiifolia* L.) populations naturalized in Turkey. S Afr J Bot 123:361–371. https://doi.org/10.1016/j.sajb.2019.03.031

Farooq M, Romdhane L, Al Sulti MKRA, Rehman A, Al-Busaidi WM, Lee D (2020) Morphological, physiological and biochemical aspects of osmopriming-induced drought tolerance in lentil. J Agron Crop Sci 206:176–186. https://doi.org/10.1111/jac.12384

Farooq S, Onen H, Ozaslan C, El-Shehawi AM, Elseehy MM (2021a) Characteristics and methods to release seed dormancy of two ground cherry (physalis) species. J Appl Res Med Aromat Plants 25:100337. https://doi.org/10.1016/j.jarmap.2021.100337

Farooq S, Onen H, Tad S, Ozaslan C, Mahmoud SF, Brestic M, Zivcak M, Skalicky M, El-Shehawi AM (2021b) The influence of environmental factors on seed germination of Polygonum perfoliatum L.: implications for management. Agronomy 11:1123. https://doi.org/10.3390/agronomy11061123

Forti C, Shankar A, Singh A, Balestrazzi A, Prasad V, Macovei A (2020) Hydropriming and biopriming improve Medicago truncatula seed germination and upregulate DNA repair and antioxidant genes. Genes (Basel) 11:242. https://doi.org/10.3390/genes11030242

Gallardo K, Job C, Groot SPC, Puype M, Demol H, Vandekerckhove J, Job D (2001) Proteomic analysis of Arabidopsis seed germination and priming. Plant Physiol 126:835–848. https://doi.org/10.1104/pp.126.2.835

Ghanbari AA, Mousavi SH, Mousapour A, Rao I (2013) Effects of water stress on leaves and seeds of bean (Phaseolus vulgaris L.). Turk J F Crop 18:73–77

Ghassemi-Golezani K, Hosseinzadeh-Mahootchy A (2009) Changes in seed vigour of faba bean (*Vicia faba* L.) cultivars during development and maturity. Seed Sci Technol 37:713–720

Gholamhoseini M, Ghalavand A, Dolatabadian A, Jamshidi E, Khodaei-Joghan A (2013) Effects of arbuscular mycorrhizal inoculation on growth, yield, nutrient uptake and irrigation water productivity of sunflowers grown under drought stress. Agric Water Manag 117:106–114

Gioria M, Pyšek P (2017) Early bird catches the worm: germination as a critical step in plant invasion. Biol Invasions 19:1055–1080. https://doi.org/10.1007/s10530-016-1349-1

Haider MU, Hussain M, Farooq M, Nawaz A (2020) Optimizing zinc seed priming for improving the growth, yield and grain biofortification of mungbean (Vigna radiata (L.) wilczek). J Plant Nutr 43:1438–1446

Hasanuzzaman M, Al Mahmud J, Anee TI, Nahar K, Islam MT (2018) Drought stress tolerance in wheat: omics approaches in understanding and enhancing antioxidant defense. In: Abiotic stress-mediated sensing and signaling in plants: an omics perspective. Springer, Singapore, pp 267–307. https://doi.org/10.1007/978-981-10-7479-0_10

Heidari M, Karami V (2014) Effects of different mycorrhiza species on grain yield, nutrient uptake and oil content of sunflower under water stress. J Saudi Soc Agric Sci 13(1):9–13

Hu Y, Chen B (2020) Arbuscular mycorrhiza induced putrescine degradation into γ-aminobutyric acid, malic acid accumulation, and improvement of nitrogen assimilation in roots of water-stressed maize plants. Mycorrhiza 30:329–339. https://doi.org/10.1007/s00572-020-00952-0

Hussain M, Malik MA, Farooq M, Ashraf MY, Cheema MA (2008) Improving drought tolerance by exogenous application of glycinebetaine and salicylic acid in sunflower. J Agron Crop Sci 194:193–199

Hussain M, Bashir W, Farooq S, Rehim A (2013) Root development, allometry and productivity of maize hybrids under terminal drought sown by varying method. Int J Agric Biol 15:1243–1250

Hussain M, Farooq S, Hasan W, Ul-Allah S, Tanveer M, Farooq M, Nawaz A (2018) Drought stress in sunflower: physiological effects and its management through breeding and agronomic alternatives. Agric Water Manag 201:152–166

Hussain HA, Men S, Hussain S, Chen Y, Ali S, Zhang S, Zhang K, Li Y, Xu Q, Liao C, Wang L (2019) Interactive effects of drought and heat stresses on morpho-physiological attributes, yield, nutrient uptake and oxidative status in maize hybrids. Sci Rep 9:3890. https://doi.org/10.1038/s41598-019-40362-7

Iqbal N, Ashraf M, Ashraf MY (2009) Influence of exogenous glycine betaine on gas exchange and biomass production in sunflower (Helianthus annuus L.) under water limited conditions. J Agron Crop Sci 195:420–426

Iqbal MN, Rasheed R, Ashraf MY, Ashraf MA, Hussain I (2018) Exogenously applied zinc and copper mitigate salinity effect in maize (Zea mays L.) by improving key physiological and biochemical attributes. Environ Sci Pollut Res 25:23883–23896

Istanbuli T, Baum M, Touchan H, Hamwieh A (2020) Evaluation of morpho-physiological traits under drought stress conditions in barley (Hordeum vulgare L.). Photosynthetica 58:1059–1067. https://doi.org/10.32615/ps.2020.041

Jatoi WA, Baloch MJ, Kumbhar MB, Khan NU, Kerio MI (2011) Effect of water stress on physiological and yield parameters at anthesis stage in elite spring wheat cultivars. Sarhad J Agric 27:59–65

Jogawat A, Yadav B, Chhaya, Lakra N, Singh AK, Narayan OP (2021) Crosstalk between phytohormones and secondary metabolites in the drought stress tolerance of crop plants: a review. Physiol Plant 172:1106–1132. https://doi.org/10.1111/ppl.13328

Kalagare VS, Ganesan NM, Iyanar K, Chitdeshwari T, Chandrasekhar CN (2021) Strategy of multiple selection indices for discrimination of potential genotypes and associated traits for yield improvement in pearl millet [Pennisetum glaucum (L.) r.Br.]. Electron J Plant Breed 12:895–906. https://doi.org/10.37992/2021.1203.124

Kapoor D, Bhardwaj S, Landi M, Sharma A, Ramakrishnan M, Sharma A (2020) The impact of drought in plant metabolism: how to exploit tolerance mechanisms to increase crop production. Appl Sci 10:5692. https://doi.org/10.3390/app10165692

Khalifa TH, Mariey SA, Ghareeb ZE, Khatab IA, Alyamani A (2022) Effect of organic amendments and nano-zinc foliar application on alleviation of water stress in some soil properties and water productivity of barley yield. Agronomy 12(3):585

Khan MB, Hussain M, Abid R, Farooq S, Jabran K (2015) Seed priming with CaCl2 and ridge planting for improved drought resistance in maize. Turk J Agric For 39:193–203. https://doi.org/10.3906/tar-1405-39

Khan MN, Zhang J, Luo T, Liu J, Rizwan M, Fahad S, Xu Z, Hu L (2019) Seed priming with melatonin coping drought stress in rapeseed by regulating reactive oxygen species detoxification: antioxidant defense system, osmotic adjustment, stomatal traits and chloroplast ultrastructure perseveration. Ind Crop Prod 140:111597. https://doi.org/10.1016/j.indcrop.2019.111597

Khan MN, Khan Z, Luo T, Liu J, Rizwan M, Zhang J, Xu Z, Wu H, Hu L (2020) Seed priming with gibberellic acid and melatonin in rapeseed: consequences for improving yield and seed quality under drought and non-stress conditions. Ind Crop Prod 156:112850. https://doi.org/10.1016/j.indcrop.2020.112850

Kobraee S, Shamsi K, Rasekhi B (2011) Soybean production under water deficit conditions. Sch Res Libr 2:423–434

Kumar B, Pandey DM, Goswami CL, Jain S (2001) Effect of growth regulators on photosynthesis, transpiration and related parameters in water stressed cotton. Biol Plant 44:475–478. https://doi.org/10.1023/A:1012408624665

Kumar S, Islam ARMT, Islam HMT, Hasanuzzaman M, Ongoma V, Khan R, Mallick J (2021) Water resources pollution associated with risks of heavy metals from Vatukoula goldmine region, Fiji. J Environ Manage 293:112868. https://doi.org/10.1016/j.jenvman.2021.112868

Kyei-boahen S, Savala CEN, Chikoye D, Abaidoo R, Kyei-boahen S (2017) Growth and yield responses of cowpea to inoculation and phosphorus fertilization in different environments. Front Plant Sci 8:646

Lafitte HR, Yongsheng G, Yan S, Li ZK (2007) Whole plant responses, key processes, and adaptation to drought stress: the case of rice. J Exp Bot 58:169–175

Lanceras J, Pantuwan G, Jongdee B, Toojinda T (2004) Quantitative trait loci associated with drought tolerance at reproductive stage in rice. Plant Physiol 135:384–399

Leggett M, Diaz-Zorita M, Koivunen M, Bowman R, Pesek R, Stevenson C, Leister T (2017) Soybean response to inoculation with *Bradyrhizobium japonicum* in the United States and Argentina. Agron J 109(3):1031–1038

Lei C, Bagavathiannan M, Wang H, Sharpe SM, Meng W, Yu J (2021) Osmopriming with polyethylene glycol (PEG) for abiotic stress tolerance in germinating crop seeds: a review. Agronomy 11:2194. https://doi.org/10.3390/agronomy11112194

Li Y, Guan K, Schnitkey GD, DeLucia E, Peng B (2019) Excessive rainfall leads to maize yield loss of a comparable magnitude to extreme drought in the United States. Glob Chang Biol 25:2325–2337. https://doi.org/10.1111/gcb.14628

Liang Y, Tabien RE, Tarpley L, Mohammed AR, Septiningsih EM (2021) Transcriptome profiling of two rice genotypes under mild field drought stress during grain-filling stage. AoB Plants 13:plab043. https://doi.org/10.1093/aobpla/plab043

Liu JX, Liao DQ, Oane R, Estenor L, Yang XE, Li ZC, Bennett J (2006) Genetic variation in the sensitivity of anther dehiscence to drought stress in rice. Field Crop Res 97:87–100. https://doi.org/10.1016/j.fcr.2005.08.019

Liu L, Bai W, Li H, Tian Y, Yuan H, Garant TM, Liu H, Zhang J, Bao A, Rowland O, Wang S (2021) ZxABCG11 from the xerophyte Zygophyllum xanthoxylum enhances drought tolerance in Arabidopsis thaliana through modulating cuticular wax accumulation. Environ Exp Bot 190:104570

Macovei A, Balestrazzi A, Confalonieri M, Carbonera D (2010) The tyrosyl-DNA phosphodiesterase gene family in Medicago truncatula Gaertn.: bioinformatic investigation and expression profiles in response to copper- and PEG-mediated stress. Planta 232:393–407. https://doi.org/10.1007/s00425-010-1179-9

Mafakheri A, Siosemardeh A, Bahramnejad B, Struik PC, Sohrabi Y (2010) Effect of drought stress on yield, proline and chlorophyll contents in three chickpea cultivars. Aust J Crop Sci 4:580–585

Maleki A, Naderi A, Naseri R, Fathi A, Bahamin S, Maleki R (2013) Physiological performance of soybean cultivars under drought stress. Bull Environ Pharmacol Life Sci 2:38–44

Manickavelu A, Kawaura K, Oishi K, Shin-I T, Kohara Y, Yahiaoui N, Keller B, Abe R, Suzuki A, Nagayama T, Yano K, Ogihara Y (2012) Comprehensive functional analyses of expressed sequence tags in common wheat (*Triticum aestivum*). DNA Res 19:165–177. https://doi.org/10.1093/dnares/dss001

Marcos FCC, Silveira NM, Marchiori PER, Machado EC, Souza GM, Landell MGA, Ribeiro RV (2018) Drought tolerance of sugarcane propagules is improved when origin material faces water deficit. PloS One 13:e0206716. https://doi.org/10.1371/journal.pone.0206716

Marthandan V, Geetha R, Kumutha K, Renganathan VG, Karthikeyan A, Ramalingam J (2020) Seed priming: a feasible strategy to enhance drought tolerance in crop plants. Int J Mol Sci 21:8258. https://doi.org/10.3390/ijms21218258

Martinez JP, Silva H, Ledent JF, Pinto M (2007) Effect of drought stress on the osmotic adjustment, cell wall elasticity and cell volume of six cultivars of common beans (*Phaseolus vulgaris* L.). Eur J Agron 26:30–38

Mehboob N, Minhas WA, Naeem M, Yasir TA, Naveed M, Farooq S, Hussain M (2022) Seed priming with boron and. Crop Pasture Sci 73:494–502. https://doi.org/10.1071/CP21377

Minhas WA, Hussain M, Mehboob N, Nawaz A, Ul-Allah S, Rizwan MS, Hassan Z (2020) Synergetic use of biochar and synthetic nitrogen and phosphorus fertilizers to improves maize productivity and nutrient retention in loamy soil. J Plant Nutr 43(9):1356–1368

Nam NH, Chauhan YS, Johansen C (2001) Effect of timing of drought stress on growth and grain yield of extra-short-duration pigeonpea lines. J Agric Sci 136:179–189

Nayyar H, Kaur S, Singh S, Upadhyaya HD (2006) Differential sensitivity of desi (small-seeded) and Kabuli (large-seeded) chickpea genotypes to water stress during seed filling: effects on accumulation of seed reserves and yield. J Sci Food Agric 2082:2076–2082

Naz S, Perveen S (2021) Response of wheat (Triticum aestivum L. var. galaxy-2013) to pre-sowing seed treatment with thiourea under drought stress. Pak J Bot 53:1209–1217

Nofouzi F (2018) Evaluation of seed yield of durum wheat (Triticum durum) under drought stress and determining correlation among some yield components using path coefficient analysis. UNED Res J 10. https://doi.org/10.22458/urj.v10i1.2023

O'Toole JC (2004) Rice and water: the final frontier. Rockefeller Foundation, Bangkok

Ogbonnaya CI, Sarr B, Brou C, Diouf O, Diop NN, Roy-Macauley H (2003) Selection of cowpea genotypes in hydroponics, pots, and field for drought tolerance. Crop Sci 43:1114–1120

Omara AED, Hafez EM, Osman HS, Rashwan E, El-Said MA, Alharbi K, Abd El-Moneim D, Gowayed SM (2022) Collaborative impact of compost and beneficial rhizobacteria on soil properties, physiological attributes, and productivity of wheat subjected to deficit irrigation in salt affected soil. Plan Theory 11(7):877

Paparella S, Araújo SS, Rossi G, Wijayasinghe M, Carbonera D, Balestrazzi A (2015) Seed priming: state of the art and new perspectives. Plant Cell Rep 34:1281–1293. https://doi.org/10.1007/s00299-015-1784-y

Parveen A, Liu W, Hussain S, Asghar J, Perveen S, Xiong Y (2019) Silicon priming regulates morpho-physiological growth and oxidative metabolism in maize under drought stress. Plan Theory 8(10):431

Porcel R, Ruiz-lozano JM (2004) Arbuscular mycorrhizal influence on leaf water potential, solute accumulation, and oxidative stress in soybean plants subjected to drought stress. J Exp Bot 55:1743–1750

Prasad PVV, Pisipati SR, Mom'cilovic I., and Ristic, Z. (2011) Independent and combined effects of high temperature and drought stress during grain filling on plant yield and chloroplast EF-Tu expression in spring wheat. J Agron Crop Sci 197:430–441

Raheem A, Shaposhnikov A, Belimov AA, Dodd IC, Ali B (2018) Auxin production by rhizobacteria was associated with improved yield of wheat (Triticum aestivum L.) under drought stress. Arch Agron Soil Sci 64:574–587. https://doi.org/10.1080/03650340.2017.1362105

Riaz M, Farooq J, Sakhawat G, Mahmood A, Sadiq MA, Yaseen M (2013) Genotypic variability for root/shoot parameters under water stress in some advanced lines of cotton (Gossypium hirsutum L.). Genet Mol Res 12:552–561. https://doi.org/10.4238/2013.February.27.4

Romdhane L, Awad YM, Radhouane L, Dal Cortivo C, Barion G, Panozzo A, Vamerali T (2019) Wood biochar produces different rates of root growth and transpiration in two maize hybrids (*Zea mays* L.) under drought stress. Arch Agron Soil Sci 65(6):846–866

Rosales-Serna R, Kohashi-Shibata J, Acosta-Gallegos JA, Trejo-López C, Ortiz-Cereceres J, Kelly JD (2004) Biomass distribution, maturity acceleration and yield in drought-stressed common bean cultivars. Field Crop Res 85:203–211

Saikia J, Sarma RK, Dhandia R, Yadav A, Bharali R, Gupta VK, Saikia R (2018) Alleviation of drought stress in pulse crops with ACC deaminase producing rhizobacteria isolated from acidic soil of Northeast India. Sci Rep 8(1):1–16

Samarah NH, Mullen RE, Cianzio SR, Scott P (2006) Dehydrin-like proteins in soybean seeds in response to drought stress during seed filling. Crop Sci 46:2141–2150

Samarah NH, Haddad N, Alqudah AM (2009) Yield potential evaluation in chickpea genotypes under lateterminal drought in relation to the length of reproductive stage. Ital J Agron 3:111–117

Sattar A, Sher A, Ijaz M, Ul-Allah S, Butt M, Irfan M, Rizwan MS, Ali H, Cheema MA (2020) Interactive effect of biochar and silicon on improving morpho-physiological and biochemical attributes of maize by reducing drought hazards. J Soil Sci Plant Nutr 20(4):1819–1826

Sen A, Puthur JT (2020) Influence of different seed priming techniques on oxidative and antioxidative responses during the germination of Oryza sativa varieties. Physiol Mol Biol Plants 26:551–565. https://doi.org/10.1007/s12298-019-00750-9

Serraj R, McNally KL, Slamet-Loedin I, Kohli A, Haefele SM, Atlin G, Kumar A (2011) Drought resistance improvement in Rice: an integrated genetic and resource management strategy. Plant Prod Sci 14:1–14. https://doi.org/10.1626/pps.14.1

Shamsi K, Petrosyan M, Noor-Mohammadi G, Haghparast R (2010) The role of water deficit stress and water use efficiency on bread wheat cultivars. J Appl Biosci 35:2325–2331

Sharma M, Delta AK, Kaushik P (2021) Glomus mosseae and Pseudomonas fluorescens application sustains yield and promote tolerance to water stress in Helianthus annuus L. Stress 1:305–316. https://doi.org/10.3390/stresses1040022

Shrestha RA, Turner NCA, Siddique KHMA, Turner DWB, Speijers JC (2006) A water deficit during pod development in lentils reduces flower and pod numbers but not seed size. Aust J Agr Res 57:427–438

Sissons M, Pleming D, Taylor JD, Emebiri L, Collins NC (2018) Effects of heat exposure from late sowing on the agronomic and technological quality of tetraploid wheat. Cereal Chem 95:274–287. https://doi.org/10.1002/cche.10027

Srivastava AK, Lokhande VH, Patade VY, Suprasanna P, Sjahril R, D'Souza SF (2010) Comparative evaluation of hydro-, chemo- , and hormonal-priming methods for imparting salt and PEG stress tolerance in Indian mustard (Brassica juncea L.). Acta Physiol Plant 32:1135–1144. https://doi.org/10.1007/s11738-010-0505-y

Stancheva I, Geneva M, Hristozkova M, Sichanova M, Donkova R, Petkova G, Djonova E (2017) Response of Vigna Unguiculata grown under different soil moisture regimes to the dual inoculation with nitrogen-fixing bacteria and arbuscular mycorrhizal fungi. Commun Soil Sci Plant Anal 48:1378–1386

Tabassum T, Ahmad R, Farooq M, Basra SMA (2018) Improving the drought tolerance in barley by osmopriming and biopriming. Int J Agric Biol 20:1597–1606. https://doi.org/10.17957/IJAB/15.0678

Thomas DT, Puthur JT (2019) Amplification of abiotic stress tolerance potential in rice seedlings with a low dose of UV-B seed priming. Funct Plant Biol 46:455. https://doi.org/10.1071/FP18258

Tůmová L, Tarkowská D, Řehořová K, Marková H, Kočová M, Rothová O, Čečetka P, Holá D (2018) Drought-tolerant and drought-sensitive genotypes of maize (Zea mays L.) differ in contents of endogenous brassinosteroids and their drought-induced changes. PloS One 13:e0197870. https://doi.org/10.1371/journal.pone.0197870

Ul-Allah S, Khan AA, Fricke T, Buerkert A, Wachendorf M (2014) Fertilizer and irrigation effects on forage protein and energy production under semi-arid conditions of Pakistan. Field Crop Res 159:62–69

Ul-Allah S, Khan AA, Fricke T, Buerkert A, Wachendorf M (2015) Effect of fertiliser and irrigation on forage yield and irrigation water use efficiency in semi-arid regions of Pakistan. Exp Agric 51(4):485–500

Ul-Allah S, Ijaz M, Nawaz A, Sattar A, Sher A, Naeem M, Shahzad U, Farooq U, Nawaz F, Mahmood K (2020) Potassium application improves grain yield and alleviates drought susceptibility in diverse maize hybrids. Plan Theory 9(1):75

Ullah H, Datta A (2018) Root system response of selected lowland Thai rice varieties as affected by cultivation method and potassium rate under alternate wetting and drying irrigation. Arch Agron Soil Sci 64(14):2045–2059. https://doi.org/10.1080/03650340.2018.1476756

Ullah A, Farooq M (2022) The challenge of drought stress for grain legumes and options for improvement. Arch Agron Soil Sci 68:1601–1618. https://doi.org/10.1080/03650340.2021.1906413

Ulzen J, Abaidoo RC, Mensah NE, Masso C, AbdelGadir AH (2016) Bradyrhizobium inoculants enhance grain yields of soybean and cowpea in northern Ghana. Front Plant Sci 7:1770

Usmani MM, Nawaz F, Majeed S, Shehzad MA, Ahmad KS, Akhtar G, Aqib M, Shabbir RN (2020) Sulfate-mediated drought tolerance in maize involves regulation at physiological and biochemical levels. Sci Rep 10:1147. https://doi.org/10.1038/s41598-020-58169-2

Valliere JM, Zhang J, Sharifi MR, Rundel PW (2019) Can we condition native plants to increase drought tolerance and improve restoration success? Ecol Appl 29. https://doi.org/10.1002/eap.1863

Varshney RK, Thudi M, Nayak SN, Gaur PM, Kashiwagi J, Krishnamurthy L, Jaganathan D, Koppolu J, Bohra A, Tripathi S et al (2014) Genetic dissection of drought tolerance in chickpea (*Cicer arietinum* L.). Theor Appl Genet 127:445–462

Vurukonda SSKP, Vardharajula S, Shrivastava M, SkZ A (2016) Enhancement of drought stress tolerance in crops by plant growth promoting rhizobacteria. Microbiol Res 184:13–24

Wahab A, Abdi G, Saleem MH, Ali B, Ullah S, Shah W, Mumtaz S, Yasin G, Muresan CC, Marc RA (2022) Plants' physio-biochemical and phyto-hormonal responses to alleviate the adverse effects of drought stress: a comprehensive review. Plan Theory 11:1620

Wang Y, Xu C, Zhang B, Wu M, Chen G (2017) Physiological and proteomic analysis of rice (Oryza sativa L.) in flag leaf during flowering stage and milk stage under drought stress. Plant Growth Regul 82:201–218. https://doi.org/10.1007/s10725-017-0252-9

Wang X, Mao Z, Zhang J, Hemat M, Huang M, Cai J, Zhou Q, Dai T, Jiang D (2019) Osmolyte accumulation plays important roles in the drought priming induced tolerance to post-anthesis drought stress in winter wheat (Triticum aestivum L.). Environ Exp Bot 166:103804. https://doi.org/10.1016/j.envexpbot.2019.103804

Wang F, Gao J, Yong JWH, Wang Q, Ma J, He X (2020) Higher atmospheric CO2 levels favor C3 plants over C4 plants in utilizing ammonium as a nitrogen source. Front Plant Sci 11:537443. https://doi.org/10.3389/fpls.2020.537443

Waraich EA, Ahmad R, Ashraf MY, Saifullah and Ahmad, M. (2011) Improving agricultural water use efficiency by nutrient management in crop plants. Acta Agric Scand Sect B Soil Plant Sci 61(4):291–304

Wassmann R, Jagadish SVK, Heuer S, Ismail A, Redona E, Serraj R, Singh RK, Howell G, Pathak H, Sumfleth K (2009) Chapter 2 climate change affecting rice production. In: Advances in agronomy, pp 59–122. https://doi.org/10.1016/S0065-2113(08)00802-X

Waterworth WM, Masnavi G, Bhardwaj RM, Jiang Q, Bray CM, West CE (2010) A plant DNA ligase is an important determinant of seed longevity. Plant J 63:848–860. https://doi.org/10.1111/j.1365-313X.2010.04285.x

Wei Y, Jin J, Jiang S, Ning S, Liu L (2018) Quantitative response of soybean development and yield to drought stress during different growth stages in the Huaibei plain, China. Agronomy 8:97

Wellstein C, Poschlod P, Gohlke A, Chelli S, Campetella G, Rosbakh S, Canullo R, Kreyling J, Jentsch A, Beierkuhnlein C (2017) Effects of extreme drought on specific leaf area of grassland species: a meta-analysis of experimental studies in temperate and sub-Mediterranean systems. Glob Chang Biol 23:2473–2481

Widuri LI, Lakitan B, Sodikin E, Hasmeda M, Meihana M, Kartika K, Siaga E (2018) Shoot and root growth in common bean (Phaseolus vulgaris L.) exposed to gradual drought stress. AGRIVITA J Agric Sci 40:442–452

Wojtyla Ł, Lechowska K, Kubala S, Garnczarska M (2016) Molecular processes induced in primed seeds—increasing the potential to stabilize crop yields under drought conditions. J Plant Physiol 203:116–126. https://doi.org/10.1016/j.jplph.2016.04.008

Yadav B, Jogawat A, Rahman MS, Narayan OP (2021) Secondary metabolites in the drought stress tolerance of crop plants: a review. Gene Rep 23:101040. https://doi.org/10.1016/j.genrep.2021.101040

Zhang X, Lei L, Lai J, Zhao H, Song W (2018) Effects of drought stress and water recovery on physiological responses and gene expression in maize seedlings. BMC Plant Biol 18:68. https://doi.org/10.1186/s12870-018-1281-x

Zheng M, Tao Y, Hussain S, Jiang Q, Peng S, Huang J, Cui K, Nie L (2016) Seed priming in dry direct-seeded rice: consequences for emergence, seedling growth and associated metabolic events under drought stress. Plant Growth Regul 78:167–178. https://doi.org/10.1007/s10725-015-0083-5

Impact of Heat Stress on Cereal Crops and Its Mitigation Strategies

Naeem Sarwar, Khuram Mubeen, Atique-ur-Rehman, Omer Farooq, Allah Wasaya, Tauqeer Ahmad Yasir, Muhammad Shahzad, Mansoor Javed, Abrar Hussain, Masood Iqbal Awan, Muhammad Dawood, and Shakeel Ahmad

Abstract

Cereals are important food crops all over the world. Cereals are consumed by human beings in one way or another way to get energy. However, due to climate change and other production and environmental effects, the yield is declining yearly. Environmental factors are affecting the phenology of cereal crops all over the world. In order to feed the burgeoning population of the globe, there is a dire need to adapt to climate change using different adaptation and mitigation strategies.

N. Sarwar (✉) · Atique-ur-Rehman · O. Farooq · A. Hussain
Department of Agronomy, Faculty of Agricultural Sciences and Technology, Bahauddin Zakariya University, Multan, Pakistan
e-mail: naeemsarwar@bzu.edu.pk

K. Mubeen
Department of Agronomy, Muhammad Nawaz Sharif University of Agriculture, Multan, Pakistan

A. Wasaya · T. A. Yasir · S. Ahmad
Department of Agronomy, Faculty of Agricultural Sciences and Technology, Bahauddin Zakariya University, Multan, Pakistan

M. Shahzad
Department of Agronomy, University of Poonch, Rawalakot, Azad Jammu and Kashmir, Pakistan

M. Javed
Department of Agronomy, University of Sargodha, Sargodha, Pakistan

M. I. Awan
Department of Agronomy, Sub-Campus Depalpur, Okara, University of Agriculture, Faisalabad, Pakistan

M. Dawood
Department of Biosciences, Comsat University, Sahiwal Campus, Sahiwal, Punjab, Pakistan

M. Ahmed, S. Ahmad (eds.), *Disaster Risk Reduction in Agriculture*, Disaster Resilience and Green Growth, https://doi.org/10.1007/978-981-99-1763-1_10

Keywords

Cereals · Climate change · Phenology · Adaptation and mitigation strategies

1 Introduction

Global warming has raised the temperature in many parts of the world, which is continuously rising due to the accumulation of greenhouse gases in the atmosphere. Rising temperature have worried crop producers as this condition create uncertainty for different cropping operations. The world population is also rising daily, which needs to boost crop production for food security, but the climate uncertainty or the elevated temperature have produced a challenging situation. We were already going deficient in cereal production, but this emerging heat stress may worsen this situation, making it difficult to maintain food security.

Wheat, maize, and rice crops are the staple food for most of the world's population. By 2050, the world population is predicted to increase to 9.8 billion, which needs a 70–100% increase in cereal crop. Boosting crop production to meet the large population for ensuring food security is a dire need of the time, but the current status revealed that we are far away from the target (Ali et al. 2021; Sarwar et al. 2022; Ray et al. 2013; Godfray et al. 2010). Current agricultural practices have further worsened the situation by decreasing the fertile or arable land to grow these crops. It is also expected that land will further decrease in the future due to climatic change and other abiotic factors (Huang et al. 2017). Many researchers have agreed that future climate change predictions will drastically affect crop production, which must be addressed to meet food security (Baker et al. 2018; Cai et al. 2018). It is also predicted that dryland will be affected more by elevated temperatures as compared to the humid region (Huang et al. 2017) Average temperature of the earth is increasing by almost 0.2 °C per decade, which reveals that our planet will be warmer in the future. Moreover, it is estimated that up to the end of the twenty-first century, world temperature will rise by 2.5 to 4.5 °C due to the accumulation of greenhouse gases (Lehner et al. 2017; Solomon 2007).

Increasing temperature induces heat stress on crop production (Sun et al. 2017). Reducing global warming can benefit many parts of the world. In Africa, stabilizing global warming can reduce one-fifth of heat stress events (Russo et al. 2019). Similarly, about 46% of heat stress events in East Asia can be decreased (Li et al. 2018). Both developing and developed world can benefit if global warming stabilized to just 1.5% rather than 2% (Russo et al. 2019). Global warming may create many issues for crop production as it will enhance respiration, reduce net carbon gain, and increase plant protection investment by introducing invasive weeds and other pests (Cai et al. 2018). The study revealed that wheat crops might reduce the crop yield by about 3 to 10% if the temperature rises by 1 °C (Sun et al. 2017).

In areas where the population is increasing up to an unsustainable level, any decrease in food production can significantly enhance the food security problem. Therefore, it is the need of the time that scientific communities from all over the world, will sit together and find some alternative crops which can sustain in this changing climate without decreasing production. The crops would also be able to release less GHGs into the atmosphere for sustainable crop production. Similarly, the community also needs crops that can fulfill the nutritional demand (Vijayan 2016). Many crops can meet the above criteria. For example, the millet crop has many desirable attributes in comparison with wheat, maize, and rice (Jain et al. 2016). Millet is a short-duration crop that can mature in 60–100 days, depending upon the cultivars and could be sown across wider ranges of weather conditions (Halilou et al. 2020).

It is also reported that millet release less greenhouse gases into the environment, which can reduce the problem of global warming. This crop can also fulfill the dietary or nutritional needs of the consumers, just like other major crops for managing food security (Jain et al. 2016). In comparison with other crops like maize or rice, millet crop is less resource intensive as it can grow on less fertile soil and also need less water input. Rice crops need lot of irrigation water for its cultivation and cannot survive under drought conditions (Yao et al. 2012). There are many other tools to enhance crop production, like breeding or some agronomic techniques, e.g., improving nutrient use efficiency, etc., while climate change affects crop production and severely induces heat and soil moisture stress (Lesk et al. 2016; Farooq et al. 2009). It is also making it difficult to control insect pest, diseases, and weeds management. Many new weed species and insects are emerging nowadays due to climate change affecting crop productivity (Porter et al. 2014). Researchers have warned that global warming is irreversible, and we need sustainable approaches to manage the environmental damage. This chapter summarizes the climate change effect, especially the high temperature on crop production and its expected management strategies.

2 Climate Change and Crop Response

2.1 Rice (*Oryzae sativa* L.)

Numerous researches revealed that the final yield of rice is significantly influenced by the growing conditions (such as irrigation, temperature, and fertilizer availability). According to the study, paddy yields under alternative wetting and drying technique was significantly lower than continuous flooding irrigation (Ali et al. 2021; Ye et al. 2013; Sarwar et al. 2022). The controlled release of nitrogen fertilizer under the alternating wetness and drying (AWD) irrigation method was also examined in the same study. In 2-year experimentations, irrigation water was reduced by 28% and 42% in wetting and drying methods compared to the flooding approach. This approach performed better with larger rice yields and biomass like root, shoot, and panicle despite reducing consumption. Especially, AWD-enhanced yield by 5.7–6.6% throughout the 2 years study. With the alternate wetting and drying irrigation method, the yield of fields fertilized by conventional urea, controlled-release blended fertilizer, and polymer-coated urea at 240 kg N/ha each rose by 62–72%,

56–67%, and 80–93%. Additionally, this technique was tested in certain other states as well that cultivate rice, such as India, Pakistan, and the Philippines, demonstrating the ability of AWD system to conserve water without sacrificing crop production (Sarwar et al. 2013a, b; Ye et al. 2013). Similarly, a study found that controlled irrigation treatments had a lower paddy yield (8516–8346 kg/ha) than alternate wetting and drying irrigation (6532–7868 kg/ha) after 2 years of experimentation (Shao et al. 2014). Furthermore, a number of studies have demonstrated that global warming may negatively affect the yields of paddy grown all over the world. The average global temperature has risen by 0.5 to 0.6 °C over the last century (Lehner et al. 2017). The plant's respiration has increased as a result of the rise in temperature, which has also increased carbon metabolism and reduced paddy output (Zhao and Fitzgerald 2013). The reproduction process may be hampered by the paddy flowers becoming sterile as a result of the higher temperatures.

According to a recent assessment from the International Food Policy Research Institute (IFPRI), climate change might decrease rice output by 10–15% which could lead to a 32 to 37% increase in market price (Li 2018).

In order to maintain yields, the study advised selecting varieties with shorter maturing and ripening times. Numerous studies have suggested that global warming will raise rice yields by 2.9–34% under highly optimistic assumptions. This is mostly because of nighttime warming and the shifting of production to regions with colder climates now but hotter ones in the 2050s and 2080s (Piao et al. 2010). However, these studies do not always consider the socioeconomic effects in the areas where rice is currently grown or the strategies that farmers have to deal with losses of paddy production. Additionally, despite a minor upsurge in production, it has been shown in numerous studies (Welch et al. 2010) that when the world's temperature rises toward the end of the century, yields will plummet dramatically. Future generations will experience food insecurity as a result of global warming if effective methods to address decreased yields are not adopted quickly.

2.2 Millet (*Pannisetum glacum*)

The deep roots of millet plants are beneficial because they may utilize the remaining nitrogen, potassium, and phosphorus in the soil. Therefore, compared to other crops, the deep-root system does not need a large amount of fertilizer. Over 50% of the millet produced worldwide is pearl millet (*Pennisetum glaucum*) (Singh et al. 2017). Pearl millet grows more slowly than conventional cereal crops and can thrive in unfavorable conditions (such as a lack of water and fertilizer) because of its physiological traits (Debieu et al. 2018).

Numerous researches revealed that rising temperatures increased millets' output in dry regions. In three distinct Chinese cities, a study found that millet yields increased on average year as the so temperature rose, from 30 to 121 kg/ha (Xifeng, Anding, and Ganzhou). Due to the province's increasing temperatures, they advise expanding millet plantations (Cao et al. 2010). Moreover, a study used a modified version of the CSM-CERES-Pearl millet simulation model to simulate the effect of temperature change on millet yield (Singh et al. 2017). According to the findings, millet output rose by 6% following a drought and by 8% following a heat simulation (upsurge from 27 to 29 °C). When drought tolerance along with heat tolerance was taken into account, the millet yield went up by 14%. Millets are also capable of growing in high and hilly areas, where cultivation of other cereals is difficult. This demonstrates that millets have the probability of being an important crop that can flourish in arid, hot, and mountainous areas of the planet with little nitrogen input (Ashraf et al. 2018). This particular crop has the greatest potential to cut carbon emissions because it has the lowest potential for global warming while also being resistant to its impacts, which include greater frequency of droughts and higher average temperatures.

2.3 Maize (*Zea maize*)

Maize yield is decreasing due to many factors associated with climate change that leads to the deficiency of irrigation water and soil nutrients (Sarwar et al. 2021). For example, modeling crop productivity showed that maize productivity in Malawi would fall by 33% by the end of the century and by 14% by mid-century (Msowoya et al. 2016) due to climate change. Similarly, global warming modeling predicted that in the northeastern part of China, the extreme temperature increase (where temperature increases by 1.32 °C) will result in a 35% drop in maize production in 2030 compared to productivity in 2008. Between 1970 and 1999, the average maize yield in the United States decreased by 2.5% as a result of global warming, and according to precipitation modeling, corn yields are expected to further decline by 20 to 50% by the 2050s, depending on the present emission scenarios (Leng and Huang 2017).

Researchers have also assessed a 2 °C increase in temperature, along with a 20% decrease in rainfall, affects maize production in Africa. Under these circumstances, they claimed that the yields dropped by around 10% (Lobell and Burke 2010). According to their findings, under both rainfed and drought situations, a loss of between 1% and 1.7% was seen in maize productivity every day of the crop was exposed to temperatures over 30 °C. Furthermore, they demonstrated that maize's capacity to withstand the high temperatures brought on by global warming depends critically on moisture (Lobell et al. 2011).

The yield sensitivity to temperature and atmospheric CO_2 was examined in a thorough study on maize yield variation was issued in 2014. Total of 23 models were applied three throughout four distinct continents (France, USA, Brazil, and Tanzania). The yields decreased by roughly 0.5 t/ha for every 1 °C increase, demonstrating a close link between temperature and yields. However, the study found that when atmospheric CO_2 levels doubled from 360 to 720 ppm, yields increased slightly (with significant uncertainty). The study concluded that catastrophic production losses in maize are possible worldwide if new management strategies are not used to prevent the rising temperature. Additionally, the anticipated increase in yields brought on by the rise in atmospheric CO_2 cannot offset these losses (Bassu et al. 2014).

2.4 Wheat (*Triticum aestivum*)

The production of wheat, one of the most prevalent grains, will be significantly affected by global climate change and extreme weather phenomena. It is grown as a staple food in many countries (Sarwar et al. 2021), so its lesser production can create a food security issue for many countries. A modeling strategy was used in France to examine winter wheat yields. This approach was based on baseline productivity and gridded meteorological data that was available from 1950 to 2015. Model projected that, during the medium term from 2037 to 2065, the wheat yield would be lowered by climate change by 3.5 to 12.9% and that, by the end of the century, winter wheat yields would be down by 14.6 to 17.2% (Gammans et al. 2017). Based on historical data from 1970 to 2000, researchers in China concluded that a 1 °C increase in temperature during the growing season would result in a 3 to 10% reduction in wheat output rates. The same study also stated that if it were not for increasing resource use (such as water, fertilizers, etc.) to cultivate crops, the temperature increase over the past two decades would have caused yields to drop by 4.5%.

Experimental studies on wheat yields between 1981 and 2009 in China revealed significant regional differences in wheat productivity. They discovered that although Southern China saw a fall of 1% to 10%, wheat productivity in Northern China increased by 1–13% (Tao et al. 2014). Since wheat is a key component of numerous food products worldwide, this could increase the price of wheat on the market. Water is one of the most crucial elements for healthy development, balanced growth, and excellent wheat yields. Winter wheat was grown using two different cultivars (Baviaans and 14SAWYT306) and three different irrigation schedules or durations, with no irrigation in the control condition. The three-schedules were (a) irrigation prior to the development of stem extension, (b) irrigation following the development of the stem extension until physiological maturity, and (c) irrigation during the crop's growth. In comparison to no irrigation treatment, the results showed that irrigation treatments considerably boosted grain production for both cultivars. The tillers, the grains/spike, the yield, the harvest index, along with the seed protein all rose in comparison to no irrigation during the course of the growth period by 20.58%, 26.07%, 42.72%, 16.71%, and 3.31%, respectively.

3 Production Status of Cereals Crops

Worldwide data for cereals production revealed that America allocated the highest area for maize crop production, followed by Asia. In America, maize was cultivated in an area of 70.1 Mha while in Asia, 63.1 Mha area was cultivated for maize crop production in the year 2016 (FAOSTAT). Similarly, America produced about 52% of world maize production after Asia, which contributed about 31%. Asia is a major continent where most of the population produces 90% of the world's rice, 43.2% millet, and 43.6% wheat. These crops were cultivated on an area of 140.5 M ha, 10.9 M ha and 100.5 M ha (Figs. 1 and 2). Data revealed that the top producer of rice was China, India, Indonesia, Bangladesh, and Vietnam. The highest rice and

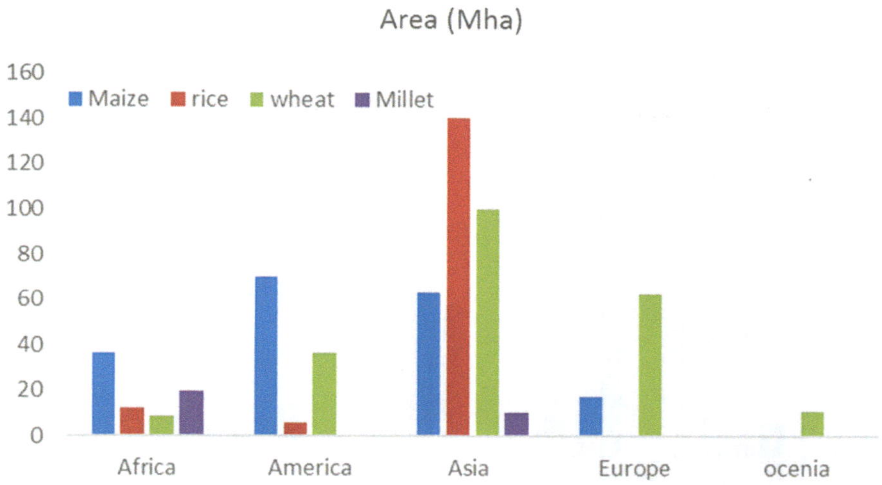

Fig. 1 Area of cereals crops in different continents

wheat production were calculated in China, followed by India. Russia, and USA were also among the top producers of wheat. Cereals production statistics revealed China is the largest cereal-producing country, followed by India. A large portion of the population (36.7%) resides in these countries, so they produce large quantities of cereals to meet food security. But China produced cereals at a higher rate as compared to India. Similarly, other Asian countries like Pakistan, Bangladesh, Nepal, and Sri Lanka produce large quantities of cereals to feed their population, but often they import cereals from neighboring countries. Growing population and recent climate change calamities have created many issues for these countries, accelerating the deterioration of natural resources (Lal 2017; Mo et al. 2017).

4 Growing Conditions for Various Cereal Crops

4.1 Rice (*Oryzae sativa* L.)

Rice is grown on flooded soil in more than 95 countries and feeds a large portion of the world's population. It is grown as the first and second staple food in many Asian countries where people are heavily dependent on rice production. Rice crops need a significant amount of irrigation water (500–600 mm) for optimum growth and yield (Ali et al. 2021). The optimum temperature should be between 22 to 31 °C, and 5–6 h of sunshine are needed for its better growth (Yao et al. 2012). Many studies reported that climatic factors like rainfall, temperature, and irrigation greatly impact rice crops. It reported that rice crops produced grain yields up to 8.23 t/ha under a continuous flooding system while reducing soil moisture reduced the crop yield. Under the current scenario, due to water shortage, farmers are looking for other rice

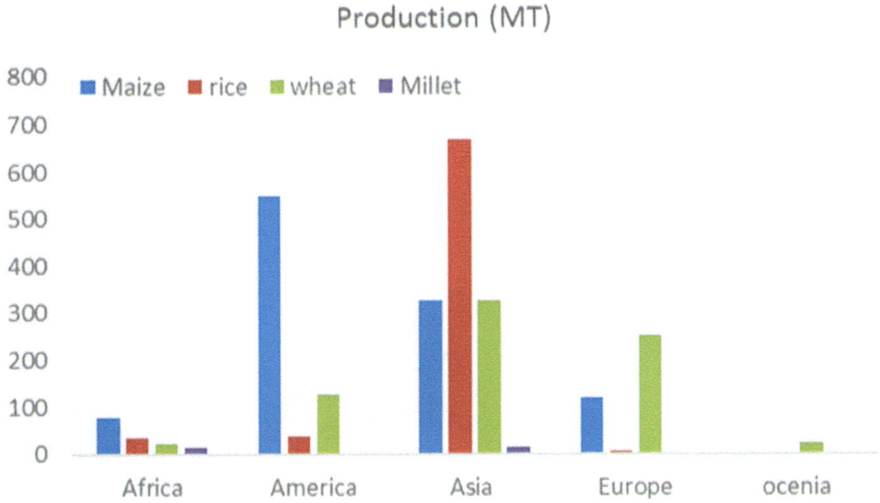

Fig. 2 Production of cereals in different continents

production systems like aerobic rice, alternate wetting, and drying to save irrigation water. These systems have the potential to save water, but crop yield is reduced linearly to water saved (Sarwar et al. 2022). Similarly, the crop grows well under 5 to 6.5 soil pH, while changing soil pH due to the application of various chemicals or flooding/drying effect the crop yield. Moreover, this crop requires 90–120 kg/ha nitrogen, 30-40 kg/ha phosphorus, and 40–60 kg/ha potassium (Santos et al. 2017). Similarly, rice crop also needs to protect from insect pest for better productivity. Due to this, many pesticides are being used on rice crop, further contaminating the soil and the environment (Lamers et al. 2011). Climate change has increased the frequency of rainfall, which creates flooding conditions in many areas of the world. Floods destroy the crops and mix up the soil-applied nutrients in the water bodies, which further leach down into the underground water. Recent floods in Pakistan have created a catastrophic conditions and destroyed a large portion of the crops.

4.2 Millet (*Pannisetum glacum*)

Millet is a very important cereal crop grown all over the world for thousands of years due to its higher nutritive value. It is grown as a major crop in different parts of the world, especially in Asian countries like India, Pakistan, China, and Africa. Different millet species are grown, especially the proso, pearl, finger, kodo, and foxtail. It is drought-resistant and can complete its life cycle in a short duration (Habiyaremye et al. 2017). It has a robust rooting system and usually needs no fertilizer. Being a C4 crop, fixes CO_2 efficiently to produce its dry matter and can withstand high temperatures due to its lower transpiration rate (Goron and Raizada 2015). It can grow under 20–35 °C and require 4–6 h of sunshine/day. It could be sown on a wide range of soil like sandy, alkaline, and acidic, while optimum pH range is 4.5 to 8. The most important factor is minimum nutrient requirement for optimum growth, which is the most favorable factor in developing countries where inputs rats are very high. Moreover, plant protection measures are very easy in this crop as most of the cultivars are disease resistant, which can provide a key role in sustainable crop production. The use of pesticides can be reduced to many folds by adopting this crop in our cropping system.

4.3 Maize (*Zea maize*)

Maize crop is also known as corn and is grown all over the world due to its wide usage (Sarwar et al. 2021). It is the oldest crop and was grown more than 7000 years ago in Mexico and spread later on in other neighboring countries (Piperno and Flannery 2001). Maize crop ranked third after wheat and rice production. It has also grown as a major crop in most developed or developing countries. Maize grain is nutritious, containing a higher percentage of carbohydrates (76–88%), protein (6–16%), fat (4–7%), and minerals (1.3%) (Sarwar et al. 2021; Woldesenbet and Haileyesus 2016). It is a warm season crop with an optimum temperature range of 11 to 30 °C and requires 5–6 h of sunshine daily. It can be cultivated on moist soil having pH 5.8 to 7 range. It requires 200 to 450 mm of rainfall for its growing season of 90 to 100 days (Yin et al. 2014). It can produce good crop yield by providing soil nutrients at 125–160 kg N/ha, 80 kg P/ha, and 85–110 kg K/ha. It is also a C4 crop that can use CO_2 and soil moisture efficiently and produce good crop yield compared to C3. These features also enable the maize plant to resist high temperatures (Woldesenbet and Haileyesus 2016; Yin et al. 2014).

4.4 Wheat (*Triticum aestivum*)

Wheat is the most extensively grown and consumed cereal crop in several areas of the world (Sarwar et al. 2021). It is classified as winter or spring wheat as per growing climatic conditions (Ngwako and Mashiqa 2013). Winter wheat germinated before freezing temperatures and remained dormant until the warm season. It also requires some weeks of cold temperatures for the initiation of flowering. Normally, it is grown in October–November and harvested in the summer season in the months of June–July. Spring wheat is sown in March–May and harvested in July–September after reaching maturity. It can complete its growth and reproductive cycle in 4 months, while winter wheat goes longer than this duration. It can grow under 3–23 °C and have 120–180 growing periods with 4–6 h of sunshine/day. Wheat can be sown on sandy loam soil and 5.5–6.5 is the optimum pH for successful growth. It requires 70–200 kg N/ha, 20–40 kg P/ha, and 80 to 100 kg K/ha but can vary a little bit as per soil condition (Hergert and Shaver 2009) (Table 1).

5 Food Security

Food availability, accessibility, consumption, and the stability of food systems make up food security (Zewdie 2014). Food security may be threatened by climate change since it has the potential to reduce agricultural yields and exacerbate competition for limited resources (Azadi et al. 2013). As the climate changes, food insecurity is becoming a problem in an increasing number of nations, as seen in Fig. 3. Figure 3 showed that high levels of food insecurity were expected in Mongolia, Mauritania, Niger, Chad, Somalia, and Namibia by the end of 2017. This appears to have been the case. By the end of the 2050s, nations like India, Zambia, Myanmar, Egypt, and Botswana will join a list of those dealing with serious food insecurity challenges, according to the World Food Programme (Fig. 4). In addition, the majority of African and Asian nations would experience severe food insecurity by the end of 2080s due to outcome of up surging temperatures and diminishing resources (Fig. 5) (Zewdie 2014).

6 Food Insecurity (2080s)

Total food quantity with satisfactory quality that is accessible for food, either by domestic production or importation, is referred to as food availability. This is one of the main criteria used to assess food security (Zewdie 2014). Climate change is currently detrimental to food security because numerous predictions point to a considerable decline in the production of various crops (Thompson et al. 2010). Numerous research revealed that climate change could lower wheat yields by 3.5% to 12.9% (Gammans et al. 2017), of maize by 34.6 to 35.4%, and of paddy by 10 to 15%.

Table 1 Required conditions for various cereal crops

Crops	Temperature (°C)	Sunshine (h/day)	Growth duration (days)	Soil pH	Soil type	References
Rice	22–31	4–6	90–120	5.0–6.5	Flooded field	Yao et al. (2012), Faostat and Production (2017)
Maize	11–30	6–7	90–110	5.8–7.0	Warm and silt loams	Faostat and Production (2017)
Wheat	−3–23	4–6	120–180	5.5–6.5	Sandy loam	Faostat and Production (2017), Acevedo et al. (2002)
Millet	20–35	4–5	60–100	4.5–8.0	Saline, sandy, clay loams	Habiyaremye et al. (2017), Goron and Raizada (2015) and Changmei and Dorothy (2014)

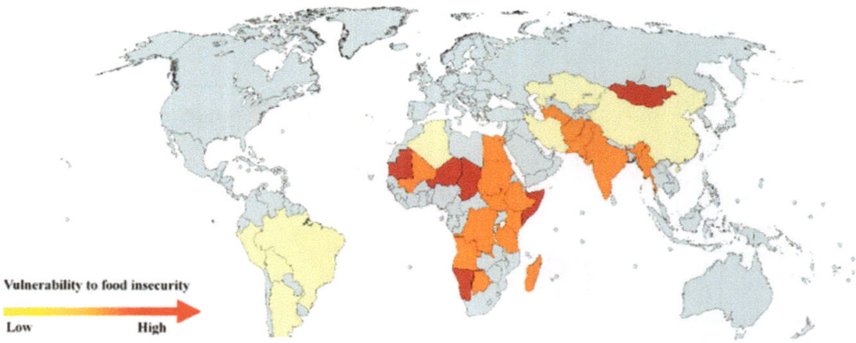

Fig. 3 Climate change and food security scenario in different regions for the year 2017. (Source: World Food Program; Wang et al. 2018)

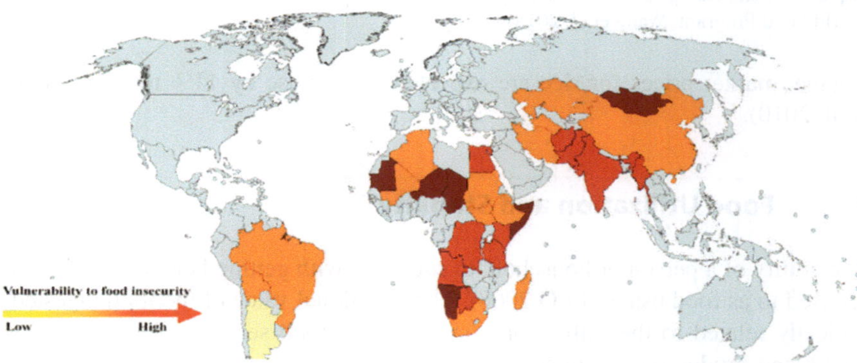

Fig. 4 Climate change and food security scenario in different regions for the year 2050. (Source: World Food Program; Wang et al. 2018)

Additionally, it is predicted that the total population would increase from 7.6 billion in 2017 to 9.8 billion in 2050, with the worldwide growth rate of the humanoid population forecast to be 1.1% annually. It will be challenging to feed this rapid population growth due to the decline in agricultural yield brought on by global warming, which could result in a highly limited food supply in the future and jeopardise food security and safety.

The ability of people, communities, and nations to obtain sufficient amounts of high-quality food is referred to as food accessibility. Access to food is impacted by a number of factors, including food costs, educational background, and property rights (Zewdie 2014). According to recent studies, the adverse impact of climate change on agricultural production led to up to a 20% rise in food costs (Trostle 2010). According to a number of projections, by the end of the 2050s, maize market prices will climb by 42 to 131% as a result of expected increases in global warming and population expansion. Due to the estimated yield drop brought on by climate

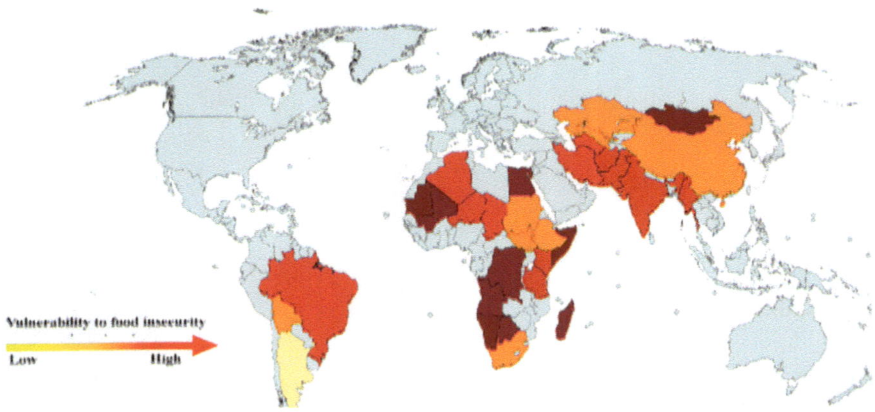

Fig. 5 Climate change and food security scenario in different regions for the year 2017. (Source: World Food Program; Wang et al. 2018)

change, market prices for rice are expected to increase by 11% to 78% (Nelson et al. 2010).

7 Food Utilization and Stability

The ability of a person or household to eat along with getting benefits from food is referred to as food usage (FAO 2008). The nutritional value of the food ingested is typically related to the utilization component of food security. The population's health and productivity may suffer significantly if food source cannot offer a balanced and nutritional diet. Numerous studies have found that food quality and nutritional content have declined due to climate change. According to a study, the cultivation temperature increased from 28–44 °C (day/night), resulting in a drop in total soluble sugars and starch in soybeans (Farooq et al. 2017). A 2–4 °C upsurge in temperature can affect starch, granular size, and seed gelatinization in wheat (Williams et al. 1995). However, climate change could also result in the mutation and emergence of new insects, diseases, and weeds, which may negatively affect the food chain, people's health, crop yield, and crop quality (Rosenzweig et al. 2001). The phrase "food systems" refers to agriculture, community, economic development, human health, and nutrition. Food, harvesting, processing, packaging, transportation, marketing, consumption, and disposal are typically included (Garnett 2011). By driving up food prices, global warming and current climate change trends can potentially tip food systems over the edge into an unsustainable state (Nelson et al. 2010). Price changes can easily affect the food security of the poor, especially in emerging nations, because they spend a sizable portion of their income on food. Maintaining the stability of food systems is difficult, particularly in light of global warming and climate change (Wheeler and Von Braun 2013).

8 Strategies to Mitigate Heat Stress for Improving Crop Production

Heat and water stresses from climate change have the potential to directly affect wheat productivity, but they can also have an indirect effect on the availability of fertilizer, diseases, and pests (Schleussner 2016). Researchers concur that the release of carbon dioxide has caused global warming, which has irreversible trends over hundreds years and may have long-term effects on the planet (Solomon et al. 2009). However, for the time being, keeping up with population expansion and limiting the effects of climate change need to increase the yields of different cereal crops and decreasing global food losses. Below is a brief discussion of some technological options for enhancing crop yields and mitigating the consequences of climate change (Kajla et al. 2015).

8.1 Breeding Approach

By the end of the century, it is predicted that global average temperatures will have increased by 2.5 to 4.5 °C. Since most cereal crops are extremely susceptible to environmental and climatic changes, maintaining an adequate temperature in growth periods is crucial for cereals in all cultivation stages. For instance, during the heading and flowering stages, high temperatures might decrease pollen viability, leading to a decrease in the number of kernels and, eventually, the yield (Stratonovitch and Semenov 2015). Breeding has always been employed to increase crop output rather than increase their resistance to extreme heat. But, there is a growing requirement to develop innovative crop kinds that can endure severe weather conditions and rising yearly temperatures while maintaining steady production rates (Fita et al. 2015). Studies have been done to find novel genotypes of high-yielding, heat-tolerant crops by extending and utilizing the genetic diversity already present (Gulluoglu et al. 2017). For instance, new kinds of cowpeas have been created that produce more grain on average when cultivated in hot climates. This was an improvement over the conventional cowpea variety, whose output dropped significantly when the nighttime temperature exceeded 20 °C. Additionally, a potato gene associated with heat resistance was chosen by three cycles of recurrent selection breeding, which increased yield by up to 37.8%. In this instance, potato types native to Europe were used in breeding to create variants that can provide superior production rates in tropical climates (Benites and Pinto 2011). In other research, it was discovered that transgenic tobacco plants with high levels of glycerol–3-phosphate acyltransferase overexpression had higher thylakoid membrane lipid saturation. Compared to wild plants, these plants recovered from heat stress more quickly (Yan et al. 2008). This demonstrated how glycerol-3-phosphate acyltransferase significantly enhances plants' capacity to withstand heat. Through molecular breeding, it is also feasible to boost the overexpression of glycerol-3-phosphate acyltransferase-related enzymes at larger cereals levels, which may raise the heat tolerance.

8.2 Agronomic Methods

Water and fertilizer use is crucial in quantifying the productivity of different crops at various growth stages. Additional water and fertilizer could augment crop outputs within a particular limit. However, because of changing climate, burgeoning population growth, intensive agricultural activity, and industrial usage, several developing nations are experiencing water scarcity problems (Olmstead 2014). For instance, since 2003, the area of China's wetlands has decreased by 9%, or 340,000 square kilometers (Reklev et al. 2018). Therefore, as part of a plan for creating sustainable agricultural practices and raising crop yield, enhancing irrigation efficiency must be given top attention. According to the International Food Policy Research Institute (IFPRI), improving irrigation effectiveness might result in a 21% increase in agricultural production (Rosegrant et al. 2014). An analysis of California's water use efficiency for various agricultural methods (Cooley et al. 2008) reported that drip irrigation had an efficiency of 88 to 90%, sprinkler irrigation had a 60 to 85% efficiency, and surface irrigation had a 70 to 90% efficiency.

Another important element that directly affects crop production is the application of fertilizer. Fertilizer use over the past few decades is thought to have contributed to a 30–60% rise in crop yield. At the moment, both wealthy and developing nations are using more fertilizer. According to the Food and Agriculture Organization (FAO), 186,900,000 tons of fertilizer were applied in 2014, an increase of 2.0% over 2013. According to estimates, the global demand for fertilizers will increase by 1.8% annually between 2014 and 2018. However, it has been demonstrated that long-term excessive fertilizer application has a negative impact on the soil as well as seed quality. Numerous researches have demonstrated excessive fertilizer can cause soil acidity (Fageria et al. 2010). In addition, in locations where urea and nitrogen fertilizers are used, ammonia volatilization and nitrogen leaching can contribute to environmental problems. Similar to this, research in China's Shouguang and Shandong provinces found that soil pH dropped up to 4.3 in 13 production years, which may have an impact on crop yields and quality (Ju et al. 2007). Therefore, the use of fertilizers in the right quantities must be prioritized to conserve resources and prevent environmental contamination.

8.3 Heat-Tolerant Crops

Millet and pearl millet are considered principal cereal grain crops that can withstand drought the best (e.g., wheat, maize, and paddy). Studies have revealed that as worldly temperature rise in Asian nations such as India and China, rice agriculture will shift to the regions that are warming up (traditionally temperate zones).

Millets might be planted in the subsequent drylands to make up for the loss of the rice harvest because of robust root systems that can withstand water stresses. Additionally, millet hardly ever requires fertilizers. Since millet is a C4 crop, it may fix carbon at lesser rates than other cereals do when it transpires. Under a broader limit for soil pH within 4.5 and 8.0, millets may adapt to a variety of soil settings, including sand, acidic and alkaline soils, and acidic soils. Research has shown that crops with C4 mechanisms, particularly maize, have many further benefits in addition to their capacity to fix carbon. First, studies have revealed that C4 crops' anticipated rate of yield decline is substantially lower than that of crops with C3 mechanisms such as wheat and rice because of their endurance for lower moisture environments. C4 crops have a 150–400% higher water use efficiency than C3 crops (Lychuk et al. 2017a, b). Second, under conditions of elevated CO_2 along with atmospheric temperature that is anticipated to come from global warming, C4 crops have a reduced photo-respiration (Lychuk et al. 2017a, b). While the yields of C3 crops would not vary significantly, some modeling studies claimed that by adopting more advanced and efficient water-saving technology in the future, the productivity of C4 crops might increase up to 38% under expected climate circumstances in a select region (Lychuk et al. 2017a, b). Although many C4 crops, like millets, are more environmentally friendly since they produce fewer greenhouse emissions than cereals, the effects of changing climate seem to have less of an impact on millet productivity. Sorghum is another drought-resistant cereal with a large and thick root system, the capacity to sustain comparatively higher limits of stomatal conductance, the ability to maintain the internal tissue water potential by osmotic adjustment, and phenological plasticity (Hadebe et al. 2017). Additionally, millet is already grown in a number of tropical nations in Africa, Asia, and to a lesser extent South America (Figs. 1, 2, and 3). Therefore, increasing the production of millet in these places might be achieved through educating local farmers, enacting regulatory variations, and implementing smart technological interventions to lighten the workload on the grower. As most of the global population currently resides in tropical environments, where changing climate could have a direct effect on crop productivity, production rates, hydrological balance, temperatures, along with quality of soil, it is important to focus on these areas (Roudier et al. 2011).

Therefore, a key tactic for minimizing the effects of changing climate, water scarcity difficulties, and food security is to increase the land area utilized to cultivate tolerant cereals.

9 Conclusions

The growth requirements of various cereal crops are outlined in this review, along with the effects of climate change on yields of different cereal crops around the world. Techniques for breeding and irrigation were suggested as answers to these problems and difficulties brought on by global warming. The yields of key grain crops will decrease due to climate change due to the rise in ambient temperatures. Therefore, in order to achieve food security goals, we must prioritize usage along with the production of food crops which could withstand ongoing climate change in order to substitute or serve as an alternative for cereal, particularly in arid and semi-arid areas of the world, which are more susceptible to food insecurity. One crop that may thrive in poor-quality soils and tolerate rising temperatures is millet. Additionally, millet cultivation helps lessen the effects of changing climate because it emits fewer GHGs than other cereals and has a lower environmental impact due to its ability to grow on marginal ground with little to no fertilizer and water input. In order to boost the market value of these crops and to encourage farmers to grow them, efforts should be made, particularly in developing nations. Based on regional variables, researchers and growers should create diversification strategies. This needs to be a component of any plan for ensuring food security in the face of changing climate along with a rising population. Additionally, it should be mentioned that there is no one answer to the problem of food insecurity; hence, a comprehensive strategy is needed to address it.

References

Acevedo E, Silva P, Silva H (2002) Wheat growth and physiology. In: Bread wheat; plant production and protection series (FAO). FAO, Rome

Ali H, Sarwar N, Muhammad S, Farooq O, Rehman AU, Wasaya A, Yasir TA, Mubeen K, Akhtar MN (2021) Foliar application of magnesium at critical stages improved the productivity of rice crop grown under different cultivation systems. Sustainability 13:4962. https://doi.org/10.3390/su13094962

Ashraf MA, Akbar A, Parveen A, Rasheed R, Hussain I, Iqbal M (2018) Phenological application of selenium differentially improves growth, oxidative defense and ion homeostasis in maize under salinity stress. Plant Physiol Biochem 123:268–280

Azadi I, Pezeshkpour P, Nasrollahi H (2013) Evaluation te effect of planting season and crop diversity of lentil (ghachsaran variety) in the dryland condition. J Annu Biol Res 4:47–50

Baker HS, Millar RJ, Karoly DJ, Beyerle U, Guillod BP, Mitchell D, Shiogama H, Sparrow S, Woollings T, Allen MR (2018) Higher CO2 concentrations increase extreme event risk in a 1.5 °C world. Nat Clim Chang 8:604–608. https://doi.org/10.1038/s41558-018-0190-1

Bassu S, Brisson N, Durand JL, Boote K, Lizaso J, Jones JW, Rosenzweig C, Ruane AC, Adam M, Baron C (2014) How do various maize crop models vary in their responses to climate change factors? Glob Chang Biol 20:2301–2320

Benites FRG, Pinto CABP (2011) Genetic gains for heat tolerance in potato in three cycles of recurrent selection. Crop Breed Appl Biotechnol 11:133–140

Cai WJ, Wang GJ, Gan BL, Wu LX, Santoso A, Lin XP (2018) Stabilised frequency of extreme positive Indian Ocean dipole under 1.5 °C warming target. Nat Commun 9:1–8. https://doi.org/10.1038/s41467-018-03789-6

Cao L, Wang Q, Deng Z, Guo X, Ma X, Ning H (2010) Effects of climate warming and drying on millet yield in Gansu province and related countermeasures. Ying Yong Sheng Tai Xue Bao 21:2931–2937

Changmei S, Dorothy J (2014) Millet—the frugal grain. Int J Sci Res Rev 3:75–90

Cooley H, Christian-Smith J, Gleick PH (2008) More with less: agricultural water conservation and efficiency in California, vol 30. Pacific Institute, Oakland, CA, p 2011

Debieu M, Sine B, Passot S, Grondin A, Akata E, Gangashetty P, Vadez V, Gantet P, Foncéka D, Cournac L et al (2018) Response to early drought stress and identification of QTLs controlling biomass Production under drought in pearl millet. PLoS One 13:e0201635

Fageria N, Dos Santos A, Moraes M (2010) Influence of urea and ammonium sulfate on soil acidity indices in lowland rice production. Commun Soil Sci Plant Anal 41:1565–1575

Faostat F, Production AC (2017) Food and Agriculture Organization of the United Nations, 2016. Rome, FAO

Farooq M, Wahid A, Kobayashi N, Fujita D, Basra SMA (2009) Plant drought stress: effects, mechanism sand management. Agron Sustain Dev 29:185–212

Farooq M, Nadeem F, Gogoi N, Ullah A, Alghamdi SS, Nayyar H et al (2017) Heat stress in grain legumes during reproductive and grain-filling phases. Crop Pasture Sci 68:985–1005

Fita A, Rodríguez-Burruezo A, Boscaiu M, Prohens J, Vicente O (2015) Breeding and domesticating crops adapted to drought and salinity: a new paradigm for increasing food production. Front Plant Sci 6:978

Food and Agriculture Organization of the United Nations (FAO) (2008) Introduction to the basic concepts of food security. FAO, Rome

Gammans M, Mérel P, Ortiz-Bobea A (2017) Negative impacts of climate change on cereal yields: statistical evidence from France. Environ Res Lett 12:054007

Garnett T (2011) Where are the best opportunities for reducing greenhouse gas emissions in the food system. Food Policy 36:S23–S32

Godfray HCJ, Beddington JR, Crute IR, Haddad L, Lawrence D, Muir JF, Pretty J, Robinson S, Thomas SM, Toulmin C (2010) Food security: the challenge of feeding 9 billion people. Science 327:812–818

Goron TL, Raizada MN (2015) Genetic diversity and genomic resources available for the small millet crops to accelerate a new green revolution. Front Plant Sci 6:157

Gulluoglu L, Bakal H, Sabagh EL, A, Arioglu H. (2017) Soybean managing for maximize production: plant population density effects on seed yield and some agronomical traits in main cropped soybean production. J Exp Biol Agric Sci 5:31–37

Habiyaremye C, Matanguihan JB, D'Alpoim Guedes J, Ganjyal GM, Whiteman MR, Kidwell KK, Murphy KM (2017) Proso millet (Panicum miliaceum L.) and its potential for cultivation in the Pacific northwest, US: a review. Front Plant Sci 7:1961

Hadebe ST, Mabhaudhi T, Modi AT (2017) Water use of sorghum (Sorghum bicolor L. Moench) in response to varying planting dates evaluated under rainfed conditions. Water SA 43:91–103

Halilou O, Assefa Y, Falalou H, Abdou H, Archirou BF, Karami SMA, Jagadish SVK (2020) Agronomic performance of pearl millet genotypes under variable phosphorus, water, and environmental regimes. Agrosyst Geosci Environ 3:e20131

Hergert GW, Shaver TM (2009) Fertilizing winter wheat. UNL-West Central Research and Extension Center, North Plate, NE, p 69101

Huang JP, Yu HP, Dai AG, Wei Y, Kang LT (2017) Drylands face potential threat under 2 degrees C global warming target. Nat Clim Chang 7:417–422. https://doi.org/10.1038/nclimate3275

Jain N, Arora P, Tomer R, Mishra SV, Bhatia A, Pathak H, Chakraborty D, Kumar V, Dubey D, Harit R (2016) Greenhouse gases emission from soils under major crops in Northwest India. Sci Total Environ 542:551–561

Ju X, Kou C, Christie P, Dou Z, Zhang F (2007) Changes in the soil environment from excessive application of fertilizers and manures to two contrasting intensive cropping systems on the North China plain. Environ Pollut 145:497–506

Kajla M, Yadav VK, Chhokar RS, Sharma RK (2015) Management practices to mitigate the impact of high temperature on wheat. J Wheat Res 7:1–12

Lal R (2017) Restoring soil and water resources and mitigating climate change in India by judicious management of agricultural and urban wastes. J Indian Soc Soil Sci 65:105–117

Lamers M, Anyusheva M, La N, Nguyen VV, Streck T (2011) Pesticide pollution in surface-and groundwater by paddy rice cultivation: a case study from northern Vietnam. Clean Soil Air Water 39:356–361

Lehner F, Coats S, Stocker TF, Pendergrass AG, Sanderson BM, Raible CC, Smerdon JE (2017) Projected drought risk in 1.5 °C and 2 °C warmer climates. Geophys Res Lett 44:7419–7428. https://doi.org/10.1002/2017GL074117

Leng G, Huang M (2017) Crop yield response to climate change varies with crop spatial distribution pattern. Sci Rep 7:1463

Lesk C, Rowhani P, Ramankutty N (2016) Influence of extreme weather disasters on global crop production. Nature 529:84

Li M (2018) Climate change to adversely impact grain production in China by 2030. IFPRI, Washington, DC

Li DH, Zhou TJ, Zou LW, Zhang WX, Zhang LX (2018) Extreme high-temperature events over East Asia in 1.5 degrees C and 2 degrees C warmer futures: analysis of NCAR CESM low-warming experiments. Geophys Res Lett 45:1541–1550. https://doi.org/10.1002/2017GL076753

Lobell DB, Burke MB (2010) On the use of statistical models to predict crop yield responses to climate change. Agric For Meteorol 150:1443–1452

Lobell DB, Schlenker W, Costa-Roberts J (2011) Climate trends and global crop production since 1980. Science 333:616–620

Lychuk TE, Hill RL, Izaurralde RC, Momen B, Thomson AM (2017a) Evaluation of climate change impacts and effectiveness of adaptation options on crop yield in the southeastern United States. Field Crop Res 214:228–238

Lychuk TE, Moulin AP, Lemke RL, Gossen BD, Leeson JY, Kirk A, Johnson EN, Olfert OO, Brandt SA, Thomas A (2017b) Effects of crop inputs, diversity, environment, and terrain on yield in an 18-yr study in the semi-arid Canadian prairies. Can J Plant Sci 97:715–730

Mo X-G, Hu S, Lin Z-H, Liu S-X, Xia J (2017) Impacts of climate change on agricultural water resources and adaptation on the North China plain. Adv Clim Chang Res 8:93–98

Msowoya K, Madani K, Davtalab R, Mirchi A, Lund JR (2016) Climate change impacts on maize production in the warm heart of Africa. Water Resour Manag 30:5299–5312

Nelson GC, Rosegrant MW, Palazzo A, Gray I, Ingersoll C, Robertson R, Tokgoz S, Zhu T, Sulser TB, Ringler C (2010) Food security, farming, and climate change to 2050: scenarios, results, policy options, vol 172. IFPRI, Washington, DC

Ngwako S, Mashiqa P (2013) The effect of irrigation on the growth and yield of winter wheat (Triticum aestivum L.) cultivars. Int J Agric Crop Sci 5:976–982

Olmstead SM (2014) Climate change adaptation and water resource management: a review of the literature. Energy Econ 46:500–509

Piao S, Ciais P, Huang Y, Shen Z, Peng S, Li J, Zhou L, Liu H, Ma Y, Ding Y (2010) The impacts of climate change on water resources and agriculture in China. Nature 467:43

Piperno DR, Flannery KV (2001) The earliest archaeological maize (Zea mays L.) from highland Mexico: new accelerator mass spectrometry dates and their implications. Proc Natl Acad Sci U S A 98:2101–2103

Porter JR, Xie L, Challinor AJ, Cochrane K, Howden SM, Iqbal MM, Lobell DB, Travasso MI, Netra Chhetri NC, Garrett K (2014) Food security and food production systems. IPCC, Geneva

Ray DK, Mueller ND, West PC, Foley JA (2013) Yield trends are insufficient to double global crop production by 2050. PLoS One 8:e66428

Reklev S, Chen K, Stanway D Fernandez C (2018) China's water squeeze worsens as wetlands shrink 9%. https://www.scientificamerican.com/article/chinas-water-squeeze-worsens-as-wetlandsshrink/. Accessed 20 Apr 2018

Rosegrant MW, Koo J, Cenacchi N, Ringler C, Robertson RD, Fisher M, Cox CM, Garrett K, Perez ND, Sabbagh P (2014) Food security in a world of natural resource scarcity: the role of agricultural technologies. IFPRI, Washington, DC

Rosenzweig C, Iglesias A, Yang X, Epstein PR, Chivian E (2001) Climate change and extreme weather events; implications for food production, plant diseases, and pests. Glob Chang Hum Health 2:90–104

Roudier P, Sultan B, Quirion P, Berg A (2011) The impact of future climate change on West African crop yields. Glob Environ Chang 21:1073–1083

Russo S, Sillmann J, Sippel S, Barcikowska MJ, Ghisetti C, Smid M, O'Neill B (2019) Half a degree and rapid socioeconomic development matter for heatwave risk. Nat Commun 10:136. https://doi.org/10.1038/s41467-018-08070-4

Santos ES, Abreu MM, Magalhães MC, Viegas W, Amâncio S, Cordovil C (2017) Nutrients levels in paddy soils and flood waters from Tagus-Sado Basin: the impact of farming system. In: Proceedings of the 19th EGU General Assembly, EGU2017, Vienna, Austria, 23–28 April 2017, p 17129

Sarwar N, Ali A, Maqsood M, Ullah E, Shahzad M, Mubeen K, Shahzad AN, Shahid MA, Ahmad A (2013a) Phenological response of rice plants to different micronutrients application under water saving paddy fields on calcareous soil. Turk J Field Crop 18:52–57

Sarwar N, Ali H, Ahmad A, Ullah E, Ahmad S, Mubeen K, Hill JE (2013b) Water wise rice cultivation on calcareous soil with the addition of essential micronutrients. J Anim Plant Sci 23:244
Sarwar N, Atique R, Omer F, Allah W, Mubshar H, Ahmed S, Shakeel A, Marian B, Samy FM, Marek Z, Shahid F (2021) Integrated nitrogen management improves productivity and economic returns of wheat-maize cropping system. J King Saud Univ Sci 33:101475
Sarwar N, Atique R, Shakeel A, Mirza H (2022) Modern techniques in rice crop production. Springer, Berlin
Schleussner CF (2016) Differential climate impacts for policy-relevant limits to global warming: the case of 15°C and 2°C. Earth Syst Dynam 7:327–351
Shao G-C, Deng S, Liu N, Yu S-E, Wang MH, She DL (2014) Effects of controlled irrigation and drainage on growth, grain yield and water use in paddy rice. Eur J Agron 53:1–9
Singh P, Boote K, Kadiyala M, Nedumaran S, Gupta S, Srinivas K, Bantilan M (2017) An assessment of yield gains under climate change due to genetic modification of pearl millet. Sci Total Environ 601:1226–1237
Solomon S (2007) The physical science basis: working group I contribution to the fourth assessment report of the IPCC, vol 4. Cambridge University Press, Cambridge
Solomon S, Plattner G-K, Knutti R, Friedlingstein P (2009) Irreversible climate change due to carbon dioxide emissions. Proc Natl Acad Sci U S A 106:1704–1709
Stratonovitch P, Semenov MA (2015) Heat tolerance around flowering in wheat identified as a key trait for increased yield potential in Europe under climate change. J Exp Bot 66:3599–3609
Sun Q, Miao C, AghaKouchak A, Duan Q (2017) Unraveling anthropogenic influence on the changing risk of heat waves in China. Geophys Res Lett 44:5078–5085. https://doi.org/10.1002/2017GL073531
Tao F, Zhang Z, Xiao D, Zhang S, Rötter RP, Shi W, Liu Y, Wang M, Liu F, Zhang H (2014) Responses of wheat growth and yield to climate change in different climate zones of China, 1981–2009. Agric For Meteorol 189:91–104
Thompson HE, Berrang-Ford L, Ford JD (2010) Climate change and food security in sub-Saharan Africa: a systematic literature review. Sustainability 2:2719–2733
Trostle R (2010) Global agricultural supply and demand: factors contributing to the recent increase in food commodity prices. DIANE Publishing, Collingdale, PA
Vijayan R (2016) Pulses: in need of more attention. Asian J Biol Sci 11:321–325
Wang J, Vanga SK, Saxena R, Orsat V, Raghavan V (2018) Effect of climate change on the yield of cereal crops: a review. Climate 2018(6):41. https://doi.org/10.3390/cli6020041
Welch JR, Vincent JR, Auffhammer M, Moya PF, Dobermann A, Dawe D (2010) Rice yields in tropical/subtropical Asia exhibit large but opposing sensitivities to minimum and maximum temperatures. Proc Natl Acad Sci U S A 107:14562–14567
Wheeler T, Von Braun J (2013) Climate change impacts on global food security. Science 341:508–513
Williams M, Shewry P, Lawlor D, Harwood J (1995) The effects of elevated temperature and atmospheric carbon dioxide concentration on the quality of grain lipids in wheat (Triticum aestivum L.) grown at two levels of nitrogen application. Plant Cell Environ 18:999–1009
Woldesenbet M, Haileyesus A (2016) Effect of nitrogen fertilizer on growth, yield and yield components of maize (Zea mays L.) in Decha district, southwestern Ethiopia. Intl J Res Granthaalayah 4:95–100
Yan K, Chen N, Qu YY, Dong XC, Meng QW, Zhao SJ (2008) Overexpression of sweet pepper glycerol-3-phosphate acyltransferase gene enhanced thermotolerance of photosynthetic apparatus in transgenic tobacco. J Integr Plant Biol 50:613–621
Yao F, Huang J, Cui K, Nie L, Xiang J, Liu X, Wu W, Chen M, Peng S (2012) Agronomic performance of high-yielding rice variety grown under alternate wetting and drying irrigation. Field Crop Res 126:16–22
Ye Y, Liang X, Chen Y, Liu J, Gu J, Guo R, Li L (2013) Alternate wetting and drying irrigation and controlled-release nitrogen fertilizer in late-season rice. Effects on dry matter accumulation, yield, water and nitrogen use. Field Crop Res 144:212–224

Yin G, Gu J, Zhang F, Hao L, Cong P, Liu Z (2014) Maize yield response to water supply and fertilizer input in a semi-arid environment of northeast China. PLoS One 9:e86099

Zewdie A (2014) Impacts of climate change on food security: a literature review in sub Saharan Africa. J Earth Sci Clim Chang 5:225

Zhao X, Fitzgerald M (2013) Climate change: implications for the yield of edible rice. PLoS One 8:e66218

Causes of Soil Erosion, Its Measurements, and Management

Omer Farooq, Muhammad Imran, Masood Iqbal Awan, Naeem Sarwar, Khuram Mubeen, Mukhtar Ahmed, and Shakeel Ahmad

Abstract

Soil erosion represents one of the main causes of degradation of agricultural land around the globe. Productivity of the agricultural land is decreasing due to the removal of the top fertile layer of soil, hence it is drawing the attention of the world. Global resources are diminishing at the rapid pace while the populations and environmental issues are gearing up at even a more pace. Factors like natural, anthropogenic, and climate change have shown that soil erosion is a serious threat to life and needs to be pro-actively looked into for sustainable solutions of various forms causing soil to erode away and pose a threat to the sustainable food security and environment. Water and wind have been recognized as the leading agencies responsible for soil erosion. While among the other factors of causing soil erosion include local climate, farming practices, soil type, vegetation, con-

O. Farooq (✉) · N. Sarwar · S. Ahmad
Department of Agronomy, Faculty of Agricultural Sciences and Technology, Bahauddin Zakariya University, Multan, Pakistan

M. Imran
Department of Environmental Sciences, COMSAT University Islamabad, Islamabad, Pakistan

M. I. Awan
Department of Agronomy, Sub-Campus Depalpur, Okara, University of Agriculture, Faisalabad, Pakistan

K. Mubeen
Department of Agronomy, Muhammad Nawaz Sharif University of Agriculture, Multan, Pakistan

M. Ahmed
Department of Agronomy, Pir Mehr Ali Shah Arid Agriculture University, Rawalpindi, Pakistan

M. Ahmed, S. Ahmad (eds.), *Disaster Risk Reduction in Agriculture*, Disaster Resilience and Green Growth, https://doi.org/10.1007/978-981-99-1763-1_11

struction (roads, housing), mining, organic matter, deforestation, overgrazing, sediments transport, urbanization, topography, surface run off, wind velocity, water velocity, and soil moisture content. Whereas, the most advanced technique for soil erosion measurement is photogrammetry which involves the application of aerial photographs coupled with image processing and Geographic information system (GIS). Agronomic measures are yet considered as the best option implemented for the control of soil erosion.

Keywords

Soil erosion · Degradation of agricultural land · Food security · Photogrammetry · Geographic information system (GIS)

1 Soil Erosion

Soils are essential to plant life, supporting ecosystems and agriculture. Climate change will affect soils, leading to changes in soil erosion, organic carbon, nutrients and alkalinity. Decreasing soil carbon due to climate change also has implications for accounting of carbon emissions from the land, which is an important avenue for all to meet its Net Zero Emissions by 2050 target. Agricultural managers may be able to limit the impact of these changes and manage their land to maximise soil health by using sustainable soil management principles. Principles include choosing appropriate crops, managing nutrient and water cycles and reforestation. Many soil properties are affected by changes in temperature and rainfall. Projected changes to our climate will therefore affect our soils (Ahmed, 2020). Degradation of our soils will have environmental impacts on our vegetation and water quality. It will also affect our agricultural production. Climate change will affect rates of soil erosion. This is because: reduced rainfall in some part of world may lead to drier topsoils and reduce soil structure, more frequent extreme weather events will bring heavy downpours. Downpours – heavy rain that falls in a short period of time – are a major cause of soil erosion and it is visible in most part of world e.g. China and Pakistan. If the rain is heavy enough, the soil cannot absorb it and water flows across the surface taking a layer of topsoil with it. The risk of soil erosion varies place to place because of changes in terrain. For example, steep terrains have a high erosion risk while flat terrains have low erosion risk. The word *erosion* is derived from the Latin word *erodere* meaning to "eat away" or "to excavate," and the term was first used in geology to explain the formation of the hollows by water, the wearing away of solid material by the action of river water. Although the term erosion was used in nineteenth century, the term soil erosion introduced later at the beginning of twentieth century, but it did not come into general use until 1930s. The term was described and defined by Bennett, Fuller, Lowdermilk, and Middleton in Anglo-American literature, Kozmenko, Pankov, Gussak, Sobolev, and Zaslavskii in Russian literature, Kuron, Schultzem, Glander, and Flegel in German and Baulig in French literature. Actually, soil erosion signifies one of the main factors causing land degradation that affects agricultural land (Pimentel 2006; Baskan et al. 2009; Angima et al. 2003). Due to the removal of the fertile topsoil part, the productivity of the agricultural land is reduced (Oldeman et al. 1990; Oldeman 1994), hence limiting stability and biodiversity of the pastures (Tilman and Downing 1994) resulting more chances of the occurrence of drought (Pimentel 2006).

2 Global Perspective on Soil Erosion

At the continental level, the baseline scenario predicts that South America will have the highest average soil erosion rate (3.53 Mg/ha/year), followed by Africa (3.51 Mg/ha/year) and Asia (3.47 Mg/ha/year). The estimated values for North America, Europe, and Oceania are significantly lower, totaling 2.23, 0.92, and 0.9 Mg/ha/year, respectively. With North America anticipated to experience the largest decline (4.8%), the latter group of continents showed an estimated decreasing trend of soil erosion induced by land use change in 2012. We forecast a marginal increase of 1% for Asia, primarily due to a discernible rise in soil erosion in the Southeast Asian nations. While South America and Africa are seeing a noticeable increase in soil erosion, China (2%) and India (0.45%), while being highly populated, show a drop in estimates of soil erosion. With an expected 10% rise in soil erosion in 2012, Africa surpasses South America to take the title of continent with the highest average soil erosion rate (3.88 Mg/ha/year) (Borrelli et al. 2017, 2022). Eekhout and de Vente, (2022) determined projected change in soil erosion as per climate zone, which was obtained from a Köppen-Geiger climate classification map with a 1 km resolution (Beck et al., 2018). The 30 Köppen-Geiger climate classes were subdivided over seven climate zones as shown in Fig. 1. Results showed that climate change is expected to lead to increased soil erosion in many locations worldwide affecting ecosystem services and human well-being. Through a systematic review of 224 modelling studies, Eekhout and de Vente, (2022) provided a global assessment of the impact of climate change on soil erosion and the adaptation potential through land use change and soil conservation. Results showed a global increasing trend in soil erosion towards the end of the 21st century, with the highest increase projected in semi-arid regions. Land use change characterized by agricultural expansion and deforestation aggravate the impact. Reforestation, agricultural land abandonment and soil conservation practices can entirely compensate the impact of climate change on soil erosion. This stresses the need for soil conservation and integrated land use planning.Number of studies per (a) land use scenario, and (b) soil conservation practice have been shown in Fig. 2. Furthermore, Projected change in soil erosion (%) (a) considering all studies, (b) per future period, and (c) per climate zone have been shown in Fig. 3. The colored boxes indicate the weighted interquantile range (25th and 75th percentiles), the black horizontal line the weighted median (50th percentile) and the whiskers extend to the weighted 10th and 90th percentiles. The jitter plot shows the projected change in soil erosion per study, considering the different study areas, periods and emission scenarios. The grey shades indicate the robustness of the studies, quantified with the normalized weight.

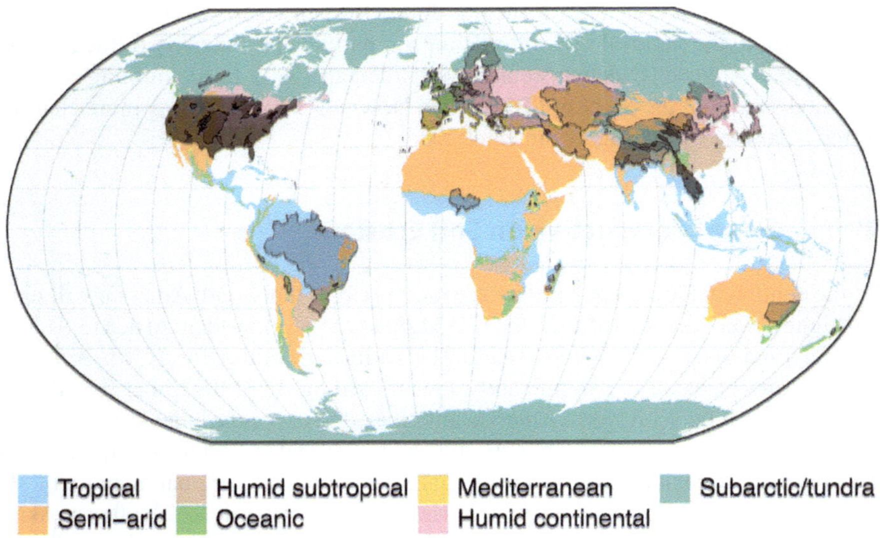

Fig. 1 Location of the 261 study areas and the climate zones (Beck et al., 2018). The study areas are projected on top of each other, where darker colours indicate repeated occurrences of the study area in the dataset (Source: Eekhout and Vente, 2022)

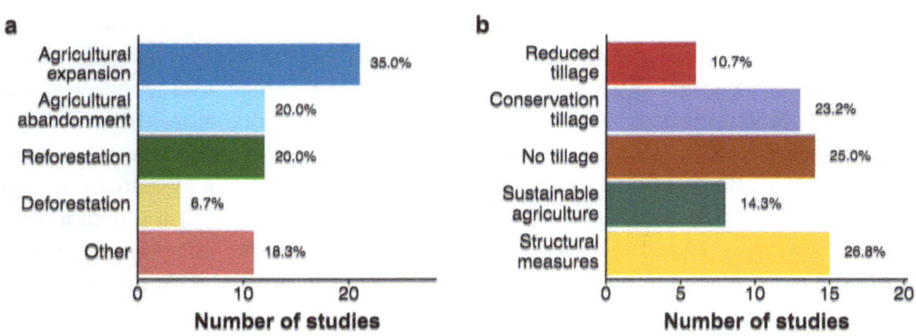

Fig. 2 Number of studies per (a) land use scenario, and (b) soil conservation practice (Source: Eekhout and Vente, 2022)

3 Classification of Soil Erosion

3.1 Erosion Intensity

Different units are used to express it. For instance, sheet and wind erosion are typically measured in terms of soil removal or loss in the unit of tons/ha or m^3/ha. In terms of kg/ha, erosion intensities are minimal. The erosion intensity for long-term erosion is also displayed as an average annual value.

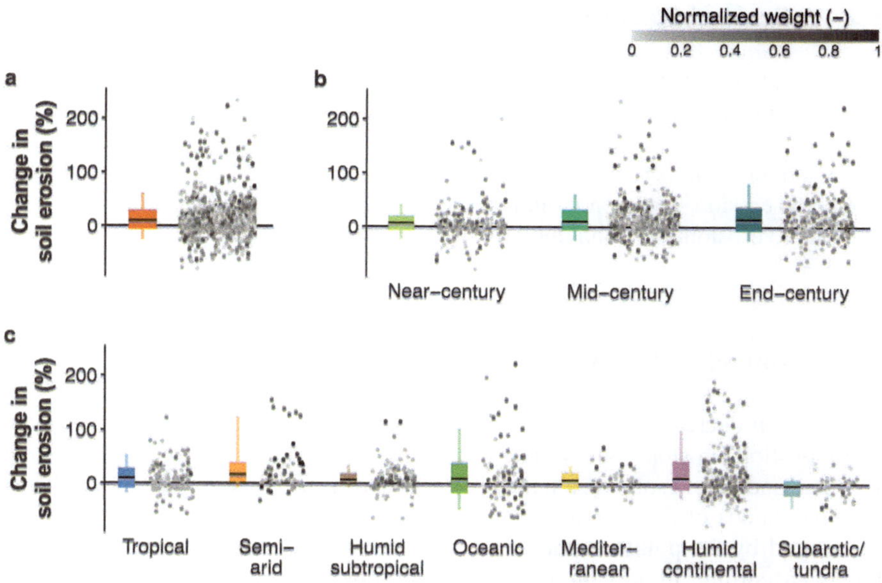

Fig. 3 Projected change in soil erosion (%) (a) considering all studies, (b) per future period, and (c) per climate zone (Source: Eekhout and Vente, 2022)

Table 1 Causes of soil erosion

Causes of erosion	Causes of intensifying gully development
Human activities (farming and land clearance), excessive grazing, deforestation, strong winds, harsh rains, and streaming water; climate (Anonymous 2022)	Cultivation, increased runoff, concentration of runoff, brought on by furrows, contour banks, rivers, etc., deficient vegetation, prolonged seepage flows, and cutting a watercourse "down" so that gullies go up the drainage lines (Anonymous 2013)

In the case of gully erosion, the length or density of gullies, i.e., length per sq. km, indicates the intensity. In addition, gully erosion is quantified using the annual length increase of gullies or the percentage of active gullies as a percentage of gully activity (i.e., total length of the gully).

Erosion height, which is defined as the depth of soil eroded by erosion in 1 year or any appropriate time, is another way to represent the intensity of erosion. This is measured in millimeters per year. If erosion occurs over a long enough length of time, it can be expressed as erosion height, average yearly change over a given number of years, or annual change as a function of time, total change in n-years, etc. Various causes of soil erosion are mentioned in Table 1.

3.2 Scale of Soil Erosion

It measures the extent of erosion.

According to scale, the erosion may be classified as under,

(a) Micro erosion—erosion in small scale
(b) Meso erosion—erosion in middle scale
(c) Macro erosion—erosion in big scale (Shweta 2022)

3.3 Impacts of Erosion

Soil erosion decreases due to the soil ecosystems and cultivable land quality and quantity. Soil erosion causes annual soil loss productivity of US $37.6 billion. Extreme unchecked soil erosion may destroy food crops, affects badly the resilience and livelihoods of the community. Damaging algal growth and pollution in water are caused by the pollutant and agrochemicals, which ultimately end up in water bodies, according to agricultural researchers. The soil particles can also obstruct water volumes and speed-up flood streams.

Some severely degraded soils are completely unusable for growing crops. Reduced water absorption capability of soil may result in water standing in the field. The planting of fresh crops may be slowed down or prevented if some areas are still under water during the planting season. Outside of farm areas, water erosion can have devastating effects: Excessive fertilizer use can have devastating effects on aquatic ecosystems and human health. The fields where it takes up soil, as well as the locations, where the dirt—along with any minerals and toxins it may contain—is deposited, can both be harmed by wind erosion, just like they can be by water erosion. In addition, it can harm people's health by lowering air quality, obstructing visibility, and making breathing harder. Here are a few instances of how wind erosion affects agricultural land. Crop injury reduced crop yields might arise from soil that is airborne due to wind sandblasting sensitive foliage and stems or even burying plants and seeds. Storms of large dust storms can arise and linger for a number of hours when dry, and loose soil particles are suspended in the air. These storms have the potential to ruin crops, kill cattle, and result in a number of major health issues for people, such as asthma attacks and dust pneumonia.

Negative operational circumstances: Dust storms make it dangerous for agricultural workers to be in the fields and can harm or impair the usage of farm equipment. Pesticide, herbicide, fertilizer, and other agricultural chemicals can become airborne due to wind and travel far outside the treatment area. This is known as chemical drift. This can be particularly problematic for farmers attempting to reduce chemical usage on their farms as well as for nearby towns where many agrochemicals are used.

3.4 Processes of Erosion

In addition to the direct effects of hydraulic action, wedging, and cavitation by waves, erosion along coastlines also results from the following processes: abrasion (corrosion), which uses sand, gravel, and larger rock fragments as tools; attrition of the rock particles themselves during this abrasive action; salt weathering or fretting; erosion by organisms (bioerosion); and chemical attack, or corrosion, which weakens the rocks and speeds up erosion. The exposure of a coast to wave attack (wave energy and exposure time), as well as the materials' resistance to erosion and weathering, determine the rates of erosion by these processes (Rampino 2005).

4 Types of Erosion

4.1 Water Erosion

Melting snow and rain causes water erosion. Barren soils free from any vegetation or after crop harvest are vulnerable to water erosion. Rainstorms cause more runoff and erosion because there is no vegetation to absorb the water, keep the soil in place, or disperse the energy of the raindrops as they fall. Rapid soil erosion may result from extreme meteorological conditions such as torrential rains, flash floods, and quick snowmelt. Forceful raindrop falling on soil surface breaks soil aggregation and eroded soil particles come within soil pores blocking water infiltration capacity of soil. It subsequently speeds up surface run off with silt. Soil vulnerability varies with erosivity of the rain, soil erodibility, slope length, and steepness.

4.1.1 Sheet and Rill Erosion

Hilly areas are characterized by sheet and rill erosion. Topsoil removal in the form of thin layer is sheet erosion, though it may not be evident immediately. Tiny water channels referred to as rills are formed due to runoff water that could be as deep as 0.3 m.

4.1.2 Gully Erosion

Gully erosion occurs if rills extend deeper than 0.3 m. Runoff volume cause erosion and transport soil particles in gullies. It hampers soil productivity, damages infrastructure. It is quite evident in hill torrent affected areas across the globe not only turning productive lands into non-productive but also making it near to impossible to rehabilitate those lands for normal agricultural activities.

4.1.3 Tunnel Erosion

Removal of subsoil is known as tunnel erosion. Seeped water through any hole in the ground causes soil to disperse and erodes subsoil developing tunnels. With the progression of tunnels, gullies develop. These soils are known as sodosols and are described as having a sodic content. When these soils' clods are exposed to water, they swiftly decompose into tiny sand, silt, and clay particles that can be easily washed away as the water permeates the subsurface.

4.1.4 Stream Bank Erosion

Floods result in development of stream bank erosion. River bank vegetation removal; sand removal from streambed lowers streambed and causes stream bank erosion.

4.1.5 Floodplains Erosion

Floods with high speeds are used to erode soils that have insufficient surface cover. The entire layer of farmed topsoil may be exposed by these erosive floods, exposing compacted subsoils. Such sites frequently have 0.1 to 0.15 m of topsoil removed. The risk of erosion on floodplains depends on:

- Flood velocities—the bigger the flood, the higher the velocities
- The orientation of crop rows and the amount of protective cover provided by crops or stubble when flooding occurs

Problems can also occur at the end of a flood event when slow moving water flows over saturated soils.

4.1.6 Mass Movement

On cleared slopes in coastal locations, there is mass movement. Earth, rock, and soil materials are suddenly and slowly (at a rate of millimeters per year) moved downward by gravity (e.g., rock falls). There are other types of mass movement, such as rock avalanches, slumping, landslips, and earthflows. A barrier like bedrock or a clay-rich soil horizon can block water entering permeable soils during periods of prolonged and severe rainfall. If this saturated soil is sitting on a rock surface that has become loosen by the accumulation of water in the soil, it may slide downslope due to its heavy weight.

4.1.7 Fluvial Erosion

This occurs when running water gouges shallow channels or deep gullies into the soil.

4.1.8 Muddy Waters

It adds a considerable amount of sediment to the river. Forest can be planted on nearby former farmland to slow the growth of the slip.

4.1.9 Slips

The transition from pasture to woodland causes slips. When the topsoil and subsoil on slopes become saturated, slips happen. It slips downhill, exposing the underlying material, unless retained by plant roots to the subsurface. Slips frequently occur if mudstone (papa) or siltstone is the underlying rock type.

4.1.10 Earth Flows

These are rare, but they can be stunning. Masses of saturated soils, which may also contain underlying rock, descend. The surface vegetative mat, however, is still present, continues to create lumps and hollows (Hales and Roering 2009).

4.1.11 Splash Erosion

Rain begins with a splash, which may be why splash erosion is also known as raindrop erosion. When a raindrop hits the ground and causes dirt fragments to splatter up and away. These particles are more susceptible to being whisked away by other forces once they are liberated. It is challenging to spot this kind of deterioration. After a very heavy downpour, you might be able to see areas where little smears of soil have been thrown up onto various surfaces. There is not much issue when the disturbed soil just returns to its original position. However, if you do nothing and the separated soil is carried away by wind or water, you will lose your topsoil.

4.1.12 Underground Erosion

Erosion on the surface and subsurface are both caused by precipitation, while subsurface erosion is not as evident as surface erosion.

4.1.13 River Erosion

Particularly in rivers with continuous water flow, usually at varied rates, this erosion takes place. Watercourses with smaller catchment areas and those with less ideal drainage characteristics are likely to experience greater erosion. In such cases, a transition between the rivers and gully is created by the topmost stream branches, which also resemble the gully.

4.2 Wind Erosion

Sandy over-grazed soils under dry period receiving strong winds cause the soil to erode. Revegetation on such soils is very much challenging. In farming regions, wind erosion typically does not pose a significant problem. The majority of Queensland's agricultural soils have a thick texture that tends to create relatively big aggregates that are too coarse to be carried by powerful winds. However, due to their low fertility and inability to hold a lot of moisture, sandy soils are susceptible to wind erosion. This restricts their use in Queensland, where only limited areas are planted for horticulture or growing vegetables (where irrigation is available).

4.3 Scalding

Scalding can happen when wind and water erosion expose salty or sodic soils by removing the top soil. Large volumes of dirt can be shifted just by the impact of raindrops. However, the movement of water or wind over the soil's surface will result in increased soil erosion, including sheet, rill, and gully erosion. In addition, erosion frequently leaves behind both fine and coarse sand by removing the lighter, smaller soil particles (such as clay and silt) first. The soil tends to seal and set hard at the surface when there are high concentrations of fine sand and low concentrations of clay, which restricts infiltration (water entering the soil). There are some other types of soil erosion, that are mention in Table 2.

Table 2 Miscellaneous forms of soil erosion

Sr. no	Form of erosion	Description	Reference
01	Scree erosion	Greywacke rocks underlie the steep or mountainous terrain where it occurs. Gravity forces debris downhill into a fan shape at high altitudes, which may be extremely impressive. If the vegetation is diminished, screes can still move, or stock can travel across their slopes	Hales and Roering (2009)
02	Glacial erosion	Typically seen in colder climates, where the mean temperature is below 0 °C (i) Action of a large mass of ice moving very slowly (ii) The protective measures including encouragement of vegetation have little effect (iii) Soil is damaged only at the edge of ice (iv) Major part of the erosion energy is dissipated during erosion of bed rock materials It mainly occurs in the form of (a) exaration (most pronounced form of glacial erosion is characterized by furrowing, cutting, ploughing, and scouring) (b) Grinding and rubbing, also referred as glacial abrasion or detersion (c) Detraction is the last phase of glacial erosion, which is a typical form of disturbance	Shweta (2022)
03	Snow erosion	In places with a permanent snow cover, there is a noticeable amount of snow erosion	
04	Organic erosion	The soil erosion caused by living organism is called organic erosion (a) phytogenic erosion (the soil erosion brought on by a plant's root system. It is also known as "root erosion" at times. In this kind of soil erosion, plant material harvesting, and external sources of soil loss in the field are replaced by the weathering process) and (b) zoogenic erosion (animals remove soil particles, especially when they migrate from one location to another in search of food or to dig for shelters. These soil particles are then carried away by wind or water. Around dam sites, this form of soil erosion is frequently seen	
05	Coastal erosion	The erosion of land due to wave action, tidal currents, or wave currents, or the removal of beach or dune sediments. The primary factors that generate waves and contribute to coastal erosion include storms, wind, or fast-moving motorized vessels. Coastal morphodynamics is the study of coastal erosion and sediment redistribution. On rocky coasts, erosion occurs if the coastline has rock layers or fracture zones with varying degrees of erosion resistance	

5 Factors Affecting Soil Erosion

It is important to understand the factors than govern the soil erosion. The understanding of these factors helps to know the danger of soil erosion and precautionary measures for minimizing the adverse effects of soil erosion. Soil erosion is mainly caused by water and wind. Soil erosion due to water as eroding agent is more severe than the winds, because water is denser than winds and hence plays an important role in shaping the earth surface. There are several natural and anthropogenic factors that control the soil erosion. Main factors for soil erosion include local climate, farming practices, soil type, vegetation, development practices (construction, roads etc.), mining, organic matter, deforestation, overgrazing, sediments transport, urbanization (active development), topography, surface run off, wind velocity, water velocity, and soil moisture content.

5.1 Vegetation and Soil Type

Land use changes with time in an area affect soil erosion. The presence of vegetation on both flat and sloppy areas provides protective cover, slows down the water movement, stabilizes the soil, and reduces the soil erosion due to binding of soil particles with the plant roots. The plant root system and roots development vary with the type of plants and soil texture. Hence, soil erosion is controlled by the type of root system and soil type due to fluctuations in soil physicochemical properties. Deep-rooted plants provide more strength to soil and consequently reduce the soil erosion. There is more soil erosion when vegetation is removed from the surface, and soil is exposed during several developmental activities (constriction, highways). In other words, anything that protects the soil from the impact of raindrops (e.g., plants, stones, residues, etc.) helps to reduce the soil erosion. Soil strength (cohesiveness) and soil drainage vary with the soil texture and ultimately control the soil erosion. In coarse soils (e.g., sand), there is improved drainage, more water infiltration, large pores (macropores), lower cohesiveness, lower water erosion, and higher soil erosion by winds. The downslope transfer of soil due to tillage operation can also be included in soil erosion.

5.2 Role of Wind Velocity and Surface Runoff

There is direct relationship of soil erosion with wind velocity and surface water runoff. The higher the wind and water velocity, the higher is the soil erosion. Hence, water velocity and wind velocity are slowed by taking special measures to control the soil erosion. Wind velocity can be reduced by planting hedges or by using trees as windbreaks along the field's boundary. Soil erosion is also induced by gravity and ice, but role of liquid water is more important in soil erosion as compared with ice. Wind velocity predominantly controls the erosion in deserts due to lower water content (drought).

5.3 Surface Topography and Slope Steepness

The physical properties of soil are considered important factors in soil vulnerability to soil erosion particularly erosion with water. Steep and unobstructed slopes are more susceptible to soil erosion by water because of higher surface runoff. While soil erosion by winds is dominant on flat land surface due to lack of obstacles and relatively slower runoff.

5.4 Construction and Urbanization

Construction and active urbanization yield more intensive soil erosion than farming practices and underdeveloped land but farming exhibits far more land and for longer time. Once construction is completed, soil erosion becomes negligible. Likewise, there is boom in soil erosion during construction of roads, shopping centers, and housing areas and a gradual decline in soil erosion is observed and ultimately becomes negligible after completion of construction.

5.5 Mining

Mining industry also contributes to soil erosion because mining destroys the landscape causing inevitable damage by removing the earth constituents including trees which are considered important to stabilize the soil thereby minimizing soil erosion. Mining leaves the erodible soil piles and broken rocks behind.

6 Soil Moisture Content, Organic Matter, and Soil Permeability

Soil moisture content, organic matter content, and permeability are crucial in soil erosion. The higher soil permeability increases soil infiltration rate and soil drainage which reduce surface runoff ultimately leading to decline in soil erosion. The lower the permeability, lower the infiltration, higher the surface runoff, and higher will be the soil erosion. It is also attributed to capillary forces in the soil. Soil erosion is reduced with increase in soil moisture content especially in flat areas. But in sloppy areas, higher water content (wet soil) makes the slope unstable, and there are more chances of slope instability leading to more soil erosion (downhill movement). The quantity and intensity of rainfall due to fluctuations in local climate is crucial in soil erosion, because it controls the soil water content. The addition of organic matter in soil improves soil aggregation, porosity, and permeability leading to a decline in water erosion.

7 Transport and Off-Road Recreational Activities

The public transport on the roads and off-road vehicles (ORVs) contribute to soil erosion. Especially, ORVs in dry areas with limited vegetation play crucial role in soil erosion. The presence of fragile plants in dry areas enhances soil erosion with ORVs due to easily destruction of these plants leaving the soil bare with intensive erosion. At recreational places, soil erosion is also triggered by walking, biking, horse-riding, and camping.

8 Measurement of Soil Erosion

Literature shows that various methods are employed for the measurement of soil erosion. The choice of a method for erosion measurement varies with the objectives of study and available resources. The method of soil erosion measurement varies with its type (rill, gully, sheet erosion) and scale (field scale, lab scale, and regional scale). Generally, erosion plots, surveys, and tracers are the techniques to measure soil erosion in an area under various meteorological conditions. These techniques have been briefly described later.

8.1 Plots

The most extensively used method for soil erosion measurement is the runoff plot (RP) method. Plots for estimation of soil erosion are with boundaries and without boundaries. The basic data on soil loss can be obtained from both types of plots effectively. Erosion plots are used for the estimation of rill and sheet erosion as kg/m^2/time. In plots, it is assumed that soil loss is uniform over that plot regardless of plot size and shape.

The water-induced soil erosion on agricultural land is predicted by using Universal Soil Loss Equation (USLE) which was developed by Wischmeier and Smith. The erosion rates on agricultural land using USLE includes characteristics of rain fall and slope (topography), soil erodibility, vegetation cover, and management practices. Contribution of erosion due to the formation of large gullies is not considered in USLE. The USLE parameters to estimate the soil loss with erosion (SL) are presented in Eq. (1) (Foth 1978).

$$\mathrm{SL} = R \times K \times L \times S \times C \times P \tag{1}$$

where SL = soil loss with erosion (tons/acre), R = rainfall factor (erosive force of rainfall), K indicates soil erodibility factor, L is the slope length factor, S represents slope gradient factor, C is the cropping management factor, and P indicates erosion control practice factor.

8.2 Surveying

8.2.1 Erosion Pins and Profilometers

Erosion pins are cost effective and intuitive approach to measure annual hill slope erosion/deposition rates. Pins (nails or steel rods) are inserted into the ground, and their reference (datum) is noted. The ground-advancement and ground-retreatment due to soil deposition and erosion are measured to estimate the respective deposition/erosion. There are generally two limitations encountered for pins application to estimate the soil erosion/deposition: (a) pins cannot be used for long-term experiments in cropped areas, and (b) small fluctuations in soil elevation are not easily detected with pins.

Profilometers are used to extract data for the topographical parameters (roughness, morphology, step height) of a surface. The surface roughness (irregularities) affects the runoff and interaction of the sediments and soil particle. There is decline in soil erosion with increase in surface roughness. Profilometers consist of frames as permanent benchmark and vertical sliding rods for mapping soil loss as contours on small areas. The stability of the permanent benchmarks is very important for effective and reliable application of profilometers. The soil swelling and disturbance due to cultivation and other agents must be considered during application of profilometers for the erosion measurement.

8.2.2 Photogrammetry

Photogrammetry uses digital photos to quantify soil erosion as compared with erosion plots in simulated rainfall and wind conditions. Aerial photographs along with remotely sensed data are used to detect fluctuations in land use and topographical mapping for the measurement of soil erosion. It is more advanced, cost effective, and efficient technique especially in inaccessible areas for the detailed information of soil erosion. Photogrammetric system is superior to typical survey by providing images and morphological characteristics simultaneously (Moritani et al. 2010; Hodge et al. 2009). Digital elevation models (DEMs) are used in combination with digital imaging technology and photogrammetry to monitor the land use changes including soil erosion over a large area. The DEMs resolution are obtained from the photograms taken before and after the rainfall using image processing software. The detail of the camera calibration, resolution, and use of images in collaboration with DEMs to monitor soil erosion (gully, rill, and sheet erosion) has been provided in Rieke-Zapp and Nearing (2005).

8.2.3 Other Survey Methods

Reconstruction of soil profile of a certain area can also be considered as indicator of soil erosion by measuring depth loss (%) of A and B horizons. The fluctuations in soil horizons are further refined with X-rays spectrographs. Exposure of tree roots can also be correlated with the soil erosion by wind and water.

8.3 Tracers

Tracers are used to follow the dynamic behavior for mapping and quantification of soil erosion on agricultural lands (Walling and Quine 1995). Radioisotopes as a tracer are artificially labelled on to soil particles for measuring the extent and source of soil erosion. The most widely developed tracer to quantify the soil erosion/deposition due to its high affinity with the soil particles is Cesium-137 isotope (137Cs). The 137Cs is a fission reaction product and is strongly adsorbed onto the soil/sediments with very limited chemical translocation. The 137Cs is accumulated in the upper soil profile (5–10 cm) of uneroded soils and is considered as reference site for 137Cs. The adsorbed 137Cs is released from the soil during erosion depending on the severity of soil erosion. The degree of soil erosion with 137Cs is estimated by comparing the eroded site with reference site of 137Cs on the same soil type (Zapata 2003). Unfortunately, at some sites, comparison of field scale soil erosion with tracer prediction showed overestimation of soil erosion with 137Cs (Evans et al. 2017). The detail of the advantages and limitations of 137Cs has been provided by Evans et al. (2017) and Zapata (2003). Sometimes, iron-59 (Fe-59) and Copper (Cu) can also be used as tracer to detect soil erosion (Loughran 1989).

9 Methods of Erosion Control

Land degradation has been a menace throughout the human history. About 33% of land is classified as degraded with moderate to high levels of degradation, and the situation is getting worse with every passing day. Among the ten identified main threats to soil functioning include soil erosion as top of the list; others being organic matter losses, waterlogging, acidification, contamination, sealing, nutrient imbalances, compaction, soil biodiversity losses, and salinization (FAO and ITPS 2015). Millions of people living on such soils will be affected, a challenge that is intensified by lack of political will and changing climate. Building and rehabilitation of the 33% degraded soils is an entry point for sustainable solution of the compound problems.

Erosion associated with intensive tillage practices and lack of suitable conservative crop rotations is degrading our precious soil resources on one hand and causing silting of reservoirs, ponds, and lakes on the other hand. Since ancient times, human civilizations recognized soil erosion associated related problems. Different effective methods were developed for controlling the runoff and erosion. For instance, terracing in Asia and South America and no till, intercropping in the pre-Columbian America. Some desert civilizations retained runoff water and eroded silt by building dams. The water so collected in depressions was then used to grow crops (Montgomery 2007). Effective runoff and erosion control without compromising soil and crop productivity demands new mindset for institutional and political reforms, considerable investment for prevention or rehabilitation, and different management practices. Erosion control efforts can work by either reducing the shear impact of water and wind, i.e., erosivity, or by keeping soil in a condition in which it is not easily eroded, i.e., erodibility. A combination of the two approaches is desirable and often advocated in the many conservation practices. In this regard, key principles are:

- do not expose the soil to water and wind erosion, i.e., by maintaining a permanent soil cover
- increase soil aggregation and infiltration by appropriate soil management practices
- minimum soil disturbance to avoid tillage erosion
- identify and focus high-risk areas, i.e., spatial measures
- focus on critical periods like pre-monsoon time, when the soil is bare and dry
- conduct economic analysis for tradeoffs between productivity, profitability, and soil health.

Erosion control methods can employ two approaches, solely or in combination: agronomic and structural. Agronomic approaches reduce erosion by necessary changes in crop management (i.e., vegetation-based) and/or soil management (soil-based). Examples of agronomic practices are reduced tillage, diversification, and cover crops. Agronomic approaches are often referred to as best management practices. Structural approaches reduce erosion by employing engineering practices in which an initial investment is made to build structures such as terraces, grassed waterways, and retention ditches. Structural approaches are often referred to as engineering, technical, or physical methods. Agronomic approach is considered as more appropriate or superior to the structural measures for following reasons (Magdoff and Van Es 2021):

- Structures in general focus containing sediments along with runoff after erosion has started, whereas agronomic management practices focus preventing erosion from happening by reducing the runoff potential
- Agronomic approach also helps building soil health and crop productivity by reducing erosion
- Agronomic management practices can be adopted at farm level and by individual farmers, where structural measures are usually adopted at the landscape level requiring government or NGOs support
- Significant advances in farm machinery and herbicides now provide alternative soil/crop management options
- Structural measures are labor-intensive and mostly expensive in terms of building and maintenance
- In most of the cases, tillage erosion is not accounted for by structures.

9.1 Agronomic Approaches

Agronomic measures achieve conservation objectives by reducing the raindrop impact through interception, increasing infiltration, reducing runoff, and hence reducing the soil erosion.

9.1.1 Contour Farming

With right implementation, contouring can reduce erosion by >50%. Contours are lines of uniform elevation and contour farming is farming around the hill, not up and down the hill. Contour farming, where all operations including sowing, ploughing are done at nearly right angles to the direction of field slope, is encouraged to reduce sheet and rill erosion, with or without terraces. Ridges and furrows formed during tillage and planting are used to intercept surface runoff by directing flow along the hill slope to increase the amount of infiltration. Contour farming can have mono- or mixed cropping but is not ideal under all conditions. It is most effective on: (a) field slope gradients between 3% and 8%; and (b) slope lengths of 30 to 120 m. If the slope is >8%, runoff water may overflow the ridges, whereas <3%, the relative benefit of contouring is diminished due to limited runoff. Grassed waterways come in handy where the contour lines are too sharp for farming equipment to plow. Other techniques to include growing bush or tree borders across the slopes (vegetative barriers), residue management, and mulching to protect the soil (Thompson and Sudduth 2017).

9.1.2 Strip Cropping

In strip cropping, farmer plants different crops in alternating strips of different levels to prevent soil erosion besides improving soil fertility. With a high amount of rainfall combined with steep slopes, strip cropping provides an extra protection layer to soil resources. The use of grassed waterways, in irregular slopes, may improve the overall usefulness for strip cropping. Rotating strips of legumes and maize will help the cereal to utilize nitrogen fixed by legumes (Raza et al. 2020; 2019a, b, c and d; Beillouin et al. 2019).

9.1.3 Reduced Tillage

Soil must be least disturbed and kept covered for ensuring reduced erosion. Transitioning to tillage systems that increase surface cover and reduce disturbance is, therefore, the single most effective approach to reducing erosion. Reduced tillage is one of the most important conservation practice of the twentieth century. Reduced tillage aims at avoiding soil inversion, rely on limited tillage depth, and maintain residues on the soil surface. Different reduced tillage approaches include shallow tillage (20–40 cm) avoiding soil inversion, conservation tillage (15–30 cm) maintaining one third of crop residues, superficial tillage (5–15 cm) maintaining minimum work depth on full surface, strip tillage (localized in sowing strips) sowing wider strips, and no tillage (localized in seeding rows) depositing seeds directly into a narrow furrow. Driving forces for the popularity of reduced tillage systems is rising fuel prices, the dust bowl phenomenon of 1930s, resource base degradation following the green revolution, and availability of new planters, improved chemistry herbicides among other add on technologies. Reduced tillage prevents soil erosion on steep slopes by: (1) reduced raindrop effect; (2) increased infiltration by leaving crop stubbles on untilled soil and anchoring the soil with roots; (3) moisture conservation especially in dry soils, and (4) improved water- and nutrient-holding capacity.

9.1.4 Cover Cropping

A cover crop, by covering the ground, is grown mainly to prevent soil erosion with the help of living vegetation/roots, which hold the soil, e.g., cowpea, oat, and ryegrass. Close-growing crops have high density and prevent erosion. The actual benefits depend on species, productive potential, and for how long it is left on the soil. Grasses like oat have extensive fibrous root system, establish quickly and hence can greatly reduce erosion. Cover crops protect the soil from splashing raindrops and scorching heat from the sun. Most of the plants used as ground cover are legumes, such as different varieties of beans and peas. Cover crops, however, are unsuitable for dry areas, i.e., less than 500 mm, as they compete for water. Under such conditions, it would be better to keep the weeds and natural vegetation (Beillouin et al. 2019; Magdoff and Van Es 2021).

9.1.5 Mulching

The word mulch is derived from a German word 'Molsch' means "easy to decay." A mulch is any material such as straw, sawdust, leaves, plastic film, loose soil, etc., that is spread or formed on the surface of the soil to protect the soil and/or plant roots from the effects of raindrops, soil crusting, freezing, evaporation, etc. Mulch prevents erosion by: (a) conserving soil structure; (b) improving aggregation; (c) reducing splash of soil particles; (d) preventing crust formation after rainfall; and (e) maintaining surface roughness (Prosdocimi et al. 2016).

9.1.6 Crop Diversification

Diversification is a process to achieve diversity. Whereas, specialized farming aims to improve the productivity, stability and delivery of ecosystem services, the crop diversification can be considered as an attempt to increase the diversity of crops through, e.g., crop rotation, multiple cropping, or intercropping. Temporal measures of diversification include: crop rotation, double to multiple cropping, catch cropping, and relay cropping. Spatial measures of diversification include: alley cropping, mixed cropping, intercropping, companion crops, and variety mixtures. The accompanying agronomic measures include reduced tillage, input management, green manuring, and livestock integration. Diversification helps to control erosion by more frequent or continuous soil cover and more diverse management strategies. This is particularly important on lands prone to degradation (Plaza-Bonilla et al. 2017).

9.1.7 Grass Strips

As a cheap alternative to terracing, the grass is planted in dense strips, up to a meter wide, along the contour. Such grass strips create barriers to minimize soil erosion and runoff. The spacing of the strips depends on the slope of the land. Optimum spacing is 20–30 cm on gentle slopes and 10–15 m on steeper slopes. The strips can be planted along ditches to stabilize, or on the rises of bench terraces in order to prevent erosion. Examples of grasses that can be used are Napier and Guinea grasses, which can also be fed to animals but require regular trimming to avoid shading of crops (Ligdi and Morgan 1995).

9.2 Structural Approaches

Structural measures achieve conservation objectives by increasing the time of concentration for runoff, allowing more infiltration, and dividing a long slope into several short ones and thus reducing the velocity of the surface runoff.

9.2.1 Retention Ditches

Large ditches, designed to catch and retain all incoming runoff until it gets infiltrated into the ground, are called as the retention or infiltration ditches. In semi-arid areas, the ditch is about 0.3–0.6 m deep and 0.5–1 m wide. On flat land, the ditches are usually spaced at 20 m with close ends to harvest water. On sloping land, the ditches are usually spaced at 10–15 m with usually open ends so that excess water can exit. Retention ditches are especially beneficial in semi-arid areas, on flat or gentle sloping land but not suitable on shallow soils or in areas prone to landslides. Retention ditches reduce erosion by retaining water but need to be maintained and de-silted (Dollinger et al. 2015).

9.2.2 Stone Lines

Stone lines along the contour is a popular, low-cost technology to slow down the runoff water speed, filter, and spread it over the field, thus enhancing water infiltration and reducing soil erosion. Such stone lines can be reinforced with crop residues, or soil to make them more stable. After a rainfall event, the soil will start to build up on the upper slope side, thus building a natural terrace overtime. The stone lines are spaced 15–30 m apart; a shorter distance is used for the steeper slopes. Stone lines are suitable on gentle slopes where annual rainfall is 200–750 mm. They are often used to rehabilitate eroded land (Wakindiki and Ben-Hur 2002).

9.2.3 Semi-Circular Bunds

These are the earthen bunds in the shape of a semi-circle with the tip of the bunds on the contour. Used for growing fruit crops, the semi-circular bunds are suitable on gentle slopes (<2%) and in areas with annual rainfall 200–750 mm. They are easy to construct, applicable to uneven terrain, and help reducing erosion (Al Mahmoud et al. 2014).

9.2.4 Planting Pits

Planting pits are the simplest form of water harvesting. Small holes are dug at a spacing of about 1 m, which catch runoff and concentrate it around the growing plant. Crops are planted in the pits. It is not necessary to follow the contour when constructing planting pits. Planting pits have been successful where annual rainfall is 200–750 mm. By tapping runoff and increasing soil moisture, they help reducing the erosion (Nyakudya et al. 2014).

9.2.5 Terracing

Terracing is an age-old art. Terraces are earthen embankments consisting of two parts: an excavated channel and a bank or ridge on the downhill side of the channel. The embankments are installed at right angles to the steepest slope to intercept the surface runoff. Bench terraces with step-like structures are made on areas of steep slope. They help to control erosion by reducing the degree and length of the slope, increasing the infiltration, and allowing improved irrigation where necessary (Thompson and Sudduth 2017).

9.2.6 Grassed Waterways

They are simple yet an effective way to reduce scouring in areas where runoff water accumulates. They help prevent surface water pollution by filtering sediments out of runoff and require only small areas to be taken out of production (Fiener and Auerswald 2003).

9.2.7 Retaining Walls

A retaining wall keeps soil, rock, and water in a place so as to prevent their washing away with the rain. Retaining walls prevent erosion caused by wind, rain, or flowing water (Youdeowei and Abam 1997).

10 Conclusion

Soil erosion draws attention to the fact that global resources are diminishing at the rapid pace while the populations and environmental issues are gearing up at even a more speed. Factors like natural, anthropogenic, and climate change have shown that soil erosion is a serious threat to life and needs to be proactively looked into for sustainable solutions of various forms causing soil to erode away and pose a threat to the sustainable food security and environment. Soil erosion due to water is more severe in shaping the earth surface than the winds due to it is denser than winds. Main factors for soil erosion include local climate, farming practices, soil type, vegetation, construction (roads, housing), mining, organic matter, deforestation, overgrazing, sediments transport, urbanization, topography, surface run off, wind velocity, water velocity, and soil moisture content. Erosion plots, erosion surveys, and radionuclides as tracers are the commonly employed techniques for soil erosion measurement depending on the erosion type and scale. The most advanced technique for soil erosion measurement is photogrammetry which involves the application of aerial photographs coupled with image processing and Geographic information system (GIS). Soil erosion must be controlled without compromising the crop productivity. Against this background, agronomic measures are considered as superior to structural measures, because they can be implemented at farm level and prevent erosion from occurring besides building soil health. However, structural measures can be used to complement the agronomic measures.

References

Ahmed M (2020) Introduction to Modern Climate Change. Andrew E. Dessler: Cambridge University Press, 2011, 252 pp, ISBN-10: 0521173159. Sci Total Environ 734:139397. https://doi.org/10.1016/j.scitotenv.2020.139397

Al Mahmoud H, Al Issam K, Awadis A (2014) Use of the universal soil-loss equation to determine water erosion with the semi-circular bund water-harvesting technique in the Syrian steppe. Int J Environ 3(2):1–11
Angima S, Stott D, O'Neill M, Ong C, Weesies G (2003) Soil erosion prediction using RUSLE for central Kenyan highland conditions. Agric Ecosyst Environ 97:295–308
Anonymous (2013) Types of erosion. https://www.qld.gov.au/environment/land/management/soil/erosion/types
Anonymous (2022) What is soil erosion? https://byjus.com/biology/soil-erosion/
Baskan O, Cebel H, Akgul S, Erpul G (2009) Conditional simulation of USLE/RUSLE soil erodibility factor by geostatistics in a Mediterranean catchment, Turkey. Environ Earth Sci 60:1179–1187
Beck HE, Zimmermann NE, McVicar TR, Vergopolan N, Berg A, Wood EF (2018) Present and future Köppen-Geiger climate classification maps at 1-km resolution. Scientific Data 5(1):180214. https://doi.org/10.1038/sdata.2018.214
Beillouin D, Ben-Ari T, Makowski D (2019) A dataset of meta-analyses on crop diversification at the global scale. Data Brief 24:103898
Borrelli P, Robinson DA, Fleischer LR, Lugato E, Ballabio C, Alewell C, Meusburger K, Modugno S, Schutt B, Ferro V, Bagarello V, Van Oost K, Montanarella L, Panagos P (2017) An assessment of the global impact of 21st century land use change on soil erosion. Nat Commun 8(1):2013. https://esdac.jrc.ec.europa.eu/themes/global-soil-erosion
Borrelli P, Ballabio C, Yang J, Robinson D, Panagos P (2022) GloSEM: high-resolution global estimates of present and future soil displacement in croplands by water erosion. Sci Data 9:406. https://esdac.jrc.ec.europa.eu/themes/global-soil-erosion
Dollinger J, Dagès C, Bailly JS, Lagacherie P, Voltz M (2015) Managing ditches for agroecological engineering of landscape. A review. Agron Sustain Dev 35(3):999–1020
Eekhout JPC, de Vente J (2022) Global impact of climate change on soil erosion and potential for adaptation through soil conservation. Earth Sci Rev 226:103921. https://doi.org/10.1016/j.earscirev.2022.103921
Evans R, Collins A, Zhang Y, Foster ID, Boardman J, Sint H, Lee M, Griffith B (2017) A comparison of conventional and 137Cs-based estimates of soil erosion rates on arable and grassland across lowland England and Wales. Earth Sci Rev 173:49–64
FAO and ITPS (2015) Status of the World's soil resources (SWSR)—Main report. Food and Agriculture Organization of the United Nations and Intergovernmental Technical Panel on Soils, Rome
Fiener P, Auerswald K (2003) Concept and effects of a multi-purpose grassed waterway. Soil Use Manage 19(1):65–72
Foth HD (1978) Fundamentals of soil science. Soil Sci 125:272
Hales TC, Roering JJ (2009) A frost "buzzsaw" mechanism for erosion of the eastern Southern Alps. New Zealand Geomorphology 107(3-4):241–253
Hodge R, Brasington J, Richards K (2009) In situ characterization of grain-scale fluvial morphology using terrestrial laser scanning. Earth Surf Process Landf 34:954–968
Ligdi EE, Morgan RPC (1995) Contour grass strips: a laboratory simulation of their role in soil erosion control. Soil Technol 8(2):109–117
Loughran RJ (1989) The measurement of soil erosion. Prog Phys Geogr 13:216–233
Magdoff F, Van Es H (2021) Building soils for better crops. The Sustainable Agriculture Research and Education (SARE) Program, US Department of Agriculture, USA, College Park, MD
Montgomery D (2007) Dirt: the erosion of civilizations. University of California Press, Berkeley, CA
Moritani S, Yamamoto T, Andry H, Inoue M, Kaneuchi T (2010) Using digital photogrammetry to monitor soil erosion under conditions of simulated rainfall and wind. Soil Res 48:36–42
Nyakudya IW, Stroosnijder L, Nyagumbo I (2014) Infiltration and planting pits for improved water management and maize yield in semi-arid Zimbabwe. Agric Water Manag 141:30–46
Oldeman LR (1994) The global extent of land degradation. In: Greenland DJ, Szabolcs I (eds) Land resilience and sustainable land use. CABI, Wallingford

Oldeman LR, Hakkeling R, Sombroek WG (1990) World map of the status of human-induced soil degradation: an explanatory note. International Soil Reference and Information Centre, Wageningen

Pimentel D (2006) Soil erosion: a food and environmental threat. Environ Dev Sustain 8:119–137

Plaza-Bonilla D, Nolot J, Raffaillac D, Justes E (2017) Innovative cropping systems to reduce N inputs and maintain wheat yields by inserting grain legumes and cover crops in southwestern France. Eur J Agron 82:331–341

Prosdocimi M, Tarolli P, Cerdà A (2016) Mulching practices for reducing soil water erosion: a review. Earth Sci Rev 161:191–203

Rampino MR (2005) Erosion processes. In: Schwartz ML (ed) Encyclopedia of coastal science. Encyclopedia of earth science series. Springer, Dordrecht. https://doi.org/10.1007/1-4020-3880-1_139

Raza MA, Feng LY, van der Werf W, Iqbal N, Khan I, Khan A, Din AMU, Naeem M, Meraj TA, Hassan MJ, Khan A, Lu FZ, Liu X, Ahmed M, Yang F, Yang W (2020) Optimum strip width increases dry matter, nutrient accumulation, and seed yield of intercrops under the relay intercropping system. Food and Energy. Security 9(2). https://doi.org/10.1002/fes3.199

Raza MA, Feng LY, van der Werf W, Iqbal N, Khan I, Hassan MJ, Ansar M, Chen YK, Xi ZJ, Shi JY, Ahmed M, Yang F, Yang W (2019a) Optimum leaf defoliation: a new agronomic approach for increasing nutrient uptake and land equivalent ratio of maize soybean relay intercropping system. Field Crop Res 244:107647. https://doi.org/10.1016/j.fcr.2019.107647

Raza MA, Feng LY, Iqbal N, Ahmed M, Chen YK, Khalid MHB, Mohi Ud Din A, Khan A, Ijaz W, Hussain A, Jamil MA, Naeem M, Bhutto SH, Ansar M, Yang F, Yang W (2019b) Growth and development of soybean under changing light environments in relay intercropping system. PeerJ 7:e7262. https://doi.org/10.7717/peerj.7262

Raza MA, Feng LY, van der Werf W, Cai GR, Khalid MHB, Iqbal N, Hassan MJ, Meraj TA, Naeem M, Khan I, Rehman S, Ansar M, Ahmed M, Yang F, Yang W (2019c) Narrow-wide-row planting pattern increases the radiation use efficiency and seed yield of intercrop species in relay-intercropping system. Food and Energy Security 8(3):e170. https://doi.org/10.1002/fes3.170

Raza MA, Bin Khalid MH, Zhang X, Feng LY, Khan I, Hassan MJ, Ahmed M, Ansar M, Chen YK, Fan YF, Yang F, Yang W (2019d) Effect of planting patterns on yield, nutrient accumulation and distribution in maize and soybean under relay intercropping systems. Sci Rep 9(1):4947. https://doi.org/10.1038/s41598-019-41364-1

Rieke-Zapp DH, Nearing MA (2005) Digital close range photogrammetry for measurement of soil erosion. Photogramm Rec 20:69–87

Shweta R (2022) Forms of soil erosion. https://www.soilmanagement.India.com/soil-erosion/forms-of-soil-erosion-8-forms-soil-science/14983

Thompson A, Sudduth K (2017) Terracing and contour farming. In: Delgado J, Sassenrath G, Mueller T (eds) Precision conservation: geospatial techniques for agricultural and natural resources conservation. ASA and CSSA, Madison, WI. https://doi.org/10.2134/agronmonogr59.c8

Tilman D, Downing JA (1994) Biodiversity and stability in grasslands. Nature 367:363–365

Wakindiki IIC, Ben-Hur M (2002) Indigenous soil and water conservation techniques: effects on runoff, erosion, and crop yields under semi-arid conditions. Soil Res 40(3):367–379

Walling D, Quine T (1995) Use of fallout radionuclide measurements in soil erosion investigations, nuclear techniques in soil-plant studies for sustainable agriculture and environmental preservation. In: Proceedings of an International Symposium held in Vienna, 17–21 October 1994

Youdeowei PO, Abam TKS (1997) Local engineering practices of erosion control in the coastal areas of The Niger Delta. Environ Geol 31(3):231–235

Zapata F (2003) The use of environmental radionuclides as tracers in soil erosion and sedimentation investigations: recent advances and future developments. Soil Tillage Res 69:3–13

Management of Crops in Water-Logged Soil

Rafi Qamar, Atique-ur-Rehman, Saad Shafaat, and Hafiz Muhammad Rashad Javeed

Abstract

Excessively water saturates the soil pores and creates waterlogging when there is indeed no or very thin coating of water present on the soil. Waterlogging typically causes changes in gene expression that affect a plant's physiology, metabolism, and anatomy. Crops respond to and adapt to waterlogging stress in a variety of ways, including the development of aerenchyma, adventitious root development, metabolism of energy, and plant-hormone signaling. One of the most damaging abiotic stresses that annually destroys 17 million km^2 of land, along with drought, is floods. Recent studies have found that increased extreme weather events, like flooding and soil waterlogging, brought on by climate change are having a substantial influence on agricultural productivity. Because of this, it is essential to understand how crops are impacted by flooding stresses and to develop better production methods that boost cropping systems' resistance and ability to endure extreme climate events. Potential management strategies that can be utilized to alleviate the stress brought on by soil waterlogging include the adoption of waterlogging-tolerant varieties, altering administration practices, improving permeability, and putting adaptive nutritional monitoring systems into

R. Qamar (✉) · S. Shafaat
Department of Agronomy, College of Agriculture, University of Sargodha, Sargodha, Pakistan
e-mail: rafi.qamar@uos.edu.pk

Atique-ur-Rehman
Department of Agronomy, Faculty of Agricultural Sciences and Technology Bahauddin Zakariya University, Multan, Pakistan

H. M. R. Javeed
Department of Environmental Sciences, COMSATS University Islamabad, Vehari Campus, Vehari, Pakistan

M. Ahmed, S. Ahmad (eds.), *Disaster Risk Reduction in Agriculture*, Disaster Resilience and Green Growth, https://doi.org/10.1007/978-981-99-1763-1_12

place. These management approaches, which may be crop- or site-specific, should be assessed for their commercial feasibility before developing future implementation strategies that enable sustainable agricultural output from waterlogged soils.

Keywords

Soil waterlogging · Abiotic stress · Physiological response · Agronomic practices · Bio drainage

1 Introduction

Water facilitates plant development and functions, making it essential to a plant's life. However, plants are put in danger by flooding or waterlogging (Normile 2008). As seen in Fig. 1, the condition known as "waterlogging" occurs when a whole or a plant portion is completely under water (Bailey-Serres et al. 2012). As a result, air pockets in the earth are simply filled, leading to wet conditions. In many plant communities around the world, soil waterlogging is an abiotic (non-living) stress which impacts species composition as well as its production (Jackson and Colmer 2005). Seasonal precipitation as a whole, as well as the differences between and among seasonal precipitation events, have changed due to climatic variations. Extremes in the availability of water have grown more severe globally in farming areas during the past 50 years (Aderonmu 2015; Bailey-Serres et al. 2012). The main causes of waterlogging in Pakistan include inadequate irrigation management techniques, a scarcity of suitable infrastructure for drainage of soils, and the use of low-quality water for irrigation purposes (Hossain 2010). Due to the threat to food security, it is

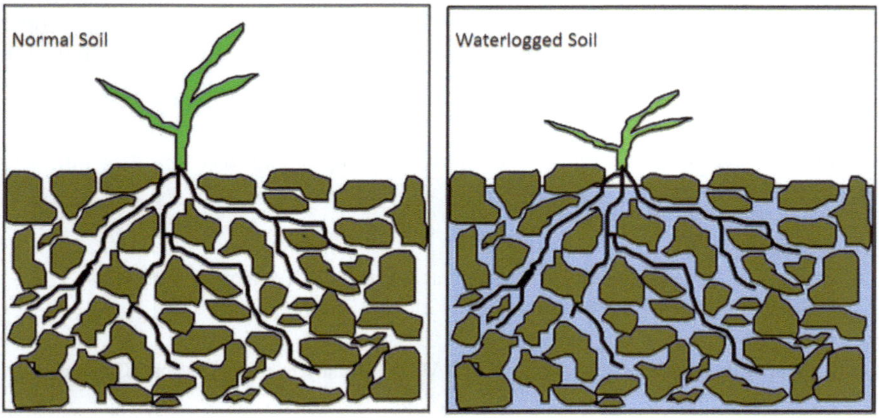

Fig. 1 Water logged condition illustration. [Source: Engineer Moid (2021), https://www.civil-click.com/waterlogging/ (source)]

urgent to find low-cost, ecologically friendly ways for managing and reclaiming these soils (Qadir and Oster 2002).

Flooding has a devastating effect on society and the environment. As terrestrial plant species including cultivated crops are susceptible to flood conditions, there is a decline in biodiversity of plants, natural species distribution, and production of food worldwide (Normile 2008). Each year, flooding damages about 17 million km^2 of land worldwide, resulting in losses in crop production and serious damage to plants (Voesenek and Sasidharan 2013). Waterlogging or serious soil drainage issues harm between 10% and 12% of the world's agricultural land (Shabala 2011). The projected annual cost of damage from severe floods that occur all over the world is much more than $74 billion (www.dartmouth.edu/~floods/Archives/2005sum.htm). According to existing fluctuations in changing climate globally, showing harsh climatic events, the National Aeronautics and Space Administration (NASA) simulation models estimated losses worth $3 billion annually in food production by 2030 (Rosenzweig et al. 2002). Pakistan has experienced exceptional monsoon conditions since June 2022, this month alone received rainfall which was 67% above average levels. As of August 27, the nation had received 2.9-fold as much rain as the 30-year average. A total of two million acres of crops and orchards have also been damaged at this point, including 1.54 million acres in Sindh, Baluchistan is 304,475 acres, and 178,186 acres in Punjab (OCHA 2022). Anatomical, physiological, and metabolic alterations are typically reported as plant responses to wet and flooding situations (Voesenek et al. 2006). Water diffusion, a mode of transportation in a biological system, is thought to be very low for terrestrial plants' survival for a long duration, which is why flooding causes damage. Essential nutrient deficits and toxicities from micronutrients like Copper (Cu), Iron (Fe), and Manganese (Mn) have an impact on plants (Setter et al. 2006). The primary source of potential energy for plant roots to absorb nutrients is aerobic respiration (Ferreira et al. 2008). These waterlogging effected roots resort to an ineffective anaerobic fermentation, using their present glucose reserves to produce the ATP they require to survive and operate. Continued hypoxia or anoxia impairs root growth and function due to reduced integrity of the membrane, hunger, and phytotoxic chemical diffusion into the root cells (Sauter 2013). Under hypoxic circumstances, the functions of shoots are compromised and may show apparent symptoms like senescence, wilting, and death because the roots are unable to transfer water and nutrients effectively (Sasidharan and Voesenek 2015). In addition, photosynthesis, carbohydrate partitioning, and the production and transport of growth regulators are all significantly impacted (Ferrer et al. 2005). Under waterlogged conditions, these physiological impedances ultimately result in a decreased crop yield.

To maintain root activity and plant survival in susceptible genotypes, waterlogged circumstances may induce and initiate crop tolerance traits or adaptation features that might enhance aeration and mitigate root hypoxia or anoxia. Plant tissues soaked with water produce ethylene (El-Esawi 2016a, b). The activation of genes related to aerenchyma production and adventitious root formation is crucial among the well-explained roles that ethylene plays in waterlogged conditions (Vidoz et al. 2010; Sasidharan and Voesenek 2015). In the shoot, the transport of auxin is reprogrammed by increased ethylene level in the stem, which causes a flow of auxin to be directed toward the submerged stem to start the growth of adventitious roots. Auxin transport inhibition reduces adventitious root development (Vidoz et al. 2010). The formation of suberin or lignin barrier, among other things, in roots in order to prevent loss of O_2, and direct its transportation to the tip of the root, were other adaptive traits displayed by resistant crops (Shiono et al. 2011).

Grain growers employ a wide range of crop management techniques to mitigate the impacts of waterlogging. Selection of crops, crop varieties that can withstand waterlogging, bio-drainage, and various agronomic techniques, like sowing season, nutrient application, engineering methods for surface and subsurface drainage, etc., and use of plant growth regulators (PGRs), are among them (Manik et al. 2019).

2 Causes of Soil Waterlogging

The oxygen concentration drops quickly in waterlogged soils because in water diffusion of a gas is several times slower than in air, causing a series of events that are detrimental to the survival of the majority of plant species (Colmer and Greenway 2011). In Asia and America, flooding is the main reason for yield losses, and waterlogging is thought to damage between 10% and 16% of the planet's cultivable soils (Yaduvanshi et al. 2012). In addition, in response to changing climate, flooding events are anticipated to occur more frequently and more intensely in every part of the planet (Westra et al. 2014). More than 21 Mha of Pakistan's 79.61 Mha total geographic area where agricultural practices take place. Almost 25% of irrigated area in Punjab province is seriously under waterlogging, but about 60% in Sindh (WAPDA 2007). Soil waterlogging in the plant-rooting zone can be caused by many variables, including the amount of water that enters the soil, the amount that flows over/through the soil's surface, and the amount of water that is absorbed by plants and other species (Kunkel 2003). Numerous factors, such as soil type, geography, meteorological circumstances, lateral ground water flows, and rising/perched water tables, can cause waterlogging (Fig. 2).

2.1 Extreme Precipitation

The frequency of heavy precipitation events and several rains is a significant factor in an increase in waterlogging or flooding (Kunkel 2003). Extremely rainy years are distinguished from dry years by the amount, frequency, size, and spacing of precipitation events (Knapp et al. 2015). The Intergovernmental Panel on Climate Change (IPCC) predicts that rising emission of greenhouse gas will probably result in more instances of extreme precipitation ahead (Cubasch et al. 2001).

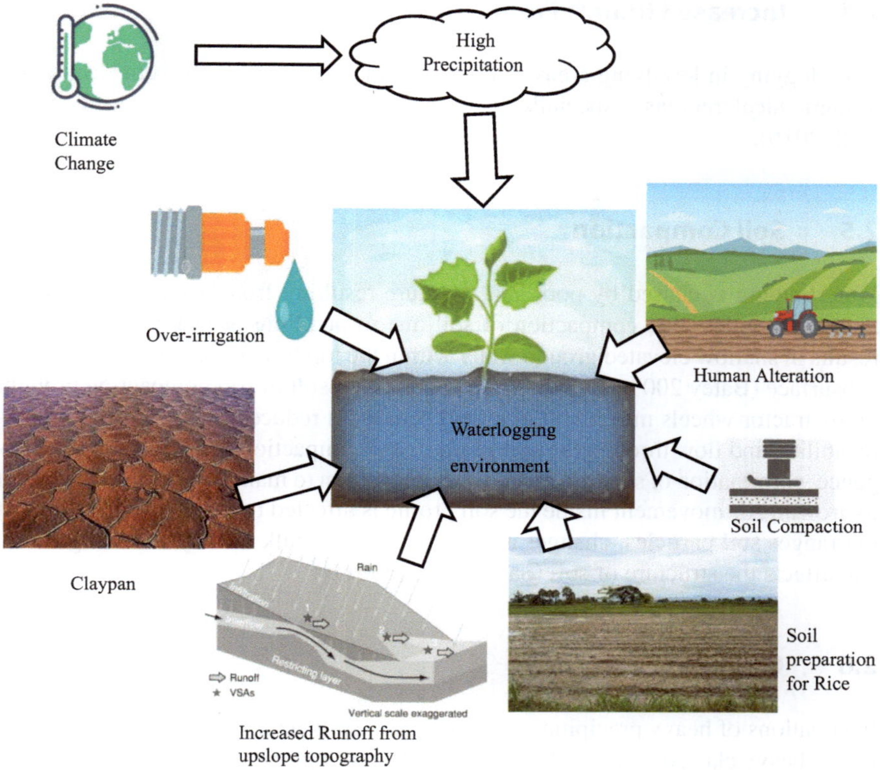

Fig. 2 Causes of waterlogging (Liu et al. 2020a)

2.2 Human Alteration in Land Use

In addition to rising precipitation frequency and severity, human modifications to stream channels, and land use are other factors contributing to rises in waterlogging (Kunkel 2003).

2.3 Over Irrigation/Rainfall after Irrigation

Soil waterlogging or floods may also be caused by over-irrigation or subsequent rains (Kirkpatrick et al. 2006). Shallow water table, compaction of soil, insufficient internal drainage as well as surface drainage are some of the issues (Kirkpatrick et al. 2006), in soils like heavy clay soils, clay pan, or duplex soil with coarse textured topsoil over compacted clay subsoil (Batey 2009).

2.4 Increased Runoff From Slope

Waterlogging in low-lying areas can result from excessive runoff from steep slope topographical regions, especially if soils there are inadequately drained (Singh et al. 2016).

2.5 Soil Compaction

Waterlogging is caused by poor soil structure resulting from natural processes or human activities, like compaction of soil due to puddling or heavy traffic, which results in shallow elevated ground water within the top few centimeters of soil or the subsurface (Batey 2009). Flooding can occur as a result of soil compaction brought on by tractor wheels movement in a field because it reduces water infiltration, permeability, and flow through the soil profile. Soil compaction can impact crop emergence, germination of seed, and its growth in addition to making roots more resistant to growth. Air movement inside the soil profile is affected by compaction because it rearranges soil particles, changes aggregate stability, bulk density, or arrangement, and affects the structure of soil (Samad et al. 2001).

2.6 Claypan

In situations of heavy precipitation or irrigation, soils with swelling–shrinking clay kinds (heavy clay soils) are vulnerable to soil waterlogging. Heavy clay soils with a high-water retention capacity and poor drainage may swell as the soil reaches its maximum water retention capacity, preventing penetration into the soil profile (Blessitt 2007). Constrictive clay subsoil horizons can be found on over 290 million ha of soil worldwide (USDA-NRCS 2006). In soils with clay pans, the subsoil horizon often suffers a fast, 100% rise in clay concentration in comparison to the soil layers above it over a small vertical distance (Motavalli et al. 2003). Depending on the topography, the claypan layer's depth could range, from 10 cm at the back slope locations to 40 cm at the front slope locations (Jiang et al. 2007).

2.7 Soil Preparation for Rice

The yield of successive non-rice crops in the rotation is negatively impacted by the breakdown of aggregates of soil also the creation of a hardpan during puddling, and these crops also demand more effort for land preparation (Kumar and Ladha 2011). Additionally, where the field had been puddled for rice, the soil infiltration rates during the wheat season are lower than they were whenever the land had been dry-drilled or maintained in no (Singh et al. 2011). Preparation of soil for rice (*Oryza sativa* L.) cultivate causes compaction of subsurface, leading to low drainage, as a result, waterlogging issues in crops like wheat in Asian countries (Samad et al. 2001).

3 Why Did Waterlogging Conditions Develop in Pakistan?

Pakistan is blessed with an abundance of water sources, including enormous rivers, tributaries, rivulets, and hill torrents, as well as significant underground water reservoirs that are known for their tall snow- and ice-covered mountain summits. The Indus irrigation system utilizes a large river and rainwater, which may help irrigate vast amounts of potentially fertile agricultural land (Aslam et al. 2015). Pakistan's economy is largely agrarian just because of that. Major crop yields, however, are much lower than any of those attained by other developing nations worldwide. Table 1 provides information on crop output (year) deficiency in the Indus Basin. Various soil, water, and management techniques, inadequate floods and spoor water management procedures, inadequate irrigation inputs of good quality water, and an insufficient drainage system could all be to blame for this (Aslam et al. 2015).

In Pakistan, irrigated agriculture is primarily limited to the Indus plains, where it has grown as a result of utilizing the main water resources the nation has to offer. Adjoining Indus Basin irrigates a total of 16 million acres. In the 1960, Indus Water Treaty, Pakistan has access to $181 \times 109\ m^3$ of water, or around 75% of the yearly available flow, from the Indus River system (Reinsch and Pearce 2005). Due to the rising depth of groundwater levels (>15 m), growers must transition, from tiny tubewells operated by diesel to powerful engines run by electricity or diesel. The majority of tubewell was driven by electricity, installations took place in the 1970s and 1980s, a time when the government offered installation cost incentives. Early in the 1990s, the government stopped providing subsidies due to rising energy costs, which caused the development of electric tubewells to stop and the number of diesel-powered tubewells to rise. Recent estimates indicate that tubewells powered by electricity are just 13%, with the remaining 85% being powered by diesel engines of various sizes (Qureshi et al. 2003). Fresh groundwater is readily available on demand, which has helped farmers attain stable and predictable yields while coping with the fluctuations in surface water supply (WAPDA 2003). To prevent a rise in the groundwater table in semiarid and arid areas, draining is seen as a complementary activity to irrigation. However, even though irrigation development has advanced significantly, Pakistan has never prioritized the building of drainage infrastructure. Due to the constant seepage over time from unlined clay canals, a wide

Table 1 Water resource draft report for the strategy study vol. 1. Islamabad (ADB 2002)

Crops	Demand	Yield	Shortage
Vegetables	14.3	9	5.3
Fruits	16.1	9	7.1
Cotton (lint)	3.5	2.7	0.8
Pulses	1.9	1.4	0.5
Sugarcane	82	46.4	35.4
Oilseed	3.3	1.5	1.8
Food-grain	50	31.5	18.5
Total	171	102.8	69.4

number of distributing channels, irrigated fields infiltration losses, groundwater levels are rising in most of the canal command regions as a result of this carelessness. In large irrigated regions, the groundwater table quickly increased within about 1.5 m of the surface of the soil (WAPDA 2007).

4 Waterlogging Stress: Physiological and Metabolic Processes in Plants

Waterlogging has been related to the number of responses shown by plants that are frequently speculative (Parent et al. 2008; Shaw et al. 2013). Oxygen transport rate in root tissues is significantly slowed down (104 times) by waterlogging in mesophytes. Ethylene, which is produced from its precursor ACC transferred from roots, regulates apoptosis induction in specific tissues and cells, nodal adventitious root formation, the creation of air chambers, metabolic variations during anaerobesis, also several other tasks (Subbaiah and Sachs 2003).

4.1 Oxygen Deprivation

Lack of oxygen caused by excessive water negatively impacts root and shoot development, photosynthesis, hydraulic conductivity, and nutrient uptake. The flow of oxygen to the soil, roughly 3 lac 20 thousand folds lesser pore spaces filled with water when compared to one filled with gas, and in water the oxygen diffusion rate compared to air is about 1/10,000th (Armstrong and Drew 2002; Colmer and Flowers 2008). Compared to air, gas diffusion in water is 104 folds slower, and O_2 deprivation is a primary barrier to waterlogging stress (Bailey-Serres and Voesenek 2008). Reduced O_2 availability slows down plant respiration and ATP synthesis, which inhibits root development (Bailey-Serres and Voesenek 2010). Decreased respiration and Adenosine Tri Phosphate production loss in wet roots are the causes of plant wilting (Sairam et al. 2008). Glycolysis uses glucose as its main fuel to provide energy for plant reproduction and growth through downstream processes including respiration (Galant et al. 2015). During respiration, glucose enters the pathway of glycolysis to create two molecules of ATP and pyruvate. Then, as a component of the TCA cycle (tricarboxylic acid), pyruvate burns to produce CO_2 and H_2O and high energy (36 ATP) in the mitochondria. Figure 3 illustrates the formation of ethanol on cytoplasm from pyruvate under hypoxic conditions, generating two ATP molecules (Sauter 2013). Waterlogged maize, rice, wheat, and barley showed energy deficiency-related restriction of root development. A study on barley and wheat for 11 days (waterlogged treatment) indicated that the growth of roots and shoots considerably decreased (Steffens et al. 2005). In comparison to plants with good drainage, wet shoots, and roots had significantly lower dry weights and root/shoot ratios (Araki et al. 2012). Waterlogging in maize slowed root senescence, which significantly reduced the roots and shoots dry weight (Ren et al. 2016a, b, c). In addition, both lowland and highland rice types' dry weight and root elongation

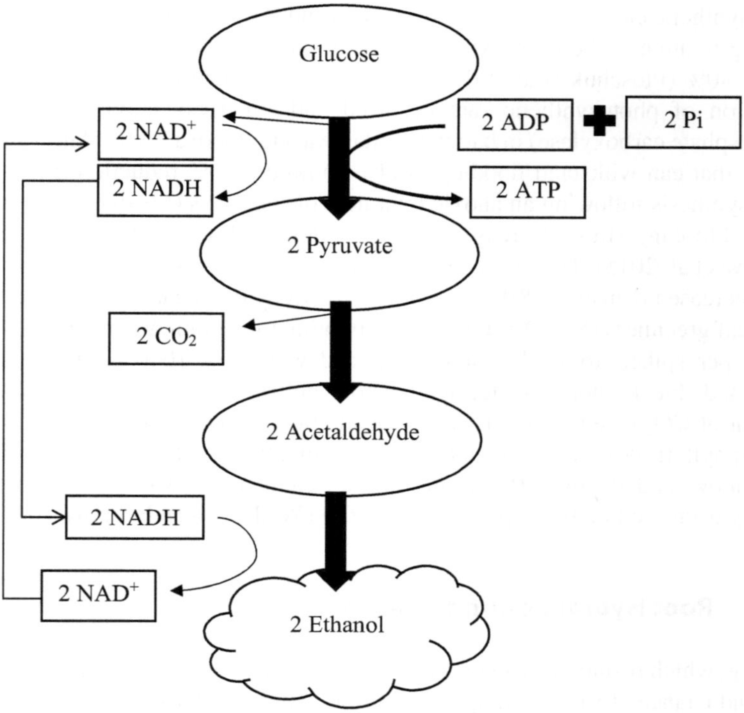

Fig. 3 Fermentation occurring in waterlogged plant roots. (Source: Poulisw 2011. Available at: http://biomhs.blogspot.com/2011/04/anaerobic-respiration-fermentation.html)

were reduced by hypoxia (Liu et al. 2020b). Waterlogged plants' poor root development also reduced their ability to absorb water and nutrients (Ren et al. 2016a, b, c).

4.2 Photosynthesis Rate

Waterlogging stress on crops reduces their photosynthetic rate because of closure of stomata, the conductance of mesophyll, degradation of chlorophyll, disruption to photosystem II, also decreased activity of photosynthetic enzymes (Ploschuk et al. 2018). Photosynthetic enzyme activity is further decreased with a prolonged waterlogging duration. Reduction in photosynthesis during flooding circumstances was shown to be caused by stomatal closure, which was found to be associated with the CO_2 exchange rate and transpiration (Irfan et al. 2010). Chlorophyll fluorescence metrics can be used to determine the various photosynthesis activities that took place in PS II including light absorption, photochemical reactions, and energy transfer (Ashraf et al. 2011). Normal leaf photosynthesis depends on the function of the chloroplast structure in mesophyll cells, which has been discovered to be damaged in waterlogged maize (Ren et al. 2016a, b, c). This damage persistently prevents photosynthetic electron transport (Yordanova and Popova 2007). After 6 h of waterlogging treatment, the photosynthesis rate of barley plants (waterlogged) initially fell by 40% (Ploschuk et al. 2018).Waterlogged treatment for 5 days, a substantial reduction of photosynthetic rate occurred and RuBisCo activity (ribulose-1.5-bisphosphate carboxylase) in barley (Yordanova and Popova 2001). Although rice is a crop that can withstand flooding, it also showed a 50% reduction in the rate of photosynthesis following an anoxic treatment of four days (Mustroph and Albrecht 2003). Flooding stress decreased soybean chlorophyll concentration by 18–34%. (Mutava et al. 2015). The maize leaf area index decreased as the period of waterlogging increased (Liu et al. 2013). Respiratory activity of wheat roots, photosynthetic rate, leaf greenness (SPAD reading), transpiration rate, grain production, number of grains per spike, stomatal conductance, and weight of 1000 grain significantly decreased due to flooding during the post-anthesis stage. Yet, the intercellular amount of CO_2 rose (Wu et al. 2012). In addition, flooded maize plants had lower chlorophyll (a + b) content and were around 20% smaller than control plants (Yordanova and Popova 2007). During the treatment for waterlogging, RuBisCo activity decreased in maize plants by 20–30% (Yordanova and Popova 2007).

4.3 Root Hydraulic Conductance

Wilting, which results from decreased root water intake and decreased root hydraulic conductance (Lp), is a frequent reaction against waterlogging stress (Herzog et al. 2016). Water absorption capacity is determined by Lp, which is connected with transpiration rate (Tan et al. 2018). Under prolonged waterlogged conditions, the death of root cells decreases Lp through erecting barriers (physical) to the flow of water (Bramley et al. 2010). Aquaporin gating and anaerobic respiration caused by a lack of oxygen are additional causes of a large shift in Lp (Tournaire-Roux et al. 2003). Energy production and cytosolic pH control aquaporin, an essential protein of membrane allowing uptake of water by the development of proteinaceous membrane pores (Aroca et al. 2012). Cellular acidosis, which is brought on by CO2 buildup through respiration and ATP depletion, and aquaporin phosphorylation, which results from these processes, control the reduction of aquaporin gating of wet plant roots (Aroca et al. 2012; Tan et al. 2018). Low-ambient oxygen and waterlogging diminish Lp in plants, but species-specific responses differ based on the water transport channel (Bramley et al. 2010).

There are three methods for transporting water:

1. Apoplastic method that is around the protoplasts.
2. Synthetic method that is by plasmodesmata.
3. Transmembrane/across the membranes.

The transmembrane system is regulated by aquaporins, whereas the apoplastic pathway depends on the structure of the root and the characteristics of the cell wall (Maurel et al. 2015). Under hypoxic conditions, lower hydraulic conductance was discovered in Arabidopsis, maize, and wheat as cellular acidosis impairs the activity of aquaporin (Tan et al. 2018). However, the primary route in other species is apoplastic, thus root Lp is not significantly affected by the decrease in the activity of aquaporin in flooding stress (Tournaire-Roux et al. 2003; Bramley et al. 2010). In addition, morphological modifications in *Oryza sativa* (rice), like the creation of barriers that prevent O_2 transport through roots, may have a deleterious impact on roots' hydraulic systems (Aroca et al. 2012).

4.4 Nutrient Absorption

Leaf chlorosis is a frequent symptom of waterlogging stress, which encourages early senescence in the leaf to remobilize N (nitrogen) to new leaves. Reduced nitrogen uptake and transport from roots results in the lower nutritional content of waterlogged shoots (Herzog et al. 2016). It accomplishes this by reduced surface, compromised function, decreased PMF (proton motive force), decreased potential of the membrane also decreased loading of xylem (Steffens et al. 2005). Particularly, wheat and barley under waterlogging stress have significantly lower amounts of magnesium (Mg), copper (Cu), phosphorous (P), potassium (K), zinc (Zn), nitrogen (N), and manganese (Mn) (Steffens et al. 2005). When compared to aerated circumstances, wheat seminal roots took up fewer nutrients from stagnant solutions (Wiengweera and Greenway 2004). Within a few minutes, the hypoxia in the barley roots' mature zone reduced net K+ uptake (Shabala and Pottosin 2014). At various phases of maize growth, Nitrogen assimilation, as well as metabolism, is reduced due to flooding stress (Ren et al. 2017). A physical barrier is created in rice in response to flooding stress in order to prevent O_2 passage form the roots, which might reduce roots nutrition absorption capacity, in contrast to waterlogging vulnerable barley, wheat, and maize that exhibited a significant loss in nutrient content (Rubinigg et al. 2002). For roots to absorb nutrients, there are three possible routes (Reichardt and Timm 2012):

1. Interception of roots' haphazard expansion into new soil areas in search of nutrients.
2. Mass flow, which represents water movement caused by evaporation and transpiration together with ion transfer to the root surface.
3. Diffusion, which is the gradient in chemical potential that encourages the flow of nutrients.

By reducing nutrient interception, in waterlogging stress reduction of growth of roots drastically reduced the potential intake of nutrients (Mancuso and Shabala 2010). The majority of roots of maize (apart from adventitious roots) were not able to take nutrients from ambient soil under waterlogging treatment for 6 days (Qiu et al. 2007). The majority of nutritional absorption relies on diffusion and is fueled by proton motive force and membrane potential, both of which are suppressed during conditions of waterlogging stress. Limited ATP supply resulted in a depolarized plasma membrane, reduced proton motive force, and impaired activity of the plasma membrane proton-pumping ATPase, all of which lowered the cytoplasmic pH. (Mancuso and Shabala 2010). With the help of plasma membrane H + -ATPase, ions that the roots have taken up are transferred to the shoot through the xylem. Waterlogged situations, inhibit xylem transport due to a decrease in H + - ATPase in the parenchyma of the xylem, which constantly lowers the shoot's nutritional content in waterlogged plants (Colmer and Greenway 2011).

5 Anatomical Adaptation

5.1 Formation of Aerenchyma

An air tissue in some plants' adventitious roots creates gaps between cells and does gaseous transportation between roots and shoots, a typical adaptive characteristic linked to the ability to withstand waterlogging (Colmer 2003). Two distinct forms of aerenchyma, schizogenous, and lysigenous, are produced when cells are separated and then lysed, respectively. The cortex of roots of the majority of cereal crops, such as wheat, maize, barley, and rice develops lysigenous aerenchyma (Yamauchi et al. 2013). Lysigenous aerenchyma for wetland plant rice develops constitutively including well soil conditions and rises in wet situations. However, the aerenchyma production in barley, maize, wheat, and other terrestrial plants is only brought on due to moisture stress (Yamauchi et al. 2011). Following 7 days of flood treatment, aerenchyma was found in mature root zones of barley in the tolerant cultivars at a distance of around 6 cm from the root apex (Zhang et al. 2015). According to a study on maize, under conditions of waterlogging, cell death began at 10 mm from the tips and was fully developed at 30–40 mm from the tips (Evans 2004). Higher root porosity and the production of aerenchyma are significant adaptive features that contribute to the ability to withstand waterlogging (Setter and Waters 2003). By starting planned death in cells of particular cell types, ROS abbreviated as reactive oxygen species and the phytohormone ethylene in gaseous form are associated with lysigenous aerenchyma formation. Due to obstruction of gas transport to the rhizosphere and the increased ethylene production caused by waterlogging stress, ethylene builds up in roots (Yamauchi et al. 2018). In response to the stress of waterlogging, antioxidant defense systems are used to combat the harmful consequences of ROS build up (Ashraf et al. 2011).

5.2 Adventitious Root Growth

Seminal root growth is inhibited in wet plants, which results in a lower root/shoot ratio. There are two main plant root types: seminal roots and adventitious roots. Comparatively to seminal roots, which only have a fully developed main root axis, adventitious roots have much more core metaxylem and cortical cell layers (Knipfer and Fricke 2011). As seen in Fig. 4, waterlogged plants frequently respond by

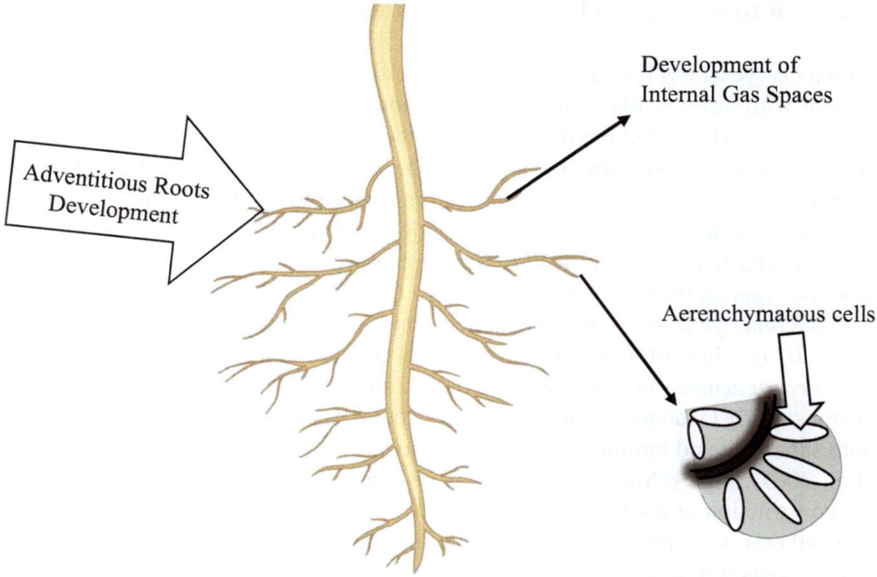

Fig. 4 Aerenchyma cells and adventitious root formation in waterlogged condition. [Kaur et al. 2020(source) (modified)]

forming adventitious roots, which can substitute the damaged seminal roots and produce more aerenchyma to increase the capacity for inner O2 delivery. In trials conducted in greenhouses, seedlings of Zea mays ssp. huehuetenangensis displayed higher adaptation toward submersion with adventitious root formation (Mano et al. 2005). The number of adventitious roots in barley genotypes (tolerant) after 21 days of waterlogging treatment was significantly higher than that in genotypes that were sensitive (Luan et al. 2018a, b). Aerenchyma was found to occupy 20–22% and 19%, respectively, of adventitious roots of wheat and barley (Ploschuk et al. 2018). A study on rice discovered that the hormone auxin gradient in root tips determines the adventitious root growth direction (Lin and Sauter 2019). Adventitious roots extend upward to get closer to the oxygen-rich water surface to help with water and nutrient absorption from the top layer of the moist soil (Jia et al. 2021; Steffens and Rasmussen 2016). Additionally, as it develops at the stem nodes, adventitious roots can shorten the distance that oxygen is transported between shoots and roots (Steffens and Sauter 2009). Epidermal cell death induced by ROS and ethylene promotes the formation of adventitious roots from the epidermis of the node (Nguyen et al. 2018a, b).

5.3 Radial Oxygen Loss (ROL) Barrier

Radial Oxygen Loss barrier, yet some other crucial response characteristic in order to deal with stresses like waterlogging in addition to aerenchyma formation. An apoplastic barrier called the ROL barrier, which is found in the outer root cell layer stops oxygen from escaping into the anaerobic environment (Yamauchi et al. 2018). In general, rice generates ROL barriers in waterlogged or stagnant conditions, whereas flooding-sensitive cereals like barley, wheat, and maize don't (Ejiri et al. 2021). Under hypoxic soil, the ROL barrier enables a plant to maintain high quantities of oxygen at the tips of its roots (Abiko et al. 2012). The development of lignified sclerenchyma and suberized hypodermis in roots regulates ROL (Watanabe et al. 2013). Light ROL barrier development was stimulated for flooding-resistant *Zea nicaraguensis* (wild maize), and lignin and suberin, found in inner and outer layers, orderly (Watanabe et al. 2017). Rice roots' basal region can be shown to have both suberized and lignified cells after two to three weeks of waterlogging (Soukup et al. 2007). Figure Microarray analysis on rice adventitious roots showed during the construction of the ROL barrier, numerous putative genes connected to suberin biosynthesis were highly elevated, while only a small number of genes related to lignin production were induced (Shiono et al. 2011). Malic acid and long-chain fatty acids are connected to the production of suberin, according to metabolite analyses

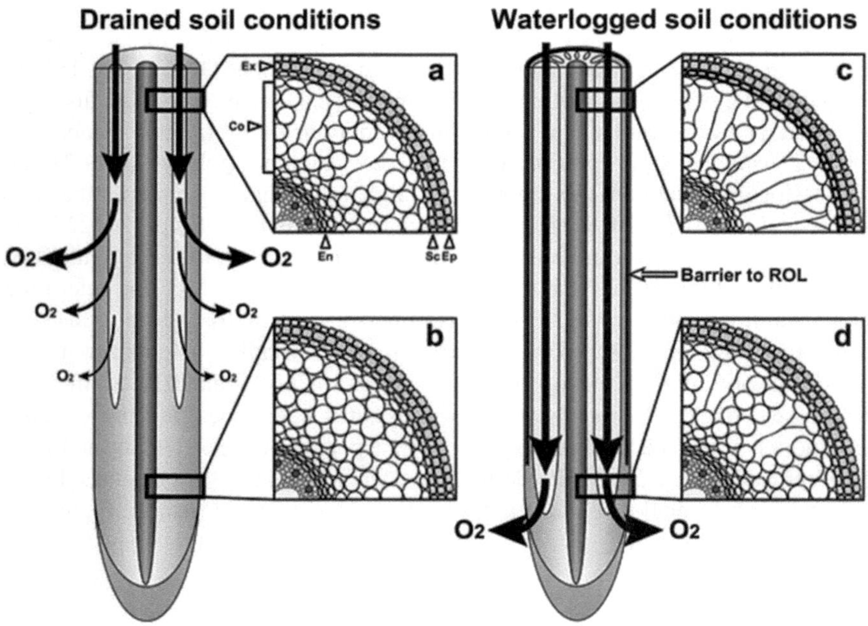

Fig. 5 Rice roots grown in both wet and drained soil exhibit different trends of radial O_2 loss (ROL) and lysigenous aerenchyma development. "Ex" stands for exodermis; "Sc" for sclerenchyma; "Co" for cortex; and "En" for endodermis. (Nishiuchi et al. 2012)

of rice adventitious roots. Malic acid and long-chain fatty acids accumulated during ROL growth (Kulichikhin et al. 2014). Lysigenous aerenchyma is continuously produced under drained soil conditions, but barriers to ROL are not established, which lowers O_2 diffusion towards the apical portion. On the other hand, lysigenous aerenchyma development is accelerated, and the construction of the barrier to ROL is stimulated in wet soil conditions, which promotes longitudinal O_2 diffusion to the root apex. Figure 5 shows how the basal region of the roots constitutively produces lysigenous aerenchyma (a) under drained circumstances of soil, typically not produced at the apical root part (b). At the basal region, lysigenous aerenchyma is formed (c) and the apical part (d) of roots in wet soil conditions. The roots' basal (a, c) compared to its apical portion, the lysigenous aerenchyma is much more developed (b, d). The O_2 availability is shown by the thickness of the arrow. In barley waterlogging tolerant cultivars, lignin deposition under waterlogging stressors greatly increased the activity of the enzyme caffeic acid o-methyltransferase (COMT), which is associated with the formation of lignin (Luan et al. 2018a).

6 Signaling and Response to the Stress of Waterlogging

6.1 Phytohormone Signaling

Under conditions of waterlogging, ethylene, an essential phytohormone that controls plant development and senescence, was shown to accumulate (Iqbal et al. 2017). The development of a plant root barrier that restricts ethylene diffusion leads to ethylene buildup (Voesenek and Sasidharan 2013). In addition, it has been discovered that the activity of two enzymes, ACC oxidase, and synthase (1-aminocyclopropane-1-carboxylic acid), increases due to waterlogging stress (Dat et al. 2004; Broekaert et al. 2006;). Ethephon, an agrochemical that releases ethylene, increased aerenchyma development at the tips of roots and prevented wilting due to waterlogging in barley after pretreatment (Shiono et al. 2019). In order to facilitate plants' movement of O_2 from shoots to roots when there is a lack of oxygen, ethylene controls the creation of gas spaces (aerenchyma) in roots (Steffens and Sauter 2009). In waterlogged maize and barley roots, a transcription level of XET expression was shown to be increased (Luan et al. 2018b). Thus, XET expression and cellulase are induced by ethylene and aid in the production of aerenchyma in roots by dissolving cell walls. Gibberellin (GA), ethylene, and abscisic acid (ABA), significantly play a role in the survival of waterlogged plants by inducing elongation of the shoot. Gibberellic acid encourages elongation between nodes through the breakdown of proteins that are growth inhibitory (Hedden and Sponsel 2015), also through the breaking of starch, releasing cell walls in order to mobilize dietary resources thus enhancing the growth of plants (Else et al. 2009). GA significantly boosts the growth of shoots whereas elongation of roots is inhibited by ABA, acting as antagonists in plants' reactions to growth stimuli (Dat et al. 2004). After 3 h of the flooded plants receiving the ethylene treatment, GA1 increased four-fold and ABA decreased by 75% in deep-water rice (Vaahtera et al. 2014). The stem node

ABA content and gene expression level in ABA production were both decreased in the adventitious roots of flooded wheat. After 3 weeks of waterlogging treatment, ABA content was observed to be reduced in the roots and leaves of varieties of barley (tolerant & sensitive), with a greater decrease in tolerant species (Luan et al. 2018a, b).

6.2 Reactive Oxygen Species Accumulation (ROS)

For crop stress conditions like droughts, salinity, freezing, and mechanical stress, ROS plays a crucial supplementary messenger role, even though they can be harmful to plants because they unrestrictedly oxidize cell components (Mittler 2002; Mhamdi and Van Breusegem 2018). Figure 6 illustrates the primary metabolic adaptations of flooding tolerance of plants as well as stress responses to

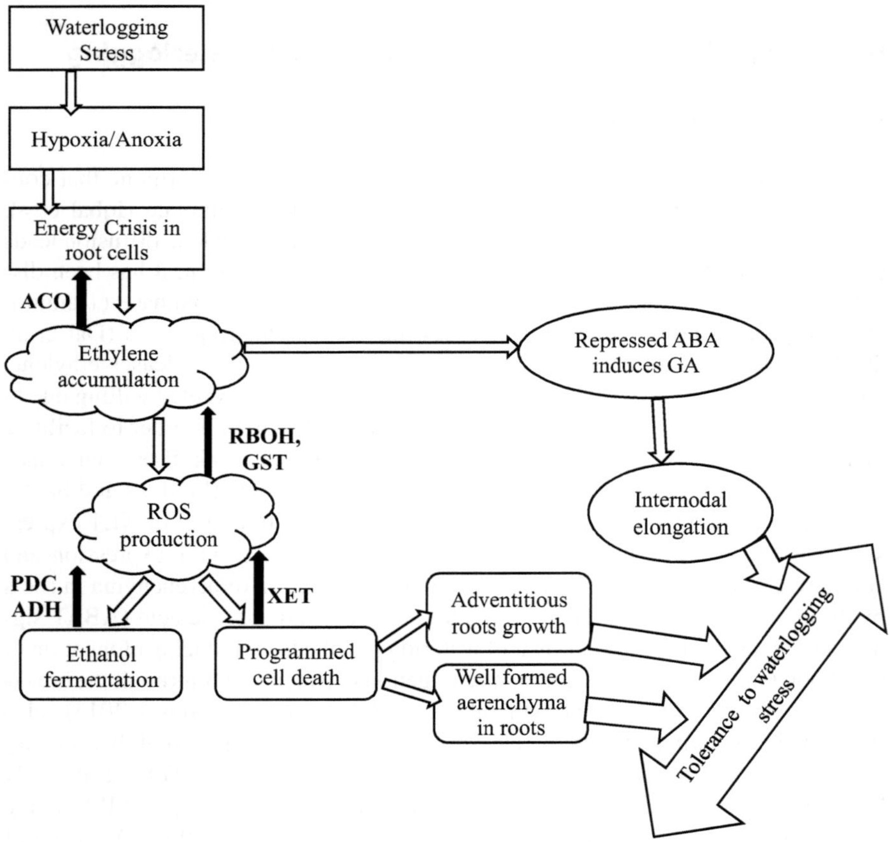

Fig. 6 Schematic diagram of the main waterlogging stress responses and metabolic adaptive traits for waterlogging tolerance in plants. PDC: pyruvate decarboxylase, ADH: alcohol dehydrogenase, RBOH: respiratory burst oxidase homolog, GST: glutathione S transferase, XET: xyloglucan endo-transglycosylase, ACO: 1-amino-cyclopropane-1-carboxylic acid oxidase (Tong et al. 2021)

waterlogging. Different organelles in plants, such as chloroplasts for photosynthesis, mitochondria for respiration, and peroxisomes for photorespiration, all participate in the metabolism of ROS (Foyer and Shigeoka 2011). The breakdown in mitochondria of the electron transport chain during oxygen deprivation results in the production of hydrogen peroxide (H_2O_2), It acts as a stimulus for epidermal cell death to generate aerenchyma, protecting plants from anaerobic environment stress (Fukao and Bailey-Serres 2004; Steffens et al. 2011; Rajhi et al. 2011). H_2O_2 treatment increased lysigenous aerenchyma production by processes of cell death in flooded *Oryza sativa* (rice) (Blokhina et al. 2001). Similarly, under wet conditions, H_2O_2 accumulation was discovered in the roots of wheat and barley (Yamauchi et al. 2014). ROS accumulation, a trigger for wheat seedlings to respond to waterlogging by controlling gene expression, is associated with fermentation of ethanol (ADH and PDC) and aerenchyma formation (Sumimoto 2008). Respiratory Burst Oxidase Homolog (RBOH), which genes for an NADPH oxidase found in the plasma membrane for the production of H_2O_2, controls the accumulation of ROS (Steffens 2014).

7 Agronomic Practices to Grow the Crop in Waterlogged Soil

Given how weather-sensitive agronomic crop production is, global climate change has an impact on the agricultural industry (Aderonmu 2015). The amount of seasonal precipitation as a whole, as well as the differences between and within seasonal precipitation events, have changed due to climatic fluctuations. Extremes in water availability, particularly waterlogging, have increased during the past 50 years in agricultural districts all over the globe (Aderonmu 2015; Bailey-Serres et al. 2012). Whenever all or a portion of a plant is submerged in water, the condition is referred to as "flooding" (Bailey-Serres et al. 2012). Figure 7 illustrates suggested procedures for various wet environments. The measures listed below can help you deal with the effects of waterlogging:

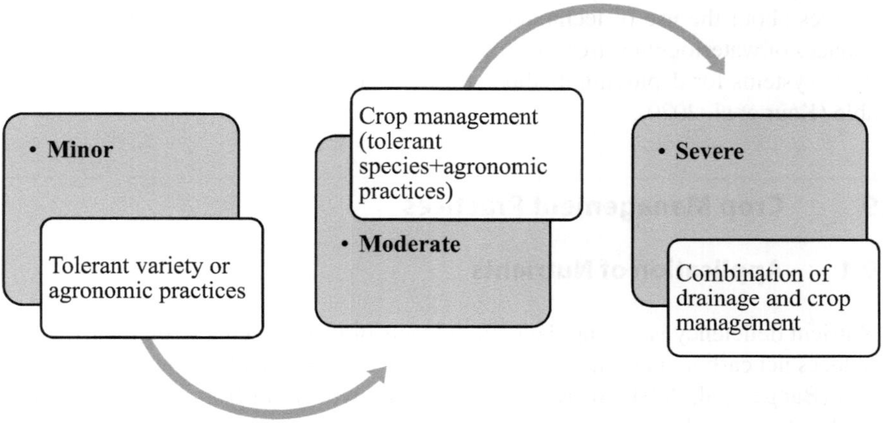

Fig. 7 Recommendations for managing soil and crops dependent on the severity of waterlogging (Manik et al. 2019)

8 Modeling of Crops and Decision-Making Systems

Numerous models, such as DRAINMOD, can simulate how crops respond to waterlogging in the soil in terms of growth and yield (Skaggs 2008). These models can be used to determine which places or circumstances will result in a decrease in yield and to evaluate the effect of management practice modifications on the reduction of flooding stress on crop plants. For instance, the Agricultural Production Systems Simulator (APSIM-Wheat) was used in order to predict impacts on wheat due to waterlogging at various dates of the plantation. It was discovered, only in places with a minimal to medium risk of waterlogging will an earlier planting date boost crop output, yet had no effect in areas that frequently experienced waterlogging (Bassu et al. 2009). However, the effectiveness of these simulation models for use in evaluating waterlogging stress depends on how well the processes are represented in them (Shaw et al. 2013). Additionally, remote sensing and GIS are utilized to locate fields' sensitive areas to soil flooding or dry conditions and can assist in the precise positioning of crops or strategies of management of nutrients thus lessening waterlogging stress. Selectively regions with the greatest nutrient losses, employment of cover crops may lower the cost of planting cover crops and result in financial savings for farmers. For producers to make decisions on the precise placement of various crop management measures to reduce the stress caused by soil waterlogging, they need decision support tools. The Right, Practice, Right, Place (RPRP) Toolbox, which consists of collection preservation strategizing tools online that connect at the local, watersheds, and field level, applying the "right practice of conservation " to the "right spot" can help increase the efficacy as well as efficiency of efforts to improve quality of the water. (McLellan et al. 2018). Using several BMPs (Best Management Practices), individually /collectively, to reduce the loss of nutrients in crop fields is evaluated using the SWAT (Water Assessment Tool) model (Merriman et al. 2019). Although these decision-support systems have been tested for identifying BMPs for improving the quality of water, yet not been examined in determining how well BMPs mitigate stresses like flooding in various situations. Crop producers can use these models as tools to help them make well-informed choices about the use of techniques of crop management for locations where the chances of waterlogging stress are high. Still no information regarding how to apply these systems for deploying methods for management at individual sites are available (Kaur et al. 2020).

9 Crop Management Practices

9.1 Application of Nutrients

Nutrient deficiency is among the main impacts of waterlogging upon plants, which reduces net carbon fixation and photosynthesis and, eventually, growth and production (Bange et al. 2004). Increased productivity will result from the application of vital nutrients, which will help to lessen the effects of abiotic pressures such as waterlogging (Noreen et al. 2018). N fertilizer applications may increase and speed up a plant's ability to adapt to waterlogging stress, including root regrowth and adventitious root growth following a flooding event. This may raise a plant's tolerance to waterlog stress. Due to low O_2 during flooding, which may prevent plants from absorbing N, its loss through leaching and denitrification may result in N deficits, decrease nitrogen availability, and restrict root function (Nielsen 2015), as seen in Fig. 8. The application of enhanced-efficiency N fertilizers, such as slow-release or controlled-release (SR/CR) fertilizers, under wet conditions, is crucial for enhancing plant growth and development (Shaviv 2001; Varadachari and Goertz 2010). By coordinating nitrogen release with crop needs, throughout growing crops, slow-release fertilizers can emit nitrogen across a long duration of time, maximizing (NUE) nitrogen use efficiency (Trenkel 2021). Externally applied fertilizers may be effective if the nutrient ions infiltrate the root architecture, enabling plants to heal from waterlogging-related damage, claims many research studies (Ashraf et al. 2011; Habibzadeh et al. 2012; Najeeb et al. 2015). Wheat (Pereira et al. 2017; Zheng et al. 2017), barley (Pang et al. 2007), corn (Kaur et al. 2018), canola (Kaur et al. 2017), and cotton (Wu et al. 2012; Li et al. 2013) are among the crops that are (Habibzadeh et al. 2012). Application of fertilizer also extends the time that the canopy is open and speeds up the development of photo-assimilates that are transferred to grain rather than straw, raising HI (harvest index) (Kisaakye et al. 2015, 2017). Additionally noted that potassium fertilizer can mitigate the negative impacts of waterlogging in a variety of crops, including rapeseed and cotton (Cong et al. 2009; Ashraf et al. 2011). In phosphorous deficiency, during a rainy growing

Fig. 8 Different soil conditions, both aerated and anaerobic, are used to demonstrate the nitrification and denitrification processes. Available at: https://civiljungle.com/waterlogging/

season, external application of different phosphorus (P) sources, such as Meat & Bone Meal (MBM) and Dairy Cow Manure (DCM), is beneficial for providing the highest yields (Ylivainio et al. 2008, 2018). Under flooded conditions, the administration of FYM (Farm Yard Manure), greatly enhanced iron, zinc and copper concentrations in grain (Masunaga and Fong 2018).

Even with high-value crops, the application of fertilizer to prevent waterlogging loss in extensive farming is limited due to the lack of research on their potential benefits for improving crop performance in waterlogged circumstances (Trenkel 2021). To prevent tissue toxicities (such as manganese) and nutrient imbalance from harming soil ecology, it is important to examine the application techniques, nutrient types, timing, and rate that are acceptable (Rochester et al. 2001; Jackson and Ricard 2003).

9.2 Plant Growth Regulator

Applying PGRs at the proper growth stage can reduce the damage that waterlogging causes to plants (Wu et al. 2018; Ren et al. 2018). Plant growth under wet conditions is improved by plant growth regulator administration (Pang et al. 2007; Ren et al. 2016a, b, c). In waterlogged barley, auxin (synthetic) 1-NAA (1-naphthalene acetic acid) encourages adventitious roots formation (Pang et al. 2007), while the external administration of cytokinin 6-BA (6-benzyl adenine) can reduce the effects of waterlogging and boost maize output (Ren et al. 2016a, b, c, 2018). By enhancing leaf photosynthesis, pre-waterlogging ABA foliar treatment enhanced the resistance of cotton plants to subsequent waterlogging-related damage (Pandey et al. 2002; Kim et al. 2018). Triazole is recognized to be a fungus-toxicants, and they also affect how plants respond to stress and regulate their growth (Rademacher 2015). For instance, paclobutrazol reduces the harm caused by waterlogging in sweet potato plants and canola (Lin et al. 2008). 5-methyl-1,2,4-triazole (3,4-b) benzothiazole (Tricyclazole) treatment reduces plant damage when there is waterlogging (Habibzadeh et al. 2013). However, there hasn't been much usage of plant growth regulators to lessen waterlogging damage due to inconsistent results at the commercial level (Manik et al. 2019).

9.3 Pretreatment with Hydrogen Peroxide

Pre-treating plants with an agent could be a successful method to boost their tolerance to various stresses as shown in Fig. 9. For instance, pretreating crops with H_2O_2 can shield them from oxidative harm brought on by waterlogging, intense light, chilly weather, salt stress, drought, and heavy metal exposure (Gechev et al. 2002; Rajaeian and Ehsanpour 2015; Andrade et al. 2018). Increases in the diameter of the stem, high accumulation of biomass, the volume of the root, and photosynthetic pigments were also brought about by H_2O_2 pretreatment (Andrade et al. 2018). H_2O_2 pretreatment resistant against waterlogging, despite substantial research being done

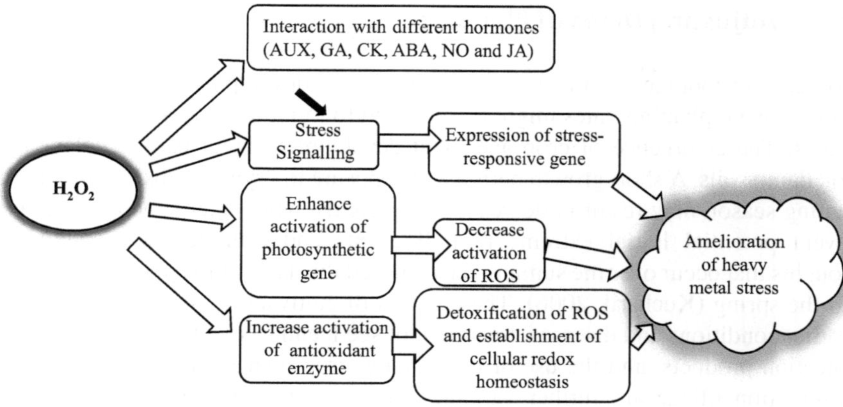

Fig. 9 Reactive oxygen species are shown to be produced by heavy metals in a schematic representation, and hydrogen peroxide is used since it is a signaling substance to activate the antioxidant system, that lowers the concentration of reactive oxygen species and safeguards plants from harm caused by heavy metal stress (Htet et al. 2019)

on treatments against both biotic and abiotic stresses, is still in its infancy (Mustafa et al. 2017; Lal et al. 2018; Ashraf et al. 2018).

9.4 Utilization of Tolerant Varieties and Species

Waterlogging tolerance is one of the most effective ways to reduce the loss brought on by flooding in available plant species (Zhou 2010; Wani et al. 2018). There are genetic variations in waterlogging resistance among several crops, including wheat and barley (Zhang et al. 2015; Huang et al. 2015; Herzog et al. 2016; Wu et al. 2018; Nguyen et al. 2018a, b). Waterlogging tolerance, however, is a dynamic condition that is regulated by a diverse range of mechanisms, including the maintenance of membrane potential (Gill et al. 2018), the control of ROS production under stress conditions, resilience to metabolites (Pang et al. 2006), toxicity of ion (Huang et al. 2018), aerenchyma formation in roots under waterlogging stress, and many quantitative trait loci (QTL) (Gill et al. 2018). The identification of genes that are associated with different tolerance mechanisms is crucial for the success of breeding programs because it enables producers to elevate tolerance genes. Depending on the region and the weather, flooding can happen at any stage of the crop's growth. Waterlogging brought on by heavy rains in the fall can delay crop harvesting, so it's critical to breeding crop varieties with traits like strong stems, superior seed quality, and reduced sensitivity to diseases and pests. It is important to create variety with tolerance both to cold and floods stress since flooding in the early part of the growing season often subjects crops to cold soil temperatures. In conclusion, it's critical to create and test novel varieties of crops that are resistant to a variety of biotic or abiotic stresses, such as heat, drought, and waterlogging stress, along with disease vectors (Kaur et al. 2020).

9.5 Adjusting Dates of Planting

In order to encourage favorable crop emergence and development during the initial spring season, planting dates might be modified to minimize waterlogging circumstances. The emergence of crop and plant development vigor can be delayed by cool, damp soils. A short growth period and exposure to dryness subsequently in the planting season may result in decreased crop yields, while later sowing dates may prevent potential initial extreme rainfall events and saturated soil conditions. Droughts that occur over the summer have pushed the dates of the plantation, sooner into the spring (Kucharik 2006). The creation of cultivars resistant to unfavorable weather conditions and diseases, treatment of seed, enhanced plantation tools, crop protection products, and the use of time-saving crop management techniques like conservation tillage are further variables that contribute to early planting dates (Kucharik 2006). By extending the time for solar radiation absorption and biomass accumulation, early planting dates enable a longer growing period and larger yields (Kucharik 2006). Crop varieties chosen for early planting ought to be resistant to a low temperature of soil that develops after or during planting because there is a danger of plant injuries due to inadequate soil temperature at these early planting dates. Changing planting dates to reduce soil waterlogging depends on when and how long the waterlogging lasts (Kaur et al. 2020).

9.6 Use of Cover Crops

By enhancing soil structure, lowering compaction, and boosting the rate of water infiltration, the use of cover crops may not just improve soil health but also reduce waterlogging (Blanco-Canqui et al. 2015). Cover crop roots can create more macropores, which will result in more water moving through the soil. Increased cover crop transpiration during the spring may potentially dry the soil in time for earlier crop planting. Through larger evapotranspiration (ET) losses, cover crops with higher water requirements and warmer springtime temperatures can assist in eliminating extra moisture from the waterlogged soils. Numerous studies have documented how cover crops can reduce soil moisture content (Monteiro and Lopes 2007; Zhang and Schilling 2006). Reed canary grass (Phalaris arundinacea) had a reduced water table and soil moisture content due to increased ET losses, which decreased groundwater recharge, according to research on the impact of land cover on these variables (Zhang and Schilling 2006). Utilizing cover crops during the winter fallow season is another possible strategy for preventing soil waterlogging. However, depending on the soil type, climate, and cover crop species employed, the impact of cover crops on the water distribution in the soil profile can be beneficial, negative, or neutral (Blanco-Canqui et al. 2015). Therefore, further research should be done on the utilization of different types of cover crops for fields that are prone to flooding or drought. Topography, which can affect nutrient and water dynamics within a field and provide variability biomass synthesis of cover crop and cash crop yields, is not always present in agricultural fields. Therefore, it is crucial to better assess how topography and cover crops interact to lessen the stress caused by waterlogging in big agricultural areas. Using spatial modeling and remote sensing techniques from the geographic information system (GIS), it may be possible to spot parts of an agricultural field that are prone to flooding (Kaur et al. 2020).

9.7 Using Conservation Tillage Techniques

Conservation tillage techniques include mulch tillage, minimal (reduced) tillage, ridge tillage, and contour tillage. The term "minimum tillage" (MT) refers to soil manipulation with only a minimal amount of plowing utilizing primary tillage tools. No-tillage (NT) refers to field cultivation with little to no soil surface disturbance. Mulch tillage involves preparing or tilling the soil in a way that allows plant wastes or other materials that cover the surface to the greatest possible amount. In ridge tillage, crops are planted in rows either on top of or along the ridges that are formed at the start of a cropping season. Variations in conservation tillage's effects on soil characteristics rely on the specific system selected. The soil qualities, particularly in the top few centimeters, have changed significantly as a result of no-till (NT) methods, that achieve high top soil coverage (Anikwe and Ubochi 2007). NT technologies are particularly successful at minimizing erosion losses, decreasing the amount of residue disturbance, and moderating soil evaporation (Lal et al. 2007). No-till soils have been linked to much more stable aggregates mostly in the soil's upper surface than tilled soils, which leads to high permeability under NT plots. Over 37–40 years of tillage operations in Gottingen, Germany, minimum tillage (MT) enhanced both levels of SOC and nitrogen inside the aggregate in the upper 5–8 cm soil depth as well as aggregate stability (Jacobs et al. 2009). In tropical and sub-humid tropics, no-till has been proven to be more beneficial in terms of water saving. Contrary to tilled plots, untilled plots hold more water (Kargas et al. 2012).

Compared to normal plowing, minimum tillage increased the soil pores (0.5–50 mm), also many elongated transmission pores (50–500 mm), which improved the soil's pore system (Pagliai et al. 2004). The upper layer of soil (0–10 cm) under NT has been observed to have a greater holding capacity for water (McVay et al. 2006). Therefore, to improve soil water storage and increase water use efficiency (WUE), the majority of research has recommended shifting to conservation tillage rather than just traditional tillage (Fabrizzi et al. 2005; Silburn et al. 2007). Table 2 lists many benefits and drawbacks of crop management techniques.

10 Adaptive Water Management

Numerous initiatives have been launched to address the issue of waterlogging since the early 1960s. Farmers made no investments in the majority of these initiatives because they were subsidized by the government. Despite significant investment, progress in resolving land degradation issues has been slow. Waterlogging issues were not as easily handled as originally thought. A high groundwater level is a

Table 2 Merits and demerits of management techniques for crops

The practice of managing the soil and crops	Merits	Demerits	Reference
Application of nutrition (particularly N)	Promoting the development and growth of plants	The right methods, nutrients, duration, and quantities should be considered for large-scale administration	Rao et al. (2002), Pang et al. (2007), Wu et al. (2012), Najeeb et al. (2015), Pereira et al. (2017), Li et al. (2013), Kaur et al. (2017, 2018), Zheng et al. (2017)
Plant growth regulators	Encourage water-logged plants' and photosynthetic ability	When applying on a wide scale, the right timing, rate, and procedures should be taken into account	Ren et al. (2016a, b, c, 2018), Habibzadeh et al. (2013)
Hydrogen peroxide pre-treatment	Defend plants from oxidative harm brought on by waterlogging	Unproven in commercial agriculture	Gechev et al. (2002), Ishibashi et al. (2011), Savvides et al. (2016)
Tolerant species and varieties	Economical for farmers	It takes time and effort to introduce waterlogging tolerance into current plant varieties	Zhou et al. (2007), Gill et al. (2018)
Modifying plant dates	Utilizing the soil's current water acts as a buffer and prevents catastrophic waterlogging incidents	Small benefit in cases of extreme waterlogging	Bassu et al. (2009), Sundgren et al. (2018), Wollmer et al. (2018)

problem for 20–30% of the population as a result of excessive surface water use (Smedema 2000).

10.1 Drainage Systems

One of the key strategies for increasing yields per available agricultural area is land drainage (Malano and Van Hofwegen 2018; Singh 2018b). The two major goals of agricultural drainage are to decrease soil submergence and open up a new areas for agriculture (Singh 2018b). When compared to irrigation, drainage is a more effective agriculture engineering solution to fight to waterlog; nevertheless, neither individual farmers nor governmental organizations have given it the same priority as irrigation. Around the world, drainage is utilized to reduce waterlogging (Milroy et al. 2009). Numerous research from North America, Europe, and England show that draining can successfully lower the water table and boost crop production (Gramlich et al. 2018). Many techniques to lessen the waterlogging problems have been proposed (Kazmi et al. 2012; Singh 2012, 2016). These techniques are explained below:

10.1.1 Surface Drainage

Surface drainage is the method of employing manufactured channels to safely remove surplus water from the surface of the ground (Ritzema et al. 2008; Ayars and Evans 2015). Surface drainage systems have been proven to be cost-effective, with cost-benefit ratios ranging from 1.2–3.2, average return rates from 20 to 58%, and payoff times of 3 to 9 years (Ritzema et al. 2008). The simplest and most affordable option is to keep the surface drains that already exist and build additional ones across edges or via depressions while considering their proper size and placement. Using cut-off drains to stop water from flowing from higher to lower paddocks is also a smart option (Palla et al. 2018). However, inadequate lateral water flow or internal soil drainage qualities frequently limit the effectiveness of surface drainage, resulting in poor drainage near the drains (Saadat et al. 2018). This means that the solutions to these issues may involve both subsurface and surface drainage.

Raised Beds

In semi-arid and arid areas, these are re-utilized to address hurdles like irrigation requirements and also waterlogging (Govaerts et al. 2007). By preserving a suitable moisture level in soil via enhanced seepage and acting as a channel to distribute irrigation water, raised seedbeds can improve crop yields (Velmurugan et al. 2016). It increases drainage, and aggregate stability and reduces bulk density (Hassan et al. 2005). To prevent compaction of soil, which promotes penetration of soil, development of roots, and surface and subsurface infiltration, traffic in the furrows should be kept to a minimum. In comparison to flat seedbed planting, raised bed planting has been shown in several studies to increase crop yields in soil that is saturated with water (Blessitt 2007). Soil structure enhancement as seen by infiltration rate and lower bulk densities in clay soil (duplex), which decreased the likelihood of waterlogging and promotes rates of runoff seen in raised beds due to the availability of furrows (Bakker et al. 2005a). As the top 15 centimeters are kept dry during planting in raised beds, as shown in Fig. 10, raised bed planting minimized waterlogging stress. Raised beds have helped to lessen the consequences caused by flooding, but it also has some demerits. These include the price of modifying and adapting machinery, the difficulty of managing drainage water, the use is restricted where the water table is too high, handling stubble and preserving fodder, firefighting and mobilizing livestock, the possibility of pesticide contamination of waterways and leaching into the water table, ineffectiveness of machinery, and weed management in furrows (Bakker et al. 2005b; Gibson 2014).

10.1.2 Subsurface Drainage

Due to the thick soil composition, compact layers, and naturally occurring or artificial hard pan as well as water flowing downhill from springs or from higher slopes, which raises the water table, poor subsurface water mobility occurs (Ward et al. 2018). Subsurface drainage reduces the water table or perched water and creates an

Fig. 10 Tackle waterlogging problematic soil, via Raised Beds (Source: Glenn McDonald 2022)

environment that is conducive to waterlogging in the root region (Christen and Skehan 2001; Xian et al. 2017). Open and pipe drains with varying drain depths and spacing constitute subsurface drainage systems (Ritzema et al. 2008). The sort of drain that should be installed often relies on the topography, the soil, and the needed drainage rate.

Horizontal Subsurface Drainage

The crop root zone is drained of excess water via horizontal subsurface drainage (Teixeira et al. 2018). The drainage system is made up of pathways of perforated pipes below the surface of the earth. Higher agricultural yields can be achieved by draining surplus soil moisture, to enhance root emergence and growth (Nelson et al. 2009, 2011). Flooded soil drainage can be improved globally through subsurface drainage technology (Nelson et al. 2012; Sharma et al. 2016). Using tiny pipes constructed of concrete that is placed at a predetermined depth, tile drainage is a form of horizontal subsurface drainage. In agricultural fields where subsurface excess water is a regular issue, tile drainage is widely used (Williams et al. 2015). In places with shallow groundwater and dense soil conditions, to strengthen the system typically gravel is utilized as a backfill material just above tile seepage (Filipović et al. 2014). This method might not be acceptable for agricultural locations where the top soils are prone to seasonal waterlogging due to inadequate hydraulic conductivity and the need to find a suitable outfall for drained water (Christen and Skehan 2001; Singh 2018a).

Tile Drainage

Agricultural fields with tile drainage may lose nitrate due to factors such as precipitation volume and time, initial soil moisture, season, tile depth, and tile distance (Drury et al. 2009). Figure 11 illustrates a study finding that this method enhanced

Fig. 11 Tile drainage advantages. Available at: https://pami.ca/beneficial-management-practices-for-agricultural-tile-drainage-in-manitoba/

moisture in the soil. In contrast to nondrained plots, it enabled soybean planting earlier at least 17 days and enhanced its output from 9% to 22%. Reducing nitrate that enters surface or groundwater systems, crop output can be increased using subsurface drainage systems that include controlled drainage and sub-irrigation (CDSI), increase the effectiveness with which nitrogen is used, and lessen the likelihood of adverse impacts on the quality of water (Drury et al. 2009; Frankenberger et al. 2004). Another possible strategy for addressing the water quality challenges while also offering crop production systems with flood and drought resilience is to connect subsurface tile-drainage to the irrigation reservoirs. However, this requires more analysis.

Vertical Subsurface Drainage

Sand compaction piles prefabricated vertical drains (PVDs), gravel piles, stone columns, and sand drains are a few examples of vertical subsurface drainage (Indraratna et al. 2005; Indraratna 2017). Compared to other subsurface drainage systems, the VD system has a few advantages. For instance, VDs are frequently chosen over other types of drainage because of their comparatively inexpensive cost of construction, also the surface drains shorter length that they provide (Christen and Skehan 2001). However, because operating a network of tube wells requires a lot of energy, the operating and maintenance expense compared to horizontal drainage is more (Food and Agriculture Organization [FAO] 2002; Prathapar et al. 2018). Vertical drainage is more effective for areas with a high water table.

Mole Drains

Subsurface drainage also includes mole drainage. In terms of design and functionality, mole drains are comparable to tile drainage as a semi-permanent solution (Dhakad et al. 2018; Tuohy et al. 2018). It is typically put in place to address issues with soil salinization and rising groundwater levels (Kolekar et al. 2014). Mole drainage depends on densely packed channels and subsoil fissures (Tuohy et al. 2015). The optimum applications for mole drains, which are put next to tile drains, are heavy soils with low permeability, such as clay (Monaghan et al. 2002; Monaghan and Smith 2004). We must put it, to the height of no more than 600 mm above the ground, forming a circle of drainage that is 40 to 50 mm in diameter (Gibson 2014). By dragging a metal object through the ground, such as a mole plow or a bullet with a blade-like foot, a mole drain can be created, as shown in Fig. 12. This method creates an open channel. The expense of installing mole drainage is less, but to maintain the integrity of the channel and improve system efficiency, the moles must be reformed every 2–5 years (Tuohy et al. 2018). To assist with drainage management in a flooded landscape and to successfully duplicate water balance and a drainage network system over a watershed, integrated drainage systems (tile and mole drainage) may be employed (Tuohy et al. 2018).

11 Strategies Adopted in Pakistan

In Pakistan, waterlogged soils have been repaired using reclamation, engineering, and bioremediation techniques. In addition to using subsurface drainage systems and industrial waste water conveyance lines, municipal surface drainage systems have been employed to remove extra water from agricultural areas. In fresh groundwater areas, Pakistan decided to construct tube wells in irrigated areas up to 14,000 in number that cover 2.6 Mha area, to lower the groundwater level to handle

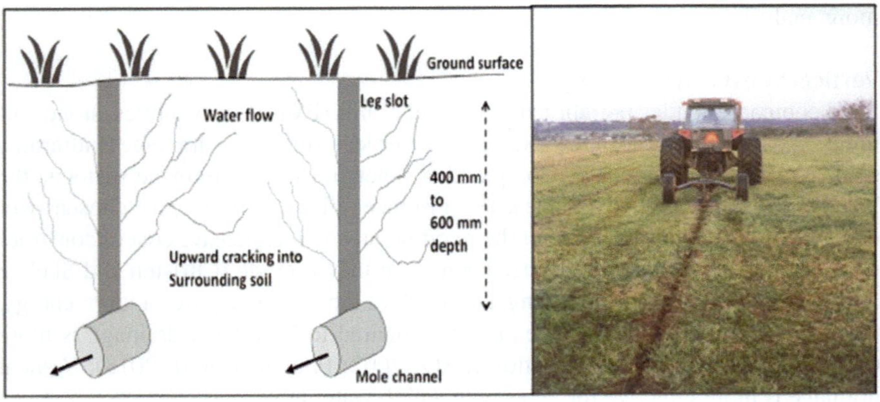

Fig. 12 Sketch of mole drainage. (Source: Don Bennett 2022). Available at: https://agriculture.vic.gov.au/livestock-and-animals/dairy/managing-wet-soils/mole-drainage-systems

waterlogging and to boost irrigation resources at the gate of the farm via blending canal and pumped ground water. The above decision was made after a thorough survey of the depth and salinity of the groundwater table in the 1950s. 63 projects costing 2 billion US$ covering an area of a total of 8 Mha were completed underneath this Salinity Control and Reclamation Projects (SCARPs) over the past 40 years (Qureshi 2011). With the help of the SCARPs initiative, the waterlogging issues were successfully managed, if not reversed, and more water was made available for irrigation. Cropping intensities increased, as a result, rising from 84 to 115% in the majority of SCARP locations. However, as time went on, rising maintenance and operational costs as well as an increase in the salinity of the groundwater being pumped diminished the effectiveness of SCARPs, causing water tables to rise and crop yield to decline. With the belief that with the passage of time drainage water quality would improve, increasing the likelihood of using drainage water for irrigation, thinking switched to horizontal (pipe) drainage systems in the middle of the 1970s. There will also be less of an issue with the disposal. In Pakistan since then, around ten significant horizontal drainage projects collectively drainage pipes of 12,600 km been constructed (Qureshi et al. 2008). In order to address this issue, Pakistan constructed a 2000 km surface drainage on the left bank of the Indus River to transport drainage water from over 500,000 acres of soil to the sea (Qureshi et al. 2008). Although the drain's initial results were highly positive, seepage quickly caused the neighboring communities to become flooded. This heightened interprovincial conflict between the Sindh and Punjab provinces led to a blockage of Punjab's drainage water's path through Sindh and ultimately into the sea. This makes Punjab's waterlogging issues worse. The advantages and disadvantages of the aforementioned adaptive management techniques for waterlogged conditions are shown in Table 3.

11.1 Bioremediation Strategies/Bio-Drainage

Scientists and engineers began considering alternate solutions that are more sustainable and cost-effective as a result of the limited effectiveness in addressing waterlogging issues despite significant investments. Using biological methods to reduce the water table is one of the possible solutions. The idea of improved evapotranspiration serves as the foundation for the utilization of bioremediation in wet environments (Ram et al. 2011). Figure 13 illustrates the fundamental idea of transpiration, absorption, and movement involved in bio-drainage. Waterlogging may be decreased by using herbaceous perennial legumes that are suited to flooding and waterlogging, like Messina (*Melilotus siculus*), lucerne (*Medicago sativa*), and Clovers (Genus: Trifolium), in cropping systems (Cocks 2001; Nichols 2018). Typically, compared to other annual crops, these deeply rooted pasture species can drain water and cause the soil to dry to greater depths (McCaskill and Kearney 2016). The suitability of different pasture species for seed production technologies, also to merge them thus providing the greatest merits have been thought of as information gaps that call for further research (Cocks 2001) because different pasture species' tolerance levels

Table 3 Merits and demerits of adaptive management practices of waterlogging condition

Crop and soil management techniques	Benefits	Drawbacks	Reference
Surface drainage	Both installations along with care are the least expensive	Less agricultural space due to open drainage; requires routine upkeep	Ritzema et al. (2008), Ayars and Evans (2015)
Raised bed system	The structure of the soil is modified	Crop acreage affects its efficacy, control of weeds in the furrows, equipment modification expense	Zhang (2005), Acuña et al. (2011), Bakker et al. (2005b, 2007)
Pipe drains	Dependable technique in extreme flooding	Expensive to install	Filipović et al. (2014), Teixeira et al. (2018)
Vertical drainage	Dependable technique for extreme waterlogging	Compared to horizontal pipe drainage systems, maintenance and operational expenses are higher	Christen and Skehan (2001), Kijne (2006), Prathapar et al. (2018)
Mole drains	Effective technique; less expensive than alternative subterranean drainage	Periodic care is required; dispersive soils will cause the integrity to be lost	Dhakad et al. (2018), Tuohy et al. (2018)
Bio-drainage	Successfully tried and tested in numerous places	Requires specialized plantation methods, thinning, pruning, and harvesting	Kapoor (2000), Lerch et al. (2017), Dash et al. (2005), Lin et al. (2011), Sarkar et al. (2018), Munoz-Carpena et al. (2018)

differ significantly from waterlogging. In order to deal with drainage congestion and environmental dangers, bio-drainage, or vertical drainage in soil water utilizing specialized forms of rapidly developing trees with a high evapotranspiration need (Kapoor 2000; Heuperman and Kapoor 2003; Sarkar et al. 2018). Trees in particular are often referred to as "biological pumps" and are crucial to the whole water cycle in a particular area. There is no need for us to:

- Stimulate soil water movement toward a pipe drain or tube well.
- Construct main and collector drains to remove water from the drainage area in bioremediation systems, which are advantageous compared to typical subsurface drainage systems.
- Run pumps to remove drained water, then transfer it to disposal facilities.
- Build disposal facilities (for example: through evaporation ponds).

Bioremediation's durable viability has been heavily debated. As an alternative to conventional field drainage techniques, it has been suggested that bioremediation

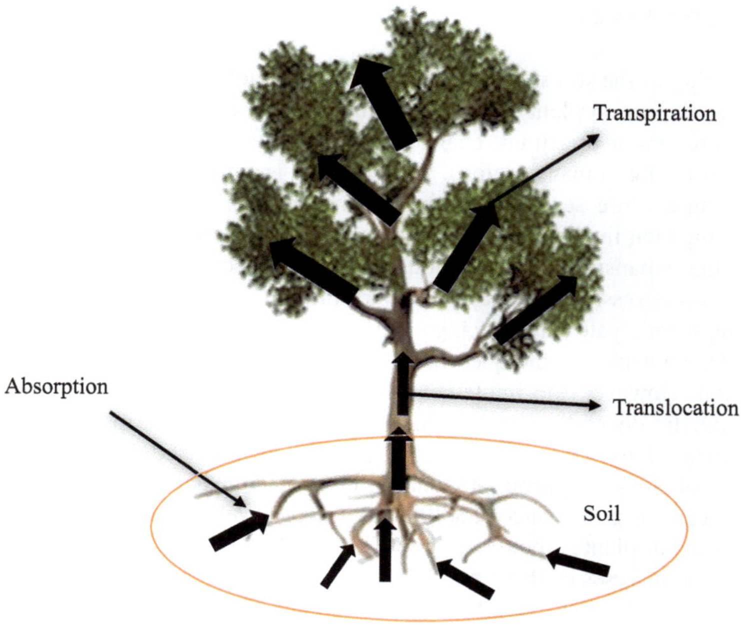

Fig. 13 Concept of biodrainage

may be employed in "parallel field drainage" designs for canal leakage interception and flooded landscape depressions (Smedema 2000). The bottom line is that this is a "pro-poor" technique that increases the revenue of struggling producers who might otherwise leave their properties fallow.

In Pakistan, waterlogged soils have been restored via bioremediation. Poplar (*Populus deltoides*), Eucalyptus, Tamarix, Gum arabic tree (*Acacia nilotica*), and mesquite (*Prosopis juliflora*), are among the trees that belong to this category. Non-woody plants, like shrubs, sedges, grasses, and herbs, can have deeply rooted systems that come into contact with groundwater like that of woody plants (Choudhry and Bhutta 2000). A recent study found that 2.5% of the 200 million irrigated agricultural trees in Punjab province are eucalyptus trees (Shah et al. 2011). The water table wouldn't be expected to be significantly affected by such a plantation until the plants occupy a sizable enough portion of the catchments so that their combined water demand equals the catchments' whole recharge. In Pakistan, the capacity of productive tree plantations to drain shallow groundwater is seen as a vital tool for controlling rising water levels.

12 Conclusion

Waterlogging of the soil significantly reduces crop yields around the world and has a negative impact on plant growth. The fluctuating precipitation patterns and temperature brought on by climate change are expected to increase crop losses owing to soil waterlogging in many regions. Waterlogging is a major problem for Pakistan's irrigated agriculture sector, depriving farmers of their productive resources and endangering their livelihoods. In general, plants can acquire some adaptive features, such as the expansion of adventitious roots or aerenchyma tissue to endure soil waterlogging stress. Commercial cultivars that are not resistant to waterlogging stress might anticipate experiencing yield losses. This article provides a summary of potential management techniques that land managers and farmers can use to increase production. However, the implementation of any management strategies will be region-specific depending on how simple it is to apply in the producers' current management plans. There are still large gaps in our knowledge of the advantages and disadvantages of appropriate management techniques for various types of soil or crop types, the governance of additional micro as well as macronutrients, and the genetic basis of plant responses to hypoxia and elemental toxicity in flooded soils. Cost-benefit analyses of these management strategies should be the main focus of future research in order to confirm their commercial feasibility and to create management plans that will encourage sustainable crop production from intermittent and variable duration waterlogged soils.

References

Abiko T, Kotula L, Shiono K, Malik AI, Colmer TD, Nakazono M (2012) Enhanced formation of aerenchyma and induction of a barrier to radial oxygen loss in adventitious roots of Zea nicaraguensis contribute to its waterlogging tolerance as compared with maize (*Zea mays* ssp. mays). Plant Cell Environ 35(9):1618–1630

Acuña TB, Dean G, Riffkin P (2011) Constraints to achieving high potential yield of wheat in a temperate, high-rainfall environment in South-Eastern Australia. Crop Past Sci 62(2):125–136

ADB [Asian Development Bank] (2002) Water resources strategy study. Draft report vol 1. Islamabad, Pakistan ADB

Aderonmu AT (2015) Assessing the impact of changing climate on agriculture in Missouri and the use of crop insurance as an adaptation strategy (1980–2010). University of Missouri-Kansas City

Andrade CA, de Souza KRD, de Oliveira Santos M, da Silva DM, Alves JD (2018) Hydrogen peroxide promotes the tolerance of soybeans to waterlogging. Scient Hort 232:40–45

Anikwe MAN, Ubochi JN (2007) Short-term changes in soil properties under tillage systems and their effect on sweet potato (Ipomea batatas L.) growth and yield in an Ultisol in South-Eastern Nigeria. Soil Res 45(5):351–358

Araki H, Hossain MA, Takahashi T (2012) Waterlogging and hypoxia have permanent effects on wheat root growth and respiration. J Agron Crop Sci 198(4):264–275

Armstrong W, Drew MC (2002) Root growth and metabolism under oxygen deficiency. In: Plant roots. CRC Press, pp 1139–1187

Aroca R, Porcel R, Ruiz-Lozano JM (2012) Regulation of root water uptake under abiotic stress conditions. J Exp Bot 63(1):43–57

Ashraf MA, Ahmad MSA, Ashraf M, Al-Qurainy F, Ashraf MY (2011) Alleviation of waterlogging stress in upland cotton (*Gossypium hirsutum* L.) by exogenous application of potassium in soil and as a foliar spray. Crop Past Sci 62(1):25–38
Ashraf MA, Akbar A, Askari SH, Iqbal M, Rasheed R, Hussain I (2018) Recent advances in abiotic stress tolerance of plants through chemical priming: an overview. Adv Seed Prim:51–79
Aslam K, Rashid S, Saleem R, Aslam RMS (2015) Use of geospatial technology for assessment of waterlogging & salinity conditions in the Nara Canal command area in Sindh, Pakistan. J Geogr Inf Syst 7(04):438
Ayars JE, Evans RG (2015) Subsurface drainage—What's next? Irrig Drain 64(3):378–392
Bailey-Serres J, Voesenek LACJ (2008) Flooding stress: acclimations and genetic diversity. Annu Rev Plant Biol 59:313
Bailey-Serres J, Voesenek LA (2010) Life in the balance: a signaling network controlling survival of flooding. Curr Opin Plant Biol 13(5):489–494
Bailey-Serres J, Lee SC, Brinton E (2012) Waterproofing crops: effective flooding survival strategies. Plant Physiol 160(4):1698–1709
Bakker DM, Hamilton GJ, Houlbrooke DJ, Spann C (2005a) The effect of raised beds on soil structure, waterlogging, and productivity on duplex soils in Western Australia. Soil Res 43(5):575–585
Bakker D, Houlbrooke D, Hamilton G, Spann C (2005b) A manual for raised bed farming in Western Australia
Bakker DM, Hamilton GJ, Houlbrooke DJ, Spann C, Van Burgel A (2007) Productivity of crops grown on raised beds on duplex soils prone to waterlogging in Western Australia. Aust J Exp Agric 47(11):1368–1376
Bange MP, Milroy SP, Thongbai P (2004) Growth and yield of cotton in response to waterlogging. Field Crop Res 88(2–3):129–142
Bassu S, Asseng S, Motzo R, Giunta F (2009) Optimising sowing date of durum wheat in a variable Mediterranean environment. Field Crop Res 111(1–2):109–118
Batey T (2009) Soil compaction and soil management–a review. Soil Use Manag 25(4):335–345
Bennett D (2022) Mole drainage in Western Australia, Department of Primary Industries and Regional development, Available at: https://www.agric.wa.gov.au/waterlogging/mole-drainage-western-australia
Blanco-Canqui H, Shaver TM, Lindquist JL, Shapiro CA, Elmore RW, Francis CA, Hergert GW (2015) Cover crops and ecosystem services: insights from studies in temperate soils. Agron J 107(6):2449–2474
Blessitt JB (2007) Productivity of raised seedbeds for soybean [Glycine max.(L.) Merr.] production on clayey soils of the Mississippi Delta. Mississippi State University
Blokhina OB, Chirkova TV, Fagerstedt KV (2001) Anoxic stress leads to hydrogen peroxide formation in plant cells. J Exp Bot 52(359):1179–1190
Bramley H, Turner NC, Turner DW, Tyerman SD (2010) The contrasting influence of short-term hypoxia on the hydraulic properties of cells and roots of wheat and lupin. Funct Plant Biol 37(3):183–193
Broekaert WF, Delauré SL, De Bolle MF, Cammue BP (2006) The role of ethylene in host-pathogen interactions. Annu Rev Phytopathol 44:393–416
Choudhry MR, Bhutta MN (2000) Problems impeding the sustainability of drainage systems in Pakistan. In: Proceedings and recommendations of the national seminar on drainage in Pakistan, pp 16–18
Christen E, Skehan D (2001) Design and management of subsurface horizontal drainage to reduce salt loads. J Irrig Drain Eng 127(3):148–155
Cocks PS (2001) Ecology of herbaceous perennial legumes: a review of characteristics that may provide management options for the control of salinity and waterlogging in dryland cropping systems. Aust J Agric Res 52(2):137–151
Colmer TD (2003) Long-distance transport of gases in plants: a perspective on internal aeration and radial oxygen loss from roots. Plant Cell Environ 26(1):17–36
Colmer TD, Flowers TJ (2008) Flooding tolerance in halophytes. New Phytol 179(4):964–974

Colmer TD, Greenway H (2011) Ion transport in seminal and adventitious roots of cereals during O2 deficiency. J Exp Bot 62(1):39–57
Cong Y, Li YJ, Zhou CJ, Zou CS, Zhang XK, Liao X, Zhang CL (2009) Effect of application of nitrogen, phosphorus and potassium fertilizers on yield in rapeseed (Brassica napus L.) under the waterlogging stress. Plant Nutrit Fertiliz Sci 15(5):1122–1129
Cubasch U, Meehl GA, Boer GJ, Stouffer RJ, Dix M, Noda A, Yap KS (2001) Projections of future climate change. In: Climate change 2001: the scientific basis. Contribution of WG1 to the third assessment report of the IPCC (TAR). Cambridge University Press, pp 525–582
Dash JP, Sarangi A, Singh AK, Dahiya S (2005) Bio-drainage: an alternate drainage technique to control waterlogging and salinity. J Soil Water Conserv India 4(3&4):149–155
Dat JF, Capelli N, Folzer H, Bourgeade P, Badot PM (2004) Sensing and signalling during plant flooding. Plant Physiol Biochem 42(4):273–282
Dhakad SS, Ambawatia GR, Verma G, Patel S, Rao KR, Verma S (2018) Performance of Mole drain system for soybean (glycine max)-wheat (Triticum aestivum) cropping system of Madhya Pradesh
Drury CF, Tan CS, Reynolds WD, Welacky TW, Oloya TO, Gaynor JD (2009) Managing tile drainage, subirrigation, and nitrogen fertilization to enhance crop yields and reduce nitrate loss. J Environ Qual 38(3):1193–1204
Ejiri M, Fukao T, Miyashita T, Shiono K (2021) A barrier to radial oxygen loss helps the root system cope with waterlogging-induced hypoxia. Breed Sci 71(1):40–50
El-Esawi MA (2016a) Nonzygotic embryogenesis for plant development. In: Plant tissue culture: propagation, conservation and crop improvement, pp 583–598
El-Esawi MA (2016b) Micropropagation technology and its applications for crop improvement. In: Plant tissue culture: propagation, conservation and crop improvement. Springer, Singapore, pp 523–545
Else MA, Janowiak F, Atkinson CJ, Jackson MB (2009) Root signals and stomatal closure in relation to photosynthesis, chlorophyll a fluorescence and adventitious rooting of flooded tomato plants. Ann Bot 103(2):313–323
Engineer Moid (2021) Waterlogging its 5 types and causes. Civilclick.com. Available at: https://www.civilclick.com/waterlogging/
Evans DE (2004) Aerenchyma formation. New Phytolog 161(1):35–49
Fabrizzi KP, Garcıa FO, Costa JL, Picone LI (2005) Soil water dynamics, physical properties and corn and wheat responses to minimum and no-tillage systems in the southern Pampas of Argentina. Soil Tillage Res 81(1):57–69
Ferreira JL, Coelho CHM, Magalhães PC, Santána GC, Borém A (2008) Evaluation of mineral content in maize under flooding. Embrapa Milho e Sorgo-Artigo em periódico indexado (ALICE)
Ferrer JLR, Magalhaes PC, Alves JD, Vasconcellos CA, Delu Filho N. Fries DD, Purcino AAC. (2005) Calcium partially relieves the deleterius effects of hypoxia on a maize cultivar selected for waterlogging tolerance
Filipović V, Mallmann FJK, Coquet Y, Šimůnek J (2014) Numerical simulation of water flow in tile and mole drainage systems. Agric Water Manag 146:105–114
Food and Agriculture Organization [FAO] (2002) Food and Agriculture Organization of the United Nations. Available at: http://www.fao.org/3/abc600e.pdf
Foyer CH, Shigeoka S (2011) Understanding oxidative stress and antioxidant functions to enhance photosynthesis. Plant Physiol 155(1):93–100
Frankenberger J, Kladivko E, Sands G, Jaynes DB, Fausey N, Helmers MJ, Brown L (2004) Drainage water management for the midwest. Agricultural and Biosystems Engineering Extension and Outreach Publications: West Lafayette, IN
Fukao T, Bailey-Serres J (2004) Plant responses to hypoxia–is survival a balancing act? Trends Plant Sci 9(9):449–456
Galant AL, Kaufman RC, Wilson JD (2015) Glucose: detection and analysis. Food Chem 188:149–160

Gechev TS, Gadjev I, Van Breusegem F, Inzé D, Dukiandjiev S, Toneva V, Minkov I (2002) Hydrogen peroxide protects tobacco from oxidative stress by inducing a set of antioxidant enzymes. Cell Mol Life Sci 59(4):708–714

Gibson G (2014) Utilising innovative management techniques to reduce waterlogging. Nuffield Australia Farming Scholars, Moama, NSW

Gill MB, Zeng F, Shabala L, Böhm J, Zhang G, Zhou M, Shabala S (2018) The ability to regulate voltage-gated K+-permeable channels in the mature root epidermis is essential for waterlogging tolerance in barley. J Exp Bot 69(3):667–680

Govaerts B, Sayre KD, Lichter K, Dendooven L, Deckers J (2007) Influence of permanent raised bed planting and residue management on physical and chemical soil quality in rain fed maize/wheat systems. Plant Soil 291(1):39–54

Gramlich A, Stoll S, Stamm C, Walter T, Prasuhn V (2018) Effects of artificial land drainage on hydrology, nutrient and pesticide fluxes from agricultural fields–a review. Agric Ecosyst Environ 266:84–99

Habibzadeh F, Sorooshzadeh A, Pirdashti H, Sanavy SAMM (2012) Effect of nitrogen compounds and tricyclazole on some biochemical and morphological characteristics of waterlogged-canola. Int Res J Appl Basic Sci 3(1):77–84

Habibzadeh F, Sorooshzadeh A, Pirdashti H, Modarres-Sanavy SAM (2013) Alleviation of waterlogging damage by foliar application of nitrogen compounds and tricyclazole in canola. Aust J Crop Sci 7(3):401–406

Hassan I, Hussain Z, Akbar G (2005) Effect of permanent raised beds on water productivity for irrigated maize–wheat cropping system. Evaluation and performance of permanent raised bed cropping systems in Asia, Australia and Mexicop, p 121

Hedden P, Sponsel V (2015) A century of gibberellin research. J Plant Growth Regul 34(4):740–760

Herzog M, Striker GG, Colmer TD, Pedersen O (2016) Mechanisms of waterlogging tolerance in wheat–a review of root and shoot physiology. Plant Cell Environ 39(5):1068–1086

Heuperman AF, Kapoor AS (2003) Biodrainage status in India and other countries. Indian National Committee on Irrigation and Drainage, New Delhi, pp 1–47

Hossain MA (2010) Global warming induced sea level rise on soil, land and crop production loss in Bangladesh. In: 19th world congress of soil science, soil solutions for a changing world, Brisbane

Htet Y, Lu Z, Trauger SA, Tennyson AG (2019) Hydrogen peroxide as a hydride donor and reductant under biologically relevant conditions. Chem Sci 10(7):2025–2033

Huang X, Shabala S, Shabala L, Rengel Z, Wu X, Zhang G, Zhou M (2015) Linking waterlogging tolerance with Mn2+ toxicity: a case study for barley. Plant Biol 17(1):26–33

Huang X, Fan Y, Shabala L, Rengel Z, Shabala S, Zhou MX (2018) A major QTL controlling the tolerance to manganese toxicity in barley (Hordeum vulgare L.). Mol Breed 38(2):1–9

Indraratna B (2017) Recent advances in vertical drains and vacuum preloading for soft ground stabilisation. In: Proceedings of 19th international conference on soil mechanics and geotechnical engineering, Seou. International Society for Soil Mechanics and Geotechnical Engineering, London, pp 145–170

Indraratna B, Rujikiatkamjorn C, Sathananthan I (2005) Analytical and numerical solutions for a single vertical drain including the effects of vacuum preloading. Can Geotech J 42(4):994–1014

Iqbal N, Khan NA, Ferrante A, Trivellini A, Francini A, Khan MIR (2017) Ethylene role in plant growth, development and senescence: interaction with other phytohormones. Front Plant Sci 8:475

Irfan M, Hayat S, Hayat Q, Afroz S, Ahmad A (2010) Physiological and biochemical changes in plants under waterlogging. Protoplasma 241(1):3–17

Ishibashi Y, Yamaguchi H, Yuasa T, Iwaya-Inoue M, Arima S, Zheng SH (2011) Hydrogen peroxide spraying alleviates drought stress in soybean plants. J Plant Physiol 168(13):1562–1567

Jackson MB, Colmer T (2005) Response and adaptation by plants to flooding stress. Ann Bot 96(4):501–505

Jackson MB, Ricard B (2003) Physiology, biochemistry and molecular biology of plant root systems subjected to flooding of the soil. Root Ecol:193–213

Jacobs A, Rauber R, Ludwig B (2009) Impact of reduced tillage on carbon and nitrogen storage of two Haplic Luvisols after 40 years. Soil Tillage Res 102(1):158–164

Jia W, Ma M, Chen J, Wu S (2021) Plant morphological, physiological and anatomical adaption to flooding stress and the underlying molecular mechanisms. Int J Mol Sci 22(3):1088

Jiang P, Anderson SH, Kitchen NR, Sadler EJ, Sudduth KA (2007) Landscape and conservation management effects on hydraulic properties of a claypan-soil toposequence. Soil Sci Soc Am J 71(3):803–811

Kapoor AS (2000) Bio-drainage feasibility and principles of planning and design. In: Role of drainage and challenges in 21st century. Vol. IV. Proceedings of the eighth ICID international drainage workshop, New Delhi, India, 31 January–4 February 2000. International Commission on Irrigation and Drainage, pp 17–32

Kargas G, Kerkides P, Poulovassilis A (2012) Infiltration of rain water in semi-arid areas under three land surface treatments. Soil Tillage Res 120:15–24

Kaur G, Zurweller BA, Nelson KA, Motavalli PP, Dudenhoeffer CJ (2017) Soil waterlogging and nitrogen fertilizer management effects on corn and soybean yields. Agron J 109(1):97–106

Kaur G, Nelson KA, Motavalli PP (2018) Early-season soil waterlogging and N fertilizer sources impacts on corn N uptake and apparent N recovery efficiency. Agronomy 8(7):102

Kaur G, Singh G, Motavalli PP, Nelson KA, Orlowski JM, Golden BR (2020) Impacts and management strategies for crop production in waterlogged or flooded soils: a review. Agron J 112(3):1475–1501

Kazmi SI, Ertsen MW, Asi MR (2012) The impact of conjunctive use of canal and tube well water in Lagar irrigated area, Pakistan. Phys Chem Earth, Parts A/B/C 47:86–98

Kijne JW (2006) Abiotic stress and water scarcity: identifying and resolving conflicts from plant level to global level. Field Crop Res 97(1):3–18

Kim Y, Seo CW, Khan AL, Mun BG, Shahzad R, Ko JW, Lee IJ (2018) Ethylene mitigates waterlogging stress by regulating glutathione biosynthesis-related transcripts in soybeans. bioRxiv, 252312

Kirkpatrick MT, Rothrock CS, Rupe JC, Gbur EE (2006) The effect of Pythium ultimum and soil flooding on two soybean cultivars. Plant Dis 90(5):597–602

Kisaakye E, Botwright Acuna T, Johnson P, Shabala S (2015) Effect of water availability and nitrogen source on wheat growth and nitrogen-use efficiency. In: 17th Australian Society of Agronomy conference, pp 1–4

Kisaakye E, Acuña TB, Johnson P, Shabala S (2017) Improving wheat growth and nitrogen-use efficiency under waterlogged conditions. In: 18th Australian agronomy conference 2017, pp 1–4

Knapp AK, Hoover DL, Wilcox KR, Avolio ML, Koerner SE, La Pierre KJ, Smith MD (2015) Characterizing differences in precipitation regimes of extreme wet and dry years: implications for climate change experiments. Glob Chang Biol 21(7):2624–2633

Knipfer T, Fricke W (2011) Water uptake by seminal and adventitious roots in relation to whole-plant water flow in barley (Hordeum vulgare L.). J Exp Bot 62(2):717–733

Kolekar O, Patil S, Rathod S (2014) Effects of different mole drain spacings on the yield of summer groundnut. Int J Res Eng Technol 3:2321–7308

Kucharik CJ (2006) A multidecadal trend of earlier corn planting in the Central USA. Agron J 98(6):1544–1550

Kulichikhin K, Yamauchi T, Watanabe K, Nakazono M (2014) Biochemical and molecular characterization of rice (Oryza sativa L.) roots forming a barrier to radial oxygen loss. Plant Cell Environ 37(10):2406–2420

Kumar V, Ladha JK (2011) Direct seeding of rice: recent developments and future research needs. Adv Agron 111:297–413

Kunkel KE (2003) North American trends in extreme precipitation. Nat Hazards 29(2):291–305

Lal R, Reicosky DC, Hanson JD (2007) Evolution of the plow over 10,000 years and the rationale for no-till farming. Soil Tillage Res 93(1):1–12

Lal SK, Kumar S, Sheri V, Mehta S, Varakumar P, Ram B, Reddy MK (2018) Seed priming: an emerging technology to impart abiotic stress tolerance in crop plants. In: Advances in seed priming. Springer, Singapore, pp 41–50

Lerch RN, Lin CH, Goyne KW, Kremer RJ, Anderson SH (2017) Vegetative buffer strips for reducing herbicide transport in runoff: effects of buffer width, vegetation, and season. J Am Water Resour Assoc 53(3):667–683

Li MF, Zhu JQ, Jiang ZH (2013) Plant growth regulators and nutrition applied to cotton after waterlogging. In: 2013 third international conference on intelligent system design and engineering applications. IEEE, pp 1045–1048

Lin C, Sauter M (2019) Polar auxin transport determines adventitious root emergence and growth in rice. Front Plant Sci 10:444

Lin KH, Tsou CC, Hwang SY, Chen LFO, Lo HF (2008) Paclobutrazol leads to enhanced antioxidative protection of sweetpotato under flooding stress. Bot Stud 49(1):9–18

Lin CH, Lerch RN, Goyne KW, Garrett HE (2011) Reducing herbicides and veterinary antibiotics losses from agroecosystems using vegetative buffers. J Environ Qual 40(3):791–799

Liu Z, Liu Z, Xiao J, Nan J, Gong W (2013) Waterlogging at seedling and jointing stages inhibits growth and development, reduces yield in summer maize. Trans Chin Soc Agric Eng 29(5):44–52

Liu K, Harrison MT, Shabala S, Meinke H, Ahmed I, Zhang Y, Zhou M (2020a) The state of the art in modeling waterlogging impacts on plants: what do we know and what do we need to know. Earth's Fut 8(12):e2020EF001801

Liu J, Hasanuzzaman M, Sun H, Zhang J, Peng T, Sun H, Zhao Q (2020b) Comparative morphological and transcriptomic responses of lowland and upland rice to root-zone hypoxia. Environ Exp Bot 169:103916

Luan H, Guo B, Pan Y, Lv C, Shen H, Xu R (2018a) Morpho-anatomical and physiological responses to waterlogging stress in different barley (Hordeum vulgare L.) genotypes. Plant Growth Regul 85(3):399–409

Luan H, Shen H, Pan Y, Guo B, Lv C, Xu R (2018b) Elucidating the hypoxic stress response in barley (Hordeum vulgare L.) during waterlogging: a proteomics approach. Sci Rep 8(1):1–13

Malano HM, Van Hofwegen PJ (2018) Management of irrigation and drainage systems–a service approach. CRC Press

Mancuso S, Shabala S (2010) Waterlogging signalling and tolerance in plants. Springer, Heidelberg. ISBN 9783642103049

Manik SM, Pengilley G, Dean G, Field B, Shabala S, Zhou M (2019) Soil and crop management practices to minimize the impact of waterlogging on crop productivity. Front Plant Sci 10:140

Mano Y, Muraki M, Fujimori M, Takamizo T, Kindiger B (2005) Identification of QTL controlling adventitious root formation during flooding conditions in teosinte (Zea mays ssp. huehuetenangensis) seedlings. Euphytica 142(1):33–42

Masunaga T, Fong JDM (2018) Strategies for increasing micronutrient availability in soil for plant uptake. In: Plant micronutrient use efficiency. Academic Press, pp 195–208

Maurel C, Boursiac Y, Luu DT, Santoni V, Shahzad Z, Verdoucq L (2015) Aquaporins in plants. Physiol Rev 95(4):1321–1358

McCaskill MR, Kearney GA (2016) Control of water leakage from below the root zone by summer-active pastures is associated with persistence, density and deep rootedness. Crop Past Sci 67(6):679–693

McDonald G (2022) Raised beds – design, layout, construction and renovation. Department of Primary Industries and Regional development, Available at: https://www.agric.wa.gov.au/waterlogging/raised-beds-_-design-layout-construction-andrenovation#skip-link

McLellan EL, Schilling KE, Wolter CF, Tomer MD, Porter SA, Magner JA, Prokopy LS (2018) Right practice, right place: a conservation planning toolbox for meeting water quality goals in the Corn Belt. J Soil Water Conserv 73(2):29A–34A

McVay KA, Budde JA, Fabrizzi K, Mikha MM, Rice CW, Schlegel AJ, Thompson C (2006) Management effects on soil physical properties in long-term tillage studies in Kansas. Soil Sci Soc Am J 70(2):434–438

Merriman KR, Daggupati P, Srinivasan R, Hayhurst B (2019) Assessment of site-specific agricultural best management practices in the upper East River watershed, Wisconsin, using a field-scale SWAT model. J Great Lakes Res 45(3):619–641

Mhamdi A, Van Breusegem F (2018) Reactive oxygen species in plant development. Development 145(15):dev164376

Milroy SP, Bange MP, Thongbai P (2009) Cotton leaf nutrient concentrations in response to waterlogging under field conditions. Field Crop Res 113(3):246–255

Mittler R (2002) Oxidative stress, antioxidants and stress tolerance. Trends Plant Sci 7(9):405–410

Monaghan RM, Smith LC (2004) Minimising surface water pollution resulting from farm-dairy effluent application to mole-pipe drained soils. II. The contribution of preferential flow of effluent to whole-farm pollutant losses in subsurface drainage from a West Otago dairy farm. N Z J Agric Res 47(4):417–428

Monaghan RM, Paton RJ, Drewry JJ (2002) Nitrogen and phosphorus losses in mole and tile drainage from a cattle-grazed pasture in eastern Southland. N Z J Agric Res 45(3):197–205

Monteiro A, Lopes CM (2007) Influence of cover crop on water use and performance of vineyard in Mediterranean Portugal. Agric Ecosyst Environ 121(4):336–342

Motavalli PP, Anderson SH, Pengthamkeerati P (2003) Surface compaction and poultry litter effects on corn growth, nitrogen availability, and physical properties of a claypan soil. Field Crop Res 84:303–318

Munoz-Carpena R, Fox GA, Ritter A, Perez-Ovilla O, Rodea-Palomares I (2018) Effect of vegetative filter strip pesticide residue degradation assumptions for environmental exposure assessments. Sci Total Environ 619:977–987

Mustafa HSB, Mahmood T, Ullah A, Sharif A, Bhatti AN, Nadeem M, Ali R (2017) Role of seed priming to enhance growth and development of crop plants against biotic and abiotic stresses. Sect Plant Sci 2(2):1–11

Mustroph A, Albrecht G (2003) Tolerance of crop plants to oxygen deficiency stress: fermentative activity and photosynthetic capacity of entire seedlings under hypoxia and anoxia. Physiol Plant 117(4):508–520

Mutava RN, Prince SJK, Syed NH, Song L, Valliyodan B, Chen W, Nguyen HT (2015) Understanding abiotic stress tolerance mechanisms in soybean: a comparative evaluation of soybean response to drought and flooding stress. Plant Physiol Biochem 86:109–120

Najeeb U, Bange MP, Tan DK, Atwell BJ (2015) Consequences of waterlogging in cotton and opportunities for mitigation of yield losses. AoB Plants 7:plv080

Nelson KA, Paniagua SM, Motavalli PP (2009) Effect of polymer coated urea, irrigation, and drainage on nitrogen utilization and yield of corn in a claypan soil. Agron J 101(3):681–687

Nelson KA, Smoot RL, Meinhardt CG (2011) Soybean response to drainage and subirrigation on a claypan soil in Northeast Missouri. Agron J 103(4):1216–1222

Nelson KA, Meinhardt CG, Smoot RL (2012) Soybean cultivar response to subsurface drainage and subirrigation in Northeast Missouri. Crop Manag 11(1):1–9

Nguyen HC, Lin KH, Ho SL, Chiang CM, Yang CM (2018a) Enhancing the abiotic stress tolerance of plants: from chemical treatment to biotechnological approaches. Physiol Plant 164(4):452–466

Nguyen TN, Tuan PA, Mukherjee S, Son S, Ayele BT (2018b) Hormonal regulation in adventitious roots and during their emergence under waterlogged conditions in wheat. J Exp Bot 69(16):4065–4082

Nichols P (2018) Yanco subterranean clover. Department of Primary Industries and Regional Development (DPIRD), Orange

Nielsen RL (2015) Effects of flooding or ponding on corn prior to tasseling. In: Corny news network. Purdue University, West Lafayette, IN

Nishiuchi S, Yamauchi T, Takahashi H, Kotula L, Nakazono M (2012) Mechanisms for coping with submergence and waterlogging in rice. Rice 5(1):1–14

Noreen S, Fatima Z, Ahmad S, Athar HUR, Ashraf M (2018) Foliar application of micronutrients in mitigating abiotic stress in crop plants. In: Plant nutrients and abiotic stress tolerance. Springer, Singapore, pp 95–117

Normile D (2008) Reinventing rice to feed the world

OCHA, Pakistan (2022) Floods response plan: 01 Sep 2022–28 Feb 2023 (Issued 30 Aug 2022), Relief web, Available at: https://reliefweb.int/report/pakistan/pakistan-2022-floods-response-plan-01-sep-2022-28-feb-2023-issued-30-aug-2022
Pagliai M, Vignozzi N, Pellegrini S (2004) Soil structure and the effect of management practices. Soil Tillage Res 79(2):131–143
Palla A, Colli M, Candela A, Aronica GT, Lanza LG (2018) Pluvial flooding in urban areas: the role of surface drainage efficiency. J Flood Risk Manag 11:S663–S676
Pandey DM, Goswami CL, Kumar B, Jain S (2002) Effect of growth regulators on photosynthetic metabolites in cotton under water stress. Biol Plant 45(3):445–448
Pang JY, Newman IAN, Mendham N, Zhou M, Shabala S (2006) Microelectrode ion and O2 fluxes measurements reveal differential sensitivity of barley root tissues to hypoxia. Plant Cell Environ 29(6):1107–1121
Pang J, Ross J, Zhou M, Mendham N, Shabala S (2007) Amelioration of detrimental effects of waterlogging by foliar nutrient sprays in barley. Funct Plant Biol 34(3):221–227
Parent C, Capelli N, Berger A, Crèvecoeur M, Dat JF (2008) An overview of plant responses to soil waterlogging. Plant Stress 2(1):20–27
Pereira EI, Nogueira ARA, da Cruz CC, Guimarães GG, Foschini MM, Bernardi AC, Ribeiro C (2017) Controlled urea release employing nanocomposites increases the efficiency of nitrogen use by forage. ACS Sustain Chem Eng 5(11):9993–10001
Ploschuk RA, Miralles DJ, Colmer TD, Ploschuk EL, Striker GG (2018) Waterlogging of winter crops at early and late stages: impacts on leaf physiology, growth and yield. Front Plant Sci 9:1863
Poulisw (2011) Anaerobic respiration (fermentation), Biology form 6, Available at: http://biomhs.blogspot.com/2011/04/anaerobic-respiration-fermentation.html
Prathapar SA, Rajmohan N, Sharma BR, Aggarwal PK (2018) Vertical drains to minimize duration of seasonal waterlogging in eastern Ganges Basin flood plains: a field experiment. Nat Hazards 92(1):1–17
Qadir M, Oster J (2002) Vegetative bioremediation of calcareous sodic soils: history, mechanisms, and evaluation. Irrig Sci 21(3):91–101
Qiu F, Zheng Y, Zhang Z, Xu S (2007) Mapping of QTL associated with waterlogging tolerance during the seedling stage in maize. Ann Bot 99(6):1067–1081
Qureshi AS (2011) Water management in the Indus Basin in Pakistan: challenges and opportunities. Mountain Res Develop 31(3):252–260
Qureshi AS, Akhtar M, Sarwar A (2003) Effect of electricity pricing policies on groundwater management in Pakistan
Qureshi AS, McCornick PG, Qadir M, Aslam Z (2008) Managing salinity and waterlogging in the Indus Basin of Pakistan. Agric Water Manag 95(1):1–10
Rademacher W (2015) Plant growth regulators: backgrounds and uses in plant production. J Plant Growth Regul 34(4):845–872
Rajaeian SO, Ehsanpour AA (2015) Physiological responses of tobacco plants (Nicotiana rustica) pretreated with ethanolamine to salt stress. Russ J Plant Physiol 62(2):246–252
Rajhi I, Yamauchi T, Takahashi H, Nishiuchi S, Shiono K, Watanabe R, Nakazono M (2011) Identification of genes expressed in maize root cortical cells during lysigenous aerenchyma formation using laser microdissection and microarray analyses. New Phytol 190(2):351–368
Ram J, Dagar JC, Lal K, Singh G, Toky OP, Tanwar VS, Chauhan MK (2011) Biodrainage to combat waterlogging, increase farm productivity and sequester carbon in canal command areas of Northwest India. Curr Sci:1673–1680
Rao R, Bryan HH, Reed ST (2002) Assessment of foliar sprays to alleviate flooding injury in corn (Zea mays L.). In: Proceedings of the Florida State Horticultural Society, vol 115, pp 208–211
Reichardt K, Timm LC (2012) Soil, plant and atmosphere. Springer, Cham. ISBN 9783030193218
Reinsch M, Pearce D (2005) Pakistan-country water resources assistance strategy: water economy running dry
Ren B, Zhang J, Dong S, Liu P, Zhao B (2016a) Effects of waterlogging on leaf mesophyll cell ultrastructure and photosynthetic characteristics of summer maize. PLoS One 11(9):e0161424

Ren B, Zhang J, Dong S, Liu P, Zhao B (2016b) Root and shoot responses of summer maize to waterlogging at different stages. Agron J 108(3):1060–1069
Ren B, Zhu Y, Zhang J, Dong S, Liu P, Zhao B (2016c) Effects of spraying exogenous hormone 6-benzyladenine (6-BA) after waterlogging on grain yield and growth of summer maize. Field Crop Res 188:96–104
Ren B, Dong S, Zhao B, Liu P, Zhang J (2017) Responses of nitrogen metabolism, uptake and translocation of maize to waterlogging at different growth stages. Front Plant Sci 8:1216
Ren B, Zhang J, Dong S, Liu P, Zhao B (2018) Exogenous 6-benzyladenine improves antioxidative system and carbon metabolism of summer maize waterlogged in the field. J Agron Crop Sci 204(2):175–184
Ritzema HP, Satyanarayana TV, Raman S, Boonstra J (2008) Subsurface drainage to combat waterlogging and salinity in irrigated lands in India: lessons learned in farmers' fields. Agric Water Manag 95(3):179–189
Rochester IJ, Peoples MB, Hulugalle NR, Gault R, Constable GA (2001) Using legumes to enhance nitrogen fertility and improve soil condition in cotton cropping systems. Field Crop Res 70(1):27–41
Rosenzweig C, Tubiello FN, Goldberg R, Mills E, Bloomfield J (2002) Increased crop damage in the US from excess precipitation under climate change. Glob Environ Chang 12(3):197–202
Rubinigg M, Stulen I, Elzenga JTM, Colmer TD (2002) Spatial patterns of radial oxygen loss and nitrate net flux along adventitious roots of rice raised in aerated or stagnant solution. Funct Plant Biol 29(12):1475–1481
Saadat S, Bowling L, Frankenberger J, Kladivko E (2018) Nitrate and phosphorus transport through subsurface drains under free and controlled drainage. Water Res 142:196–207
Sairam RK, Kumutha D, Ezhilmathi K, Deshmukh PS, Srivastava GC (2008) Physiology and biochemistry of waterlogging tolerance in plants. Biol Plant 52(3):401–412
Samad A, Meisner CA, Saifuzzaman M, van Ginkel M (2001) Waterlogging tolerance. In: Reynolds MP, Ortiz-Monasterio JI, McNab A (eds) Application of physiology in wheat breeding, pp 136–144. ISBN:970-648-077-3
Sarkar A, Banik M, Ray R, Patra SK (2018) Soil moisture and groundwater dynamics under bio drainage vegetation in a waterlogged land
Sasidharan R, Voesenek LA (2015) Ethylene-mediated acclimations to flooding stress. Plant Physiol 169(1):3–12
Sauter M (2013) Root responses to flooding. Curr Opin Plant Biol 16(3):282–286
Savvides A, Ali S, Tester M, Fotopoulos V (2016) Chemical priming of plants against multiple abiotic stresses: mission possible? Trends Plant Sci 21(4):329–340
Setter TL, Waters I (2003) Review of prospects for germplasm improvement for waterlogging tolerance in wheat, barley and oats. Plant Soil 253(1):1–34
Setter TL, Khabaz-Saberi H, Waters I, Singh KN, Kulshreshtha N, Sharma SK (2006). Review of waterlogging tolerance in wheat in India: involvement of element/microelement toxicities, relevance to yield plateau and opportunities for crop management. In: International symposium on balanced fertilization. Ludhiana, India, pp 22–25
Shabala S (2011) Physiological and cellular aspects of phytotoxicity tolerance in plants: the role of membrane transporters and implications for crop breeding for waterlogging tolerance. New Phytol 190(2):289–298
Shabala S, Pottosin I (2014) Regulation of potassium transport in plants under hostile conditions: implications for abiotic and biotic stress tolerance. Physiol Plant 151(3):257–279
Shah AH, Gill KH, Syed NI (2011) Sustainable salinity management for combating desertification in Pakistan. Int J Water Res Arid Environ 1(5):312–317
Sharma PC, Kaledhonkar MJ, Thimmappa K, Chaudhari SK (2016) Reclamation of waterlogged saline soils through subsurface drainage technology
Shaviv A (2001) Advances in controlled-release fertilizers
Shaw RE, Meyer WS, McNeill A, Tyerman SD (2013) Waterlogging in Australian agricultural landscapes: a review of plant responses and crop models. Crop Past Sci 64(6):549–562

Shiono K, Ogawa S, Yamazaki S, Isoda H, Fujimura T, Nakazono M, Colmer TD (2011) Contrasting dynamics of radial O2-loss barrier induction and aerenchyma formation in rice roots of two lengths. Ann Bot 107(1):89–99

Shiono K, Ejiri M, Shimizu K, Yamada S (2019) Improved waterlogging tolerance of barley (Hordeum vulgare) by pretreatment with ethephon. Plant Prod Sci 22(2):285–295

Silburn DM, Freebairn DM, Rattray DJ (2007) Tillage and the environment in sub-tropical Australia—tradeoffs and challenges. Soil Tillage Res 97(2):306–317

Singh A (2012) Development and application of a watertable model for the assessment of waterlogging in irrigated semi-arid regions. Water Resour Manag 26(15):4435–4448

Singh A (2016) Hydrological problems of water resources in irrigated agriculture: a management perspective. J Hydrol 541:1430–1440

Singh A (2018a) Managing the salinization and drainage problems of irrigated areas through remote sensing and GIS techniques. Ecol Indic 89:584–589

Singh A (2018b) Salinization of agricultural lands due to poor drainage: a viewpoint. Ecol Indic 95:127–130

Singh Y, Singh VP, Singh G, Yadav DS, Sinha RKP, Johnson DE, Mortimer AM (2011) The implications of land preparation, crop establishment method and weed management on rice yield variation in the rice–wheat system in the Indo-Gangetic plains. Field Crop Res 121(1):64–74

Singh G, Williard KW, Schoonover JE (2016) Spatial relation of apparent soil electrical conductivity with crop yields and soil properties at different topographic positions in a small agricultural watershed. Agronomy 6(4):57

Skaggs RW (2008) DRAINMOD: a simulation model for shallow water table soils

Smedema L (2000) Irrigation-induced river salinization: five major irrigated basins in the arid

Soukup A, Armstrong W, Schreiber L, Franke R, Votrubová O (2007) Apoplastic barriers to radial oxygen loss and solute penetration: a chemical and functional comparison of the exodermis of two wetland species, Phragmites australis and Glyceria maxima. New Phytol 173(2):264–278

Steffens B (2014) The role of ethylene and ROS in salinity, heavy metal, and flooding responses in rice. Front Plant Sci 5:685

Steffens B, Rasmussen A (2016) The physiology of adventitious roots. Plant Physiol 170(2):603–617

Steffens B, Sauter M (2009) Epidermal cell death in rice is confined to cells with a distinct molecular identity and is mediated by ethylene and H2O2 through an autoamplified signal pathway. Plant Cell 21(1):184–196

Steffens D, Hutsch BW, Eschholz T, Losak T, Schubert S (2005) Water logging may inhibit plant growth primarily by nutrient deficiency rather than nutrient toxicity. Plant Soil Environ 51(12):545

Steffens B, Geske T, Sauter M (2011) Aerenchyma formation in the rice stem and its promotion by H2O2. New Phytol 190(2):369–378

Subbaiah CC, Sachs MM (2003) Molecular and cellular adaptations of maize to flooding stress. Ann Bot 91(2):119–127

Sumimoto H (2008) Structure, regulation and evolution of Nox-family NADPH oxidases that produce reactive oxygen species. FEBS J 275(13):3249–3277

Sundgren TK, Uhlen AK, Lillemo M, Briese C, Wojciechowski T (2018) Rapid seedling establishment and a narrow root stele promotes waterlogging tolerance in spring wheat. J Plant Physiol 227:45–55

Tan X, Xu H, Khan S, Equiza MA, Lee SH, Vaziriyeganeh M, Zwiazek JJ (2018) Plant water transport and aquaporins in oxygen-deprived environments. J Plant Physiol 227:20–30

Teixeira DL, de Matos AT, de Matos MP, Miranda ST, Vieira DP (2018) Evaluation of the effects of drainage and different rest periods as techniques for unclogging the porous medium in horizontal subsurface flow constructed wetlands. Ecol Eng 120:104–108

Tong C, Hill CB, Zhou G, Zhang XQ, Jia Y, Li C (2021) Opportunities for improving waterlogging tolerance in cereal crops—physiological traits and genetic mechanisms. Plan Theory 10(8):1560

Tournaire-Roux C, Sutka M, Javot H, Gout E, Gerbeau P, Luu DT et al (2003) Cytosolic pH regulates root water transport during anoxic stress through gating of aquaporins. Nature 425(6956):393–397
Trenkel ME (2021) Slow-and controlled-release and stabilized fertilizers: an option for enhancing Nutrient use effiiency in agriculture. International Fertilizer Industry Association (IFA)
Tuohy P, Humphreys J, Holden N, Fenton O (2015) Mole drainage performance in a clay loam soil. In: NJF congress: Nordic view to sustainable rural development, 25, Riga (Latvia), 16–18 Jun 2015. NJF Latvia
Tuohy P, O'Loughlin J, Fenton O (2018) Modeling performance of a tile drainage system incorporating mole drainage. Trans ASABE 61(1):169–178
USDA-NRCS (2006) Land resource regions and major land resource areas of the United States, the Caribbean, and the Pacific Basin. USDA Handbook 296. Retrieved from https://www.fertilizer.org/images/Library_Downloads/2010_Trenkel_slow%20release%20book.pdf
Vaahtera L, Brosché M, Wrzaczek M, Kangasjärvi J (2014) Specificity in ROS signaling and transcript signatures. Antioxid Redox Signal 21(9):1422–1441
Varadachari C, Goertz HM (2010) Slow-release and controlled-release nitrogen fertilizers. Indian Nitrogen Group, Society
Velmurugan A, Swarnam TP, Ambast SK, Kumar N (2016) Managing waterlogging and soil salinity with a permanent raised bed and furrow system in coastal lowlands of humid tropics. Agric Water Manag 168:56–67
Vidoz ML, Loreti E, Mensuali A, Alpi A, Perata P (2010) Hormonal interplay during adventitious root formation in flooded tomato plants. Plant J 63(4):551–562
Voesenek LACJ, Sasidharan R (2013) Ethylene–and oxygen signalling–drive plant survival during flooding. Plant Biol 15(3):426–435
Voesenek LACJ, Colmer TD, Pierik R, Millenaar FF, Peeters AJM (2006) How plants cope with complete submergence. New Phytol 170(2):213–226
Wani SH, Choudhary M, Kumar P, Akram NA, Surekha C, Ahmad P, Gosal SS (2018) Marker-assisted breeding for abiotic stress tolerance in crop plants. In: Biotechnologies of crop improvement, vol 3. Springer, Cham, pp 1–23
WAPDA (2007) Waterlogging, salinity and drainage situation. SCARP Monitoring Organization, Water and Power Development Authority, Lahore
WAPDA (Water and Power Development Authority) (2003) Salinity and Reclamation Department. SCARP Monitoring Organization, Lahore
Ward A, Sharpley A, Miller K, Dick W, Hoorman J, Fulton J, LaBarge GA (2018) An assessment of in-field nutrient best management practices for agricultural crop systems with subsurface drainage. J Soil Water Conserv 73(1):5A–10A
Watanabe K, Nishiuchi S, Kulichikhin K, Nakazono M (2013) Does suberin accumulation in plant roots contribute to waterlogging tolerance? Front Plant Sci 4:178
Watanabe K, Takahashi H, Sato S, Nishiuchi S, Omori F, Malik AI, Nakazono M (2017) A major locus involved in the formation of the radial oxygen loss barrier in adventitious roots of teosinte Zea nicaraguensis is located on the short-arm of chromosome 3. Plant Cell Environ 40(2):304–316
Westra S, Fowler HJ, Evans JP, Alexander LV, Berg P, Johnson F, Roberts N (2014) Future changes to the intensity and frequency of short-duration extreme rainfall. Rev Geophys 52(3):522–555
Wiengweera A, Greenway H (2004) Performance of seminal and nodal roots of wheat in stagnant solution: K+ and P uptake and effects of increasing O2 partial pressures around the shoot on nodal root elongation. J Exp Bot 55(405):2121–2129
Williams MR, King KW, Fausey NR (2015) Drainage water management effects on tile discharge and water quality. Agric Water Manag 148:43–51
Wollmer AC, Pitann B, Mühling KH (2018) Nutrient deficiencies do not contribute to yield loss after waterlogging events in winter wheat (Triticum aestivum). Ann Appl Biol 173(2):141–153
Wu QX, Zhu JQ, Liu KW, Chen LG (2012) Effects of fertilization on growth and yield of cotton after surface waterlogging elimination. Adv J Food Sci Technol 4(6):398–403

Wu H, Xiang J, Chen HZ, Zhang YP, Zhang YK, Zhu F (2018) Effects of exogenous growth regulators on plant elongation and carbohydrate consumption of rice seedlings under submergence. J Appl Ecol 29(1):149–157

Xian C, Qi Z, Tan CS, Zhang TQ (2017) Modeling hourly subsurface drainage using steady-state and transient methods. J Hydrol 550:516–526

Yaduvanshi NPS, Setter TL, Sharma SK, Singh KN, Kulshreshtha N (2012) Influence of waterlogging on yield of wheat (Triticum aestivum), redox potentials, and concentrations of microelements in different soils in India and Australia. Soil Res 50(6):489–499

Yamauchi T, Rajhi I, Nakazono M (2011) Lysigenous aerenchyma formation in maize root is confined to cortical cells by regulation of genes related to generation and scavenging of reactive oxygen species. Plant Signal Behav 6(5):759–761

Yamauchi T, Shimamura S, Nakazono M, Mochizuki T (2013) Aerenchyma formation in crop species: a review. Field Crop Res 152:8–16

Yamauchi T, Watanabe K, Fukazawa A, Mori H, Abe F, Kawaguchi K, Nakazono M (2014) Ethylene and reactive oxygen species are involved in root aerenchyma formation and adaptation of wheat seedlings to oxygen-deficient conditions. J Exp Bot 65(1):261–273

Yamauchi T, Colmer TD, Pedersen O, Nakazono M (2018) Regulation of root traits for internal aeration and tolerance to soil waterlogging-flooding stress. Plant Physiol 176(2):1118–1130

Ylivainio K, Uusitalo R, Turtola E (2008) Meat bone meal and fox manure as P sources for ryegrass (Lolium multiflorum) grown on a limed soil. Nutr Cycl Agroecosyst 81(3):267–278

Ylivainio K, Jauhiainen L, Uusitalo R, Turtola E (2018) Waterlogging severely retards P use efficiency of spring barley (Hordeum vulgare). J Agron Crop Sci 204(1):74–85

Yordanova RY, Popova LP (2001) Photosynthetic response of barley plants to soil flooding. Photosynthetica 39(4):515–520

Yordanova RY, Popova LP (2007) Flooding-induced changes in photosynthesis and oxidative status in maize plants. Acta Physiol Plant 29(6):535–541

Zhang S (2005) Soil hydraulic properties and water balance under various soil management regimes on the Loess Plateau, China, vol 2005, no 2005, p 126

Zhang YK, Schilling KE (2006) Effects of land cover on water table, soil moisture, evapotranspiration, and groundwater recharge: a field observation and analysis. J Hydrol 319(1–4):328–338

Zhang X, Shabala S, Koutoulis A, Shabala L, Johnson P, Hayes D, Zhou M (2015) Waterlogging tolerance in barley is associated with faster aerenchyma formation in adventitious roots. Plant Soil 394(1):355–372

Zheng W, Liu Z, Zhang M, Shi Y, Zhu Q, Sun Y, Geng J (2017) Improving crop yields, nitrogen use efficiencies, and profits by using mixtures of coated controlled-released and uncoated urea in a wheat-maize system. Field Crop Res 205:106–115

Zhou M (2010) Improvement of plant waterlogging tolerance. In: Waterlogging signalling and tolerance in plants. Springer, Berlin, Heidelberg, pp 267–285

Zhou MX, Li HB, Mendham NJ (2007) Combining ability of waterlogging tolerance in barley. Crop Sci 47(1):278–284

Climate Change Impact on Mangrove Forests in Pakistan

Muhammad Rafique Khan, Sajid Rashid Ahmad, and Shakeel Ahmad

Abstract

Mangroves are shrubs and trees mostly found in tropical and sub-tropical regions along the coastlines. Mangroves being halophytes could survive in soil with high salinity and low oxygen. There are significant number of mangroves species and genera usually found in dense and thickets. In Pakistan mangrove forests are in Sindh and Balochistan provinces in coastal areas along Arabian sea. Mangrove ecosystem provides breeding space, habitat and niches to several flora and fauna. This ecosystem makes available number of goods (timber and timber products, food and livelihoods, recreation and sports, employment and business) and services (clean water and healthy environment, prevention of shoreline erosion and protection from storms, carbon sequestration and carbon sinks). Additionally, that very ecosystem does contribute to local and national economy. All this demand sustainability of mangrove ecosystem which has been threatened by anthropogenic activities (cutting of trees, clearing of forest areas for agriculture, housing, urbanization, industrialization) and climate change (global warming, rising sea level, high salinity, storm surges). Things if not controlled timely may cause biodiversity loss, unemployment, unproductive local community exposed to the risk of extreme weather conditions. This situation warrants to adopt holistic but inclusive approach while taking all stakeholders on board and framing

M. R. Khan (✉) · S. R. Ahmad
College of Earth and Environmental Sciences, Punjab University Lahore, Lahore, Pakistan

S. Ahmad
Department of Agronomy, Faculty of Agricultural Sciences and Technology, Bahauddin Zakariya University, Multan, Pakistan

M. Ahmed, S. Ahmad (eds.), *Disaster Risk Reduction in Agriculture*, Disaster Resilience and Green Growth, https://doi.org/10.1007/978-981-99-1763-1_13

'Disaster Risk Reduction Policy' and undertaking best conservation practices to save mangrove forests of Pakistan.

Keywords

Sustainable ecosystem · Halophytes · Ecotourism · Habitat and niche · Carbon sequestration and carbon sinks · Biodiversity · Food web · Economic gains · Community participation · Holistic approach

1 Introduction

Mangroves are mostly found in tropical and subtropical coastal areas. These are primarily shrubs and trees of varying height depending on climatic conditions, soil type, water salinity, tidal inundation time. Such coastal forests can be found in 118 countries and territories, though nearly 75% of their area occurs in just 15 countries. They are most often found straddling the equator between 25° North and South latitude. About 42% of the world's mangroves are found in Asia, with 21% in Africa, 15% in North and Central America, 12% in Australia and the islands of Oceania, and 11% in South America (NASA (2022) Earth observatory, 'Mapping Mangroves by Satellite').

As mangrove trees are halophytes – salt tolerant trees, therefore they could survive and thrive in soils having characteristics of high salinity and low oxygen- conditions under which it is very difficult for other plants to survive (Fig. 1). Moreover, mangroves grow in dense forests along coast lines or in intertidal zones. The coastal ecology sustainability is mainly indebted to the contributions of mangroves (Baig and Iftikhar 2005). Globally, there are 60 true mangrove species in 27 genera and 20 families (Sengupta 2010).

Socio -economic, ecological and environmental benefits derived from the mangrove forests make them an essential component of shoreline environment. They are the best shield against cyclones, tsunamis and storm surges as they break the energy of devastating waves and provide a buffer zone between coastline dwellers and the storms besides preventing coastal erosion. Present day rapid urbanization, economic growth and concerns about environmental sustainability have compelled everybody including governments, planners, developers, scientists, and coastal dwellers to realize the value of mangroves owing to their remarkably diverse but very important advantages to the sustainable development.

2 Salient Services Provided by Mangrove Forests

Major contributory services of mangroves including economic goods and services are listed below:

Fig. 1 Pictorial view of Mangrove forests in Pakistan

2.1 Mangroves Ecosystem's Contributions

Ecological as well as economic importance of Mangroves Ecosystem cannot be undermined as mangroves help maintaining coastal environment on sustainable basis. Benefits accrued from this special ecosystem are discussed herein:

2.2 Mangrove-Seagrass-Coral Reef Continuum

Mangroves are also known as mainstay of any coastal ecosystem. The classical phenomenal functional correlation between mangroves, seagrass beds, and coral reefs is highly remarkable (Fig. 2). This marvel prodigy poses itself as an essential integrant for the sustainability of that ecosystem. When mangroves stop the eroded soil particles from flowing to sea then Seagrass beds work as filters to contain mud and silt from destroying the coral reefs whereas coral reefs in turn provide shield to the seagrass beds and mangroves from strong waves while diluting waves energy. So, without mangrove forests, this incredibly productive ecosystem might have been unproductive rather would likely collapse.

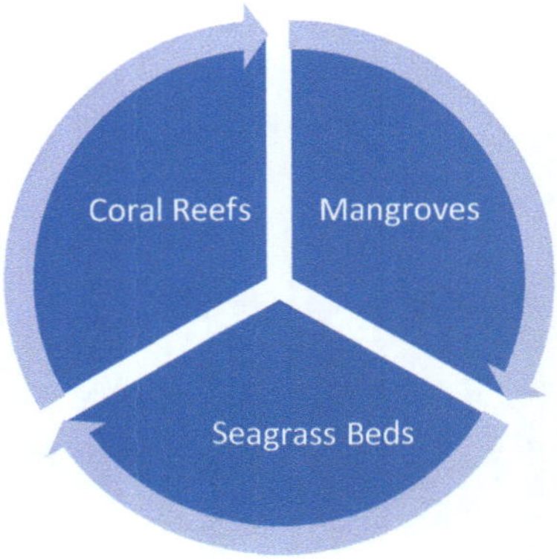

Fig. 2 Functional correlation

2.3 Breeding and Nursery Places

Mangroves swamps do provide a congenial breeding place for a lot many fish and crustaceans etc. Nearly 100 species of fish have so far been recorded from mangroves in Pakistan, of which 46 species were in fingerling or young stages while 52 in sub-adult or adult stages. In fact, more than 75% of commercially caught fish may inhabit mangroves at some point of their life (Dey 2020; Sadilyan et al. 2010; Sahu et al. 2016; World Economic Forum 2019; The Editors of Encyclopaedia Britannica 2018). Because intertidal zones are quite nutrients-rich areas therefore, many fish and crustacean's species select these places as their breeding grounds. Again, marine life such as barracuda, tarpon, and snook also find shelter in the roots of Mangroves forests and spend their juvenile period of their life. This is the area wherein these juveniles can easily find ample food for their growth and development while hiding themselves from predators. In this way mangroves and seagrasses provide foraging opportunities to young ones as well, in addition to food and shelter. Afterwards these juveniles move into open sea as adults after completing their 'nursey and kindergarten' age. This interesting phenomenon can easily be understood in the given illustration (Fig. 3).

2.4 Habitat and Niches

Variations in biotic and abiotic characteristics promote various types of niches and habitats for macrofauna within the intertidal zone. Mangrove forests are important habitats that support a unique assemblage of organisms. Mangrove forests not only provide habitat for thousands of species of flora and fauna at all levels but also serve as forest food webs ranging from bacteria to Bengal tigers. On the other hand, mangrove forests do serve as niches- a subset of habitat, for many species with their own exclusive functioning role in the energy chain and remarkable impact on the environment.

Fig. 3 Breeding area, nursery and kindergarten

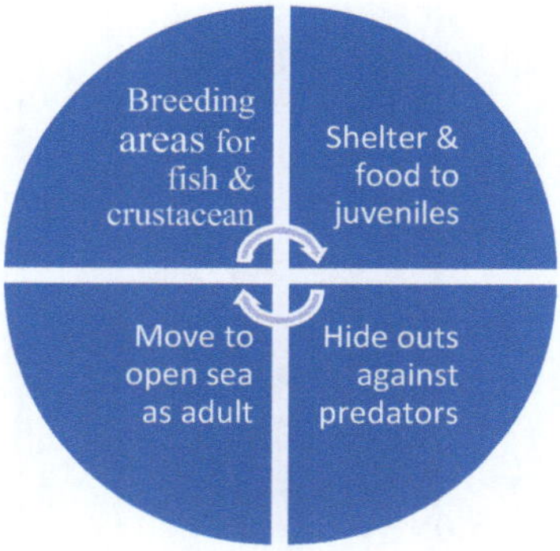

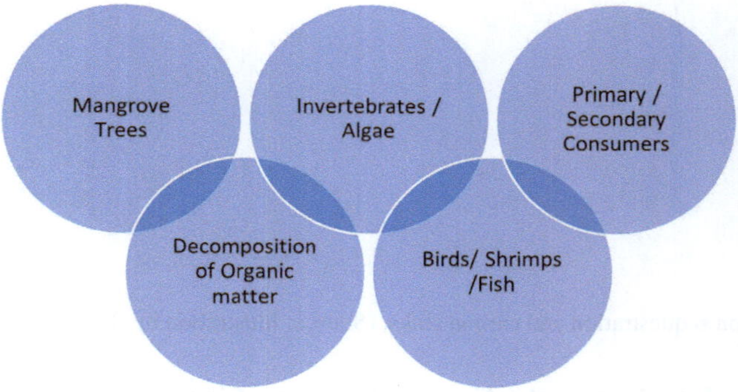

Fig. 4 Mangroves food web

2.5 Mangrove Food Web

Food web in Mangrove ecosystem not only supports the species concerned in their survival but also makes the very system sustainable (Fig. 4). Mangroves, being autotrophs, produce their own food as primary producers through the process of photosynthesis. There are many herbivore species in marine ecosystem which consume mangrove leaves, bark or fruits in the ecosystem. Whereas, on the other hand, fallen leaves of mangrove forest become the very basis for this incredible and productive mangroves food web. Essential nutrients become available on decomposition of these organic materials which are consumed by invertebrates and algae. These in turn feed many organisms like birds, sponges, worms, anemones, jellyfish, shrimp, and young fishes. Water tides also help shifting nutrients to different mudflats, coral reefs and estuaries providing food to oysters etc. These primary consumers then feed secondary and tertiary consumers in turn at multiple levels.

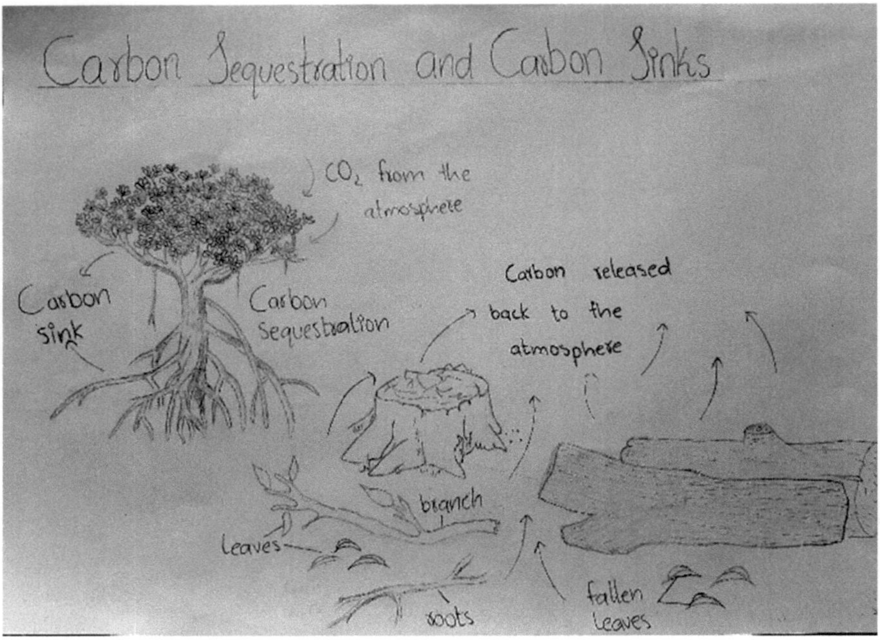

Fig. 5 Carbon sequestration and carbon sinks. (Source: Illustration by Rahma Rafique Rana)

2.6 Carbon Sequestration and Carbon Sinks

Mangrove forests do help fighting against global warming by reducing atmospheric carbon dioxide (CO_2). They consume a lot of atmospheric CO_2 in photosynthesis and thereby sequestering carbon in the process. Likewise, mangroves store that carbon dioxide in the trees and soil, so they serve as carbon sinks also. However, cutting of forests cause release of carbon but regrowth of forest trees again starts carbon sequestration (Fig. 5). Whereas wood logs and fallen leaves, branches, or dead roots on decomposition release carbon very slowly and may take years. In this way, mangroves provide necessary support to combat climate change impacts while helping in reducing the overall "carbon footprint" prerequisite for environmental and developmental sustainability.

3 Status of Mangrove Forests of Pakistan

Mangrove forests constitute a significant part of green cover of Pakistan. They are located/situated in coastal areas of Sindh and Balochistan Provinces along Arabian sea. Mangroves play an important role in marine ecosystem sustainability in Pakistan. They not only provide food, shelter and security to the inhabitants of

Table 1 The key features of Mangrove forests

Sr. No.	Category	Description
1.	Sustainable environment	• Carbon sink and carbon sequestration • Prevent soil and water pollution
2.	Habitat and niche	• Serve as essential habitat for flora and fauna • Preserve biodiversity and prevent it from extinction
3.	Coastline defenders	• Preventing erosion and stabilizing coast lines • Protecting the local community from storms surges, cyclones, tsunami
4.	Economic services	• Provision of wood and wood products for energy, construction and household use • Commercial fish and other sea food
5.	Clean water	• Improving water quality by filtering pollutants and trapping dirt • Water management
6.	Breeding zone	• Breeding place for fish and crustaceans • Shelter to juvenile marine life
7.	Ecotourism	• Recreation like bird and wildlife watching • Boating, hiking and fishing

mangrove habitat but also sources of livelihood, employment, economic prosperity, shield against cyclones and storm surges and wellbeing to the coastal dwellers. Mangroves are a rich source of ecotourism and adventure sports. Indus delta mangroves provide habitat to many fish, crustaceans, birds, mammals, reptiles, amphibians, etc. Their role in environmental sustainability cannot be undermined as they contain soil erosion, and water pollution while stopping the sedimentation load being flown over to the sea on one hand and saving agriculture and arable land while stopping the saltwater intrusion on the other hand. Moreover, mangroves provide employment to hundreds of thousands of unemployed residents of local community attached with fisheries profession and are source of billions of rupees export earning much needed foreign exchange besides local consumption leading to contributing the local economy and national growth. The key features of Mangrove forests are presented in Table 1.

4 Indus Delta Mangroves

In the Sindh province, mangroves are found in the Indus Delta, Karachi harbor and Sandspit. The Indus Delta extends from Korangi Creek in the west to Sir Creek in the east, whereas Sandspit is a small locality in the west of Karachi city. Indus originated from northern part. Indus delta consists of creeks. Mudflats, sand dunes, estuaries, marshes and bays spread over an area of 600,000 hectares (ha). These mangrove forests are under the management of three different organizations namely Sindh Forest Department (SFD), Port Qasim Authority (PQA) and Sindh Board of Revenue (SBR).

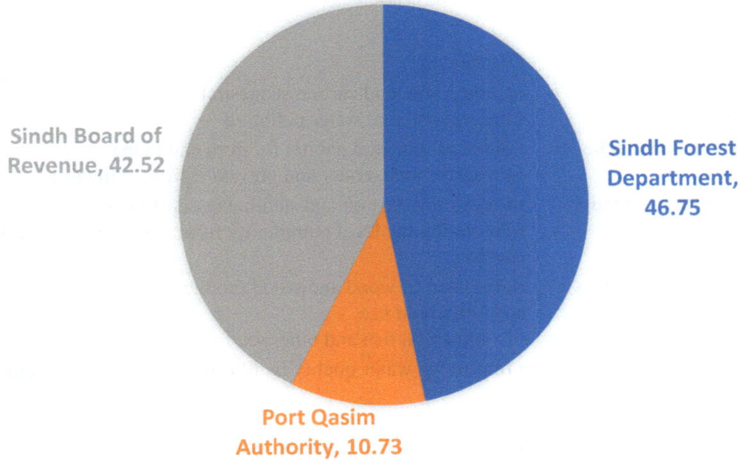

Fig. 6 Percentage area with different bodies

Table 2 Indus delta mangroves management

Sr.#	Name of Department	Area (in ha)	% age
1	Sindh Forest Department	2,80,470	46.75
2	Port Qasim Authority	64,400	10.73
3	Sindh Board of Revenue	2,55,130	42.52
Total		600,000	

Source: Forest Department, Government of Sindh

Mangroves under the management of SFD and PQA that constitute 57.48% of total mangroves at Indus deltaic region have been declared by Government of Sindh as "Protected Forests" under Forest Act, 1927 whereby, cutting of trees is prohibited without permission.

This graph demonstrates that major area of mangroves of Indus Delta is under the control and management of SFD followed by SBR and PQA respectively (Fig. 6). These organizations are responsible for conservation as well as afforestation of mangroves falling under their area of responsibility (Table 2).

5 Species Diversity

Species diversity of Mangroves of Indus delta along with their presence in percentile can be seen from the table below:

This table depicts that 90% of Indus delta mangrove forests consist of Avicennia Marina which is the dominant mangrove species followed by Rhizophora Mucronata with 8% population (Table 3).

Table 3 Mangroves species diversity in Indus delta

Sr.#	Mangrove Species	% age cover
1	Avicennia Marina	90
2	Rhizophora Mucronata	08
3	Aegiceras Corniculatum	1.5
4	Ceriops Tagal	0.5

Source: Forest Department, Government of Sindh

Table 4 Mangroves forest in Balochistan (hectarage)

Sr.#	Location	Area (in ha)	% age
1	Sonmiani khor	4280.040	75.27
2	Kalmat khor	933.030	16.41
3	Sahidi khor	20.000	0.35
4	Sawar khor	1.600	0.03
5	Shabi & Ankara creeks	228.000	4.01
6	Jiwani	223.290	3.93
Total		5685.960	

Source: Forest Department, Government of Balochistan

6 Balochistan Mangroves

Mangrove forests in Balochistan are located along the coastline at scattered locations mainly at Sonmiani Khor, Kalmat Hor, shabi & Ankra Creeks and Jiwani. Mangroves covered area also keeps on changing slightly due to variance in intertidal zones. However, mangrove cover in Balochistan is as under (Table 4):

This table shows that out of total mangroves consisting of 5685.960 ha in Balochistan, major part of mangroves exists at Sonmiani Khor with an area of 4280.040 ha followed by Kalmat Khor with an area of 933.030 ha which is a second major establishment of mangroves forest in Balochistan whereas Shabi & Ankara Creeks as well as Jiwani do have significant mangroves spread over 288.000 and 223.290 ha respectively.

This chart illustrates that 75.27% of mangroves forest of Balochistan are located at Sonmiani Khor area followed by 16.41% at Kalmat Khor and 4.01% of forest is available at Shabi & Ankara Creeks (Fig. 7).

7 Deteriorating Factors

Forest departments of Sindh and Balochistan with the essential aid of local communities have attempted to conserve the mangroves. Although, forest departments and Pakistan Navy also been increasing the mangrove coverage by planting new saplings through different projects and campaigns yet mangrove forests in Pakistan are quite under stress and prone to deterioration due to anthropogenic activities

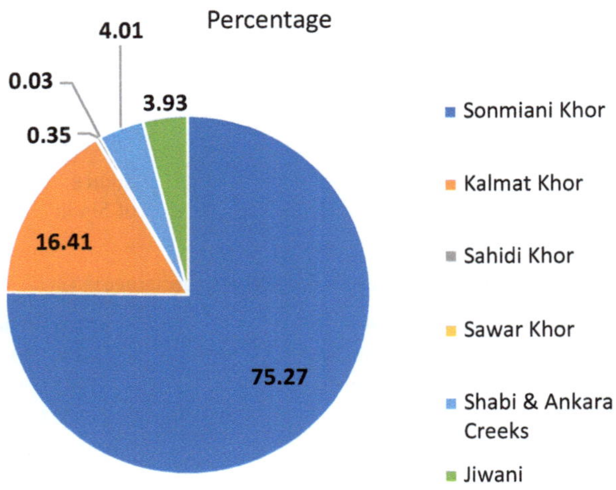

Fig. 7 Mangroves area in percentile

Fig. 8 Destruction of Mangrove forests in Pakistan

coupled with climate change impacts (Fig. 8). Some of those glaring causes are enumerated below:

- Cutting of forest trees for fuel, wood products, fodder, construction etc.
- Ill planned urbanization and construction thereof.
- Clearing of forest lands for agriculture, housing or other industrial ventures.
- Extreme weather conditions like cyclones, windstorms or storm surges etc.
- Environmental pollution caused by anthropogenic activities.

- Disposal of untreated wastes, effluences into water sources feeding these mangroves on their way to sea.
- Poor implementation of mangrove conservation Policy and Practices.
- Inefficiency and lack of capacity and resources at the part/disposal of those responsible for management of mangroves.
- Absence of political will and non-participation of civil society, community organizations and citizens in mangroves conservation.
- Lack of awareness on the part of those whom mangroves serve the most.

8 Manifestations of Climate Change

In the past 20 years, the world has lost almost 50 per cent of its mangrove forests, making them one of the most endangered landscapes. It is essential to recover them and to use them as a shield against a tsunami and as a resource to secure optimal socio-economic, ecological and environmental benefits (Osti et al. 2009). Occupying a harsh margin between land and sea, most mangrove plants and associated organisms are predisposed to be either resilient or resistant to most environmental change (Alongi 2015). Like all other fields of life, climate change is also severely affecting mangrove ecosystem and things related to that marine ecological stratum. This impact can very easily be visualized in the following areas:

8.1 Biodiversity Loss

Change of environment caused by the destruction or change of any component of mangrove ecosystem pose a challenge to the survival of its inhabitants thereof. Therefore, foremost impact of climate change becomes visible in the form of biodiversity loss. Some of the species, owing to their inability to cope with the new environment, become endangered whereas if the situation prolongs enough some species might got extinct. Several species are listed as vulnerable or endangered on the International Union for Conservation of Nature (IUCN) Red List of Threatened Species (Rodrigues et al. 2006).

Destruction of mangrove forests has caused the loss of habitat for several fish, crustacean and wildlife species that live and breed in mangroves.

8.2 Decline in Area

The rising level of carbon dioxide and other industrial gases in the atmosphere may lead to global warming with an accompanying rise in sea-level (Field 1995). Sea-level rise, first, affects the habitat and niches provided by the mangroves. Secondly rise in sea level also influences the growth of mangroves whereas rise in temperature is again crucial for density of the thickets. All these factors would lead to decline in mangroves area resulting into non availability of mangrove ecosystem products and services consequently.

8.3 Mangrove Swamp Destruction

Climate change has reportedly various multidimensional components of varied nature and scale with different level effects on the ecosystems (Mitra 2013). Extreme weather conditions like windstorms, cyclones, tsunami and storm surges destroy mangrove swamps. Many mangrove saplings and trees got damaged or uprooted under these conditions.

8.4 High Salinity Impact

Although, mangroves have the ability to survive in high saline conditions and low oxygen levels yet up to a certain threshold. A notable decrease of gastropods (snails, slugs, etc.) was reported and, similarly, migratory water birds' populations were also shown to decline, due to increases in salinity. As the change in climatic conditions affect salinity in water and soils which in turn impact the diversity of mangroves. High saline conditions also affect the marine life as well.

8.5 Reduction in Fresh Water Flow

As the climate change affects the weather patterns and in case of drought water is hardly availabe. Again, climate change phenomenon coupled with extreme saline conditions have also led to a 90 per cent reduction in freshwater flow (Dasgupta and Shaw 2013). Therefore, with the decrease in freshwater availability afforestation of mangrove forest in wetlands with high salinity and low oxygen is quite challenging.

8.6 Environmental Pollution

Polluted environment including water and air pollution has become a big challenge especially for the megacities. Improper and untreated disposal of wastes and effluents of all kinds into water sources discharging into sea as well as solid wastes dumped into mangroves severely affects the mangroves growth and functionality of mangrove ecosystem. Besides this, oil spills in coastal waters again badly impact the mangrove ecosystem thereby affecting the goods and services accrued from that ecosystem.

9 Implications of Loss of Mangroves

Climate change coupled with environmentally non-friendly and unsustainable human activities not only becoming challenge for afforestation of mangrove forests but severely threatening the survival of mangrove ecosystem. Loss of mangroves would expose the coastal dwellers to extreme weather events which are occurring more frequently and with more intensity. Secondly, mangrove ecosystem goods - timber and timber products, livelihoods, ecotourism, foods, fisheries, etc. as well as ecosystem services - healthy environment, coastline protection, shield against storms, clean water, etc. would either be non-available, or their quality would get compromised. Thirdly, economic gains including contributions in local and national economy whether through consumption or exports would be squeezed. Fourthly, it would cause unemployment of hundreds of thousands of people connected with fisheries and other professions relating to goods and services provided by mangrove ecosystem. Fifthly, mangrove loss may impact the economic productivity of coastal community and may lead to their food security situation. Lastly, services rendered by mangroves in carbon sequestration and carbon sinks much required for climate change mitigation would be non-available in case of no mangroves.

All situations, thus warrants the conservation of mangroves forests on sustainable basis in addition to afforestation of mangroves wherever wetlands are available to get the benefits attached with the mangrove ecosystem.

10 Conservation and Restoration of Mangroves Not an Option but a Way Forward

There is dire need for preservation and restoration of mangrove forests given their role as a specialized marine ecosystem. As the optimal productivity of mangrove forests and ecosystem gets affected by a large range of anthropogenic and climate change factors therefore, an inclusive framework for understanding complexity of interactions and their management thereof needs to be evolved. With this end in view following suggestions are vital to face the challenges posed by the climate change vis-à-vis mangroves conservation:

1. First and foremost, there is need to devise a proper *Disaster Risk Reduction Policy* by the government while involving all stakeholders including local community, NGOs, leaders of civil society, scientists/environmentalists/experts on the subject and all relevant departments/organizations and agencies responsible for implementation.
2. Create awareness among the people in general and local dwellers about the pivotal role of mangroves and mangrove ecosystem.

3. Local community needs to be educated and trained as well by imparting necessary skills and knowledge to conserve and restore mangrove forests leading to mangrove ecosystems' conservation.
4. Well planned and suitable land-use plans and best farming practices for the coastal areas which complement conservation of mangroves may be developed.
5. Local community participation in the conservation of mangroves and to stop illegal cutting and encroachment on mangrove areas is critical for the success. Therefore, a robust community participatory mechanism may be devised.
6. Development policies and plans for coastal areas/cities urbanization should address the issues of coastal communities and should include the component of mangrove preservation.
7. As the survival and growth of mangrove forests are affected by salinity level therefore, factors responsible for salinity rise required to be studied and appropriate measures may be taken to cap the salinity in a range suitable for mangroves thrive.

As flourishment of mangrove forests is directly linked with sustainability of mangrove ecosystem hence, their survival and conservation are the linchpin for getting mangrove ecosystem products and services on sustainable basis. It is imperative to all stakeholders including public authorities, leaders of civil society, community organizations, citizens and coastal dwellers to play their part to achieve the objective of mangrove conservation and to relieve the mangroves from the stress being posed to them by human activities and climate change.

References

Alongi DM (2015) The impact of climate change on mangrove forests. Curr Clim Chang Rep 1(1):30–39

Baig SP, Iftikhar UA (2005) Are the mangroves for the future. Empirical evidence of the value of Miani Hor Mangrove ecosystem as the basis for investments

Dasgupta R, Shaw R (2013) Cumulative impacts of human interventions and climate change on mangrove ecosystems of south and Southeast Asia: an overview. J Ecosyst:379429

Dey K (2020) India's blue economy net getting bigger! Country ranks third in fisheries and second in aquaculture. Financial Express, 14 Feb 2020. https://www.financialexpress.com/opinion/indias-blue-economy-net-getting-bigger-country-ranks-third-in-fisheries-and-second-in-aquaculture/1867607/

Field CD (1995) Impact of expected climate change on mangroves. In: Asia-Pacific symposium on mangrove ecosystems. Springer, Dordrecht, pp 75–81

http://mangrove.org/video/mangroves.html

https://www.amnh.org/explore/videos/biodiversity/mangroves-the-roots-of-the-sea/why-mangroves-matter

Mitra A (2013) Impact of climate change on mangroves. In: Sensitivity of mangrove ecosystem to changing climate. Springer, New Delhi, pp 131–159

NASA (2022) Earth observatory, mapping mangroves by satellite. https://earthobservatory.nasa.gov/images/47427/mapping-mangroves-by-satellite

Osti R, Tanaka S, Tokioka T (2009) The importance of mangrove forest in tsunami disaster mitigation. Disasters 33(2):203–213

Rodrigues AS, Pilgrim JD, Lamoreux JF, Hoffmann M, Brooks TM (2006) The value of the IUCN red list for conservation. Trends Ecol Evol 21(2):71–76

Sadilyan S, Thiyagesan K, Nagarajan R et al (2010) Salinity rise in Indian mangrove- a looming danger for coastal biodiversity. Curr Sci 98(6):754–756

Sahu SC, Kumar M, Ravindranath NH (2016) Carbon stocks in natural and planted mangrove forests of Mahanadi mangrove wetland, East Coast of India. Curr Sci 110(12):2234–2241

Sengupta R (2010) Mangroves: soldiers of our coasts. TERI, New Delhi, p 32

The Editors of Encyclopaedia Britannica (2018) Acanthaceae. Encyclopedia Britannica. https://www.britannica.com/plant/Acanthaceae

World Economic Forum (2019) Future of the environment. https://www.weforum.org/agenda/2019/02/5-reasons-to-protect-mangrove-forests-for-the-future/

Climate Change, Flash Floods and Its Consequences: A Case Study of Gilgit-Baltistan

Mehwish Aslam, Rifat Hayat, Nelum Pari, Aashir Sameen, and Mukhtar Ahmed

Abstract

The fundamental factor influencing the development and expansion of glacier lakes and glacier lake outburst floods (GLOF) is climate change. The climatic variables like temperature have been used to analyze hydro-meteorological risks at the local level, and their results put impacts on stream flow patterns, soil degradation, and drought conditions occurred. Wild weather in the Gilgit-Baltistan (Hindukush, Karakorum and Himalaya HKH) has a detrimental effect on the social infrastructure. The actions have substantially changed the situation; yet, they have had a detrimental effect on the environment, the socioeconomic progress of the populace, and the main sources of livelihood, including the cattle and forestry industries. The inability to maintain an accurate inventory, categorization, and susceptibility profile of glacial lakes and newly developed GLOFs is one of the issues that Pakistani disaster planning and risk reduction efforts face. The FAO agricultural assessment report states that the floods destroyed primary infrastructure, including tube wells, water channels, household storage units, houses, animal pens, individual seed stocks, fertilizers, and agricultural machinery, and caused unprecedented scale damages to agriculture crops, livestock, fisheries, and forestry.

M. Aslam · R. Hayat (✉)
Institute of Soil and Environmental Sciences, PMAS-Arid Agriculture University, Rawalpindi, Pakistan
e-mail: hayat@uaar.edu.pk

N. Pari
SDGs Academy, Islamabad, Pakistan

A. Sameen · M. Ahmed
Department of Agronomy, Pir Mehr Ali Shah Arid Agriculture University, Rawalpindi, Pakistan

M. Ahmed, S. Ahmad (eds.), *Disaster Risk Reduction in Agriculture*, Disaster Resilience and Green Growth, https://doi.org/10.1007/978-981-99-1763-1_14

Keywords

Glacier Lake Outburst Floods (GLOF) · Hydro-meteorological risks · Stream flow patterns · Soil degradation · Drought · Gilgit-Baltistan

1 Introduction

The biggest challenge of the twenty-first century is now climate change (Ahmed 2020). The global weather pattern, including trends in temperature, precipitation, and humidity, has evolved over the past 50 years. Different nations all across the world are being negatively impacted by climate change. It significantly impacts agriculture and increases the risk to food security. One of the top 10 nations most impacted by changes in climatic temperatures and weather patterns is Pakistan. Wide-ranging effects include decreased agricultural output, ongoing droughts, coastal erosion, and higher-than-average rainfall. One of the main contributors to the current flood scenario in the nation is climate change, which has resulted in the destruction of thousands of acres of land, the displacement of millions of people, and the loss of life.

Each and every Pakistani province, including Gilgit-Baltistan, is susceptible to geological and meteorological risks (Ahmad et al. 2011). The glaciated mountains of the Upper Indus Basin of Pakistan are warming more quickly than the rest of the nation. It has caused abrupt and erratic glacier changes that have advanced glacial lakes and increased the risk of Glacier Lake Outburst Floods (GLOF) in Pakistan's mountain systems. Northern part of Pakistan, Gilgit-Baltistan is a province having 14 districts. This year two main districts, Hassanabad Hunza and Ghazir, are been hit by Climate change. The GLOF episode caused the destruction of homes, orchards, powerhouses, and fiber optics in the vicinity, hundreds of trees, agricultural land, and two hydro-power plants were also impacted by the storm.

Glacial lakes in high mountains around the world are growing in number and size as a result of glacier retreats brought on by rising temperatures. Particularly, the Hunza River Basin's glacial ice has been shrinking, going from 44.02% in 1989 to 34.99% in 2010 (Baig 2018). Due to a rate of temperature increase coupled with deglaciation, glacier lake outburst floods (GLOFs) pose severe dangers in high mountain glaciated habitats since the past 10 years (Hongyu 2020). Glaciers are melting quickly as a result of rising temperatures, adding to the 33 fragile lakes that have already developed and resulting in the creation of more than 3000 lakes. 33 are thought to be in danger of bursting, putting the lives of 7.1 million people in danger from glacial lake outburst floods (GLOF). These unexpected occurrences have the potential to release millions of cubic meters of water and debris, endangering lives, homes, livestock, and the way of life in isolated mountain communities. Due to a rate of temperature increase coupled with deglaciation, glacier lake outburst floods (GLOFs) pose severe dangers in high mountain glaciated habitats since the past 10 years (Senese et al. 2018).

According to reports, the Rakaposhi Valley in Gilgit-Nagar District is particularly susceptible to climate changes related risks, which include droughts, erosion, cloud bursts, landslides, glacial lake outburst floods (GLOFs), flash floods, avalanches, and extreme weather events (WWF 2015). Over the past 30 years, extreme weather events have frequently impacted the area's key agricultural crops, pastures, infrastructure, and way of life (Ali et al. 2019; Chettri et al. 2008a, b; Convention on the Conservation of European Wildlife and Natural Habitats 2010; Immerzeel et al. 2012; International Centre for Integrated Mountain Development 2009; Nawaz et al. 2009; Nogues-Bravo et al. 2017; Qureshi 2017; Shrestha et al. 2015). In Gilgit-Baltistan, forestry (11%) and livestock (41%) follow agriculture as the top three sources of farm revenue (World Bank 2010). The people are agro-pastoralists, highly dependent on mountain farming and livestock herding to earn bread and butter for their families. Consequently, agriculture is regarded as the principal economic activity and irrigation as the primary component of livelihood security. The socio-hydrological interactions are shaped by various water-related disasters, water distribution systems, socioeconomic developments, and other external developments (Parveen et al. 2015; Nüsser 2017). Glacial lake is a lake with origin in a melting glacier and accumulation of water up to a certain limit leads to the sudden release of water and debris known as Glacial Lake Outburst Floods (GLOFs). 2500 glacial lakes are formed in HKH and 52 of them are considered as potentially dangerous. Glacial Lake area larger than 0.02 km^2 is considered as dangerous (Ashraf et al. 2012) (Fig. 1).

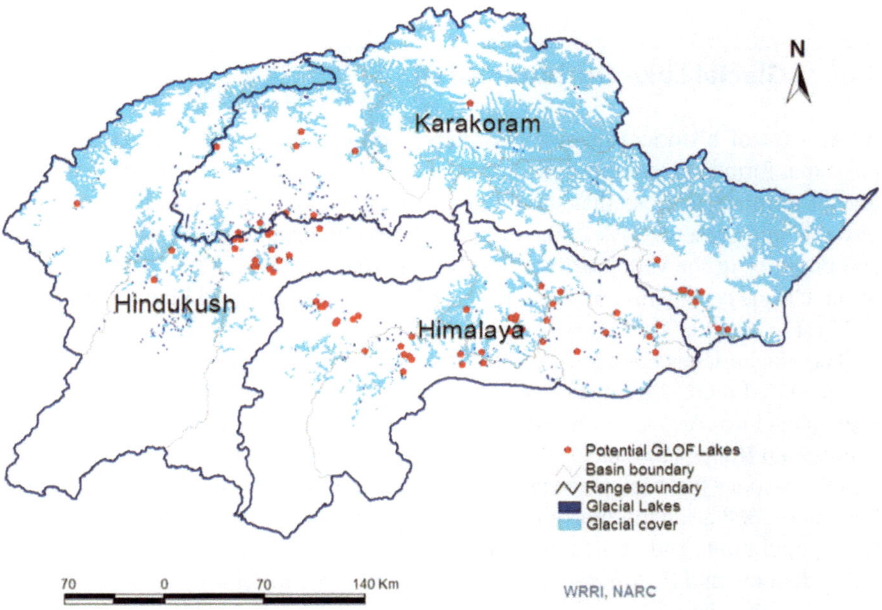

Fig. 1 Potentially dangerous lakes in HKH—Gilgit-Baltistan Pakistan. (Ashraf et al. 2012)

2 Glacial Lake Formation in Karakoram Range (Gilgit-Baltistan)

Since the second century, global warming has hastened the melting of glaciers in the HKH region, which has led to the emergence of several glacial lakes. A glacial lake is a lake that was formed when a glacier melted. Proglacial lakes are appearing more frequently as a result of retreat, while existing lakes are growing in size and volume. Moraines that are frequently dammed to create lakes sometimes are not stable. Glacial lake outburst floods (GLOFs), which can have catastrophic downstream repercussions, can be caused by dam failure. Once these lakes reach a particular level of water accumulation, there is a sudden release of water that results in GLOFs (Glacial Lake Outburst Floods), which inflict loss of life, property damage, and the devastation of several resources. The development of meltwater lakes on the lower portions of numerous glaciers has been one of the consequences of recent atmospheric warming in the Himalayan region (Ashraf et al. 2012). GLOFs are potentially dangerous because meltwater gets blocked inside a glacier. It can burst anytime. We cannot see it coming and that is what makes them a bigger problem than the larger glacial surface lakes. Glaciers in Karakoram are active, maintaining their natural balance. Meltwater build up inside and not on surfaces. These trapped lakes can burst anytime when glaciers travel tens or hundreds of meters down the valley a year (https://www.dawn.com/news/330034). It is hard to work when scale of the problem is hidden. A number of villages in Gojal tehsil are home to many glaciers including Batura and Passu glaciers and are prone to these kinds of flooding (Fig. 2).

3 Glacial Lake Outburst (GLO)

As a result of historical or ongoing glacier action, glacial lakes are created. The lakes that form behind a loosely consolidated end-moraine dam beneath a retreating glacier terminal is those that should be of the most concern. These lakes could be unstable and present a risk to the infrastructure of hydropower as well as to people and property in the valley below (ICIMOD 2016a, b). 25% of the GLOF lakes are of the Cirque type, while 62% are of the End Moraine Dammed kind (Ashraf et al. 2012) (Fig. 3).

The glacial lakes connected to glaciers like Supraglacial, Valley, Cirque, and/or dammed by Lateral Moraine or end Moraine with an area larger than 0.02 km^2 have been taken into consideration and they have been defined as major glacial lakes. The Karakoram Range has had 35 devastating outburst floods in the past 200 years. Five GLOF episodes that took place in the Hunza basin of the Karakoram Range between 2007 and 2008 are particularly frightening because they significantly impacted the local populations and constituted a threat to the future. Better hazard assessment, risk reduction, and GLOF mitigation are required given the circumstances. The situation demands better hazard assessment, risk reduction, mitigation of GLOF hazards, and adoption of a suitable early warning system (Fig. 4).

Fig. 2 Inventory of Glaciers, Glacial Lake and Glacial Lake Outburst Floods Monitoring and Early Warning System in Hindukhush-Himalayan Region, Nepal (Source: Pradeep K. Mool Samjwal R. Bajracharya Shrad P. Joshi)

Submerged –glacial lake **Super glacial lake** **Moraine Dammed Lake**

Fig. 3 Types of lakes

The physical conditions and criteria for identification of potential GLOF lakes, as mentioned by Mool et al. (2001), Bajracharya et al. (2011) and ICIMOD, may include:

- The threat of a glacial lake reaching a breaching point is increased by the rising water levels in moraines-dammed glacial lakes.
- The characteristics of the damming material and the mother glacier can be used to determine the potentially harmful status of lakes that have been dammed by moraines.

Fig. 4 Glacial lake outburst. (Source: Kreutzmann 1994, modified by Iturrizaga 2005)

- The activity of supraglacial lakes, or lakes that have developed over glacial surfaces: over time, clusters of sparsely spaced supraglacial lakes around glacier tongues combine to form larger, potentially hazardous lakes.
- Cirque lakes even smaller than 0.1 km^2 that are connected to steep hanging glaciers at a distance of less than 0.5 km are thought to be potentially deadly.
- A lake with a moraine dam that has already breached, closed, and then replenished with water may do so again.

4 Remote Sensing (RS) and Geographic Information System (GIS)

Early literature has demonstrated that remote sensing techniques can be used to study glacier dynamics in remote areas with difficult topography, such as the Hindukush-Karakoram-Himalaya (HKH). When compared to contemporary approaches like RS and GIS, conventional and traditional methods of glacier monitoring are exceedingly time consuming and frequently unfeasible, especially in remote places. Due to the paucity of ground observation, the distant locations, and the lack of funding for research, the glaciers of Pakistan's HKH regions are among the least studied. However, since many years, the region has experienced glacial lake outburst floods due to the melting of these glaciers in reaction to climate change. Understanding the evolution of glaciers requires more than just systematic long-term field observations. Due to the area's diverse climate patterns and difficult topography, the HKH region. In the last few decades, technological development has made it possible for people to see glaciers like never before. The best option for monitoring glaciers in high mountain locations is remote sensing (Paul et al. 2004). A common technique for glacier mapping and monitoring is the automatic recognition of glaciers and GIS-based extraction of glacier bodies. We can now map and monitor glaciers in distant areas of HKH and other glaciers across the world, which were previously too difficult to monitor on a broad scale, with the aid of remote sensing and using GIS as a tool (Bishop et al. 1998; Kulkarni et al. 2007; Racoviteanu et al. 2008).

Berthier et al. (2006) used remote sensing data to supervise changes in glacier elevation and mass balance in Western Himalaya, by comparing a digital elevation model (DEM) of 2004 from SPOT5 satellite to the 2000 STRM DEM and resulting suggested that glaciers are experiencing rapid loss in ice accumulated area with specific mass balance of −0.7 to −0.8 m/yr. w.e (water equivalent). Ali et al. (2017) suggested that correct estimation about snow and ice within a basin facilitate the mass balance assessment on the observations of accumulation area ratio obtained through remote sensing. Monitoring the extent of glaciers and their mass balances are key elements to analyze the impact of climatic disturbances on the health of glaciers (Paul et al. 2004).

5 Study Area

The Hindukush-Karakoram-Himalayas (HKH) region, which runs from Afghanistan to Myanmar, is made up of the hilly region of Northern Pakistan, which is located in its western portion (Fig. 5). The research focuses on Gilgit-Baltistan (GB), province of Pakistan it covers an area of 72,971 km^2. It has three divisions and 14 districts; Baltistan Division (Ghanche, Skardu, Shigar, Kharmang, Roundu), Gilgit Division (Ghizer, Gupis–Yasin, Gilgit, Hunza, Nagar), Diamer Division (Astore, Diamer, Darel, Tangir), and is extremely mountainous area it shares borders with Afghanistan, China, and Indian-administrated Kashmir. The Karakoram range starts in the Wakhan Corridor in Afghanistan and extends into Ladakh in India, the Aksai Chin region of China, and most of Gilgit-Baltistan in Pakistan.

The Karakoram range is occupied by six districts: Ghizer (11,700 km^2), Gilgit (4000 km^2), Hunza (11,200 km^2), Nagar (3000 km^2), Shigar (9000 km^2), and Ghanche (8200 km^2), as well as a small piece of Kharmang (2800 km^2). The entire districts of Astore (5200 km^2) and Skardu (7000 km^2) are located within the Himalayan range. Approximately 7000 km^2 of the Diamer districts are located in the Hindu-Kush range and the Karakorum. Shigar, Hunza, and Ghizer are the next two largest districts in Gilgit-Baltistan (Baig et al. 2021).

Due to its location in the southern part of the Karakoram region, the region has extremely cold weather, with winters lasting around 9 months out of the year. The summer season is hot but brief, with highs of 40 degrees; as a result, avalanches and

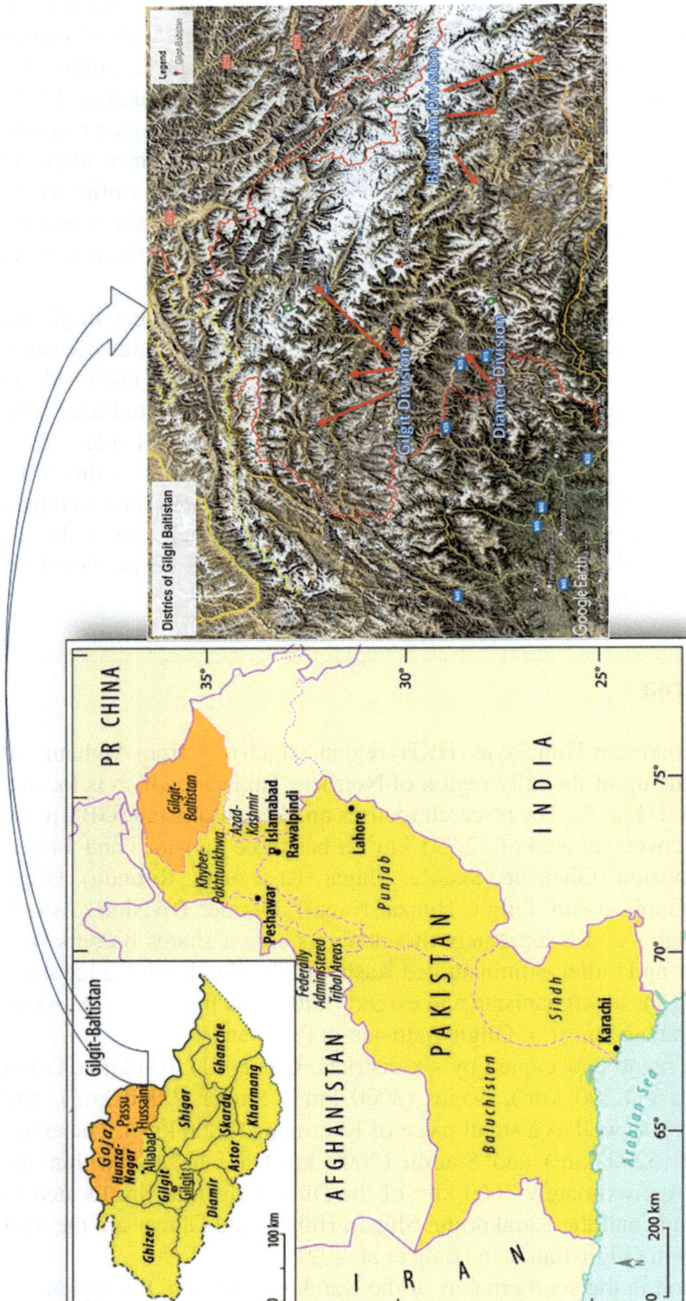

Fig. 5 Map of study site

landslides are frequent. Some common glaciers present in Gilgit-Baltistan, Pakistan are Baltoro glacier, Chimmik glacier, godwin-austen glacier, Hainablak, Hispar glacier, Kutiah glacier, Miar glacier, Panmah glacier and Passu glacier (Baig et al. 2021). The Gilgit River enters the Indus River close to the settlements of Juglot and Bunji. It originates from Shandur Lake. Gilgit experiences less precipitation, ranging between 120 and 240 mm annually. Water from glacier snowmelt is used in agriculture and other industries. The Katpana and Sarfaranga deserts in Gilgit, which have a climate similar to a cold desert, are among the highest cold deserts in the world.

6 Methodology

To record how the community feels about climate change and its effects, a qualitative research approach was used. Focus group discussions (FDGs), structured and semi-structured interviews, and other social survey techniques were employed during fieldwork in 2015 to gather initial data. A random sample of households was interviewed in-person. Key stakeholders such as herders, elderly villagers, school instructors, and youth, including both women and men, participated in the interviews and FGDs as respondents.

The approximate time and place of GLOF occurrences have been assembled from a number of reports released by public and commercial entities as well as printed media. Table 1 provides a list of noteworthy events.

7 Results and Discussions

Climate change is devastating the Himalayan glaciers, according to a 2021 article in the Journal of Natural Science Advances. The primary risk is that the lakes and rivers formed by the melting glaciers can overflow and produce flash floods. 33 out of the 3000 glacial lakes are catastrophic in nature. Faster glacier melting as a result of rising temperatures causes flash floods and glacial lake eruptions. Floods were brought on by a combination of heavy rain and glacier melting in the district Hunza and Ghizer recently, 2022. Due to upcoming Glacial Lake Outburst Floods, more glaciers are anticipated (GLOF).

Floods brought on by climate change might significantly damage inland structures. The majority of responders claimed that climate change is to blame for flood outburst. Gilgit Baltistan may see flash floods as a result of glaciers melting. There are various effects of flooding which are as such;

8 Impact of Flood on Biodiversity

Mountainous ecosystems are vulnerable to climate change and are more fragile (Munoz et al. 2015). Mountainous and coastal environments are where the majority of the world's biological diversity hotspots are found. Because of their harsh

Table 1 Overview of glacial lake outburst floods in the HKH from 1884 to 2012. (Source: Kreutzmann 1994, modified by Iturrizaga 2005, Din et al. 2014)

Year	Date/ season	Glacier/ locality	Loss	Source
1884	–	Khurdopin	Considerable damage to lands at Altit and Ganesh	
1893	July	Khurdopin	Damage to lands at Altit	
1901	–	Shimshal	Breaking of a dam, bridge at Ganesh destroyed	
1904	–	Shimshal	Damage to terraces at Shimshal after emptying of two–year old lake (1902)	
1905	Second August	Khurdopin	Destruction of Chalt Bridge and Gilgit–Chalt road damaged; loss of fields at Passu and Shimshal	
	–	Malungutti/ Khurdopin	Rise of river level of 30 ft. at Bunji	
	–	Khurdopin	Damage of 7 houses at Shamets and to Hunza–Nager bridge at Ganesh	
1906	11/12 August	Shimshal	Damage to bridges at Askurdas, Tashot, and Chamogah Rise of river level in Chilas 36 ft., in Chalt 55 ft. Fields, houses and bridges destroyed at Passu, Hussaini, Gulmit and Ganesh 3 houses, 35 fields and 3 watermills and orchards destroyed Nomal–Chalt road damaged	
1907	Summer	Khurdopin/ Malungutti	Slow drainage in 11 days	
	–	–	Bonfire system (Puberanj) to warn the settlements in the Hunza valley	Local information's
1922	–	Shimshal	Loss of farms	
1923	–	Shimshal	–	Kreutzmann (1994)
1927	June	Khurdopin	Damages to bridges and farms at Shimshal	
1941	–	Shimshal	–	
1944	–	Shimshal	Damage to terraces at Passu	
1957	–	Shimshal	None	
1959	21 August	Shimshal	Damage to bridge between Nagar and Baltit	

(continued)

Table 1 (continued)

Year	Date/ season	Glacier/ locality	Loss	Source
1960–1964	–	Shimshal	Loss of farms and terraces at Shimshal in consecutive years Damage to bridges at Shimshal and Passu Damage to irrigation channel at Nomal	
1976	–	Shimshal	None	
1978	–	Shimshal	Damage to terrace at Passu	
1980	28 June	Khurdopin/ Yukshin Gardan	None	Local information
2000	11 June	Khurdopin Yukshin Gardan	None	
2005	25-july	Sost/Gupis	–	Din et al. (2014)
2007	5-April	Ghulkin	–	Din et al. (2014)
2008	6-January	Ghulkin	–	Din et al. (2014)
2008	2-April	Passu	–	Din et al. (2014)
2008	22-May	Ghulkin	–	Din et al. (2014)
2008	24-May	Ghulkin	–	Din et al. (2014)
2008	14/15-July	Ghulkin	–	Din et al. (2014)
2009	26-March	Ghulkin	–	Din et al. (2014)
2012	08-July	Sost/Gupis	–	Din et al. (2014)

environmental slopes, varied habitats, abundance of water, and biodiversity, mountains are among the world's most fragile ecosystems and one of the most important "experimental areas of nature." Mountains also offer services that have a measurable economic worth. One-fifth of the earth's continental areas are covered by mountain ecosystems, which are referred to be "storehouses of world biodiversity." Due to the fact that these regions are home to numerous globally endangered and threatened species as well as endemic ones, climate change is seen as a serious threat to mountain biodiversity. The Himalayas, Karakorum, and Hindu Kush (HKH) are hotspots with a wealth of species, gene pools, and ecosystems that are significant globally. Threats to Himalayan species are expected as a result of human-caused climate change. Due to their vertical (altitudinal) dimension, these mountain ecosystems are significantly impacted by the climate (ICMOD 2008, 2009). The Himalayas, Karakorum, and Hindu Kush (HKH) are regions with a diversity of species, gene pools, and ecosystems that are noteworthy worldwide. Hazards to Himalayan species are anticipated as a result of human-caused climate change. Due to their vertical (altitudinal) dimension, these mountain ecosystems are adversely affected by the climate. Mountain habitats are habitats to some of the most endangered and indigenous species in the world (Behera 2012). Ecosystems on mountains are more susceptible to little climate change. Due to the fact that a small shift in the climate can cause both snow and ice to melt into water. Many species can cope with

Fig. 6 Damage due to flood

climate change by adapting, changing their geographic ranges, modifying how prevalent they are, or even extinction. As per Convention on the Conservation of European Wildlife and Natural Habitats numerous animal species are compelled to leave their natural habitats as a result of climate change (2010). Flooding is a direct consequence of glacial melting that is exacerbated by global warming. Because wetlands serve as a fundamental life support system for diverse species, their loss is a massive concern. Flooding is a direct consequence of glacial melting that is intensified by global warming. Because wetlands serve as a crucial life support system for diverse biodiversity, their loss is a massive concern (Malik et al. 2010). Climate change is more likely to affect the Himalayas, Karakorum, and Hindu Kush mountains (Khan et al. 2015). According to study, Pakistan's most significant water sources the Karakorum, Himalayas, and Hindu Khush are more impacted by climate change. As a result, more research on alpine ecosystems is urgently needed, as well as management plans and strategies for purposes of restoration. Numerous studies have demonstrated that the mountain biota (tree line) is rising in many mountain systems as a result of global climate change (El-Keblawy 2014).

9 Damage to Agriculture Land, Crop Productivity and Livestock

The links between floods and food security are extremely important, especially in developing nations where the availability of food can be seriously threatened by flood events that have negative direct and indirect effects on agriculture and reduce food availability, access, utility, and stability at the local, national, and international levels (Pacetti et al. 2017) (Fig. 6). Agriculture activity depends heavily on climate,

Fig. 7 Loss of fertile land due to flooding

hence any change in the climate has an impact on plant and animal productivity. During and after flood events, floods can cause damage to food produce as well destroy potential for good crop yields (Shongwe et al. 2014).

Thus, climate change can affect food production through flood directly and indirectly. Soil erosion is a result of all of these landslides and floods brought on by climate change. All of these tree removals and vegetation losses result in a loss of fertile topsoil and subsurface soil, which has an adverse effect on regional agriculture (Fig. 7). Gilgit-Baltistan experienced an increase in soil erosion from 7.54 to 20.25 tons per hectare per year on average in 2005 to 9.06 to 29.69 tons per hectare per year in 2015. The Shigar Valley is most significantly affected (Environmental Issue 2022).

The FAO agricultural assessment report states that the floods destroyed primary infrastructure, including tube wells, water channels, household storage units, houses, animal pens, individual seed stocks, fertilizers, and agricultural machinery, and caused unprecedented scale damages to agriculture crops, livestock, and forestry. Since a few decades ago, hydrometeorological dangers have had a significant impact on agriculture.

In addition to agriculture, crops such as wheat, potatoes, maize, corn, deciduous fruits, nuts, and vegetables also suffer harm, costing the impoverished people a great deal of money. Food crop production in the area is badly impacted by both types of flooding in distinct ways. Although flash floods more frequently have an adverse impact on food crop output, these impacts are frequently much less severe than those of large floods brought on by extended periods of heavy rainfall. Gilgit-Baltistan has experienced eight negative consequences of floods on food crop cultivation, including poor seed development, insect and disease infestation, crop death, soil nutrient loss, crop rotting, stunted crop growth, low yields, and plant felling and washing away. Due to heavy rainfall, wetness, and mouldy weather, crops rot on farms in wet environments with poor sunlight.

Aside from that, flooding in the area is brought on by water spilling which impacts people, agriculture, and livestock. These outcomes reduce food availability and agricultural production in rural markets and communities. Food security is therefore compromised. Flooding occurrences also have an influence on the road infrastructure, which is used to transport food to markets and storage facilities. Food security is put at danger as a result of the physical damage of the region's already insufficient food processing and storage facilities as well as its transportation infrastructure. Livestock is the second most important but high-risk source of income after agriculture. Avalanches and landslides typically cause 75% damage to cow barns. Animal deaths are frequently caused by sickness and a lack of food during floods (Ali et al. 2015a, b).

10 Damages to the Irrigation Channels

In Gilgit-Baltistan, forestry (11%) and livestock (41%) follow agriculture as the top three sources of farm revenue (World Bank 2010). As a result, agriculture is regarded as the principal economic activity and irrigation as the primary component of livelihood security. The basic classification of irrigation in the high mountains is socio-hydrology, which combines the social and natural sciences on multiple scales

Fig. 8 Infrastructure damage due to flooding

and offers a flexible and sophisticated approach to dealing with a number of water dangers (Nüsser 2017). However, the irrigation system faces a number of difficulties. The irrigation system has seen recurrent impacts from threats like floods, glacial lake outburst floods (GLOFs), erosion, and sedimentation, which have decreased its capacity (Fig. 8). On the other hand, the glacier and snow environment has been significantly impacted by climate change, which has had a negative influence on freshwater flows, water supply, and irrigation. For instance, the Hunza valley has experienced regular disruptions to the crucial irrigation infrastructure and decreased irrigation water supply as a result of flash floods, landslides, and sinking of moraine areas brought on by increased glacier melt (Parveen et al. 2015). Furthermore, future predictions indicate that both the intensity and frequency of extreme occurrences will rise (Wijngaard et al. 2017). It would probably endanger livelihoods based on agriculture. 2% or less of the total area is arable, and the majority of the remaining 0.73 hectares is used for subsistence farming. Only 1.2% of Gilgit-Baltistan's land is cultivated, which is less than the total amount of arable land. This is probably a result of irrigation water's poor accessibility and placement, which is substantially higher or lower than the water sources. Gilgit-Baltistan has about 2% of cultivable wasteland that can be planted with crops using sophisticated irrigation techniques (IFAD 2015). However, bed furrow or ridge furrow systems were favored by the villagers over flood and basin irrigation because they were effective in terms of water conservation, worker convenience, and agricultural yield enhancement. The unavailability of modern irrigation techniques like drip and spray irrigation in the community was probably caused by a lack of awareness of their advantages.

Fig. 9 Infrastructure damage due to flooding

11 Damage Infrastructure and Shelters

Destruction of road infrastructure due to flooding was extensive. Recently, flooding destroys Hassanabad Bridge (Fig. 9). Sudden temperature increases and extreme weather in Pakistan's northern regions in the Hunza area of Gilgit Baltistan, Pakistan, a Glacial Lake Outburst Flood (GLOF) has devastated and washed away the strategically significant Hassanabad Bridge on the Karakoram Highway. The Hassanabad Bridge served as a crucial connection between Pakistan's northern regions and the rest of the nation, hence its location was extremely significant. The Karakoram Highway, which links Gilgit Baltistan with China, only had one bridge as its primary source. The Hunza River has a branch stream at Hassanabad that receives water from the Shishpar Glacier, which is about 10 km above the city. It is located between Aliabad and Murtazabad, over a branch of the Hunza River. Moreover, this GLOF episode caused destruction of over 52 homes that have been harmed by this glacier lake outburst-induced erosion. Along with 22 homes, power-houses, and fiber optics in the vicinity, and two hydro-power plants were also impacted by the storm. Shisper Glacier in Hassanabad, Hunza, experienced a glacier lake that burst because the heat in the highlands hastened ice melt. The heat wave was connected to the climate problem by scientists. According to one study, climate change increased the likelihood of a heat wave by 30 times.

References

Ahmed M (2020) Introduction to Modern Climate Change. Andrew E. Dessler: Cambridge University Press, 2011, 252 pp, ISBN-10: 0521173159. Sci Total Environ 734:139397. https://doi.org/10.1016/j.scitotenv.2020.139397

Ahmad F, Kazmi SF, Pervez T (2011) Human response to hydro-meteorological disasters: a case study of the 2010 flash floods in Pakistan. J Geograp Reg Plann 4(9):518–524

Ali S, Li D, Congbin F, Khan F (2015a) Twenty first century climatic and hydrological changes over upper Indus Basin of Himalayan region of Pakistan. Environ Res Lett 10(1):014007

Ali F, Babar K, Garee K, Yawar A, Ejaz H, Ambar M, Karim A, Rizwan K (2015b) Hazard vulnerability risk assessment of district Gilgit, Gilgit-Baltistan, Pakistan. Modern Environ Sci Engineer 1(5):255–268

Ali A, Shukla A, Romshoo SA (2017) Assessing linkages between spatial facies changes and dimensional variations of glaciers in the upper Indus Basin, Western Himalya. Geomorphology 284:115–129

Ali S, Eum H-I, Cho J, Dan L, Khan F, Dairaku K, Shrestha ML, Hwang S, Nasim W, Khan IA (2019) Assessment of climate extremes in future projections downscaled by multiple statistical downscaling methods over Pakistan. Atmos Res 222:114–133

Ashraf A, Naz R, Roohi R (2012) Glacial lake outburst flood hazards in Hindukush, Karakoram and Himalayan ranges of Pakistan: implications and risk analysis. Geomet Nat Hazard Risk 3(2):113–132

Baig SU (2018) Spatio-temporal analysis of glacial ice area distribution of Hunza River basin, Karakoram region of Pakistan. Hydrol Process 32:1491–1501

Baig SU, Muheeb UR, Nirmeen NJ (2021) District-level disaster risk and vulnerability in the Northern mountains of Pakistan. Geomantic, Natural Hazards and Risks, 12

Bajracharya S, Maharjan S, Shrestha F (2011) Glaciers shrinking in Nepal Himalaya. 10.5772/25172

Behera SK (2012) Impact of climate change on mountains ecosystems of India: special reference to the Himalayas. Int Soc Environ Bot, p 18

Berthier E, Arnaud Y, Vincent C, Remy F (2006) Biases of SRTM in high-mountain areas: implications for the monitoring of glacier volume changes. Geophys Res Lett 33(L08502):1–5

Bishop MP, Shroder JF, Hickman BL, Copland L (1998) Scale dependent analysis of satellite imagery for characterization of glacier surfaces in the Karakoram Himalaya. Geomorphology 21:217–232

Chettri N, Shakya B, Sharma E (2008a) Biodiversity conservation in the Kangchenjunga landscape. International Centre for Integrated Mountain Development (ICMOD), Kathmandu, Nepal

Chettri N, Shakya B, Sharma E (2008b) Biodiversity conservation in the Kangchenjunga landscape. International Centre for Integrated Mountain Development (ICMOD), Kathmandu

Convention on the Conservation of European Wildlife and Natural Habitats (2010) Impact of climate change on mountain biodiversity in Europe

Din K, Tariq S, Mahmood A, Rasul G (2014) Temperature and precipitation: GLOF triggering indicators in Gilgit-Baltistan, Pakistan. Pakistan J Meteorol 10(1)

El-Keblawy A (2014) Impact of climate change on biodiversity loss and extinction of endemic plants of arid land mountains. J Biodivers Endanger Species 1:2–20

Environmental Issue (2022) Effect of climate change on Gilgit Baltistan of Pakistan. Environ Sci

Hongyu D (2020) Lake inventory and potentially dangerous glacial lakes in the Nyang Qu Basin of China between 1970 and 2016. J Mt Sci 17:851–870

ICIMOD (2016a) Impact of climate change on the cryosphere, Hydrological Regimes and Glacial Lakes of the Hindukhush Himalayas

ICIMOD (2016b) Impact of climate change on the cryosphere, hydrological regimes and glacial lakes of the Hindukhush Himalayas

Immerzeel WW, Pellicciotti F, Shrestha AB (2012) Glaciers as a proxy to quantify the spatial distribution of precipitation in the Hunza Basin. Mount Res Develop 32(1):30–38

International Centre for Integrated Mountain Development (2009) Mountain biodiversity and climate change. Kathmandu, Nepal

International Fund for Agricultural Development (IFAD) (2015) Economic transformation initiative Gilgit-Baltistan programme, Islamic Republic of Pakistan Design Report

Iturrizaga L (2005) New observations on present and prehistorically glacier–dammed lakes in the Shimshal valley (Karakoram Mountains). J Asian Earth Sci 25(4):545–555

Khan MA, Gul B, Khan H (2015) Impact of climatic change on flora of high altitudes in Pakistan, vol 6. Springer, pp 361–381

Kreutzmann H (1994) Habitat conditions and settlement processes in the Hindu Kush Karakoram. Petermanns Geogr Mitt 138:337–356

Kulkarni AV, Bahuguna IM, Rathore BP, Singh SK, Randhawa SS, Sood RK, Dhar S (2007) Glacial retreat in Himalaya using Indian remote sensing satellite data. Curr Sci 92(1):69–74

Malik M, Imran A, Naseer N, Mahmood R, et al (2010) Wetlands as indicators of climate change. Nature. A quarterly magazine of WWF-Pakistan, vol 34, pp 1–32

Mool PK, Bajracharya SR, Joshi SP (2001) Inventory of glaciers, glacial lakes, and glacial lake Outbust floods: monitoring and early warning systems in the Hindu Khush Himalayaan Regions-Nepal, Kathmandu. ICIMOD. ISBN 92 9115331, 363PP

Munoz M, Faz A, Mermut R (2015) Soil carbon reservoirs at high-altitude ecosystems in the Andean plateau. In: Climate change impacts on high-altitude ecosystems. Springer, pp 135–153

Nawaz MA, Shadie P, Zalaria V (2009) Central Karakorum conservation complex, Draft Management Plan (IUCN), Gilgit-Baltistan

Nogues-Bravo D, Araujo MB, Errea MP, Martinez-Rica JP (2017) Exposure of global mountain systems to climate warming during the 21st century. Glob Environ Chang 17:420–428

Nüsser M (2017) Socio-hydrology: a new perspective on mountain waterscapes at the nexus of natural and social processes. Mt Res Dev 37(4):518–520

Pacetti T, Caporali E, Rulli MC (2017) Floods and food security: a method to estimate the effect of inundation on crops availability. Advan Water Res 110:494–504

Parveen S, Winiger M, Schmidt S, Nüsser M (2015) Irrigation in upper Hunza: evolution of socio-hydrological interactions in the Karakoram, northern Pakistan. Erdkunde 69(1):69–85

Paul F, Huggel C, Kääb A (2004) Combining satellite multispectral image data and a digital elevation model for mapping debris-covered glaciers. Remote Sens Environ 89(4):510–518

Qureshi MA (2017) Glacier status during the period 1973–2014 in the Hunza Basin, Western Karakoram. Quat Int 444:125–136

Racoviteanu AE, Arnaud Y, Williams MW, Ordonez J (2008) Decadal changes in glacier parameters in the cordillera Blanca, Peru, derived from remote sensing. J Glaciol 54(186):499–510

Senese A, Maragno D, Fugazza D, Soncini A, Agata D, Azzoni RS, Minora U, Ul-Hassan R, Khan MA, Rana AS (2018) Inventory of glaciers and glacial lakes of the Central Karakoram National Park (CKNP—Pakistan). J Maps 2018(14):189–198

Shongwe P, Masuku MB, Manyatsi AM (2014) Cost benefit analysis of climate change adaptation strategies on crop production system: a case of mpolonjeni area development programme in Swaziland. Sustain. Agric Res Center Sci Educ 3:5539

Shrestha M, Koike T, Hirabayashi Y, Xue Y, Wang L, Rasul G, Ahmad B (2015) Integrated simulation of snow and glacier melt in water and energy balance-based, distributed hydrological modeling framework at Hunza river basin of Pakistan Karakoram region. J Geophys Res 120(10):4889–4919

Wijngaard RR, Lutz AF, Nepal S, Khanal S, Pradhananga S, Shrestha AB, Immerzeel WW (2017) Future changes in hydro-climatogical extremes in the upper Indus, Ganges, and Brahmaputra River basins. PLoS One 12(12):1–26

World Bank (2010) Pakistan, Gilgit-Baltistan economic report, broadening the transformation. World Bank

WWF (2015) Climate change vulnerability and capacity assessment of Hoper Valley, World Wide Fund for Nature Pakistan, Gilgit

Conservation Agriculture a Sustainable Approach for Disaster Risk Reduction in Rice Wheat Cropping System of Pakistan

Sajid Ali, Adnan Zahid, Ammara Fatima, Memoona Aziz, Hamza Maqsood, Aamir Shezhad, Saqaina Younas, and Bushra

Abstract

Globally, climatic vulnerabilities and natural hazards adversely affect crop productivity and food security. Agriculture is an economic activity that produces food as well as livelihood for human beings. Still, it is highly dependent and vulnerable to climatic and natural hazards like drought, flood, increasing of global and regional temperature, uneven rains, insect infestation, and biological diseases. Rice–wheat cropping system is one of the most important cropping systems composed of 2.2 mha, and many people are linked for their livelihood. The climatic vulnerabilities and hazards such as drought, climate and temperature, reduction in water availability, global warming, smog, soil health reduction, flood, and heat wave. The yield and productivity in this region are badly affected by natural adversaries, which is a source of poverty. It is predicted that the near future food insecurity issue can prevail in Pakistan due to these hazards. Due to climatic and natural hazards, a decline in agricultural productivity affects the farming community, manufacturing, and business sectors through a multiplier effect. Conservation agriculture is the emerging sustainable tool for a reduction in natural and climatic vulnerabilities. Conservation agriculture enhances water productivity, soil health improvement, drought reduction, and overcome the effect of global warming.

S. Ali · A. Zahid (✉) · M. Aziz · H. Maqsood · A. Shezhad · S. Younas · Bushra
Department of Agronomy, Faculty of Agricultural Sciences, University of the Punjab, Lahore, Pakistan
e-mail: adnan.iags@pu.edu.pk

A. Fatima
Department of Environmental Science, Lahore College for Women University, Lahore, Pakistan

M. Ahmed, S. Ahmad (eds.), *Disaster Risk Reduction in Agriculture*, Disaster Resilience and Green Growth, https://doi.org/10.1007/978-981-99-1763-1_15

Keywords

Climatic vulnerabilities · Natural hazards · Food security · Rice–wheat cropping system · Agricultural productivity · Conservation agriculture

1 Wheat Rice Cropping System Significance

The wheat–rice cropping system (RWCS) is among the biggest cropping systems in World. In Asia, RWCS has an extensive history. Since 700 AD in China, this system has been practiced in Pakistani Punjab since 1920 (Chauhan et al. 2012). RWCS in Southeastern Asia has 24 million hectares. There is nearly15.5 million hectares area in South Asian countries, of which India, Pakistan, Bangladesh, and Nepal have 10, 4.25, 0.8, and 0.5 million hectares, respectively (Chauhan et al. 2012). About 1.1 million farm familie's living is dependent on RWCS (Khaliq et al. 2019). The total area under cultivation is 22.54 million ha, out of which 4.25 million ha is under RWCS, accounting about 19% (FAO 2004, 2023). Punjab has a 2.8 million ha area under RWCS, which covers nearly 66% of RWCS and 12% of the total agricultural area of Pakistan. In a nutshell, the total cultivated area of Pakistan is 56.2% in Punjab, similarly, 53% of the total agricultural gross domestic product (Pakistan Bureau of Statistics 2020). Agriculture is the biggest employer in the province, especially in rural areas where it employs 60% of the workforce (GOP 2020). Figure 1 depicted RWCS in Punjab, and Pakistan is mainly prevailing in district Sheikhupura, Gujranwala, Sialkot, Hafizabad, Nankana Sahib, and Narowal (Crop Reporting Services 2020).

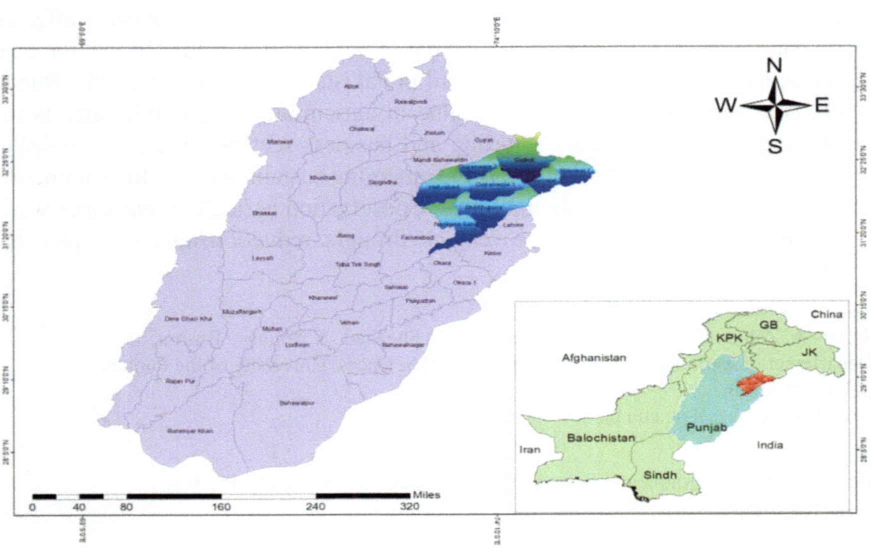

Fig. 1 Rice–wheat cropping system areas in Punjab, Pakistan

2 Disasters and Its Impacts

Globally intensity and incidence of extreme weather and climatic events have augmented due to climate change (Ahmed 2020; IPCC 2018). Agriculture sector provides livelihood to almost half of the population worldwide and is the food source to over seven billion people worldwide (Abdullah et al. 2005; Khan et al. 2015). The agriculture is an economic activity that produces food as well as livelihood for human beings. Still, it is highly dependent and vulnerable to climatic and natural hazards like drought, flood, increasing of global and regional temperature, uneven rains, insect infestation, and biological diseases. Food security at the global and regional levels is under severe threat in future climatic projections (Ahmed et al. 2022; Field et al. 2014; Ndamani and Watanabe 2017).

Variation in precipitation and increased surface temperature due to climatic hazards pose serious consequences to developing countries like Pakistan (Ali and Erenstein 2017; IPCC 2018). Due to scarce resources, less adaptation, and dry topographical background, Pakistan is highly vulnerable to natural calamities (Schilling et al. 2013). In the past two decades, Pakistan has faced a rise in frequency, degree, and harshness of natural and climatic hazards, including floods, droughts, water shortage, extreme temperature, hailstorms, uneven rains, and high existence of pests and diseases (Smit and Skinner 2002; Ali and Erenstein 2017). In weather and climatic risks of 2009–2010, 2010–2011, 2012 and 2015, Pakistan was graded as the 29th, 16th, 12th, and eighth, respectively, among greatest vulnerable countries (Khan and Fee 2014; Abid et al. 2016), and ranked seventh among the highest prone areas of the biosphere (Kreft et al. 2013). In Pakistan, consecutive floods of 2010, 2011, 2012 and 2014 and severe droughts from 1999 to 2003 are common examples of climatic and natural hazards (Abid et al. 2017). Risk awareness, attitude, and adaptation measures are vital factors affecting farm investment, management decisions, and productivity. Farmers seem that severe climatic and weather changes can pose a gigantic risk to the agricultural sector in the forthcoming period (Khan et al. 2015), so their livelihood and food security.

The agriculture sector in Pakistan is under extreme threat to climatic and natural hazards related risks. Major cereal and food crops face a significant decline in yield due to climatic events, which causes a huge loss to food security. In the past few years, rice is a staple food in Pakistan and is extremely susceptible to life-threatening climatic risks (Ahmad et al. 2015; Ali and Erenstein 2017). The production of rice in Pakistan from the past few years has reduced by 20% due to fluctuating patterns of monsoon rains and depleting water resources (Ahmad et al. 2015). Furthermore, the Rice–Wheat cropping zone, along with some other districts of Punjab, is badly affected by recent killer floods. In addition, rice yield could reduce by 26% from 2040to 2070 and 35% from 2071 to 2100, if the temperature continue rising in Pakistan (Ahmad et al. 2015). Likewise, rice production would decline by 6% with decreasing average precipitation, and it will increase 29% net irrigation water requirement (Ali and Erenstein 2017).

The farming communities' food and livelihood sources in Pakistan largely depend on agriculture, so it is essential to provide knowledge about climatic and natural hazards to concerned stakeholders. Understanding climatic hazards and their impacts on agricultural production system are fundamental in smart decisions-making process and hence developing effective management strategies to cope with these calamities and hazard. Strategies to manage risk like adaptation and extenuation help the growers to minimize or avoid the possible damages to their final yield (Frank et al. 2010; Moghariya and Smardon 2014). Farmers' livelihood, directly or indirectly linked with climatic and natural hazards, affects their intention to the perceived risks. Risk attitude responds to uncertain conditions (Gattig and Hendrickx 2007).

Risk perception and risk attitude are usually used in decision making and policy planning regarding climatic vulnerabilities. Under risk conditions, the attitude of the farmers determines their choice to realize the risk on the production of crops. Likewise, the perception of risk influences their aims to retort the professed costs, which eventually leads to actual management of risk (Sjoberg et al. 2004).

3 Disasters in Rice–Wheat Cropping System

Many farming families are facing severe loss each year by crises and disasters. The impact of a disaster upon various cropping systems determines by its frequency, nature, duration, and intensity. On a global scale, all these disasters include climatological, biological, hydrological, and geographical are causing economic loss of about 175 billion US$ per year during past decade, especially in 2011 and 2017, when loss reaches upto US$ 300 billion (FAO 2021). Natural disasters are becoming more severe since 1980 and hitting almost every continent on earth. The overall loss caused by these disasters to farming is much higher.

Rice–wheat cropping system (RWCS) of our region (South Asia) is facing water and labour shortage and turning less profitable as these resources are reduces. During past decade extreme weather and climate events are observed which causes huge loss to farming systems, given their major dependence on weather and climate.

3.1 Unpredictable Risks of Climate and Weather

The Rice–Wheat cropping zone of Punjab, Pakistan, is vulnerable to unpredictable risks related to climate and weather. Many studies (Bryan et al. 2013; Harvey et al. 2014; Abid et al. 2016; Akhtar et al. 2019) have elaborated the possible threat of climate change to the agriculture sector. The socioecological landscape of Pakistan is under serious threat of climatic hazards (Ali and Erenstein 2017). Goals of accomplishing food security and poverty reduction are jeopardized with the current weather and climatic abnormalities (Ali and Erenstein 2017). Many studies revealed that uneven rainfall in RWCS resulted in the decline in yield and production. Khan et al. (2020a), who reported that when the rice cultivation starts in the RWCS, the monsoon season mostly starts late. At this time there is a need for irrigation water and farmers rely on this seasonal rain. So, wheat and rice production may fall due to uneven and short rainfall; as a result, availability as well as rural farm income and food security decreases. Time series data of rainfall and temperature reported by the Pakistan metrological department also revealed a decline in rainfall and an increase in temperature in the RWCS. Crop health and yield have significantly deteriorated with biological risks like weed germination, pest and insect attacks, and crop diseases. In Pakistan, the crop sector, mainly cereal production, is victim to many biological risks (Usman et al. 2012; Khaliq et al. 2019; Khan et al. 2020b).

3.2 Drought Risk

In Pakistan, every four out of 10 years, a drought was observed, which had a catastrophic effect on all sectors and, so, on people's livelihoods and food security (Durrani et al. 2021). For the past two decades, drought has been prevalent, and farmers have perceived it consistently. Economic dependence on crops and livestock, shrinking surface water resources, reduction of groundwater, and inadequate supply of electricity have further augmented their vulnerability to drought. Socioeconomic impacts of drought are loss of employment and reduced the production of crops and livestock. An increase in social crimes, drop out of schoolchildren, migration to other places, and impacts on health and festivals are social impacts of the drought. Decrease in intensity of rainfall, rising temperature, and nonclimatic factors are environmental impacts due to the prevalence of drought. Farmers' understanding of risk perception about drought may help develop policy making to design suitable intervention strategies for its mitigation (Durrani et al. 2021). Natural hazards like drought, floods, rainfall fluctuations, and temperature are also causing major risk to the food security at local, national, and regional level (Privara and Privarova 2019). Flood hazards increased food prices due to disequilibrium of demand and supply of food, and abridged crop production is negatively affecting low-income and poor households' food access. Food prices are further rising due to regional as well as global population expansion (Abbas et al. 2017; Ahmad and Afzal 2019). As global warming patterns of rainfall and temperature are changing, so, disturbing crop production (Abbas and Mayo 2021). Pakistan has been facing extreme floods for the past two decades due to natural and climatic vulnerabilities which are declining productivity of food crops and deteriorating rural livelihood (Ahmad et al. 2020).

3.3 Risk of Global Warming

There are many studies that investigated the impact of rainfall and temperature on crop yield by applying input-output production function (Brown and Rosenberg 1997; Yuan 2011; Mahmood et al. 2012; Zhang et al. 2015; Abbas et al. 2016). The study of Lobell and Asner (2003) used Cobb–Douglas type production function to analyse climate change impact on the yield of wheat-rice crops because of global warming in Australia. Rice yield was reduced due to intensification in minimum temperature. Overall, temperature and rainfall patterns severely affected the crops yield. Efficient and effective use of agricultural resources to gain the best and highest output is called productivity. Farmers are facing frustration, social and economic losses due to unfavourable environmental alterations. The severe climatic alterations badly affected the productivity of cash and food crops in Pakistan that have caused a huge loss to community, food and raw material availability (Ahmad and Afzal 2020a). Moreover, Pakistan's agricultural productivity and rural livelihood in Pakistan have been adversely influenced due to regional and global climatic dynamics during the last 20 years. Farm losses at the farm level can be abridged by implementing timely and effective climate change adaptation measures (Ahmad and Afzal 2020b). Global warming is altering the pattern of temperature and rainfall, which significantly influences the growing phases of crops (Sridevi and Chellamuthu 2015). Parry et al. (2013) stated that discrepancy in rainfall and temperature has a negative effect on rice development stages subsequently, production of rice declined. Quantity of plants and degree of tillering in rice are badly affected by high maximum and minimum temperatures, decreasing the crops yield at critical tillering and stem elongation stages (Mahbubal Aslam et al. 1985). Furthermore, rice production decreases due to maximum temperature at the propagative stage resulting abridged crop duration (Dabi and Khanna 2018). In Punjab Mahmood et al. (2012) stated that rice productivity is negatively affected by the increase in rainfall pattern during the propagative and maturing stages.

3.4 Reduction in Water Availability

A large amount of water is used to maintain flooding in rice fields. Rice grown by traditional practices requires approximately 1800 mmH_2O during a complete rice season. Moreover, around 50 mm water is needed to grow nursery before transplantation. The actual water that is being applied by farmers, however, is far more than actual crop requirement, particularly where rice is grown on light-textured soils in India and Pakistan (Timsina and Connor 2001).

Availability of sufficient irrigation water has made the RW system in the NW IGP a typical example of a highly productive system in non-ideal rice soils (porous, coarse, and highly permeable). Four decades of RW cropping have depleted the water resources in this region to a great extent. In the Indian Punjab alone, there is an annual shortage of about 1.2 M ha meters of water (Hira 2009). The excess demand for water is being met through overexploitation of groundwater, leading to a decline in the water table.

3.5 Risk of Soil Health

Crop production removes various essential nutrients from the surface of soil, depending on production and nutrient-supplying capacity of the soil, which in turn is influenced by soil type, soil organic matter content, amount of nutrients applied, and removal or recycling of crop residues in the soil. The rice wheat cropping system has not only mined few major nutrients (N, P, K, and S) but also has caused a nutrient imbalance in the soil, further causing soil quality degradation. Soils are mostly deficient in nitrogen, phosphorus, and potassium are most extensive. It is about 25, 3.9, and 27.5 kg of N, P, and K have been removed from soil for producing one ton wheat. Similarly, rice grains production depletes about 20.2 kg nitrogen, 4.8 kg phosphorus, and 25.5 kg potassium from soil. Low soil organic matter is present in the IGP and are being consistently eliminating of their finite reserve of nutrients by field crops (Yadvinder-Singh and Bijay-Singh 2004). One of the prominent causes of nutrient shortage from soil is their mining under rice wheat farming. More nutrients are mined from soil with cropping than their addition through fertilizers. Compared with wheat rice required more quantities of micronutrients. Zinc deficiency has become prominent in India and Pakistan, and it has become the third most limiting nutrient after N and P, particularly in soils with high pH and those irrigated with poor-quality water.

Nutrients are mostly lost through leaching, emit as gases and fixed by the soil, these are the common reasons for their low efficiency. These loses play their part in soil degradation and deteriorate water quality which ultimately affects environment. There is a need to tackle the emerging concerns of soil organic matter degradation and reduced nutrient supplying capacity of the soils in the RW system. Inclusion of short duration legumes between wheat and rice, balanced application of nutrients, returning rice crop residues to the soil after harvest, and the possibility of introducing some microbes for fast decomposition of residues to facilitate wheat sowing may help to restore soil fertility.

3.6 Disaster by Crop Residues Burning

Rice and wheat residues which are totalled 36 M tons, out of these 36 M tons about 15 M tons residues are in situ burned annually, causing a 5 M tons loss of C equivalent to a CO_2 load of about 18.3 M tons per year and a loss of 69,000 tons of N in Indian Punjab. One ton of wheat residue contains 4.8 kg N, 0.7 kg P, and 9.8 kg K, whereas 1 ton of rice residues contains 6.1 kg N, 0.8 kg P, and 11.4 kg K. Burning of rice straw causes gaseous emission of 70% CO_2, 7% CO, 0.66% CH_4, and 2.09% N_2O (Sharma 1998).

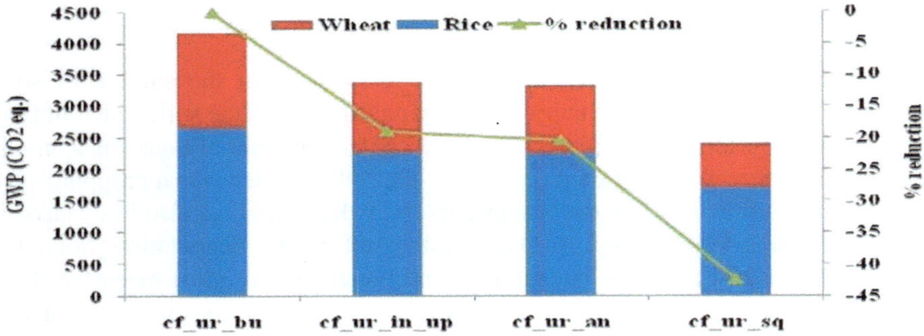

Fig. 2 Rice–wheat cropping system global warming potential under various management practices. (**ncf_ur_bu**: continuous flooding, urea application and straw burnt; **cf_ur_in_up:** continuous flooding, urea application and straw incorporated in upland crop; **cf_ur_an**: continuous flooding, urea application and straw fed to cattle; **cf_ur_sq**: continuous flooding, urea application and straw sequestered as construction material)

3.7 Smog

Crop residue burning produced smoke which cause air pollution with harmful gases, which can clog the lungs and cause breathing problems. Asthmatic people have great difficulty in breathing under these conditions. The peak in asthmatic patients in Indian hospitals are according to the residue burning estimates (Yadvinder-Singh and Bijay-Singh 2004). Smoke particles can remain in the atmosphere up to several weeks. Smoke can also create haze that impairs visibility. Thus, burning of crop residues not only results in a loss of organic matter and nutrients but also causes atmospheric pollution due to the emission of toxic and greenhouse gases such as CO, CO_2, and CH_4 (methane) that pose a threat to human and ecosystem health (Fig. 2).

3.8 Floods Disaster

Sustainability of wheat–rice cropping system is at risk in Southeast Asia as far as food security of continuously increasing population is concerned. Atmosphere warming of this region is already reported in studies (Prabhjyot-Kaur and Hundal 2001) and various climate models also predicted that (IPCC 2001). Under these conditions, more respiration than photosynthesis in field crops causing yield reduction, less assimilate production of water use, less dry matter production, more spikelet sterility, and reduction in crop sink capacity at its grain filling causing low crop yields (Surinder et al. 2012; Tohru and Naoya 2004).

In current year 2022, Pakistan has faced more than 60% rainfall in just 3 weeks of the entire normal monsoon season. Natural calamities like floods, glacial lake outbursts and landslides have occurred as a result of these heavy rainfalls affecting

Baluchistan, South Punjab, KPK, and Sindh provinces of Pakistan. Previously, The floods in 2010 have increased poverty and badly affected the poor population in flood affected areas. These areas which are affected severely by floods are still constantly suffering from educational and socioeconomic hazads and lagging behind those areas which are not affected by floods. Infrastructure breakage and loss of livelihood sources worked as icing on cake. Mostly, small holders and unskilled labourers are badly affected by these floods. They are livings already below poverty line and are among the most vulnerable communities. Agriculture is the sector which is affected the most by these floods and other natural hazards. Reconstructions and damages cost in the flood affected areas is estimated 430 billion rupees according to *Pakistan Flood Assessment Report 2010.* Most of the major crops specially rice, cotton and sugarcane were ready to harvest. Again in 2013, major provinces of Pakistan, Punjab, Sindh and Baluchistan cropped area was badly affected by floods and it is estimated that 0.42 million hectares area was faced this challenge. Likewise in 2014, floods affected only Punjab crops area that is estimated 0.98 million hectares. One million acres of land with standing crops were destroyed, according to National Disaster Management Authority (NDMA). Major destruction was reported in Muzaffargarh, Jhang, Sargodha and Multan Districts. SPARCO reports estimated around 218,000 tons rice production was lost as a result of 2014 floods. Besides this, loss of agricultural tools, seed stocks and land erosion are also major destruction caused by heavy floods.

3.9 Heat Wave Disaster

Agriculture sector has faced a new major challenge which is global warming and alleviating its damaging effects is now major talk in the town. Amazon fires are causing damage each year in the months of June to November. Number of blazes reached highest in 9 years during 2019 and 1.9 gigatons of carbons emissions are estimated which raises a global crises. Which caused long wave solar radiations to strike directly to ground without any difficulty. These long waves changes into shoot wave lengths after hitting the ground and greenhouse gases like CH_4, CO_2, N_2O absorbs and hinders these short wave length radiations to go out of the earth environment which results in the enhancement of the global temperature known as greenhouse effect. RWC systems produce heavy crop left outs and their burning is the common practice among farmers for timely sowing of the successive crop. Flames caused by these burnings results into the huge production of greenhouse gases, hydrocarbons, and aerosols which are causing major damage to the atmospheric gases composition. For instance, 69%, 6.9%, 0.68% of carbon and 2.09% of N emitted as carbon dioxide, carbon monooxide, methane, and nitrous oxide upon these rice straw burning (Niveta et al. 2014). This negative practice might have caused radiation imbalance directly or indirectly. These effects become so severe that they could result into the globally or regionally increase in aerosols, acid deposition, tropospheric ozone levels, and depleted ozone layer which give us protection

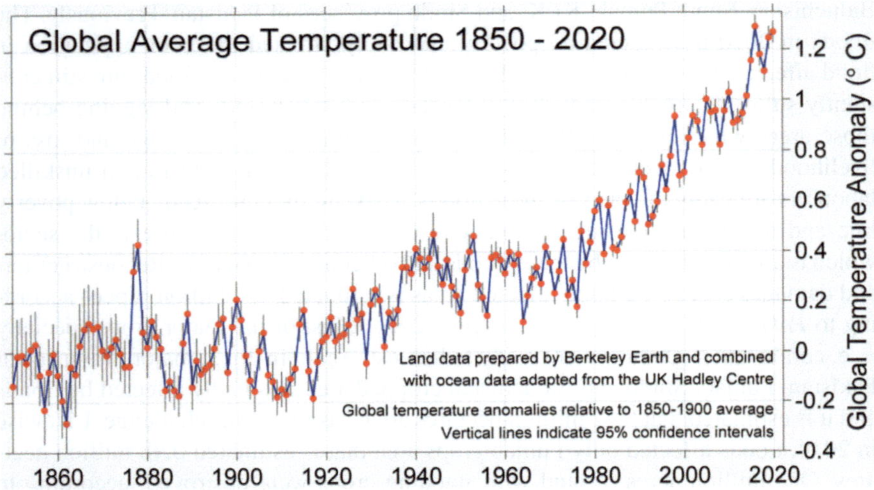

Fig. 3 Global temperature anomalies from 1850 to 1900

from damaging sunrays. Hence, various sustainable ways must be adopted on a large scale to control the production of these harmful gases and one of these techniques is drilling of wheat seeds directly into rice stubbles without burning those. Furthermore, global warming is reducing wheat crop yield as reported by Express tribune (June 6, 2022). The effects of these Climatic alterations have started to appear in the farming system of Pakistan. Current year 2022, wheat crop is badly affected by these sudden changes in the temperature due to climate change. Wheat grain filling period reduced to 4–5 days due to these temperature changes which is normally continues for 10–12 days, and the reduction in weight and size of wheat grain occurred due to these changes. It is expected that the important wheat crop will face 4–6% yield loss this year (Fig. 3).

4 Farmer's Perception Regarding Climatic and Natural Disasters

The study of Usman et al. (2022) conducted a survey on farmers' perception about the impact of climactic and natural vulnerabilities on the food production in the Rice–Wheat cropping system and resulted indicated that farmers are well aware of the possible impacts of these events on crops, income, and ultimately food security. There was the highest impact of the flood on food crops production followed by hailstorms, biological diseases, insect infestations, wind storms in summer, extreme heat heatwaves, early rains, late rains, extreme cold, wind storms in winter, droughts, typhon, smog, fog, too much rains, frost, and humidity.

There was a positive but insignificant impact of drought on the profitability of rice–wheat cropping system (RWCS). The variable of the flood was negatively and significantly affecting the food production. Pakistan has faced severe floods during the past two decades. These floods had affected a large area under food crops. Heavy rains are beneficial for rice crop; however, it affects the wheat crop adversely. The effect of heavy rains was found to be positive and insignificant. Uneven rains have negative effects on the production of food crops. Early and late rains have negative and significant effects. In Pakistan food crops, particularly wheat, has been badly affected at the harvesting stage due to rain. Furthermore, extreme hot and extreme cold variables have negative and insignificant impacts on the dependent variable. They significantly decline food production. Extreme cold, wind storm in winter, fog, smog, and typhon have a negative but insignificant impact on the outcome variables.

Extreme heat was affecting crop production negatively. It causes more evaporation, and thus, plants require more irrigation which accelerates the production cost. Similarly, heat waves and wind storm in winter has a significant and negative impact on the productivity of food crops. It has been experienced that hailstorms ruined the food crops brutally in Pakistan. Wheat and rice production are declined with hailstorms, particularly when these crops are at ripening and harvesting stages. The attack of insects and recently locust breakdown has brutally affected the Pakistan's agriculture and food security. Insect infestation was found to affect crop production negatively and significantly. In Pakistan, the agriculture sector is the main victim of biological diseases and a huge amount of food production is prone to these diseases (Fig. 4).

5 Conservation Agriculture as a Disaster Risk Reduction in Rice-Wheat Cropping System

5.1 History of No-Till/CA Farming in South Asia

Deforestation for building housings by urbanization and degradation of fertile agricultural land for other uses resulted in deterioration of soil and water quality which affects the natural habitats. Moreover, the traditional farming practices, for example, uncontrol/intensive tillage/ploughing and burning of crop stubbles in the field, have caused loss of soil organic carbon as carbon dioxide which results into soil erosion and other problems associated with it. Thus, a modern agricultural system known as conservation agriculture (CA) which is a sustainable alternate system of land-usage that minimizes the disasters effects of calamities and enhances social balance, stabilizes economic growth, protects soil health, cause environmental conservation, and alleviates climatic alterations. These common things are crucial for sustainable farming and improvement of the long-term living standards of small landholders' and their poverty alleviation. In the past, conservation agriculture in south Asia commences with directly wheat sowing in the rice stubles in Indian and Pakistani Punjab in the 1980s decade (Hafeez-Ur-Rehman et al. 2019). During 1990s, Coservation agriculture program is started by CIMMYT in south Asian states viz. Pakistan, India, Bangladesh, Pakistan, Nepal, China, and Afghanistan.

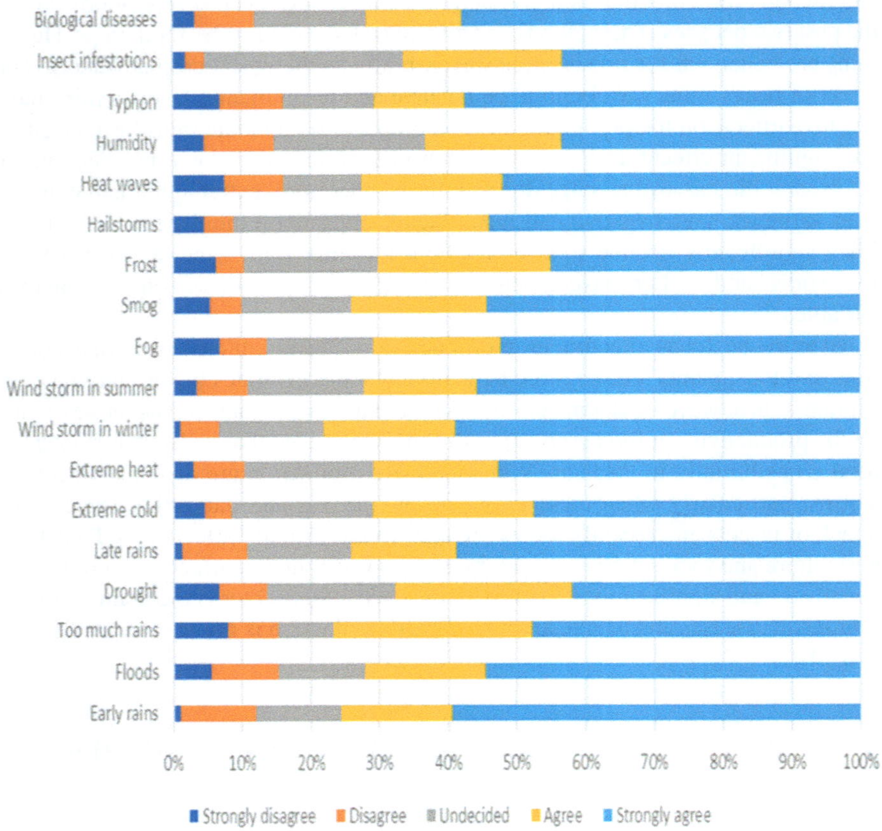

Fig. 4 Farmers' perceptions regarding natural hazards. (Source: Usman et al. 2022)

Latterly, to cope with the rice–wheat (RW) farming practices in the Himalayan hills and Indo-Gangetic Plains, Wheat-Rice Consortium (WRC) was inaugurated by CGIAR in mid 1990s. Similarly, the Conservation Agriculture Network for Southeast Asia (CANSEA) was established in 2009 to gather similarities among south Asian institutions and countries to enhance the research activities performed by the different programs, and to gather resources and means (Somasundaram et al. 2020). In south Asia, on about 5.5 million hectares, wheat is cultivated under zero tillage because various implements are now under practice and farmers become more aware of wheat cultivation under this system. Traditional tillage is commonly practiced among farmers for cultivation of other crops. The introductory steps taken by Professional Alliance for Conservation Agriculture (PACA) are developing common practices of conservation agriculture. It has enhanced more interest among farmers for higher adoption of conservation agriculture techniques. Later on, Indian Council of Agriculture Research has initiated Consortia-Research Platform on Conservation Agriculture to practicing best conservation agriculture technologies by focusing mainly on major field crops under distinctive ecologies, viz. started from rainfed to entire irrigated systems across India (Annual Report 2016–17 of CRP on CA). Various challenges in the implementation of conservation agriculture techniques on a huge scale are also present, such as weed control, nitrogen management, residue management, farmers' mind-set, climate change, socio-economic issues, policy interventions, irrigation, and less incentives for carbon credits.

5.2 Spread of Conservation Agriculture in South Asia

In the Rice–Wheat cropping system of India (325.1 Mha of land with area under cultivation is 142 Mha), the no tillage practices are adopted by farming community mostly in the rice-wheat cropping system (RWCS) (<5 Mha), and also in rainfed areas where pigeon pea, maize, and cotton are mostly cultivated (Kassam et al. 2009). Similarly in Bangladesh, in 2015 conservation agriculture practices (minimum tillage practices) were adopted on around 27,615 hectare area (Poddar et al. 2017). Resource conservation technologies which are based on conservation agriculture are being successfully practiced over almost 3.95 Million hactare in south Asia, reported earlier. Thus, Kassam et al. (2019) found that area under conservation agriculture in Pakistan, India, and Bangladesh is 0.6 Mha, 1.5 Mha, and 0.0015 Mha, respectively; the valid information about total area where conservation agriculture is being practiced in other South Asian countries' is lacking. Concluded that no clear information is available about the spread of conservation agriculture in Asian region. However, owing to its intangible and tangible benefits, the adoption of conservation agriculture in Asia has enhanced since 2000 (Erenstein and Farooq 2007). Similarly, in Central Asian states, development of conservation agriculture has been occurred between 2003 and 2018 for almost 5 years particularly in Kazakhstan where conservation agriculture is being practiced on almost 10.45 Million hactares under minimum tillage, especially in dry northern areas. 10% of of this, 1.6 Million hactares conservation agriculture area was considered conservation agriculture with continuous no tillage that made Kazakhstan prominent with highest percentage of cropped area occupied by conservation agriculture and placed it within top ten countries worldwide. Similar to this, China started conservation agriculture at the end 1990s (i.e., 1.35 Mha) and expanded it continuously (Wani et al. 2016).

6 Conservation Agriculture Role as a Disaster Risk Reduction

6.1 Improvement in Water Use and Efficiency

Conservation agriculture enhance H_2O productivity by improving soil water holding capacity, increasing infiltration, decreasing soil evaporation for later stomatal transpiration (Dillaha et al. 2010). He et al. (2009) reported that more ultimate infiltration rate in the field under no tillage was depends on residue retention on the surface, less disturbance to the water holding pores and improves aggregate stability. Das et al. (2018) found that the permanent narrow bed with residue and permanent broad bed with residue in a conservation agriculture-based wheat-maize farming system resulted in higher water-use efficiency and accumulation of more carbon in soil with higher sequestration potential, besides giving sustainable production over the years Sapkota et al. (2015) reported that in rice–wheat production system, direct-seeded rice with previous wheat residue retention led to 52% more water saving than puddled traditional rice. They investigated that increased water productivity in zero tillage-based production systems was mainly due to requirement of less water to produce same or even more than conservation tillage-based production systems and one presowing irrigation for wheat germination could be saved under zero tillage by residual moisture retention after harvest of rice. Generally, 22–26% conserving in irrigation water can be achieved from the zero tillage practice for wheat. Bhale and Wanjari (2009) found an improvement in water content available under conservation tillage, especially in the surface horizon due to improved soil structure and crop residue mulch, which increased more use of water by crops and enhanced WUE.

Weed management is a major concern in adoption of conservation agriculture, despite having its excellent effects on soil conservation, it gained little fame among farmers worldwide. Weed management is major issue for those farmers who are ready to adopt this technique. Most of the issues in conservation agriculture can be resolved and have no serious validation for well managed conservation agriculture techniques. These issues should be anticipated, especially for the early years, until the soil seed bank produced during those years when tillage was performed.

6.2 Improvement in Soil Health

Physical properties of soil also got favor from conservation agriculture practices. Root growth of crops, water holding capacity, soil moisture contents, MWD, AFP, hydraulic conductivity, to reduce soil bulk density, metric flux potential, and regulation of soil temperature can be properly managed by the crop residues usage with zero tillage. Moreover, the investigation of long-term conservation agriculture practices is strongly recommended to various agro-ecologies for their long-scale implementation in the north eastern areas of India.

Impact of various conservation agriculture techniques on soil carbon and soil organic matter was confirmed from various studies (Judy et al. 2002). Improvement in soil organic carbon, from the implementation of conservation agriculture practices, no or reduced tillage was found (Beukes and Swanepoel 2017), covering soil with mulch (Muzangwa 2016) and multiple cropping (Arnold et al. 2016). Minimum or less tillage had the many positive effects on improved soil organic matter or soil orgaic carbon (Esmeraldo 2017) as comparison to some other conservation agriculture practices. Although with some positive effects of conservation agriculture treatments on soil organic carbon, few studies revealed that little improvements

(Cheesman et al. 2016), low (du Preez et al. 2011), and having no such importance as compared to traditional systems (Njaimwe et al. 2016).

Although there are many studies are available on the role of conservation agriculture on nutreient status and fertility of soil ($n = 13$), positive role of conservation agriculture on nutrient status of soil was found by few of these studies. Loke et al. (2012) and Dube et al. (2012a, b) investigated improved build-up of sulfur (S), phosphorus (P), or zinc (Zn) in soil layers. Similar to this, increased soil organic carbon also improved the availability of some other nutrients, like as potassium (K), calcium (Ca), and phosphorus (P) (Graham et al. 2002). Conservation agriculture also enhanced the soil nitrogen contents (Loke et al. 2012; Swanepoel et al. 2014). However, soil nitrogen contents were not altered by changing tillage practices (Maali and Agenbach 2003), Legume-based cropping systems are especially positively affect the nitrogen contents (Maali and Agenbag 2004; Swanepoel et al. 2014). No till soil was positively correlated with improved soil fertility (Kutu 2012), retention of previous crop residues (Muzangwa 2016) and conservation agriculture practices (Wiltshire and du Preez 1993). These alterations not always have impact on yields significantly (Loke et al. 2012; Nciizah et al. 2015a, b). Soil acidification enhancement was liked with the more soil organic carbon (Graham et al. 2002; Sosibo et al. 2017). It is reported that soil pH was also increased with conservation agriculture practices like no tillage and stubble management (Loke et al. 2012).

Soil strength which is measured as resistance of penetrometer (PR), soil aggregate stability and soil compaction was also reported in several studies. Reduced tillage increased the penetrometer resistance of soil (Swanepoel et al. 2014). These alterations mostly happened in the soil upper layer as they are related with compaction (Swanepoel et al. 2014). Similar to these studies some other researchers also measured an increase in soil bulk density under no tillage soils that could be due to the absence of ploughing or tillage pan that remove the compaction from top soil layer. Low soil organic carbon or biomass production was observed from these compacted layers (Lampurlanés and Cantero-Martínez 2003). Increased soil strength also cause reduction of roots elongation, which ultimately hinders crop production and overall yield, thus have negative effect on the success of conservation agriculture (Swanepoel et al. 2014). Aggregate stability improvement of soil usually linked with the capacity of holding water, water infiltration and soil erosion reduction (Lampurlanés and Cantero-Martínez 2003). A study revealed that oat cause reduction in aggregate stability and grazing vetch improved it (Mupambwa and Wakindiki 2012). In some other studies, with cover crops cultivation and under no or reduced tillage increased aggregate stability was happened (Kidson 2014; Esmeraldo 2017; Njaimwe et al. 2016).

Under no tillage with intercropping system water holding capacity of soil and water that is available for plant usage also increased (Mupambwa and Wakindiki 2012; Mzezewa et al. 2011). Similar to this, Knot (2014) found that cultivation of cover crops also improved water infiltration and reduced water runoff (Bennie and Hensley 2001; Tesfhuney et al. 2013) and also reduced soil loss (Mchunu et al. 2011).

Soil organic matter supported a diversity of organisms in soil as beneficial to soil biology (Graham and Haynes 2006). Tillage practices in soil or application of chemicals have negatively impact the soil biology (Badenhorst 2016). These effects could be harmful for a whole year or showed their impact immediately (Swanepoel et al. 2015). Soil organic matter also cause biological changes. Conservation agriculture practices that affected soil organic matter, latterly also impacted biology of soil (Mukumbareza et al. 2015). For instance, no tillage increased glomalin levels, abundance of earthworms as well as nematode population in soil as it improved soil organic matter (Koch 2017; Mcinga et al. 2017). Farming systems, particularly those with residue retention and cover crops, resulted in more enzyme activity and microbial biomass (Mukumbareza et al. 2015). However, not all changes were positive in the biology of soil.

6.3 Land Productivity Improvement

Improving soil fertility and conservation of water are the major benefits of conservation agriculture. The conservation agriculture principles not only improve soil quality and enhance supply of water but it should also altered soil physical, biological, and chemical processes and soil properties. Under climate change scenario, to maintain suitable productivity and achieving yield goals these improvements are fundamental. Several studies investigated both minimum tillage and no-tillage by altering biological activities, physical characteristics of topsoil layer (Dennis et al. 1994; Balabane et al. 2005; Riley et al. 2005). Reduced tillage helps in accumulation of soil organic matter on top layer of soil and vertical stratification of carbon (Hernanz et al. 2002; Moreno et al. 2006), it results in enhanced fertility of soil and cause greater production. Furthermore, reduced tillage and no tillage can also control soil erosion that could be a great benefit for higher crop productivity. Tebrügge and Düring (1999) and Hernanz et al. (2002) found that these treatments could help in improving the soil stability and reduce surface sealing of soil. Soil fertility and ultimately soil productivity improves by these changes.

6.4 Drought Reduction Through CA

"Green water," which is known as water present in the soil, is required for more than 80% cultivated farming land for its proper production. In rainfed agriculture, cultivation depends only on green water. The major sources for food production in the near future will be of Green water. Blue water sources are gaining attention, but 15% of total rainfall is used wisely for crop production. In future, it might be compulsory to enhance the productivity of green water resources because water bodies either present on surface or under ground are limited and it will be hard to improve the area under irrigation.

After wheat, Rice is one of the most important food crops worldwide and provides calories to about half of the human population. It is a good source of 15% of protein, 27% of dietary energy, and 3% of fat to human population. It is cultivated under different ecologies from rainfed uplands to deep water on an area is about 167 million hectares yielding 770 million tons and averaging 4.60 tons/ha in the world (FAO 2023). Among abiotic stresses drought is the most damaging stress which negatively affected 23 million hectares of rainfed rice only in Southeast Asia. In rainfed ecologies, drought has been recognized as the major cause of low rice productivity.

Among the principles of conservation agriculture, permanent soil organic cover is a major principle. Intensive tillage can reduce the water infiltration in soil profile because of surface crust and tillage pan formation, but the residues present on the surface of soil can improve water storage (Lampurlanés and Cantero-Martínez 2003). Furthermore, overland flow velocity can also be reduced by the cultivation of cover crops, so improved soil structure (Susama et al. 2008). The drought impacts could be reduced by increasing the effect of rainfall as the increased water availability and improved soil structure improved biopores in the soil profile (Ben Moussa-Machraoui et al. 2010). Thus, in dry areas, for reducing yield losses due to climate change, permanently covered soil with organic matter is an excellent approach.

Conservation agriculture also has prominent effect on soil evaporation. Tillage cause more dryness as it moves wet soil into the outer most layer (Hatfield et al. 2001). Manipulation of soil surface layer through tillage, improved soil moisture evaporation more than those areas which are untilled. The soil surface receives energy is mainly influenced by residue cover and canopy. Greb (1966) determined that crop residues minimizes evaporation of water from soil by decreasing soil temperature, more absorbance of water vapors into residues, residue thickness has greater significance for reducing evaporation of water. Evaporation can be reduced by 34–51% by the presence of residues on the soil surface (Sauer et al. 1996). Atmospheric evaporative potential also helps in the calculating rate of evaporation, in addition to thickness of residues (Tolk et al. 1999).

More soil moisture is retained under conservation agriculture system and more water is available for crop cultivation which is a beneficial effect of CA. Azooz and Arshad (1996) determined in British Columbia that greater water contents in the soil under minimum tillage compared with traditional mouldboard plough. Mulching/residues retention helped in conserving moisture of soil in a long dry season without rainfall at various experimental sites of Zimbabwe (Mupangwa et al. 2007). Traditional low yielding varieties also increased the impact of drought stress as losses enhanced many folds. Cultivation of high yielding rice cultivars which are drought tolerant, play pro-active and important role in ensuring food security among farming communities in the areas affected by drought.

References

Abbas S, Mayo ZA (2021) Impact of temperature and rainfall on rice production in Punjab, Pakistan. Environ Dev Sustain 23:1706–1728

Abbas S, Khalida K, Ali Z (2016) Green economic growth: an opportunity for sustainability and poverty alleviation, HKH, Pakistan. Sci Int 28:3715–3720

Abbas G, Ahmad S, Ahmad A, Nasim W, Fatima Z, Hussain S (2017) Quantification the impacts of climate change and crop management on phenology of maize based cropping system in Punjab, Pakistan. Agric For Meteorol 247:42–55

Abdullah A, Gillani W, Naveed S, Amanullah K, Kashif H (2005) Computerized farm guide: using ICT for better dissemination of agriculture extension information. Proc Int Symp Intel Info Tech Agric 3:14–16

Abid M, Scheffran J, Schneider UA, Ashfaq M (2016) Farmers' perceptions of and adaptation strategies to climate change and their determinants: the case of Punjab province, Pakistan. Earth Syst Dyn 6:225–243

Abid M, Schilling J, Scheffran J, Zulfiqar F (2017) Climate change vulnerability, adaptation and risk perceptions at farm level in Punjab, Pakistan. Sci Total Environ 547:447–460

Ahmad D, Afzal M (2019) Household vulnerability and resilience in flood hazards from disaster-prone areas of Punjab, Pakistan. Nat Hazards 99:337–354

Ahmed M (2020) Introduction to modern climate change. Andrew E. Dessler: Cambridge University Press, 2011, 252 pp, ISBN-10: 0521173159. Sci Total Environ 734:139397. https://doi.org/10.1016/j.scitotenv.2020.139397

Ahmad D, Afzal M (2020a) Flood hazards and factors influencing house-hold flood perception and mitigation strategies in Pakistan. Environ Sci Pollut Res 27:1–13

Ahmad D, Afzal M (2020b) Climate change adaptation impact on cash crop productivity and income in Punjab province of Pakistan. Environ Sci Pollut Res 27:30767–30777

Ahmad A, Ashfaq M, Rasul G, Wajid SA, Khaliq T, Rasul F, Saeed U, Rahman MH, Hussain J, Baig IA, Naqvi AA, Bokhari SAA, Ahmad S, Naseem W, Hoogenboom G, Valdivia RO (2015) Impact of climate change on the rice–wheat cropping system of Pakistan. In: Daniel H, Rosenzweig C (eds) Handbook of climate change and agroecosystems: the agricultural model intercomparison and improvement project (AGMIP) integrated crop and economic assessments. World Scientific, ASA, CSSA, and SSSA, St. Louis, MO, pp 219–258

Ahmad D, Afzal M, Rauf A (2020) Environmental risks among rice farmers and factors influencing their risk perceptions and attitudes in Punjab, Pakistan. Environ Sci Pollut Res 27:21953–21964. https://doi.org/10.1007/s11356-020-08771-8

Ahmed M, Hayat R, Ahmad M, Ul-Hassan M, AMS K, Ul-Hassan F, Ur-Rehman MH, Shaheen FA, Raza MA, Ahmad S (2022) Impact of climate change on dryland agricultural systems: a review of current status, potentials, and further work need. Int J Plant Prod. https://doi.org/10.1007/s42106-022-00197-1

Akhtar N, Saqib Z, Khan MI, Martin M, Atif S, Rai Z (2019) A bibliometric analysis of contemporary research regarding industrial symbiosis: a path towards urban environmental resilience. Appl Ecol Environ Res 17:1159–1221

Ali A, Erenstein O (2017) Assessing farmer use of climate change adaptation practices and impacts on food security and poverty in Pakistan. Clim Risk Manag 16:183–194. https://doi.org/10.1016/j.crm.2016.12.001

Arnold N, Pearson M, Cornelius C, Muchaonyerwa P, Isaiah W (2016) Tillage and crop rotation effects on carbon sequestration and aggregate stability in two contrasting soils at the Zanyokwe Irrigation Scheme, Eastern Cape province, South Africa. South Afr J Plant Soil 33:1–8

Azooz RH, Arshad MA (1996) Soil infiltration and hydraulic conductivity under long-term no-tillage and conventional tillage systems. Can J Soil Sci 76:143–152

Badenhorst H (2016) Mesofaunal assemblages in soil of selected crops under diverse cultivation practices in central South Africa,with notes on collembola occurrence and interactions. MSc thesis, University of the Free State, South Africa

Balabane M, Bureau F, Decaens T, Akpa M, Hedde M, Laval K, Puget P, Pawlak B, Barray S, Cluzeau D, Labreuche J, Bodet JM, Le Bissonnais Y, Saulas P, Bertrand M, Guichard L, Picard D, Houot S, Arrouays D, Brygoo Y, Chenu C (2005) Restauration de fonctions et propriétés des sols de grande culture intensive : effets de systèmes de culture alternatifs sur les matières organiques et la structure des sols limoneux, et approche du rôle fonctionnel de la diversité biologique des sols. GESSOL/projet Dmostra. Rapport final, 119pp

Bennie ATP, Hensley M (2001) Maximizing precipitation utilization in dryland agriculture in South Africa — a review. J Hydrol 241:124–139

Ben Moussa-Machraoui S, Errouissi F et al (2010) Comparative effects of conventional and no-tillage management on some soil properties under Mediterranean semi-arid conditions in northwestern Tunisia. Soil Till Res 106(2):247–253

Beukes DJ, Swanepoel CM (2017) The effects of conservation tillage practices and fertiliser management on soil structural properties at an experimental farm. South Afr J Plant Soil 34:19–26

Bhale VM, Wanjari SS (2009) Conservation agriculture: a new paradigms to increase resource use efficiency. Indian J Agronomy 54:167–177

Brown RA, Rosenberg NJ (1997) Sensitivity of crop yield and water use to change in a range of climatic factors and CO2 concentrations: a simulation study applying EPIC to the Central USA. Agric For Meteorol 83:171–203

Bryan E, Ringler C, Okoba B, Roncoli C, Silvestri S, Herrero M (2013) Adapting agriculture to climate change in Kenya: household strategies and determinants. J Environ Manage 114:26–35

Chauhan BS, Mahajan G, Sardana V, Timsina J, Jat ML (2012) Productivity and sustainability of the rice–wheat cropping system in the Indo-Gangetic Plains of the Indian subcontinent: problems, opportunities, and strategies. Adv Agron 117:315–369

Cheesman S, Thierfelder C, Eash NS, Kassie GT, Frossard E (2016) Soil carbon stocks in conservation agriculture systems of Southern Africa. Soil Tillage Res 156:99–109

Crop Reporting Services Government of Punjab (2020) Punjab crops data. http://www.crs.agripunjab.gov.pk/

Dabi T, Khanna VK (2018) Effect of climate change on rice. Agrotechnology 7:2–7

Das TK, Sonaka G, Dinesh S, Kamlika G (2018) Potential of conservation agriculture for ecosystem services: a review. Indian J Agric Sci 89:1572–1579. https://doi.org/10.56093/ijas.v89i10.94578

Dennis P, Thomas MB, Sotherton NW (1994) Structural features of field boundaries which influence the overwintering densities of beneficial arthropod predators. J Appl Ecol 31:361–370

Dillaha Theo, Shenk Cheryl, Moore Keith (2010) Conservation agriculture and ecosystem services. ASABE - 21st century watershed technology: improving water quality and environment 2010. 10.13031/2013.29412

Dube E, Chiduza C, Muchaonyerwa P (2012a) Biomass production weed suppression, nitrogen and phosphorus uptake in white oat (Avena sativa L.) and grazing vetch (Vicia dasycarpa L.) cove crop bicultures under an irrigated no-till system. South Afr J Plant Soil 29:135–141

Dube E, Chiduza C, Muchaonyerwa P (2012b) Conservation agriculture effects on soil organic matter on a Haplic Cambisol after four years of maize-oats and maize-grazing vetch rotations in South Africa. Soil Tillage Res 123:21–28

Durrani H, Syed A, Khan A, Tareen A, Durráni NA, Khwajakhail BA (2021) Understanding farmers' risk perception to drought vulnerability in Balochistan, Pakistan. AIMS Agric Food 6:82–105

du Preez CC, van Huyssteen CW, Mnkeni PNS (2011) Land use and soil organic matter in South Africa 1: A review on spatial variability and the influence of rangeland stock production. South Afr J Sci 107:27–34

Erenstein O, Farooq U (2007) A survey of factors associated with the adoption of zero tillage wheat in the irrigated plains of South Asia. Exp Agric 45(2):133–147. https://doi.org/10.1017/S0014479708007448

Esmeraldo MQ (2017) Effects of tillage practices on some key soil parameters: A case study in the KwaZulu-Natal Midlands, South Africa. MSc thesis, Stellenbosch University, South Africa

FAO (2004) FAOSTAT, 2004. Database for database for wheat crop, Pakistan. Rome, Italy. Retrieved from http://www.fao.org/faostat/en/#data/QC. Accessed 3 Aug 2023

FAO (2021) Global agriculture towards 2050. High level expert forum—how to feed world in 2050. Food and agriculture organization. Rome, Italy. http://www.fao.org/fileadmin/templates/wsfs/docs/Issues_papers/HLEF2050_Global_Agriculture.pdf

Field RD, Kim D, LeGrande AN, Worden J, Kelley M, Schmidt GA (2014) Evaluating climate model performance in the tropics with retrievals of water isotopic composition from Aura TES. Geophys Res Lett 41(16):6030–6036. https://doi.org/10.1002/2014GL060572

Frank JR, Mungroo R, Ahmad Y, Wang M, De Rossi S, Horsley T (2010) Toward a definition of competency-based education in medicine: a systematic review of published definitions. Med Teach 32(8):631–637

Graham MH, Haynes RJ, Meyer JH (2002) Changes in soil chemistry and aggregate stability induced by fertilizer applications, burning and trash retention on a long-term sugarcane experiment in South Africa. Eur J Soil Sci 53:589–598

Graham MH, Haynes RJ (2006) Organic matter status and the size, activity and metabolic diversity of the soil microbial community in the row and inter-row of sugarcane under burning and trash retention. Soil Biol Biochem 38:21–31

Greb BW (1966) Effect of surface-applied wheat straw on soil water losses by solar distillation. Soil Sci Soc Am Proc

GoP (2020) Land cover atlas of Pakistan: the Punjab Province. FAO, SUPARCO

Hafeez R, Ahmad N, Masood A, Muhammad I, Mubshar H, Shakeel A, Muhammad F (2019) Direct seeding in rice: Probl. Prospects. https://doi.org/10.1007/978-981-32-9151-5_11

Harvey CA, Rakotobe ZL, Rao NS, Dave R, Razafimahatratra H, Rabarijohn RH, Rajaofara H, MacKinnon JL (2014) Extreme vulnerability of smallholder farmers to agri cultural risks and climate change in Madagascar. Phil Trans R Soc B 369:20130089. https://doi.org/10.1098/rstb.2013.0089

Hatfield J, Sauer T, Prueger J (2001) Managing soils to achieve greater water use efficiency: a review. Agron J 93:271–280

He TH, Sonaka G, Das TK, Dinesh S, Kamlika G (2009) Potential of conservation agriculture for ecosystem services: A review. Indian j agr sci 89:1572–1579. https://doi.org/10.56093/ijas.v89i10.94578

Hernanz JL, Arrúe JL, Cantero C (eds) (2002) Creación de una red temática sobre laboreo de conservación. Informe final, Acción Especial CICYT, AGF96-1613-E, ETSIA, UPM, Madrid, 180p

Hira GS (2009) Water management in Northern States and the food security of India. J Crop Improv 23(2):136–157. https://doi.org/10.1080/15427520802645432

IPCC (2018) IPCC Special Report 1.5 - summary for policymakers. In Global warming of 1.5 °C. An IPCC Special Report on the impacts of global warming of 1.5 °C above pre-industrial levels and related global greenhouse gas emission pathways, in the context of strengthening the global response to the threat of climate change. 10.1017/CBO9781107415324

Judy G, Edward S, Joan M, Mark G (2002) Gender differences in investment strategies: an information processing perspective. Int J Bank Market 20. https://doi.org/10.1108/02652320210415953

Kassam A, Friedrich T, Derpsch R (2019) Global spread of Conservation Agriculture. International Journal of Environmental Studies 76(1):29–51. https://doi.org/10.1080/00207233.2018.1494927

Kassam A, Friedrich T, Shaxson F, Pretty J (2009) The spread of Conservation Agriculture: justification, sustainability and uptake. International Journal of Agricultural Sustainability 7(4):292–320. https://doi.org/10.3763/ijas.2009.0477

Khan JA, Fee L (2014) Cities and climate change initiative-abridged report: Islamabad Pakistan, climate change vulnerability assessment. United Nations Human Settlements Programme (UN-Habitat). http://www.fukuoka.unhabitat.org/programmes/ccci/pdf/Islamabad(Pakistan)_23_February_2015_FINAL(5th_revision).pdf. Accessed 7 Oct 2015

Khan M, Mahmood HZ, Damalas CA (2015) Pesticide use and risk perceptions among farmers in the cotton belt of Punjab, Pakistan. Crop Protection 67:184–190. https://doi.org/10.1016/j.cropro.2014.10.013

Khan MB, Saleem H, Shabbir MS, Huobao X (2020a) The effects of globalization, energy consumption and economic growth on carbon dioxide emissions in South Asian countries. Energy Environ. https://doi.org/10.1177/0958305X20986896

Khaliq T, Gaydon D et al (2019) Analyzing crop yield gaps and their causes using cropping systems modelling–A case study of the Punjab rice-wheat system, Pakistan. Field Crops Res 232:119–130

Khan I, Lei H, Shah AA et al (2020b) Climate change impact assessment, flood management, and mitigation strategies in Pakistan for sustainable future. Environ Sci Pollut Res 28:29720–29731. https://doi.org/10.1007/s11356-021-12801-4

Kidson MV (2014) Comparison of soil erosion under no-till and conventional tillage systems in the high rainfall Mlondozi area, Mpumalanga province, South Africa. MInstAgrar thesis, University of Pretoria, South Africa

Knot J (2014) Promoting conservation agriculture: commercial farmers in the eastern Free State. PhD thesis, University of the Free State, South Africa

Koch S (2017) Effect of time, tillage and cropping systems on glomalin production by arbuscular mycorrhiza in Western Cape soils. Paper presented at the Combined Congress, Klein Kariba, 23–26 January 2017

Kreft S, Eckstein D, Junghans L, Kerestan C, Hagen U (2013) Global climate risk index 2014. Who suffers most from extreme weather events? vol 1A. Germanwatch e.V., Bonn, pp 31–32

Kutu FR (2012) Effect of conservation agriculture management practices on maize productivity and selected soil quality indices under South Africa dryland conditions. Afr J Agric Res 7:3839–3846

Gattig A, Hendrickx L (2007) Judgmental discounting and environmental risk perception: dimensional similarities, domain differences, and implications for sustainability. J Soc Issues 63(1):21–39. https://doi.org/10.1111/j.1540-4560.2007.00494.x

Lampurlanés J, Cantero-Martínez C (2003) Soil bulk density and penetration resistance under different tillage and crop management systems and their relationship with barley root growth. Agron J 95:526–536

Lobell DB, Asner GP (2003) Climate and management contributions to recent trends in US agricultural yields. Science 299:1032–1032

Loke PF, Kotze E, du Preez CC (2012) Changes in soil organic matter indices following 32 years of different wheat production management practices in semi-arid South Africa. Nutr Cycl Agroecosyst 94:97–109

Mahbubal Aslam SM, Islam MT, Muhsi AAA (1985) Effect of light and night temperature on some cultivars of rice (Oryza sativa L.). Indian J Plant Physiol 28:385–394

Maali SH, Agenbach AG (2003) Effect of soil tillage, crop rotation and nitrogen application rates on grain yield of spring wheat (Triticum aestivum L.) in the Swartland wheat producing area of the Republic of South Africa. South Afr J Plant Soil 20:111–118

Maali SH, Agenbag GA (2004) Effect of soil tillage, crop rotation and nitrogen application rates on plant-N content of spring wheat (Triticum aestivum L.) in the Swartland wheat-producing area of the Republic of South Africa. South Afr J Plant Soil 21:255–260

Mahmood N, Ahmad B, Hassan S, Bakhsh K (2012) Impact of temperature adn precipitation on rice productivity in rice-wheat cropping system of Punjab province. J Anim Plant Sci 22(4):993–997. ISSN: 1018-7081

Mchunu CH, Lorentz S, Jewitt G, Manson A, Chaplot V (2011) No-till impact on soil and soil organic carbon erosion under crop residue scarcity in Africa. Soil Sci Soc Am J 75:1502–1511

Mcinga S, Muzangwa L, Mnkeni PNS (2017) Effects of tillage, crop rotation, and residue management on earthworm diversity and abundance in the Alice Jozini ecotope in the Eastern Cape, South Africa. Poster presented at the Combined Congress, Klein Kariba, 23–26 January 2017

Moghariya DP, Smardon R (2014) Rural perspectives of climate change: a study from Saurastra and Kutch of Western India. Public Underst Sci 23:660–677

Moreno F, Murillo JM, Pelegrín F, Girón IF (2006) Long-term impact of conservation tillage on stratification ratio of soil organic carbon and loss of total and active CaCO3. Soil Till Res 85:86–93

Mukumbareza C, Muchaonyerwa P, Chiduza C (2015) Effects of oats and grazing vetch cover crops and fertilisation on microbial biomass and activity after five years of rotation with maize. South Afr J Plant Soil 32:189–197

Mupambwa HA, Wakindiki IIC (2012) Winter cover crop effects on soil strength, infiltration and water retention in a sandy loam Oakleaf soil in Eastern Cape, South Africa. South Afr J Plant Soil 29:121–126

Mupangwa W, Twomlow S, Walker S, Hove L (2007) Effect of minimum tillage and mulching on maize (Zea mays L.) yield and water content of clayey and sandy soils. Phys Chem Earth 32:1127–1134

Muzangwa L (2016) Effects of conservation agriculture components on soil carbon sequestration, carbon dioxide fluxes, enzyme activities and crop yields in two ecotopes in the Eastern Cape Province, South Africa. Ph.D. thesis, University of Fort Hare, South Africa

Mzezewa J, Gwata ET, van Rensburg LD (2011) Yield and seasonal water productivity of sunflower as affected by tillage and cropping systems under dryland conditions in the Limpopo province of South Africa. Agric Water Manag 98:1641–1648

Nciizah AD, Matlou MC, Wakindiki IIC (2015a) Weed assessment report: Mveso CATs project phase 2. Pretoria: department of rural development and land reform

Nciizah AD, Matlou MC, Wakindiki IIC (2015b) Weed assessment report: Candu CATs project phase 2. Pretoria: department of rural development and land reform

Ndamani F, Watanabe T (2017) Determinants of farmers' climate risk perceptions in agriculture—A rural ghana perspective. Water 9:210

Niveta J (2014) Emission of air pollutants from crop residue burning in India. Aerosol Air Qual Res 14. https://doi.org/10.4209/aaqr.2013.01.0031

Njaimwe AN, Mnkeni PNS, Chiduza C, Muchaonyerwa P, Wakindiki IIC (2016) Tillage and crop rotation effects on carbon sequestration and aggregate stability in two contrasting soils at the Zanyokwe Irrigation Scheme, Eastern Cape province, South Africa. South Afr J Plant and Soil 33:317–324

Parry S, Marsh TJ, Kendon MC (2013) 2012: from drought to floods in England and Wales. Weather 68(10):268–274

PBS (2020) Pakistan bureau of statistics. Population census, Government of Pakistan, Islamabad

Poddar S, Narkhede P, Kumar V, Kumar A (2017) PSO aidedadaptive complementary filter for attitude estimation. J Intell Robot

Prabhjyot-Kaur, Hundal SS (2001) Global climate change vis-à-vis crop productivity. In: Jha MK (ed) Natural and anthropogenic disasters. Springer, Dordrecht. https://doi.org/10.1007/978-90-481-2498-5_18

Privara A, Privarova M (2019) Nexus between climate change, displacement and conflict: Afghanistan case. Sustainability 11:1–19

Riley HCF, Bleken MA, Abrahamsen S, Bergjord AK, Bakken AK (2005) Effects of alternative tillage systems on soil quality and yield of spring cereals on silty clay loam and sandy loam soils in the cool, wet climate of central Norway. Soil Till Res 80:79–93

Sapkota TB, Jat ML et al (2015) Climate change adaptation, greenhouse gas mitigation and economic profitability of conservation agriculture: some examples from cereal systems of Indo-Gangetic plains. J Integr Agric 14(8):1524–1533

Sauer TJ, Hatfield JL, Prueger JH (1996) Over-winter changes in radiant energy exchange of a corn residue-covered surface. Agr Forest Meteorol 85:279–287

Schilling J, Vivekananda J, Khan MA, Pandey N (2013) Vulnerability to environmental risks and effects on community resilience in mid-West Nepal and south-East Pakistan. Environ Nat Resour Res 3:27–45

Sharma K, Saini AL, Nawab S, Ogra JL (1998) Feeding behaviour and forage nutrient utilization by goats on a semiarid reconstituted silvipasture. Asian-Aust J AnimSci 11(4):344–350

Sjoberg L, Moen BE, Rundmo T (2004) Explaining risk perception. An evaluation of the psychometric paradigm in risk perception research. Rotunde publikasjoner Rotunde 84:55–76

Smit B, Skinner M (2002) Adaptation options in agriculture to climate change: a typology. Mitig Adapt Strateg Glob Chang 7:85–114. https://doi.org/10.1023/A:1015862228270

Somasundaram J, Nishant S, Ram D, Rattan L, Monoranjan M, Anandkumar N, Kuntal H, Ranjeet C, Ashis B, Ashok P, Chaudhari S (2020) No-till farming and conservation agriculture in south asia – issues, challenges, prospects and benefits. Crit Rev Plant Sci 39. https://doi.org/10.1080/07352689.2020.1782069

Sosibo NZ, Muchaonyerwa P, Visser L, Barnard A, Dube E, Tsilo TJ (2017) Soil fertility constraints and yield gaps of irrigation wheat in South Africa. South Afr J Sci 113:9

Sridevi V, Chellamuthu V (2015) Impact of weather on rice – a review. Int J Appl Res 1(9):825–831

Surinder J, Harsimran K, Shibendu R, Tripathi R, Vashisht B, Bal S (2012) Mitigating future climate change effects by shifting planting dates of crops in rice–wheat cropping system. Reg Environ Chang 12. https://doi.org/10.1007/s10113-012-0300-y

Susama S, Anchal D, Kumar LN (2008) Efficacy of vegetative barriers for rehabilitation of degraded hill slopes in eastern India. Soil Till Res. 99:98–107. https://doi.org/10.1016/j.still.2008.01.004

Swanepoel CM, Marais M, Swart A, Habig J, Koch S, Sekgota WM, Mampana R, Trytsman G, Beukes DJ (2014) Final report: quantifying effects of CA practices on soil and plant properties. ARC-ISCW report no. GW/A/2014/37. Pretoria: agricultural Research Council–Institute for Soil, Climate and Water

Swanepoel PA, du Preez CC, Botha PR, Snyman HA, Habig J (2015) Assessment of tillage effects on soil quality of pastures in South Africa with indexing methods. Soil Res 53:274–285

Tebrügge F, Düring RA (1999) Reducing tillage intensity – a review of results from long-term study in Germany. Soil Till Res 53:15–28

Tesfhuney WA, van Rensburg LD, Walker S (2013) In-field runoff as affected by runoff strip length and mulch cover. Soil Till Res 131:47–54

Timsina J, Connor DJ (2001) Productivity and management of rice–wheat cropping systems: issues and challenges. Field Crops Res 69:93–132

Tohru K, Naoya U (2004) High temperatures during the grain-filling period do not reduce the potential grain dry matter increase of rice. J Agron 96. https://doi.org/10.2134/agronj2004.0406

Tolk JA, Howell TA, Evett SR (1999) Effect of mulch, irrigation, and soil type on water use and yield of maize. Soil Till Res 50:137–147

Usman K, Ullah I et al (2012) Integrated weed management through tillage and herbicides for wheat production in rice-wheat cropping system in Northwestern Pakistan. J Integr Agric 11(6):946–953

Usman M, Anwar S, Yaseen MR, Makhdum MSA, Kousar R, Jahanger A (2022) Unveiling the dynamic relationship between agriculture value addition, energy utilization, tourism and environmental degradation in South Asia. J Public Aff e2712. https://doi.org/10.1002/pa.2712

Wani IA, Gani A, Tariq A, Sharma P, Masoodi FA, Wani HM (2016) Effect of roasting on physicochemical, functional and antioxidant properties of arrowhead (Sagittaria sagittifolia L.) flour. Food Chem 197:345–352

Wiltshire GH, du Preez CC (1993) Long-term effects of conserva tion practices on the nitrogen fertility of a soil cropped annually to wheat. South Afr J Plant Soil 10:70–76

Yadvinder-Singh, Bijay-Singh et al (2004) Long-term effects of organic inputs on yield and soil fertility in the rice-wheat rotation. Soil Sci Soc Am J 68:845–853

Yuan Z (2011) Analysis of agricultural input-output based on Cobb–Douglas production function in Hebei Province, North China. Afr J Microbiol Res 5:5916–5922

Zhang P, Zhang J, Chen M, M. (2015) Economic impacts of climate change on Chinese agriculture: the importance of relative humidity and other climatic variables. Soc Sci Res Netw 2598810:1–78

Forestry a Way Forward for Disaster Risk Reduction in Agriculture

Irfan Ahmad, Muhammad Asif, Haroon Ur Rashid, Salman Ahmed, Shakeel Ahmad, Abdul Jabbar, Zainab Shahbaz, and Zoha Adil

Abstract

Emergency Event Database (EM-DAT) reported 432 devastating events during 2021, related to natural hazards, which is much higher than the last year's calamities (i.e. 357). Amongst all, Asia is most affected continent with 40% sufferings of all continents along with 66% of the total people affected. These catastrophes include drought, earthquake, floods, extreme temperature, storms, landslides, wild fires, etc. These disasters impair socioeconomic and environmental growth of all the developing countries across the globe. Besides posing serious threats to sustainable development, such events also amplify the issue of food security by affecting cropped area. Most of the scientists believe that these catastrophes have become more frequent than ever before only because of severe impact of climate change.

Natural forests, plantations, and linearly planted trees besides delivering numerous ecosystem services play a critical role in combating the serious threats

I. Ahmad (✉) · M. Asif · H. U. Rashid · Z. Shahbaz
Department of Forestry and Range Management, University of Agriculture, Faisalabad, Pakistan
e-mail: irfanahmad@uaf.edu.pk

S. Ahmed
Department of Plant Pathology, University of Sargodha, Sargodha, Pakistan

S. Ahmad
Department of Agronomy, Faculty of Agricultural Sciences and Technology, Bahauddin Zakariya University, Multan, Pakistan

A. Jabbar
Department of Agronomy, University of Agriculture, Faisalabad, Pakistan

Z. Adil
Department of Plant Pathology, University of Agriculture, Faisalabad, Pakistan

M. Ahmed, S. Ahmad (eds.), *Disaster Risk Reduction in Agriculture*, Disaster Resilience and Green Growth, https://doi.org/10.1007/978-981-99-1763-1_16

of climate change, which are the fundamental causes of mentioned disasters. Serious efforts are indispensable to build disaster and climate resilient agricultural systems that will be capable of improving the food security of targeted areas. Forests protect the top fertile layer of the soil from erosion by reducing run off and enhancing seepage, percolation, infiltration, and ultimately recharge but also protect different agricultural crops from high-velocity winds and flash floods. These trees besides providing protection also improve fertility of the soil by addition of organic matter.

Generally, forestry programs across the globe are launched to address the issues like climate change, biodiversity conservation, carbon sequestration, global warming, drought, energy crises, etc. The idea of enhancing area under forests is considered as an effective approach to tackle all the above-mentioned issues. This chapter encompasses the role of our forests in mitigating all these catastrophes.

Keywords

Disasters · Forestry · Agroforestry · Agriculture · Climate change

1 Introduction

Disasters affect three major pillars of sustainable development, i.e., social, environmental, and economic. Such calamities and tragedies are happening frequently than anticipated and are beyond people's control (Liu-Lastres et al. 2020). The effects of disasters vary based on their nature, severity, people's response and recovery (Djalante et al. 2013; Kusumastuti et al. 2021). Disasters are mostly categorized as natural and unnatural. Natural calamities include tsunamis, volcano eruptions, storms, floods, earthquakes, and wildfires (Muir et al. 2020; Fig. 1). Un-natural disasters, on the other hand, include toxic pollutants, transport incidents, explosions, and so on (Yulianto et al. 2020; Fig. 1). Wildfires, flooding, and landslides are sometimes linked to the human actions (Gore and Fischer 2014). Disasters have struck almost all regions of the world. In addition, anthropogenic or human-induced climate change that results in droughts or uneven rainfall will be considered a disaster (Pratiwi et al. 2020). Disasters affect many sectors, including infrastructure, economics, livelihoods, and social welfare (Bakkour et al. 2015). When disaster strikes an agricultural based economy, its agriculture sector becomes the one to be highly impacted as, the country relies heavily on this sector to drive its economy (Sun et al. 2010). A natural disaster that smashes agriculture can have a devastating effect on the lives of farmers and associated communities, because this sector has perceived as the major livelihood source for many people (Mottaleb et al. 2013; Helmi and Sasaoka 2018; Fig. 2). In order to formulate mitigation strategies, abundant studies have already been conducted to understand the impact of disasters on different sectors (Shrestha et al. 2019; Yulianto et al. 2020; Rozaki 2020), but being the most

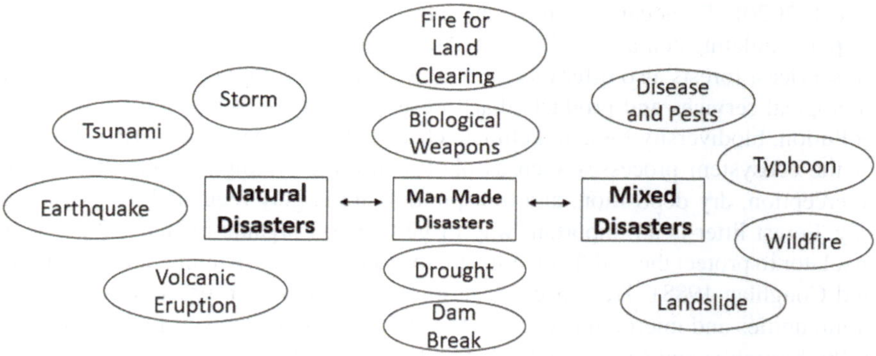

Fig. 1 Illustrations of catastrophes broken down into its 'element' and degree of human involvement

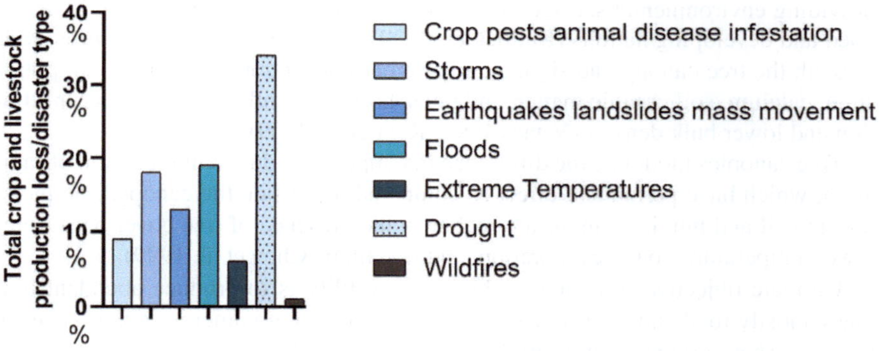

Fig. 2 Total crop and livestock production loss by region and per disaster in least developed and lower middle-income countries, 2008–2018

vulnerable segment, agriculture is still ignored (Helmi et al. 2019; Kusumastuti et al. 2021). Serious efforts are indispensable to build disaster and climate resilient agricultural systems that will be capable of improving the food security of targeted areas, even in the era of escalating threats (FAO 2021).

2 Forests and Nature Conservation

Forests are the most abundant bionetwork in nature. Forests and trees supply a plenty of ecological services that assist in shaping healthy living environments and in refurbishing degraded biomes. Besides, tangible products, natural forests, and forest plantations assist in alleviating floods, drought stresses, and act as barrier to noise and dust, maintain water quality by purifying water, improve soil fertility and control erosion which is considered as a major cause of land degradation (FAO and

UNEP 2020). Trees can minimize the adverse effects of climate change and can help in regulating contagious diseases (Ahmad et al. 2019). Other than environmental services, forests also offer recreational, social, and cultural attributes. However, ecological services and products that forests provide are at risk by deforestation, pollution, biodiversity loss, and climate change (Karjalainen et al. 2010), but most of the ecosystem processes such as photosynthesis, cycling of nutrients, rainfall interception, dry deposition, and transpiration are largely regulated by forest canopy. Forest litter is an important way of getting soil organic matter and act as an insulator to protect the soil from moisture loss and changes in temperature (Running and Coughlan 1988). Tree species can affect decomposition process as microbial communities and microclimate in the forest floor regulate the soil fertility (Hobbie 1996; Sariyildiz and Anderson 2003; Mitchell et al. 2007).

Forests protect and maintain the environment through ecological (e.g., biodiversity maintenance, bio-geo-chemical cycles of water, C and N), financial (e.g., natural resources such as timber and energy source) and social functions. Trees are providing environmental services as trend of social forestry is increasing in developed and developing nations (Bullock et al. 2011). In comparison to pastures, soils beneath the tree canopy had significantly higher concentrations of nitrogen, potassium, calcium, soil organic matter, and phosphorus, as well as higher water infiltration and lower bulk density (Vetaas 1992; Kahi et al. 2009).

Tree canopies modulate the difference in temperature due to interception of radiation, which have permanent effect as compared to outside the canopy. The moisture in soil and humidity in air are higher in the presence of tree cover because of lower temperature, lower evaporation, and tree litter (Chen et al. 1999).

Ultimate objective of achieving high soil fertility is to produce abundant biomass, mostly food and fiber. Because of the interaction of numerous components in the ecosystem, organic matter produced as a result of litter fall varies from species to species (Kaye et al. 2000). The pivotal significance of diverse plant types varies across and within plant species, particularly in gymnosperms and angiosperms. Furthermore, trees alter the soil's chemistry profile (Reich et al. 2005), and such changes in soil quality are influenced by a variety of tree species (Dijkstra and Smits 2002).

3 Disasters; Major Threat to Agriculture

Annually various natural disasters, including hurricanes, floods, forest fires, volcanic eruptions, earthquakes, and windstorms are considered as major threats to agricultural productivity. As agriculture relies on biotic and abiotic factors and among these, climate and water availability are more important. So, it is definitely affected by natural events and calamities (EPQA, US Environment protection agency). Most common impacts of disasters and misfortunes on agriculture include reduction in yield of crops, mortality of livestock, amplified susceptibility to diseases, damage to infrastructure, and irrigations systems, etc. (Putra et al. 2021).

These impacts can have enduring effects on agricultural output including cereal crops, vegetables, fruits plants, and tree growth. Well-timed and effective planning about how to prepare for and rehabilitate from natural disturbances and disasters will definitely decrease their long-lasting effects on agriculture and allied disciplines (FAO 2021).

Drought and waterlogging have been recognized among the most severe threats to agriculture. Waterlogging is characterized by short- or long-term water saturation of soil pores (Alaoui-Sossé et al. 2005). If there is a surplus of water compared to its optimal needs, it will be considered as waterlogging. Waterlogging can be of different types as riverine flood waterlogging, oceanic flood waterlogging, seasonal waterlogging, perennial waterlogging, and sub-soil waterlogging. Salinity and waterlogging are major degradation issues responsible for the low yield in irrigated and agricultural regions of the world. According to UNESCO and FAO, half of the existing irrigated lands of the globe are more or less under the threat of secondary salinization, alkalization, and waterlogging problems. Poor drainage in irrigated lands is a major factor responsible for raising the water table (Bruinsma 2009).

Drought stress occurs when water that is accessible to the ground is concentrated, and in some regions, the atmospheric conditions results in everlasting evaporation which causes loss of water (Akhtar and Nazir 2013). Globally, 82% of all drought impact is concentrated in the agriculture sector (FAO 2021).

Pests and threatening diseases of crops and livestock are also a major danger for this sector. During the last decade or so, such natural adversities affected about 9% of agricultural crops and livestock production. During 2020–2021, desert locust crisis in Punjab and Sindh provinces of Pakistan and in east Africa exacerbated the role of such disasters as it substantially reduced the yield of major crops (Raza et al. 2015; FAO 2021).

In arid and semi-arid areas, salinization of soil and ground water is one of the major limiting factors for crop and forest growth (Kaur et al. 2002). Since last two to three decades, vast area of arable agriculture land has been degraded due to the presence of soluble salts in excess than the normal range (Rego and Ferreira 2011). Around the globe, more than 100 countries in different climatic zones have been affected by soil salinity, which is still spreading unabated (Saddiqui et al. 1984; Massoud et al. 1988).

Intergovernmental Panel on Climate Change (IPCC) assessed that the current global mean temperature of earth surface was about 0.42 °C to 0.54 °C during 1961–1990 (IPCC 2007). During next two decades, 12 years from 1995–2006 were reported as the warmest years. More proof about global warming comes from the scrutiny of the natural systems like glaciers (Hanna et al. 2006), coral reefs and atolls (Hughes et al. 2003), polar, and alpine ecosystems (Ackley et al. 2003). Recently observed climate change has affected the life of poor communities who depend on semi-subsistence agriculture for their existence (Slingo et al. 2005; IPCC 2007; Nelson et al. 2009). Maximum of the current research has focused on the effects of climate change in agriculture along with workable strategies for the revival of agricultural productivity (Hijmans 2003; Jones and Thornton 2003). Deforestation, urbanization, industrialization, and various other factors are considered responsible for such abrupt climate changes. Some major causes of change in climate are carbon emission in the atmosphere and greenhouse gases. Due to emission of gases, ozone layer is depleting continuously. Deforestation and change in land use pattern takes place due to emission of different greenhouse gases. In the recent decades, changes in precipitation pattern and rise in global temperature have been observed due to severe climate changes (Nema et al. 2012).

An abrupt, disastrous occurrence that substantially impairs a community's or society's capability to perform and results in losses to people, property or the climate that are more than what the community or society might reasonably expect to be able to recover from using its assets (Hon'ble et al. 2019). A natural disaster can have human causes, although nature frequently causes them. Disasters tend to arise from the forces of nature, but sometimes humans are responsible too. Natural and (to a lesser extent) man-made disasters can be distinguished by the elements such as Earth, Water, Wind, Fire, and Biota, the geographical and chronological scopes of their effects, and the degrees to which they are triggered (McElwee et al. 2020).

The human reaction can be broken down into three stages: before, during, and after. As a long-term strategy, knowing about potential dangers and taking steps to mitigate them as well as avoiding potentially dangerous situations altogether, is essential. Resilience, or the ability to recover from adversity, has both tangible and immaterial (uplifting) elements, the former of which include such strategies as running away, hiding, and avoiding detection quickly and effectively. Disasters are becoming increasingly difficult to classify in terms of their 'natural' and 'man-made' origins. Yet, these distinctions are still important for determining how to respond to disasters and what kind of protection. Recent earthquakes in Lombok provide evidence against designating the destruction as a "national disaster," as this would have severe effects on foreign tourism and slow down the rehabilitation program (Field and Barros 2014).

3.1 Devastating Floods

Floods are hefty water bodies causing serious damage to all the living organisms. Longer spell of heavy rain or over burdening of water reservoirs results in floods normally. These have direct as well as indirect impacts upon all the living beings. Property damage or its complete loss, agricultural crops, and livestock loss along with causalities are few examples of direct impacts, whereas, temporary accommodation and communal break ups are the indirect impacts of this disaster (Johnson 2008). The 20th century emerged with more frequent disasters including flood, like never before. Principle reason seems to be change in land use patron by the human beings for their personal benefits (Crooks and Davies 2000; De Roo et al. 2001; Costa 1975; Zang 2001).

3.2 Reasons and Impacts of Flood Disasters

Reasons for more frequent floods were investigated, and data were analyzed from 1908 to 1990. It was observed that rise in temperature and more precipitation were more frequent than ever before. Similar conclusions were observed for the period during 1950–2007. It was observed that rise in temperature significantly increased the flood hazards during the mentioned years. Amplified temperature caused the glaciers and snowpack to melt with more frequent storms and flood risks at basin level. Ultimately, this flooding poses serious economic threats to the economy of a country. Changing climate was considered as the key reason for such disasters across the globe (Wang and Bill 2007; Zang 2001). New conflicts were raised among different 126 flood affected developing countries during 1985–2009. So the impacts of these disasters were more pronounced in developing countries as compared to developed countries (Ghimire et al. 2003).

Asian countries are more susceptible to climate change and among all these, Pakistan is at the top. Ultimate consequences of climate change include insect attack, water scarcity, and food security accompanied by disasters like flood. These floods affect crop productivity threaten the economy of developing countries alarmingly (Abid et al. 2016). Historically Pakistan has faced pervasive and frequent flooding that has not only effected agriculture but also caused substantial damage to infra structure along with huge loss of lives (Sardar et al. 2008; Abid et al. 2016). Unexpected and heavy rainfalls in Pakistan normally results in floods, and this flood water flows through Pakistan after damaging everything. Pakistan has faced devastating floods since 2001 and serious decline in its economy. The flood in 2010 swamped 70,238 km^2 and affected about 884,715 families (Haq et al. 2010). About six million people, 1500 km of road network, 382 km of railway tracks, 498 km^2 of forests, and 16,440 km^2 of agricultural land were affected by flood in 2011. Flood of 2012 smashed 22 districts of Pakistan with huge economic loss (Haq et al. 2012; Memon et al. 2015). Even in 2022, the damage was on higher side (due to floods) in all the provinces of Pakistan.

From Fig. 3, we can see that in 2022 about 116 districts were affected by flood out of which 66 were officially declared calamity hit. About 218,000 houses were completely destroyed, 452,000 houses were damaged, two million acres cropped area was impacted, and 794,000 livestock loss accompanied with more than 937 death tolls was reported. Nationwide 2.87 times higher rain fall was reported than the average 30 years. Furthermore, the situation is going to worsen, as more rain fall is expected (OCHA 2022). Floods pose stern socio-economic intimidation each year, across the globe by affecting at least 20 million people (Kellens et al. 2013). Meteorologists believe that, this climate change will lead to added flood hazards in coming years. Flood of 2015 in UK attracted the policy makers towards ecosystem services of vegetation in reducing flood risks which casted about £5 billion (KPMG 2015) as more frequent floods are expected in future.

Apart from of reason, floods have a thoughtful effect on people and their socio-economic standards. Annually, more than three million people lose their residences and more than 60 million face serious economic crises due to these floods (WCD

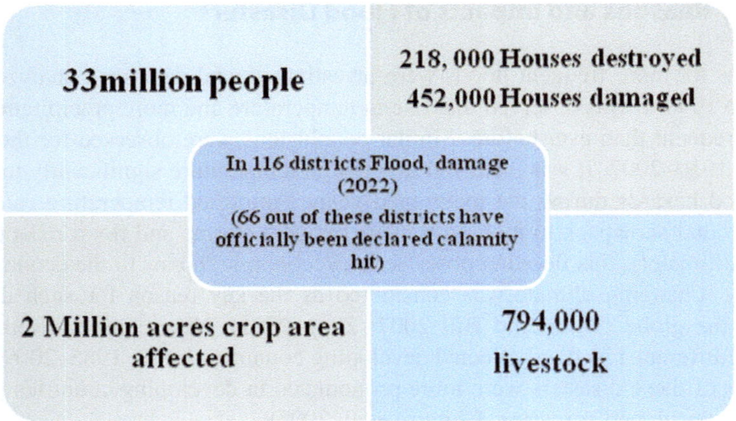

Fig. 3 Damage caused by flood in Pakistan during 2022

2000). Since the beginning, devastating floods have seriously damaged the residences of dwellers in the catchment areas of respective rivers. It can be seen throughout the history that huge economic losses resulted in untold sufferings of millions of people only due to these floods though they have a little positive impact as well. These most devastating natural hazards caused the largest amount of deaths and property damage across the globe (CEOS 2003). Red Cross reported that floods between 1971 and 1995 affected more than 1.5 billion people globally, together with 318,000 persons lost their lives and more than 81 million were left homeless (IFRCRCS 1997). About 3723 sufferers around the world were reported in 2003, due to floods (Munich Re 2003). A 10 years comparative investigation of disasters, revealed that frequency of floods has increased three times since 1950 to 1998, along with ten times higher economic losses during the mentioned period (Loster 1999). Data recorded for various natural disasters during the last few decades indicate that there is huge number of average deaths due to these devastating floods.

Successive settlement in the catchments of Grant River at different times has increased run off and erosion to many folds only due to the change in land use pattern by the community (Knox 1972). Piereira (1973) conducted a study to interrogate the prominent effects of change in land use pattern. A well-planned change was taken into count for 8 years to study its impacts upon downstream flow at Wagon Wheel Gap, Colorado. Flow from two catchments was compared during successive 8 years, and after that one area was clear felled and monitored again on regular basis. It was reported that run off was increased, peak flows was increased, lag time was reduced, and over all yield of water was increased up to 17% in clear felled area. Similar findings were reported by Bosch and Hewlett (1982).

Urbanization and modernization has put incredible pressure upon actual landscape of the natural forests of Limburg (Netherlands). This expansion of urbanization in the hilly area of the mentioned province has deteriorated the soil health of cropped areas, increased surface flow, reduced recharge and increased erosion hazards (Leenaers et al. 1989).

Anthropogenic activities contribute a lot in the phenomenon of climate change by altering hydrological conditions of an area and changing the time and size of floods (Hornbeck et al. 1970; Knox 1972, 1977; Potter 1991; Orbock-Miller et al. 1993) accompanied by the alteration in extent and type of erosion (Douglas 1967; Piest and Spoomer 1968; Costa 1975; Renfroe 1975). Afforestation and consistent landscape were considered most appropriate for sustainable forest management and retaining moisture contents of the soil (FAO 2003). Recent developments in urbanization near river banks have resulted in most devastating floods (Handmer 2000; Van der Sande et al. 2003; Water Directors 2003; Zang 2001). Alteration in land use pattern significantly affects the relationship among run-off, sediments and rain fall or soil water holding capacity (Yang and Yu 1998; Handmer 2000; Kim et al. 2002).

3.3 Flood Risk Mitigating Measures

There are numerous and repeatedly clashing investigations about flood risk and its mitigation measures. Unexpected and acute weather conditions and lack of control over the weather is often used to give reason for weak flood protection policy and decision-making. Moreover vast size, heterogeneity and lack of honourship of catchments disrupted by the habitation are mounting potential threats for flooding. This eventual uncertainty creates serious hurdles in decision- and policy-making ultimately reducing the effectiveness of different mitigation measures (O'Connell et al. 2004).

Normally, there are three types of flood mitigation approaches, which include hard engineering, soft engineering, and passive procedures. Hard measures are expensive, time consuming and often include construction of embankments, dams, and water reservoirs, etc., like these were weighed down in floods of December 2015 in the North West of England. Soft measures comprise improved land management practices, improved water, and drainage system along with improved vegetation cover (Wilkinson et al. 2010). These hard measures were overwhelmed in flood affected areas of Northern England during 2015 thus shifted the attention of researchers and policy makers towards soft measures. This has paved the way for new alternative soft measures. These inclined focuses comprise extensive support for proper management of the catchment areas, reducing velocity of flowing water, and improved land use pattern than conventional ones (Dixon et al. 2016; Harrabin 2016; The Guardian 2016). Recent researches has revealed that poor management along with conventional vegetation cover exhibited reduced lag time and more discharge which ultimately increases flood hazards (CIRIA 2013).

Accordingly, the idea of enhancing area under forests is considered as an effective approach to reduce run off, increasing lag time to reduce discharge and for improving the overall conditions of catchment areas. Generally, forestry programs across the globe are launched to address the issues like climate change, biodiversity conservation, carbon sequestration, energy crises, etc. Trees are the best source to regulate the flow of water and to purify the environment. Likewise, regeneration of barren areas is the best approach to enhance interception, to reduce run off, to amplify infiltration and percolation and to increase recharge. Forests play a critical role in shelter belts and in soil conservation when used on stream banks, etc. Role of trees was evaluated against extreme floods to protect community. It was observed that vegetation significantly contributed towards the mentioned aspects (Johnson 2008; Dixon et al. 2016).

3.4 Role of Forests in Flood Control

Ecosystem services of trees cannot be denied especially during flood disasters. Their role has been evaluated by various researchers during different decades. But still, there are many aspects related to this issue needed to be explored. With recent developments in urbanization has put tremendous pressure in catchments. This escalating urbanization has created the risk of vegetation depletion and degraded the infiltration capacity of the soil to absorb the rain water. It is important to introduce the trees and enhance forest cover to deal with the flood risks (Hernandez et al. 2000).

Since decades, the role of forests has been evaluated to manage excessive run off and to reduce downstream flow of different systems. Tree cover reduce flood risks by reducing the velocity of flowing water thus the discharge of the water channel is also reduced (Bosch and Hewlett 1982). Riverine forests play a critical role in different hydrological processes of an area which ultimately affect other related geomorphic processes occurring at particular sites. Vegetation along river banks plays critical role in soil conservation by reducing the velocity of flowing water (Gilman 2002; Hernandez et al. 2000). Remarkable impact of vegetation on velocity of flowing water has been investigated by various researchers. Cultivation of lands has an opposite effect upon the same processes (Robinson et al. 2003).

A case study of Brazil revealed that dense vegetation is less affected by flood disasters as compare to patchy or scanty vegetation. Flood plains of various regions were investigated, as these were hit by floods annually. After thoroughly probing into the issue, it was assessed that areas had more trees faced less erosion hazards as compare to the areas had less vegetation (Kurzatkowski et al. 2015). Similarly in Switzerland, it was reported that 100% vegetated area had 1/3 to 1/2 less flooding as compare to 30% vegetated area (Burger 1943). Likewise, in another study, 700 hectares of degraded lands were regenerated situated well inside the catchments, and it was concluded that though the volume of the flooding was same but the peak flows were reduced to only 15%, the lag time was also reduced from 1.5 to 8 h, when compared with previous situations. In another study of cascading mountains, it was concluded that clear felling ^^^^silvicultural system can has adverse impacts on flood peaks by enhancing the speed of flowing water from 1.1 $m^3/s/km^2$ (Anderson 1958).

Although the considerable amount of water is used by trees in catchments but still, it is crystal clear that forests can reduce the risk of flood disasters. A large quantity of rainfall is intercepted by the canopies of trees due to which the momentum of rain droplets is changed considerably. This change in momentum helps to reduce surface run off, increased seepage, increased infiltration, and percolation which ultimately affect the peak flows of floods (Bosch and Hewlett 1982; Robinson et al. 2003). Moreover, the biological wastes or debris deposited the trees on river banks in riverine forests can help to manage a balanced ecosystem by supporting different species of flora and fauna and by avoiding soil erosion. This deposited debris can help in sedimentation in minute areas and can provide habitat to different organisms. These ideas can only be realized by restocking/regenrating the riverine forests, catchments, and floodplains to their capacity (Bosch and Hewlett 1982).

A comprehensive, holistic, and integrated approach is need of the time to manage such flood disasters for this new paradigm. Downstream and upstream areas of a catchment (may be a forest, range areas, pasture, or an urban area) involve variety of land uses. So we can conclude that for the accomplishment of sustainable forest management (SFM) approach, the involvement of all government agencies across the board should brought together with the involvement of all other concerned quarters.

4 Forests Mediated Disaster Prevention and Mitigation in Agriculture Sector

Disaster risk reduction (DRR) management needs to take into consideration how farmers perceive disaster management. Disaster incidents can increase people's awareness of disaster risk reduction and develop a favorable perception of disaster prevention and mitigation (Peng et al. 2018). People's understanding about DRR and perception of disaster prevention and mitigation can both be enhanced by their experiences with disasters (Agussabti et al. 2020). Disaster mitigation is the strategy used to decrease the consequences of calamities, including those that occur before, during, and after. People need to take action in response to the calamities they are experiencing. They need to adopt different approaches to tackle various natural disasters for minimizing their affects (Harrowell and Özerdem 2019).

Natural forests, plantations, and linearly planted trees deliver numerous ecosystem services that have a central role to play in addressing the fundamental causes of disasters. In order to tackle these issues such concepts need to be highlighted at global level (FAO 2021). Forests encompass an essential part of the resilience of societies and their livelihoods to disasters, menaces and catastrophes and can assist in tackling primary causes of food uncertainty and poverty. It has been reported that 1–1.7 billion people depend on forest resources for their livelihoods. Yet, wood remains a main source of energy for cuisine for about 2.4 billion people. Moreover, nearly 80 percent of an unprecedented 79.5 million displaced people around the world depend on fuelwood and charcoal, for culinary and heating. The function of trees in disaster management, conservation of soil and controlling soil erosion is critical as is persuading farmers to plant trees on their farmland. The forest ecosystem regulates runoff and reduces the rate of siltation in downstream aquatic environments, extending reservoir, and dam shelf lives (Tripathi et al. 2009). Foresters and horticulturists are creating awareness among masses for the promotion of gardens, forest parks, orchards, irrigated plantation, and agroforestry including the services of trees in soil reclamation (Carnus et al. 2006).

4.1 Forestry as a Climate Change and Disaster Risk Reduction Approach in Urban Agriculture

Climate change and rural/urban development are inextricably intertwined. Moreover, half of the world's population lives in cities, which is expected to climb upto 70% by 2050 (Joshi et al. 2019). Cities consume up to 80% of the energy generated globally and account for more than 70% of energy-related global greenhouse gases (GHGs), with both statistics predicted to escalate. It is anticipated that developing countries, particularly fast-growing areas in Asia and Africa, would account for about 90% of the rise in CO_2 emissions from energy use (IEA 2008). Furthermore, poor waste management across many places contributes to CFC and methane emissions (United Nations HABITAT NAP 2019).

Growing areas already have to deal with many problems, like ensuring people have suitable accommodation, infrastructure, economic security, and nutrient-rich, cost-effective, and socially acceptable food to the community. They also have to deal with climate change and the risk of climate-related catastrophes, which are some of the world's major environmental concerns today. Nevertheless, urban communities also have the ability to take massive action, as they are home to the majority of the world's population and the vast majority of the world's cultural and social capital. Many local governments during the past decade have recognized their ability to affect climate change in both causes and effects. They are now helping national and global climate change policies. These problems can be solved by building resilient communities and managing issues related to agriculture and forestry (Dubbeling et al. 2019; Fig. 4).

5 Impact of Disasters on Rural Economy

The majority of farmers in agriculture-based countries are small landholders with very less or no technical education accompanied by poor productivity connected, particularly in developing countries, when it comes to preventing poverty (Sun et al. 2010). Because of natural disasters, instability in agricultural incomes upsets the balanced use of agricultural inputs (Mottaleb et al. 2013; Helmi and Sasaoka 2018). Disasters affect people more deeply if they have low incomes (Viganò and Castellani 2020). There are different characteristics associated with each type of disaster that affect agriculture, mostly in relation to the loss of crops. Agriculture is affected by disasters to different degrees according to their severity (Guarnacci 2016). Agricultural insecurity and rural economic inequality are strongly related to poverty-reduction strategies especially in developing countries (Gignoux and Menéndez

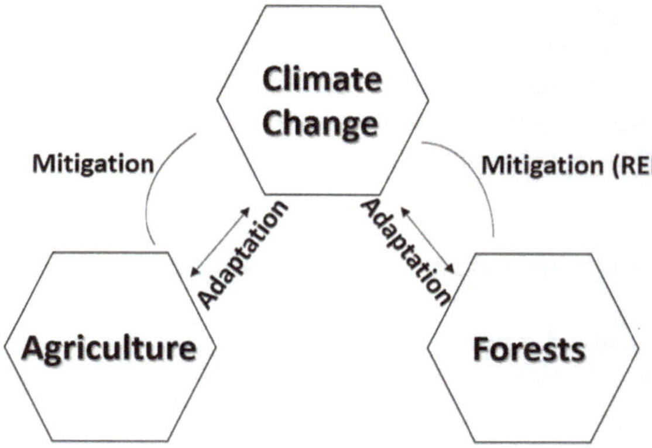

Fig. 4 Role of forests in climate change mitigation

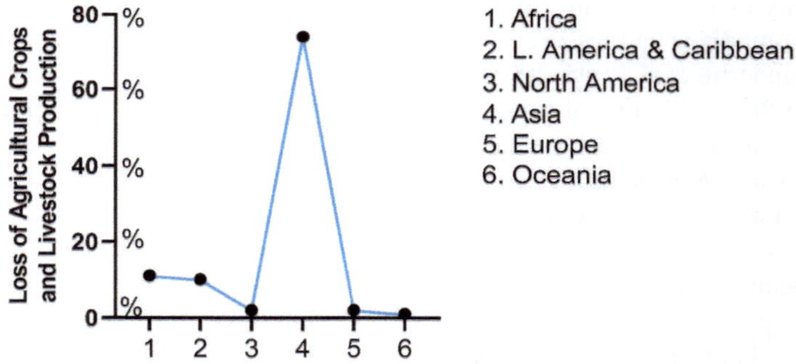

Fig. 5 Regional distribution of crop and livestock production loss due to various disasters, 2008–2018

2016). Landslides frequently occur around the world, and human actions are crucial in regulating the stability of the slopes (Garcia-Chevesich et al. 2021). Landslides occur because of steep hills, rock properties, and poor soil characteristics, which make the soil vulnerable to erosion and affect agricultural productivity (Sitorus and Pravitasari 2017). The consequences of disasters on agriculture are strongly related to crop losses (Putra et al. 2021; Fig. 5). Floods, even if they have a little impact on agricultural land, are likely to cause crop losses (Pratiwi et al. 2020). The greater the size of the flood, the greater the impact on agriculture. Flood has a direct impact on food security and agriculture. This tragedy is also linked to climate change, as it affects processes like unexpected rain, etc. (Shrestha et al. 2019; Dzulkarnain et al. 2019). Climate change, because of both natural destruction and human profiteering has become a heated topic across the world (Suryanto et al. 2020). Agriculture is the industry most influenced by climate change and global warming. It causes irregular precipitation, which makes it challenging for farmers to estimate sowing dates and manage supply of water. Aside from floods climate change also causes drought accompanied many other issues. Drought reduces or eliminates productivity and can cause conflict, particularly with regard to water consumption (Mursidi 2017). In addition to protecting human life, disaster risk reduction management also protects food security by guaranteeing that the amount of food produced will not be significantly squeezed by crop losses due to natural disasters (Udmale et al. 2014).

5.1 Why Forests Are Important for Global Poverty Alleviation?

As the history of mankind forests are considered very vital to the well-being of the poor (Byron and Arnold 1999; Angelsen and Wunder 2003; Vedeld et al. 2004) and provide some monetary benefits (Ruíz Pérez et al. 2004). Forests provide us free ecological goods and services and act as a form of "subsidization of nature (Anderson et al. 1991). They also support agricultural areas, help in maintenance of soil fertility on fallow lands utilized in shifting cultivation farming cycles, and in the availability of sweet, clean water through active watersheds. Products made from these commodities and resources are consumed and sold in a wide range of households (Sunderlin et al. 2005, 2008). There are numerous ways in which forests contribute to the livelihoods of rural poor people. People living in and around forests rely on them for food, fuel, fodder, construction materials, medicines, and a variety of other substantial items (Byron and Arnold 1999). The poor, rural community must diversify their income sources to reduce their risk exposure (Ellis 2000). Forest resources serve as a safety net in the case of a disaster, such as a death in the family, or a societal calamity, such as failing agriculture, drought, flood, storm, or conflict and earthquake (Angelsen and Wunder 2003; Takasaki et al. 2004). Forests can serve as a kind of savings (Chambers et al. 1993) as well as natural protection (Pattanayak and Sills 2001; McSweeney 2004). The open-access nature of large tracts of public forest in developing nations ensures the poor community to use these forest resources to fulfil their bonafide needs without any difficulty (Sunderlin et al. 2008).

There is an apparent overlap in China between counties classified as critically impoverished and countries having substantial natural resources (Zhou and Veeck 1999). The majority of poor tribal people are positioned near forests in India. Forests provide a source of income to around 275 million people in India's rural areas; forest dwellers, which are mainly tribal, and are among the poorest and most-deprived people in India (World Bank 2006; Khare et al. 2000). Approximately, half of India's 350 million impoverished people reside in three states with little natural, physical, social, and human capital, and those in forest-based economies endure the highest poverty; 84% of India's tribal ethnic minority lives in forested areas (Mehta and Shah 2003). The incidence of poverty, as measured by the headcount ratio, is greater than the all-India figures for the majority of forest-based livelihoods. Except for Maharashtra, the prevalence of poverty is substantially lower in semi-arid regions (Shah and Guru 2004).

5.1.1 "Agroforestry Tree Products (AFTPs): Targeting Poverty Reduction and Enhanced Livelihoods

It's now widely accepted that agroforestry is an alternative paradigm for rural development. Instead of high-input monocultures that produce a limited set of staple food crops, agroforestry emphasizes species-rich, low-input agricultural techniques, including indigenous tree crops (Leakey 2001; Leakey and Tchoundjeu 2001; Haider 2019). This alternative approach, according to the United Nations Millennium Development objectives and environmental accords, tackles a variety of global concerns (Garrity 2004). There are several challenges related to deforestation, land degradation, unsustainable crop practices, biodiversity loss, rising hunger, poverty, malnutrition, and increased risk of climate change. In the last 10–15 years, agroforestry domestication strategies, approaches, techniques, along with commercialization, and marketing of agroforestry products (AFTPs), have become one of the pillars of this paradigm shift (Simons and Leakey 2004). Domesticating indigenous trees has the potential to become new cash crops by promoting different agroforestry practices.

6 Social Forestry an excellent Approach towards Nature Conservation

6.1 Social Forestry a Remarkable Solution for Sustainability and Disaster Management

Forestry can be classified into various branches based on climatic variations, practices, and uses. Among all, divisions social forestry is recognized as a leading subdivision concerning to unexpected and speedy environmental distresses (Sands 2013).

Social forestry is the management and practice of tree planting on various lands with the help of local communities and non-governmental organizations (Sands 2013). It is being practiced for improvement in the social fabric of the local communities. It has the capacity to enhance the ecological and socio-economic conditions of the community by promoting multipurpose trees. It is a ground reality that forests can be protected and managed by involving local people and government officials of that territory. It is observed that people with miserable socioeconomic conditions can improve their status and living standard by involving in management of forests on sustained basis (Baig et al. 2008).

Social forestry can be further divided into four types such as farm forestry, community forestry, urban forestry and agroforestry (Sands 2013). Farm forestry is growing of trees by farmers individually on farmlands to meet their domestic needs of wood for fuel wood, forage, and timber. Community forestry is stated as increasing forest area by tree plantation practices with the collaboration of local communities as well as Governmental and non-Governmental organizations and with the involvement of various stakeholders for the betterment of communities. Urban forestry involves practice of planting trees along roadsides, canal banks, railway tracks, and along play grounds for increasing tree/forest cover for amenity value and tackling noise and pollution hazards (Urban forestry project papers). Agroforestry is the practice of tree plantation with agricultural crops on the same unit of land for improving the socio-economic conditions of the farmers (Bargali and Bargali 2009). Social forestry will improve microclimate by avoiding the effects of direct exposure to solar radiation which is a major cause of high temperature (Beer et al. 1998). The elimination of trees can increase soil temperature about 4°C and reduce relatives' humidity at 2 m above ground by about 12% (Belsky et al. 1993).

Agricultural production can be increased and safeguarded from natural disasters by planting trees in many ways (Kiepe and Rao 1994). The ecosystem and soil structure is enhanced by multipurpose trees and shrubs and also considered as a source of human food, medicine, and animal feed in agroforestry systems. Because of the considerable economic advantage, the farmers are pinched to multipurpose trees in arid or semi-arid regions (Oyewole and Carsky 2001).

6.1.1 Agroforestry a Realistic Approach for Disaster Risk Reduction

Agroforestry has great potential and can uplift socioeconomic conditions of communities and ensure environmental sustainability in various climate zones of the world (Garrett et al. 2000; Alavalapati et al. 2001; Nair 2001). Sagacious use of multipurpose trees in agroforestry systems and practices can play a key role in lessening exposure to disasters which can be helpful for effective land use planning. Such practices can similarly help in retaining or restoring the wetlands and embankments of rivers and streams in the agricultural landscape that will ultimately reduce disturbances, including soil erosion. Through its alleviating effects on global climate change, agroforestry practices also help to counter the ongoing aggregating trends in disaster prevalence as a result of climate change (Van Noordwijk et al. 2019).

Social forestry practices and agroforestry systems are very useful in maintaining the soil efficiency at optimum level for long term. When we compare the agricultural crops with trees for soil reclamation and productivity the leguminous trees adopted in agroforestry systems, fix nitrogen, which is very important component for enhancement of soil properties. Leaf litters of the trees add essential micronutrients in soil. The productivity of land increases when agricultural crops are grown in combination with trees on the same land. The yield of agricultural crops is higher in the soil where trees grow easily than in the soil where trees cannot be easily grown (Chaturvedi 1981; Verinumbe 1987). In developing countries, shifting cultivation and agricultural expansion is a major factor responsible for land deterioration which ultimately affecting the food productivity. Land degradation can be considered a major reason for decline in food production. Planting trees on farm lands in the form of agroforestry improve cropping pattern/practices. The potential of such practices will enhance the cultivated land and will definitely play an important role in climate change mitigation and managing other environmental hazards (Mbow et al. 2014).

Farmers in disaster-prone areas frequently use agroforestry, which combines crops with trees. This technique can assist in preventing soil erosion and protecting crops (Orduño Torres et al. 2020; Rozaki et al. 2021). This farming approach can also be utilized as a disaster mitigation strategy for earthquake and landslide disasters. Agroforestry may rehabilitate the vegetation in the periphery of the forests that are prone to fires while simultaneously providing food through the crops, giving farmers the option to either use the produce themselves or sell in the market (Gross 2015). To avoid natural disasters like forest fires, landslides, and floods, it is important to manage natural resources, such as establishing and maintaining of tree crops (Mehring et al. 2011). Planning for flood disaster mitigation while considering the climate change impacts is important (Shrestha et al. 2019). Farming practices as mitigation strategies against natural hazards include wise use of crop varieties, fertilizers, and pesticides (Dzulkarnain et al. 2019; Budhathoki et al. 2020). The disaster insurance for crops is a better way to moderate the adverse effects of calamity before it happens (Viganò and Castellani 2020). Usually, risk-financing mechanisms are commonly used worldwide as a strategy. In agriculture insurance, there are several challenges, including the lack of experience, limited products and data and insufficient financial capacity (Alam et al. 2020). Livelihoods need to be restored because farmers or people need to earn money and think about how they will survive after the disaster (Sina et al. 2019). Agricultural-related institutions and farming communities play an important role in this situation by providing assistance and support. Diversity of livelihoods is a good way to secure economic situation, and variety of employments provide greater resilience against natural disasters (Rubiyanto and Hirota 2021). In comparison with other sectors, agriculture provides a better source of livelihood (Singha and Maezawa 2019). Government support, such as supplying of seeds or other inputs to reinstate the agricultural activities, is their major responsibility in such conditions (Aksa 2020).

The following products, such as fruit and nuts, fuel wood, lumber, drugs, animal feed, green manure, gum, resins, spices, and increase in income, are the major outcomes of agroforestry, which help small farmers (Raj et al. 2014). The poor, especially those who reside in rural areas, rely on nature for a variety of basic necessities like food, fuel, shelter, and medicines (Jhariya and Raj 2014). The production of dairy, honey bees, silk worms, and selection of appropriate plants for the manufacturing gum and resin in agroforestry systems is providing alternatives for livelihood (Dhyani and Handa 2013). Agroforestry approaches are becoming a realistic alternative for reducing the negative consequences of climate change (Singh et al. 2013). As a result, it is the most suitable practical solution for addressing the population's expanding needs. According to a global initiative of agroforestry records, it has the potential to: supplement multiple products, ecological restoration, sequester carbon and reduce adverse climatic effects; maintain soil fertility, which leads to quality and quantity production; reduce nutrient loss and soil erosion; improve microclimate of area by lowering soil temperature; and provide resistance from diseases, insect, etc., due to variety in crops (World Agro-Forestry Centre 2010).

Agroforestry has huge potential for reducing poverty in developing counties and improving living standards of people. Moreover, there is need to increase productivity on sustained basis without compromising the quality (Gaspar et al. 2007). Agroforestry has been recognized as a valid solution for meeting the demands of exploding population and also mitigating the adverse effects of climate change

(Dinesh et al. 2017). Various agroforestry systems and practices including integration of nitrogen fixing trees with fallow periods and intercropping trees with shrubs and crops can lead towards improved crop productivity and carbon sequestration in tropical and subtropical regions (Hall et al. 2005; Thorlakson and Neufeldt 2012). Carbon sequestration potential of agroforestry systems can fluctuate due to the adoption of diverse agroforestry practices (e.g., woodlots, farm forestry, shelterbelts, etc.) under various environmental conditions (Nair and Nair 2014; Nawaz et al. 2018).

Growing of woody trees in the farmlands of Punjab, Pakistan, is a tradition but it is practiced without any appropriate information and methodology. Practice of agroforestry on scientific basis and the suitability maps are very essential. A productive agroforestry system requires proper planning and management for maximizing the profit of agricultural lands. (Saralch et al. 2007).

Agroforestry practices that are focused on ecological, social and economic needs of farmers include alley cropping, woods cultivating, riparian forest buffers, agrisilvicultural, silvopastoral, agrisilvopastoral, and windbreaks. Naturally, agroforestry practices can diminish erosion, enhance water quality, microclimates, improve nutrient cycling and provide habitat to the wildlife (Workman et al. 2003; Bargali and Bargali 2009). No doubt, agroforestry can mitigate climate change and protect biodiversity from natural disasters but farmers will take interest only if they are convinced of its economic benefits.

7 Conclusion

Various natural disasters, including hurricanes, floods, forest fires, volcanic eruptions, earthquakes, and windstorms, are considered as major threats to agricultural productivity. The most common impacts of disasters and misfortunes on agriculture include reduction in the yield of crops, mortality of livestock, amplified susceptibility to diseases, and damage to infrastructure and irrigation systems. These impacts can have enduring effects on agricultural output, including cereal crops, vegetables, fruits, and tree growth.

Disaster mitigation is the strategy used to decrease the consequences of calamities, including those that occur before, during, and after such incidents. People need to take action in response to the calamities they are experiencing. They need to adopt different approaches to tackle various natural disasters for minimizing their affects.

Forests and trees supply plenty of ecological services that assist in shaping healthy living environments and in refurbishing degraded biomes. Trees can minimize the adverse effects of climate change and help to increase the agricultural crop productivity. The function of trees in disaster management and in soil conservation is critical for persuading farmers to plant trees on their farmland. The forest ecosystem regulates runoff and reduces the rate of siltation in downstream aquatic environments, extending reservoir, and dam shelf lives. So, we can conclude that agroforestry practices can help to counter the ongoing aggregating trends in disaster prevalence because of climate change.

References

Abid M, Schilling J, Scheffran J, Zulfiqar F (2016) Climate change vulnerability, adaptation and risk perceptions at farm level in Punjab, Pakistan. Sci Total Environ 547:447–460

Ackley S, Wadhams P, Comiso JC, Worby AP (2003) Decadal decrease of Antarctic sea ice extent inferred from whaling records revisited on the basis of historical and modern sea ice records. Polar Res 22(1):19–25

Agussabti A, Romano R, Rahmaddiansyah R, Isa RM (2020) Factors affecting risk tolerance among small-scale seasonal commodity farmers and strategies for its improvement. Heliyon 6:12

Ahmad I, Atiq M, Nawaz MF, Ahmed S, Asif M, Gull S, Tanvir MA, Aabdullah M, Azhar MF, Rajput NA (2019) Prediction of dieback disease of *Dalbergia sissoo* (shisham) based upon environmental factors and tree age. Appl Ecol Environ Res 17(3):6483–6495

Akhtar I, Nazir N (2013) Effect and waterlogging and drought stress in plants. Int J Water Res Environ Eng 2:34–40

Aksa FI (2020) Wisdom of indigenous and tacit knowledge for disaster risk reduction. Indones J Geogr 52:418–426

Alam ASAF, Begum H, Masud MM, Al-Amin AQ, Filho WL (2020) Agriculture insurance for disaster risk reduction: a case study of Malaysia. Int J Disaster Risk Reduct 47:101626

Alaoui-Sossé B, Gérard B, Binet P, Toussaint ML, Badot PM (2005) Influence of flooding on growth, nitrogen availability in soil, and nitrate reduction of young oak seedlings (Quercus robur L.). Ann Forest Sci 62:593–600

Alavalapati J, Nair P, Barkin D (2001) Socioeconomic and institutional perspectives of agroforestry. In: Palo M, Uusivuori J, Mery G (eds) World forests, markets and policies. World forests, vol 3. Springer, Dordrecht. https://doi.org/10.1007/978-94-010-0664-4_5

Anderson HW (1958) Rain-snow flood sources, meteorologically defined (Abstract). Bull Am Meteorol Soc 39:174–175

Anderson AB, May PH, Balick MJ (1991) The subsidy from nature: palm forests, peasantry, and development on an Amazon frontier. Columbia University Press, New York, NY

Angelsen A, Wunder S (2003) Exploring the forest–poverty link: key concepts, issues and research implications. CIFOR occasional paper number 40. Center for International Forestry Research, Bogor

Baig MB, Ahmad S, Khan N, Ahmad I, Straquadine GS (2008) The history of social forestry in Pakistan. Int J Soc For 1(2):167–183

Bakkour D, Enjolras G, Thouret JC, Kast R, Mei ETW, Prihatminingtyas B (2015) The adaptive governance of natural disaster systems: insights from the 2010 mount Merapi eruption in Indonesia. Int J Disaster Risk Reduct 13:167–188

Bargali K, Bargali SS (2009) Acacia nilotica: a multipurpose leguminous plant. J Nat Sci 7:11–19

Beer J, Muschler R, Kass D, Somarriba E (1998) Shade management in coffee and cacao plantations. Agrofor Syst 38(3):139–164

Belsky AJ, Mwonga SM, Duxbury JM (1993) Effects of widely spaced trees and livestock grazing on understory environments in tropical savannas. Agrofor Syst 24:1–20

Bosch JR, Hewlett JD (1982) A review of catchment experiments to determine the effect of vegetation changes on water yield and evapotranspiration. J Hydrol 55:3–22

Bruinsma J (2009) The resource outlook to 2050: by how much do land, water and crop yields need to increase by 2050. In: Expert meeting on how to feed the world in 2050. FAO and ESDD, Rome. ftp://ftp.fao.org/docrep/fao/012/ak542e/ak542e06.pdf. Accessed 14 Dec 2020

Budhathoki NK, Paton D, Lassa AJ, Zander KK (2020) Assessing farmers' preparedness to cope with the impacts of multiple climate change-related hazards in the Terai lowlands of Nepal. Int J Disaster Risk Reduct 49:101656
Bullock JM, Aronson J, Newton AC, Pywell RF, Rey Benayas JM (2011) Restoration of ecosystem services and biodiversity: conflicts and opportunities. Trends Ecol Evol 26:541–549
Burger H (1943) The water economy in the Sperbel and Rappen watersheds from 1927-28 to 1941-1942. mitt Schweiz Anstalt fur Forestl Versuchsw 23 (in German)
Byron RN, Arnold JEM (1999) What futures for the people of the tropical forests? World Dev 27:789–805
Carnus JM, Parrotta J, Brockerhoff E, Arbez M, Jactel H, Kremer A, Walters B (2006) Planted forests and biodiversity. J For 104:65–77
CEOS (2003) The use of earth observing satellites for hazard support, assessments and scenarios. Final report of the CEOS Disaster Management Support Group (DMSG).
Chambers R, Leach M, Conroy C (1993) Trees as savings and security for the rural poor. Gatekeeper series number 3. International Institute for Environment and Development, London
Chaturvedi AN (1981) Poplar for planting. Uttar Pradesh Department Bull No 50. Uttar Pradesh Department, Lucknow, p 27
Chen J, Saunders SC, Crow TR, Naiman RJ, Brosofske KD, Mroz GD, Brookshire BL, Franklin JF (1999) Microclimate in forest ecosystem and landscape ecology: variations in local climate can be used to monitor and compare the effects of different management regimes. Bioscience 49:288–297
CIRIA (2013) Land use management effects on flood flows and sediments—guidance on prediction (C719). CIRIA, London, England
Costa JE (1975) Effects of agriculture on erosion and sedimentation in the Piedmont Province, Maryland. Geol Soc Am Bull 86:1281–1286
Crooks S, Davies H (2000) Assessment of land use change in the Thames catchment and its effect on the flood regime of the river. Phys Chem Earth B Hydrol Oceans Atmos 26:583–591
De Roo A, Odijk M, Schmuck G, Koster E, Lucieer A (2001) Assessing the effects of land use changes on floods in the Meuse and Oder catchment. Phys Chem Earth B Hydrol Oceans Atmos 26:593–599
Dhyani SK, Handa AK (2013) Area under agroforestry in India: an assessment for present status and future perspective. Indian J Agrofor 15(1):1–11
Dijkstra FA, Smits MM (2002) Tree species effects on calcium cycling: the role of calcium uptake in deep soils. Ecosystems 5(4):385–398
Dinesh D, Campbell BM, Bonilla-findji O, Richards M (2017) 10 best bet innovations for adaptation in agriculture: a supplement to the UNFCCC NAP technical guidelines. CGIAR Research Program on Climate Change, Agriculture and Food Security (CCAFS), Wageningen
Dixon SJ, Sear DA, Odoni NA, Sykes T, Lane SN (2016) The effects of river restoration on catchment scale flood risk and flood hydrology. Earth Surf Process Landf 41:997–1008
Djalante R, Holley C, Thomalla F, Carnegi M (2013) Pathways for adaptive and integrated disaster resilience. Nat Hazards 69:2105–2135
Douglas I (1967) Man, vegetation, and the sediment yield of Rivers. Nature 215:925–928
Dubbeling M, van Veenhuizen R, Halliday J (2019) Urban agriculture as a climate change and disaster risk reduction strategy. Field Actions Sci Rep J Field Actions 20:32–39
Dzulkarnain A, Suryani E, Aprillya MR (2019) Analysis of flood identification and mitigation for disaster preparedness: a system thinking approach. Procedia Comput Sci 161:927–934
Ellis F (2000) Rural livelihoods and diversity in developing countries. Oxford University Press, Oxford
FAO (2003) Sustainable use and management of freshwater resources: the role of forests. In: Perlis A, Wearne L, Moore B (eds) State of World's forests. Center for International Forestry Research Food and Agriculture Organization of the United Nations, Rome
FAO (2021) The impact of disasters and crises on agriculture and food security: 2021. FAO, Rome. https://doi.org/10.4060/cb3673en

FAO, UNEP (2020) The State of the World's Forests 2020. Forests, biodiversity and people. FAO and UNEP, Rome, p 214. http://www.fao.org/3/ca8642en/ca8642en.pdf

Field CB, Barros VR (2014) Climate change 2014—impacts, adaptation and vulnerability: regional aspects. Cambridge University Press, New York

Garcia-Chevesich P, Wei X, Ticona J, Martinez G, Zea J, Garcia V, Alejo F, Zhang Y, Flamme H, Graber A, Santi P, McCray J, Gozales E, Krahenbuhl R (2021) The impact of agricultural irrigation on landslide triggering: a review from Chinese, English, and Spanish literature. Water 13:1–17

Garrett HE, Rietveld WJ, Fisher RF (2000) North American agroforestry. In: An integrated science and practice. American Society of Agronomy (ASA), Madison, WI, p 402

Garrity D (2004) World agroforestry and the achievement of the millennium development goals. Agrofor Syst 61:5–17

Gaspar P, Mesías FJ, Escribano M, Ledesma ARD, Pulido F (2007) Economic and management characterization of dehasa farms: implications for their sustainability. Agrofor Syst 71:151–162

Ghimire SK, Higaki D, Bhattarai TP (2003) Study of land use and stream planform dynamics using aerial photographs: a case study in Siwalik Hills of Nepal. Unpublished report. Department of Regional Environment Science, Hirosaki University, Japan

Gignoux J, Menéndez M (2016) Benefit in the wake of disaster: long-run effects of earthquakes on welfare in rural Indonesia. J Dev Econ 118:26–44

Gilman K (2002) Modelling the effect of land use change in the upper Severn catchment on flood levels downstream. English nature report no. 471

Gore T, Fischer TB (2014) Uncovering the factors that can support and impede post-disaster EIA practice in developing countries: the case of Aceh Province, Indonesia. Environ Impact Assess Rev 44:67–75

Gross M (2015) A fire with global connections. Curr Biol 25:R1107–R1109

Guardian T (2016) Planting more trees can reduce UK's flood risk, research shows. The Guardian. http://www.theguardian.com/ environment/2016/mar/11/planting-more-trees-can-reduce-uk-flooding-research-shows

Guarnacci U (2016) Joining the dots: social networks and community resilience in post-conflict, post-disaster Indonesia. Int J Disaster Risk Reduct 16:180–191

Haider S (2019) Agroforestry as a tool for sustainable rural development: a mini-review. PSM Biol Res 4:BR-2019-021

Hall NM, Kaya B, Dick J et al (2005) Effect of improved fallow on crop productivity, soil fertility and climate-forcing gas emissions in semi-arid conditions. Biol Fertil Soils 42:224–230. https://doi.org/10.1007/s00374-005-0019-8

Handmer JW (2000) Flood hazard and sustainable development. In: Parker D (ed) Floods. Routledge, London, pp 276–286

Hanna E, McConnell J, Das S, Cappelen J, Stephens A (2006) Observed and modeled Greenland ice sheet snow accumulation in 1958–2003 and links with regional climate forcing. J Clim 19(3):344–358

Haq M, Said R, Rahmatullah J, Mohammad J (2010) Pakistan—flood 2010 monitoring using MODIS data. In: The 9th international workshop of the CAS-TWAS-W Forum (2010 CTWF) on climate and environmental changes and challenges for developing countries, Beijing, China

Haq M, Akhtar M, Muhammad S, Paras S, Rahmatullah J (2012) Techniques of remote sensing and GIS for flood monitoring and damage assessment: a case study of Sindh province, Pakistan. Egypt J Remote Sens Space Sci 15:35–141

Harrabin R (2016) Tree planting 'can reduce flooding'. BBC, London. http://www.bbc.co.uk/news/science-environment-35777927

Harrowell E, Özerdem A (2019) Understanding the dilemmas of integrating post-disaster and post-conflict reconstruction initiatives: evidence from Nepal, Sri Lanka and Indonesia. Int J Disaster Risk Reduct 36:101092

Helmi A, Sasaoka M (2018) Dealing with socioeconomic and climate-related uncertainty in small-scale salt producers in rural Sampang, Indonesia. J Rural Stud 59:88–97

Helmi H, Basri H, Sufardi S (2019) Flood vulnerability level analysis as a hydrological disaster mitigation effort in Krueng Jreue sub-watershed, Aceh besar, Indonesia. Jàmbá 11:a737

Hernandez M, Scott N, Miller David C, Goodrich Bruce F, Goff William G, Kepner Curtis M, Edmonds, Bruce JK (2000) Modeling runoff response to land cover and rainfall spatial variability in semi-arid watersheds. Environ Monit Assess 64:285–298

Hijmans RJ (2003) The effect of climate change on global potato production. Am J Pot Res 80(4):271–279

Hobbie SE (1996) Temperature and plant species control over litter decomposition in Alaskan tundra. Eool Monogr 66:503–522

Hon'ble P, Patil PP, Kumar S, Deore SJ, Bhatt EB, Patil SU, Jadhav PD, Sonawane MJ (2019) Disaster management: new challenges and solutions

Hornbeck JW, Pierce RS, Federer CA (1970) Streamflow changes after Forest clearing in New England. Water Resour Res 6:1124–1132

Hughes TP, Baird AH, Bellwood DR, Card M, Connolly SR, Folke C, Grosberg R, Hoegh-Guldberg O, Jackson JB, Kleypas J, Lough JM (2003) Climate change, human impacts and the resilience of coral reefs. Science 301(5635):929–933

IFRCRCS (1997) World disasters report 1997. Oxford University Press, Oxford

IPCC (2007) Impacts, adaptation and vulnerability. Contribution of working group II to the fourth assessment report of the intergovernmental panel on climate change. In: Parry ML, Canziani OF, Palutikof JP, van der Linden PJ, Hanson CE (eds) Climate change 2007. Cambridge University Press, Cambridge, 976pp

Jhariya MK, Raj A (2014) Human welfare from biodiversity. Agro bios Newsletter 12:89–91

Johnson R (2008) The role of catchment land use planning in flood risk management. In: Paper presented at a Workshop on Flood Management in Local Planning, Austria/Slovenia

Jones PG, Thornton PK (2003) The potential impacts of climate change on maize production in Africa and Latin America in 2055. Glob Environ Chang 13(1):51–59

Joshi A, Arora A, Amadi-Mgbenka C, Mittal N, Sharma S, Malhotra B, Grover A, Misra A, Loomba M (2019) Burden of household food insecurity in urban slum settings. PLoS One 14(4):e0214461

Kahi CH, Ngugi RK, Mureithi SM, Ngethe JC (2009) The canopy effects of Prosopis juliflora (DC) and Acacia tortilis (HAYNE) on harbaceous plant species and soil phsico-chemical properties in Njempts, Kenya. Trop Subtrop Agroecosystems 10(3):441–449

Karjalainen E, Sarjala T, Raitio H (2010) Promoting human health through forests: overview and major challenges. Environ Health Prev Med 15(1):1–8. https://doi.org/10.1007/s12199-008-0069-2

Kaur B, Gupta SR, Singh G (2002) Bio amelioration of a sodic soil by silvopastoral systems in North-Western India. Agrofor Syst 54:13–20

Kaye JP, Resh SC, Kaye MW, Chimner RA (2000) Nutrient and carbon dynamics in a replacement series of *eucalyptus and Albizia* trees. Ecology 81(12):3267–3273

Kellens W, Terpstra T, De Maeyer P (2013) Perception and communication of flood risks: a systematic review of empirical research. Risk Anal Int J 33:24–49

Khare A, Sarin M, Saxena NC, Palit S, Bathla S, Vania F, Satyanarayana M (2000) Joint forest management: policy, practice, and prospects. Policy that works for forest and people Seri policy ber 3. World Wide Fund for Nature, New Delhi, and International Institute for Environment and Development, London

Kiepe P, Rao MP (1994) Management of agroforestry for the conservation and utilization of land and water resources. Outlook Agric 23:17–25

Kim Y, Engel BA, Lim KJ, Larson V, Duncan B (2002) Runoff impacts of land-use change in Indian River lagoon watershed. J Hydrol Eng 7:245–251

Knox JC (1972) Valley alleviation in southwestern Wisconsin. Ann Am Assoc Geogr 77:224–244

Knox JC (1977) Human impacts on Wisconsin streams. Ann Am Assoc Geogr 67:323–342

KPMG (2015) Flooding economic impact will breach £5bn. KPMG. https://home.kpmg.com/uk/en/home/media/press-releases/2015/12/flooding-economic-impact-will-breach-5bn.html

Kurzatkowski D, Leuschner C, Homeier J (2015) Effects of flooding on trees in the semi-deciduous transition forests of the Araguaia floodplain, Brazil. Acta Oecol 69:21–30
Kusumastuti RD, Arviansyah A, Nurmala N, Wibowo SS (2021) Knowledge management and natural disaster preparedness: a systematic literature review and a case study of East Lombok, Indonesia. Int J Disaster Risk Reduct 58:102223
Leakey RRB (2001) Win: Win landuse strategies for Africa: 2. Capturing economic and environmental benefits with multistrata agroforests. Int For Rev 3:11–18
Leakey RRB, Tchoundjeu Z (2001) Diversification of tree crops: domestication of companion crops for poverty reduction and environmental services. Exp Agric 37:279–296
Leenaers H, Rang MC, Schouten CJ (1989) Variability of the metal content of flood deposits. Environ Geol Water Sci 11:95–106
Liu-Lastres B, Mariska D, Tan X, Ying T (2020) Can post-disaster tourism development improve destination livelihoods? A case study of Aceh, Indonesia. J Destin Mark Manag 18:100510
Loster T (1999) Flood trends and global change. Geoscience Research Group, Munich Reinsurance Company
Massoud FI, Abrol IP, Yadav JSP (1988) Salt-affected soils and their management. Food and agriculture Organization of the United Nations, Rome. http://www.fao.org/docrep/x5871e/x5871e00. Accessed 03 Nov 2008
Mbow C, Smith P, Skole D, Duguma L, Bustamante M (2014) Achieving mitigation and adaptation to climate change through sustainable agroforestry practices in Africa. Curr Opin Environ Sustain 6:8–14
McElwee P, Calvin K, Campbell D, Cherubini F, Grassi G, Korotkov V, Le Hoang A, Lwasa S, Nkem J, Nkonya E, Saigusa N (2020) The impact of interventions in the global land and Agri-food sectors on Nature's contributions to people and the UN sustainable development goals. Glob Chang Biol 26(9):4691–4721
McSweeney K (2004) Forest product sale as natural insurance: the effects of household characteristics and the nature of shock in eastern Honduras. Soc Nat Resour 17:39–56
Mehring M, Seeberg-Elverfeldt C, Koch S, Barkmann J, Schwarze S, Stoll-Kleemann S (2011) Local institutions: regulation and valuation of forest use-evidence from Central Sulawesi, Indonesia. Land Use Policy 28:736–747
Mehta AK, Shah A (2003) Chronic poverty in India: incidence, causes and policies. World Dev 31:491–511
Memon AA, Muhammad S, Rahman S, Haq M (2015) Flood monitoring and damage assessment using water indices: a case study of Pakistan flood-2012. Egypt J Remote Sens Space Sci 18:99–106
Mitchell RJ, Campbell CD, Chapman SJ, Osler GHR, Vanbergen AJ, Ross LC, Cameron CM, Cole L (2007) The cascading effects of birch on heather moorland: a test for the top down control of an ecosystem engineer. J Ecol 93:540–554
Mottaleb KA, Mohanty S, Hoang HTK, Rejesus RM (2013) The effects of natural disasters on farm household income and expenditures: a study on rice farmers in Bangladesh. Agric Syst 121:43–52
Muir JA, Cope MR, Angeningsih LR, Jackson JE (2020) To move home or move on? Investigating the impact of recovery aid on migration status as a potential tool for disaster risk reduction in the aftermath of volcanic eruptions in Merapi, Indonesia. Int J Disaster Risk Reduct 46:101478
Munich Re (2003) Topics geo: annual review of catastrophes. Munich Re, Munich
Mursidi A (2017) Management of disaster drought in Indonesia. J Terap Manaj Dan Bisnis 3:165
Nair PKR (2001) Agroforestry. In: Our fragile world: challenges and opportunities for sustainable development, forerunner to the Encyclopedia of Life Support Systems, EOLSS Chapter 1, vol 25. EOLSS, pp 375–393
Nair PKR, Nair VD (2014) 'Solid–fluid–gas': the state of knowledge on carbon-sequestration potential of agroforestry systems in Africa. Curr Opin Environ Sustain 6:22–27. https://doi.org/10.1016/j.cosust.2013.07.014

Nawaz MF, Yousaf MTB, Yasin G, Gul S, Ahmed I, Abdullah M, Rafay M, Tanvir MA, Asif M, Afzal S (2018) Agroforestry status and its role to sequester atmospheric CO2 under semi-arid climatic conditions in Pakistan. Appl Ecol Environ Res 16(1):645–661

Nelson GC, Rosegrant MW, Koo J, Robertson R, Sulser T, Zhu T, Ringler C, Msangi S, Palazzo A, Batka M, Magalhaes M, Valmonte-Santos R, Ewing M, Lee D (2009) Climate change: impact on agriculture and costs of adaptation.

Nema P, Nema S, Roy P (2012) An overview of global climate changing in current scenario and mitigation action. Renew Sust Energ Rev 16(4):2329–2336

O'Connell PE, Beven KJ, Carney JN, Clements RO, Ewen J, Fowler H, Tellier S (2004) Review of impacts of rural land use and management on flood generation: impact study report (R&D technical report no. FD2114/TR). Defra/EA

Orbock-Miller S, Ritter DF, Kochel RC, Miller JR (1993) Fluvial responses to land-use changes and climatic variations within the Drury Creek watershed, southern Illinois. Geomorphology 6:309–329

Orduño Torres MA, Kallas Z, Ornelas Herrera SI (2020) Farmers' environmental perceptions and preferences regarding climate change adaptation and mitigation actions; towards a sustainable agricultural system in México. Land Use Policy 99:105031

Oyewole BD, Carsky RJ (2001) Multipurpose tree use by farmers using indige-nous knowledge in sub-humid and semi-arid northern Nigeria. Int Tree Crop J 11:295–312

Pattanayak SK, Sills EO (2001) Do tropical forests provide natural insurance? The microeconomics of non-timber forest product collection in the Brazilian Amazon. Land Econ 77:595–612

Peng Y, Zhu X, Zhang F, Huang L, Xue J, Xu Y (2018) Farmers' risk perception of concentrated rural settlement development after the 5.12 Sichuan earthquake. Habitat Int 71:169–176

Piereira HC (1973) Land use and water resources in temperate and tropical climates. Cambridge University Press, London

Piest RF, Spoomer RG (1968) Sheet and gully erosion in the Missouri Valley Loessial region. Trans Ame Soc Agric Engin 11:850–853

Potter KW (1991) Hydrological impacts of changing land management practices in a moderate sized agricultural catchment. Water Res Res 27:845–855

Pratiwi EPA, Ramadhani EL, Nurrochmad F, Legono D (2020) The impacts of flood and drought on food security in Central Java. J Civ Eng Forum 6:69

Putra AN, Nita I, Jauhary MRA, Nurhutami SR, Ismail MH (2021) Landslide risk analysis on agriculture area in pacitan regency in east java Indonesia using geospatial techniques. Environ Nat Resour J 19:141–152

Raj A, Jhariya MK, Pithoura F (2014) Need of agroforestry and impact on ecosystem. J Plant Dev Sci 6:577–581

Raza MM, Khan MA, Ahmad I, Bajwa AA, Aslam HMU, Ullah BA, Riaz K (2015) Forest pathogens and diseases under changing climate—a review. Pak J Agric Res 28(3):318–337

Rego SS, Ferreira MM (2011) Estresse hídrico e salino na germinação de sementes de Anadenanthera colubrina (Veloso) Brenan. J Biotec Biodivers 2(4):37–42

Reich PB, Oleksyn J, Modrzynski J, Mrozinski P, Hobbie SE, Eissenstat DM, Tjoelker MG (2005) Linking litter calcium, earthworms and soil properties: a common garden test with 14 tree species. Ecol Lett 8(8):811–818

Renfroe GW (1975) Use of erosion equations and sediment-delivery ratios for predicting sediment yield. In: Present and prospective technology for predicting sediment yields and sources, vol 40. U.S. Department of Agriculture ARS-S, Washington, pp 33–45

Robinson M, Cognard-Plancq AL, Cosandey C, Davidd J, Durande P, Fuhrerf HW, Zollner A (2003) Studies of the impact of forests on peak flows and baseflows: a European perspective. For Ecol Manag 186:85–97

Rozaki Z (2020) COVID-19, agriculture, and food security in Indonesia. Rev Agricult Sci 8:243–260

Rozaki Z, Rahmawati N, Wijaya O, Mubarok AF, Senge M, Kamarudin MF (2021) A case study of agroforestry practices and challenges in Mt. Merapi risk and hazard prone area of Indonesia. Biodiversitas 22:2511–2518

Rubiyanto CW, Hirota I (2021) A review on livelihood diversification: dynamics, measurements and case studies in montane mainland Southeast Asia. Rev Agricult Sci 9:128–142

Ruíz Pérez M, Belcher B, Fu M, Yang X (2004) Looking through the bamboo curtain: an analysis of the changing role of forest and farm income in rural livelihoods in China. Int For Rev 6:306–316

Running SW, Coughlan FC (1988) A general model of forest ecosystem processes for regional applications. Ecol Model 42:125–154

Saddiqui KM, Khan JA, Yasin SM (1984) Eucalyptus camaldulensis DEHN, its growth properties and utilization. Bulletin number 4. Pakistan Forest Institution, Peshawar

Sands R (ed) (2013) Forestry in a global context. CABI, Wallingford, pp 215–240. 9781780641584

Saralch HS, Singh B, Chauhan SK, Puri S, Panwar P (2007) Promising agroforestry practices in Punjab. Agroforestry: systems and practices. New India Publishing Agency, New Delhi

Sardar MS, Tahir MA, Zafar MI (2008) Poverty in riverine areas: vulnerabilities, social gaps and flood damages. Pak J Life Soc Sci 6:25–31

Sariyildiz T, Anderson JM (2003) Interactions between litter quality, decomposition and soil fertility: a laboratory study. Soil Biol Biochem 35:391–399

Shah A, Guru B (2004) Poverty in remote rural areas in India: a review of evidence and issues. CPRC-IIPA working paper number 21. Indian Institute of Public Administration, New Delhi

Shrestha BB, Perera EDP, Kudo S, Miyamoto M, Yamazaki Y, Kuribayashi D, Sawano H, Sayama T, Magome J, Hasegawa A, Ushiyama T, Iwami Y, Tokunaga Y (2019) Assessing flood disaster impacts in agriculture under climate change in the river basins of Southeast Asia, vol 97. Springer, Dordrecht, p 157

Simons A, Leakey RRB (2004) Tree domestication in tropical agroforestry. Agrofor Syst 61:167–181

Sina D, Chang-Richards AY, Wilkinson S, Potangaroa R (2019) A conceptual framework for measuring livelihood resilience: relocation experience from Aceh, Indonesia. World Dev 117:253–265

Singh NR, Jhariya MK, Raj A (2013) Tree crop interaction in agroforestry system. Readers Shelf 10:15–16

Singha U, Maezawa S (2019) Production, marketing system, storage and future aspect of potato in Bangladesh. Rev Agricult Sci 7:29–40

Sitorus SRP, Pravitasari AE (2017) Land degradation and landslide in Indonesia. Sumatra J Disaster Geogr Geogr Educ 1:61–71

Slingo JM, Challinor AJ, Hoskins BJ, Wheeler TR (2005) Food crops in a changing climate. Philos Trans R Soc Lond B Biol Sci 360(1463):1983–1989

Sun M, Chen B, Ren J, Chang T (2010) Natural disaster's impact evaluation of rural households' vulnerability: the case of Wenchuan earthquake. Agric Agric Sci Procedia 1:52–61

Sunderlin WD, Arild A, Brian B, Paul B, Robert N, Levania S, Sven W (2005) Livelihoods, forests, andconservation in developing countries: an overview. World Dev 33(9):1383–1402

Sunderlin W, Dewi S, Puntodewo A, Müller D, Angelsen A, Epprecht M (2008) Why forests are important for global poverty alleviation: a spatial explanation. Ecol Soc 13(2):24

Suryanto GE, Daerobi A, Susilowati F (2020) Crop insurance as farmer's adaptation for climate change risk on agriculture in Surakarta residency-Indonesia. Int J Trade Glob Mark 13:251–266

Takasaki Y, Barham BL, Coomes OT (2004) Risk coping strategies in tropical forests: floods, illnesses, and resource extraction. Environ Dev Econ 9(2):203–224

Thorlakson T, Neufeldt H (2012) Reducing subsistence farmers' vulnerability to climate change. Agricult Food Security 1:15. https://doi.org/10.1186/2048-7010-1-15

Tripathi OP, Pandey HN, Tripathi RS (2009) Litter production, decomposition and physicochemical properties of soil in 3 developed agroforestry systems of Meghalaya, Northeast India. Afr J Plant Sci 3(8):160–167

Udmale P, Ichikawa Y, Manandhar S et al (2014) Farmers' perception of drought impacts, local adaptation and administrative mitigation measures in Maharashtra state, India. Int J Disaster Risk Reduct. https://doi.org/10.1016/j.ijdrr.2014.09.011

United Nations HABITAT NAP (2019) Addressing urban and human settlement issues in National Adaptation Plans—a supplement to the UNFCCC technical guidelines on the National Adaptation Plan Process Nairobi. United Nations Human Settlements Programme (UN-habitat)
Van der Sande CJ, de Jong SM, De Roo APJ (2003) A segmentation and classification approach of IKONOS-2 imagery for land cover mapping to assist flood risk and flood damage assessment. Int J Appl Earth Obs 4:217–229
Van Noordwijk M, Hairiah K, Tata HL, Lasco L (2019) How can agroforestry be part of disaster risk management? In: van Noordwijk M (ed) Sustainable development through trees on farms: agroforestry in its fifth decade. World Agroforestry (ICRAF) Southeast Asia Regional Program, Bogor, pp 251–267
Vedeld P, Angelsen A, Sjaastad E, Kobugabe Berg G (2004) Counting on the environment: forest incomes and the rural poor. Environmental economics series number 98. World Bank, Washington, DC
Verinumbe I (1987) Crop production on soil under some forest plantations in the Sahel. Agrofor Syst 5(2):185–188
Vetaas OR (1992) Micro-site effects of trees and shrubs in dry savannas. J Veg Sci 3:337–344
Viganò L, Castellani D (2020) Financial decisions and risk management of low-income households in disaster-prone areas: evidence from the portfolios of Ethiopian farmers. Int J Disaster Risk Reduct 45:101475
Wang S, Bill W (2007) Pluralism in the economics of sustainable forest management. For Policy Econ 9:743–750
Water Directors (2003) Core Group on flood protection of the water directors (Europe): best practices on flood prevention, protection and mitigation. European initiative on flood prevention
WCD (2000) Dams and development: a new framework for decision-making. Earthscan, London
Wilkinson ME, Quinn PF, Benson I, Welton P (2010) Runoff management: mitigation measures for disconnecting flow pathways in the Belford burn catchment to reduce flood risk. Paper presented at the third international symposium of the British hydrological society: managing consequences of a changing global environment. British Hydrological Society, Newcastle University. https://research.ncl.ac.uk/proactive/belford/papers/BelfordBHS1.pdf
Workman SW, Bannister ME, Nair PKR (2003) Agroforestry potential in the southeastern United States: perceptions of landowners and extension professionals. Agrofor Syst 59:73–83
World Agro-forestry Centre (2010) Transforming lives and landscapes. World Agro-forestry Centre, Nairobi, pp 1–5
World Bank (2006) India: unlocking opportunities for forest-dependent people in India. Report number 34481-IN. World Bank, Washington DC
Yang TC, Yu PS (1998) The effect of land-use change on the design hydrograph. J Hydrol Changing Environ 3:207–216
Yulianto E, Utari P, Satyawan IA (2020) Communication technology support in disaster-prone areas: case study of earthquake, tsunami and liquefaction in Palu, Indonesia. Int J Disaster Risk Reduct 45:101457
Zang Y (2001) Economics of transaction costs saving forestry. Econ Ecol 36:197–204
Zhou L, Veeck G (1999) Forest resource use and rural poverty in China. For Econ 4:80–92

Risks of Deserts Locust and Its Mitigation

Tauseef Khan Babar

Abstract

The ancient migratory pest "Desert Locust" is one of the natural disasters that the generation of twenty-first century is still grappling with. Locusts belong to the family Acrididae of grasshoppers, which includes most short-horned grasshoppers. Globally, more than 10,000 grasshopper species are dispersed throughout tropical, temperate grassland, and desert areas. Of these, locusts are a group of 18–21 species with the ability to swarm over long distances. Desert locust, *Schistocerca gregaria* (Forksal), is the most dominant species of these with about 10 subspecies distributed in Europe, Asia, Africa, and Australia. Taxonomically, the locust and grasshopper are indistinguishable. But primary dissimilarity is that whether grasshopper species under favorable environment conditions forms a swarm (gregarious) or not (solitary). Often considered to be the "most important and dangerous of all migratory pests" which is known to invade 60 countries of the world. Interestingly, the proportion of crops damaged by locusts around the world is less than 0.2% but due to its voracious feeding, rapid reproduction, and extensive flight range, it can deprive a farmer from his sustenance in a single morning. Therefore, due to theses inherent features, the locusts have left indelible imprints in thoughts, views, scientific literature, along with arts of several cultures. Desert locust has three developmental stages (eggs, hoppers, adults), and its life cycle usually takes 2–4 months depending on the weather and ecological conditions. Although, the desert locust has been found in the Horn of Africa, since biblical times yet its intense outbreak in the recent past (2019–2020) is being linked with anthropogenic climate change and the increased frequency

T. K. Babar (✉)
Department of Entomology, Faculty of Agricultural Sciences and Technology, Bahauddin Zakariya University, Multan, Pakistan
e-mail: tauseefkhan@bzu.edu.pk

M. Ahmed, S. Ahmad (eds.), *Disaster Risk Reduction in Agriculture*, Disaster Resilience and Green Growth, https://doi.org/10.1007/978-981-99-1763-1_17

of extreme weather events such as El Niño and La Niña. Other factors like poor governance and the political uncertainties of these countries also intensified the plague and placed particular pressure on the already devastated nations. Unfortunately, the last locust outbreak (2019–2020) was the largest seen in the past 25, 30, 50, and 70 years in Somalia and Ethiopia, Pakistan, Iran, and Kenya, respectively, which posed serious economic, social, and environmental challenges. Overall, it affected 20 million persons in six (Ethiopia, Kenya, Somalia, South Sudan, Uganda, Tanzania) of the eight East African countries and put them at risk of acute food insecurity. The World Bank estimated that East Africa and Yemen alone suffered damages totaling US$8.5 billion. All over the world, maximum control measures implemented over years depended upon the application of conventional insecticides which are generally neurotoxic for locust. However, on account of their hazardous effects, there is an urge to invest for alternate controlling techniques which are ecofriendly as well as sustainable, like microbial insecticides. Other environmentally safe tools include georeferencing, global positioning systems, insect growth regulators, botanicals, and semiochemical traps. However, most of these are at infancy stages or not readily available in market. Herein, this chapter a review about the historical aspects, biology, and physiology of desert locusts is being presented. Furthermore, this chapter also sheds light on how climate change played a role in the in the irruption of the locust crisis, repercussions of the locusts invasion, and the potential of novel tools like remote sensing and the fungal-based biopesticides in its management is discussed.

Keywords

Biopesticides · Conventional insecticides · Desert Locust · El Niño and La Niña · Locust outbreak · Natural disaster · Remote sensing · Socio-economic effects

1 Introduction

Since the dawn of civilization, human beings have been vulnerable to the unforeseen and unpredictable natural disasters for their survival. These calamities include sudden-onset of events such as droughts, earthquakes, floods, mudslides, tsunamis, and wild fires (Alexander 1993; Sorokin and Horowitz 2017). Even the contemporary world includes the catastrophic effects of the ancient migratory pest "Desert Locust" among these (FAO 2021). Locusts belong to the family Acrididae of grasshoppers, which includes most short-horned grasshoppers (Ellis 1950; Uvarov 1944). Globally, greater than ten thousand species of grasshopper are dispersed in tropical, temperate grasslands, along with desert regions. Locusts are group of 18–21 species with the ability to swarm over long distances (Peng et al. 2020). *Schistocerca gregaria* Forksal, is dominant species of these having around 10 sub-species dispersed in continents like Asia, Africa, Australia, and Europe (Çiplak 2021; Fontana et al. 2017). Importantly, in terms of crops and food supply population, the species symbolize major and maximum damaging group of insects (Bennett 1975). Due to the extensive migration and rapid increase in their numbers, *S. gregaria* is a key worry while *Locusta migratoria* Linnaeus, is most extensive (Cease et al. 2015; Pandey et al. 2021). Different locust species found around the world are listed in Table 1.

Often considered to be the "most important and dangerous of all migratory pests" in the world, as it can deprive a farmer from his sustenance in a single morning (Cressman 2016). A swarm of the desert locust (*S. gregaria*) quantifying in a square kilometer contains 80 million adults, has the ability to eat foodstuff for 35,000 persons/day (Fig. 1) (Cressman 2013). After every 8 weeks, a new generation of locusts emerges, with an average population increase of 20-times each generation (FAO 2021), and it spreads to new areas, disrupting the food supply, and upending livelihoods that requires substantial resources to address (Yuga and Wani 2022; Zhang et al. 2019). Unfortunately, the locust swarms often brought devastation and famine to entire nations. For example, locust plague of the Roman colonies in 125 BC created a famine which resulted in the death of 0.8 million people (Uvarov 1944). Similarly, the voracious pest in 1958 in Ethiopia ruined 167,000 tons of grain that might be sufficient for feeding one million citizens in a year (Steedman 1988).

Locusts and grasshoppers dispersed in the whole world, however, increasing trends are reported towards equator and flourish in subtropical as well as tropical states (Chen et al. 2020; Çiplak 2021). All the locusts are grasshoppers; however, all the grasshoppers are not locusts. Taxonomically, the locust and grasshopper species are indistinguishable. Grasshopper species under favorable environment conditions forms a swarm (gregarious) or not (solitary) (Mohamed et al. 2018; Pflüger and Bräunig 2021; Simpson et al. 1999). However, the locusts in the gregarious forms are voracious feeders and a worldwide threat to the food availability (Lecoq 2003; Xu et al. 2021). The desert locust invades more than 60 countries of the world which constitutes more that 20% of earth's dry-land of North Africa, Middle East, and Indian subcontinent (Fig. 2) (Chen et al. 2020; Harmon 2009). Although, the locusts inhabitate remote as well as sparsely populated zones; however, their spread through

Table 1 Different locust species distributed across the World (Usmani and Usmani 2018)

Common name	Scientific name	Country
African migratory locust	*Locusta migratoria migratorioides*	Africa
Australian plague locust	*Chortoicetes terminifera*	Australia
Bombay locust	*Patanga succincta*	South West-South East Asia
Brown locust	*Locustana pardalina*	South Africa
Italian locust	*Calliptamus italicus*	Western Europe -Central Asia
Moroccan locust	*Dociostaurus maroccanus*	North West Africa-Asia
Oriental migratory locust	*Locusta migratoria manilensis*	South-East Asia
Red locust	*Nomadacris septemfasciata*	East Africa
Tree locusts	*Anacridium sp.*	Africa, Mediterranean, near east

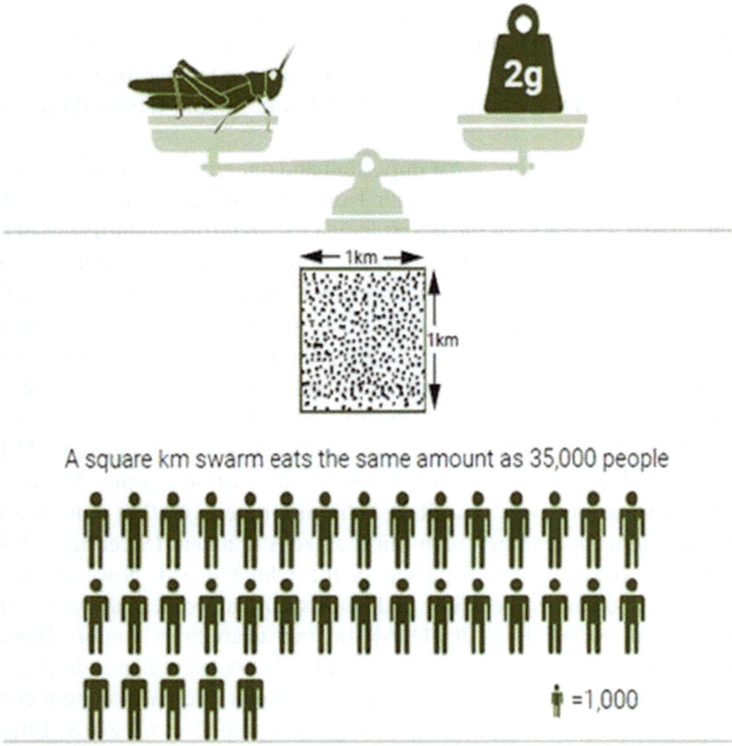

Fig. 1 Locusts consume equivalent of their own weight in a day and 1 sq. km of locusts can consume the same amount of food consumed by 35,000 people (UN 2021)

the continents is a principle factors responsible for outbreaks. Therefore, there are major obstacles in monitoring and controlling locust populations (Usmani and Usmani 2018). It is worth noting that global percentage of crops spoiled by locusts is less than 0.2% (Wewetzer et al. 1993), while comparatively 30–40% yield losses are reported every year due to insect pests (Lewis et al. 1997).

2 Locusts in the World History

Because of the degree of destruction and remarkable behavior of migrant swarms, the locusts left ineffaceable marks from ancient times in thoughts, views, scientific literature, along with arts of several cultures around the world (Dominy and Fannin 2021; Kevan 1978). Historically, a tablet of stone in pre-cuneiform writings for the first time in Assyrian era (ca. 3200 BC) mentioned their damage (Barton 1913; Horne 1917). Since then, the evidence has multiplied over the centuries with many books, religion-based text, travelers' testimonies, missionaries, and ecologists (Cressman 2016). Even some cultures believe that locusts created the world and

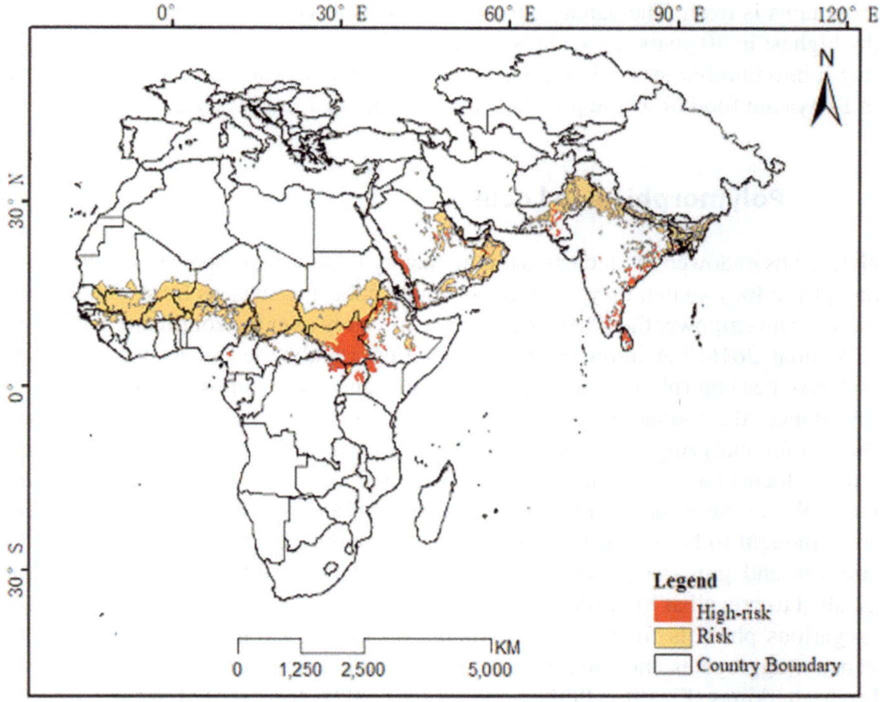

Fig. 2 Desert locusts migrate from east to west between these areas in Africa and Asia between10°N and 30°N. Since the range of at-risk areas changes mainly from March to October, the risk areas are likely the main activity areas for desert locusts in summer; the high-risk areas are the areas where desert locusts are common in winter (Chen et al. 2020)

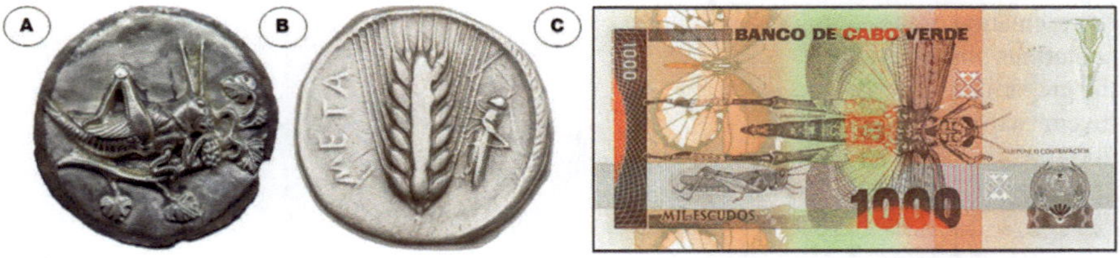

Fig. 3 Historical coins and currency notes showing images of the desert locust (Marino-Perez 2022)

ruled the fire (Lecoq and Pierozzi 1995). In the past, the locusts were also used for political and military drives (Peloquin 2013). Similarly, in the archives of few earliest medallions and coins illustrating the interactions among locusts and crops such as barley were found (Fig. 3a, b) (Dominy and Fannin 2021). For example, to portray the furious act of the desert locust, Cape Verde, an African country issued a banknote depicting gregarious adults of *S. gregaria* consuming maize stalk, an important food crop for the country (Fig. 3c) (Dominy and Fannin 2021). Even now, the trauma is fresh. The damage to the economies of East Africa and North India is the highest in 70 years. Its scale is so great that news reports have a hard time putting it into familiar terms. Economists' estimate that a swarm of 200 billion locusts in Kenya ate food of whole populace of Germany in 1 day (FAO 2021).

3 Polymorphism in Locusts

Nature has endowed the locusts with the faculty to alter their appearance, behavior, and physiology so that they can transform into cohesive-swarms from solitary individuals and empower them for long distance traveling over the continents and oceans (Cressman 2016; Latchininsky et al. 2016). Importantly, the locusts upon crowding undergo these morphological and behavioral changes (Latchininsky et al. 2011). In this stance, the founder of Anti-Locust Research Centre "Sir Boris Uvarov" was the first for analyzing and presenting "Theory of Phase Polymorphism" which states that the locust has two phenotypes or phases, and an individual can switch between these phases based on population or density (Uvarov 1921). Initially, these phases were thought to be two distinct species, but later on recognized and nominated as solitaria and gregaria (Uvarov 1977; Uvarov 1944). Solitarious phase is distinguished to prevail at lower density, and has cryptic green or tan color (Fig. 4a). The gregarious phase is found when favorable ecological circumstances favor higher density; aggregates and migrates *en masse*, along with combinations of dark and bright markings (Fig. 4b) (Pflüger and Bräunig 2021; Rogers et al. 2014).

Physiologically, it has been identified that changing of serotonin, an mediator of neuronal plasticity, is responsible for this behavioral shift at the onset of mass migration (Anstey et al. 2009). The locust population density determines the

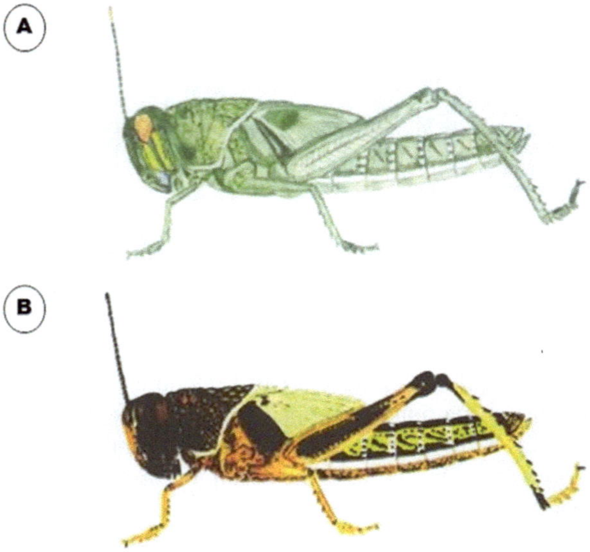

Fig. 4 Pictorial representation of the solitarious phase (**a**) and the gregarious phase (**b**) juveniles (Sword et al. 2000)

transition from disorder individual movement to a highly aligned collective motion (Fig. 5) (Buhl et al. 2006; Rogers et al. 2014). Typically, during the favorable circumstances of drought followed by fast growth of vegetation, locust's brains start releasing serotonin that triggers changes in series, and started breeding abundantly (Figs. 5 and 6) (Morgan 2014; Tanaka and Nishide 2013). Nymphs being wingless form bands those later on mature to adults with wings forming swarms. While moving around the field strips in a speedy manner both bands and swarms damage the crops. Adults being powerful fliers can travel greater distances and devour mainly green vegetation, where swarms settle. Groups are being acting as cohesive-units while moving across landscapes, mainly moving downhill; however, making way round barriers while merging with new bands. Although attraction among insects appears to be primarily visual; olfactory cues are also involved, allowing them to travel tens of kilometers in few weeks, bands seem to rely upon the sun for navigation. They stop marching for feeding at regular intervals (Usmani and Usmani 2018). Low-density solitary locust plays a serious part in grassland ecosystems functioning through nutrient recycling, determining plant-community structures, and serving as a source of food for several animals however gregarious locusts swarm migrates longer distances, destroys 80–100% of crops as well as pastures, exposes bare soil for erosion, and affects the livelihood of human beings (Latchininsky et al. 2011; Le Gall et al. 2019). A worldwide distribution of the different phases of the desert locust is shown in Fig. 7.

Fig. 5 Transformation of solitarious phase of a locust into the gregarious phase. The process begins with the unintentional gathering of the solitarious phase locusts especially in search of food during the dry spells when the vegetation dies off and leaves minimal areas of green within the desert. Within hours of crowding, a boost of serotonin in the central nervous system (CNS) spurs behavioral changes such as rapid movements, sociability, increased self-grooming, and a more varied appetite. The returning rains provide moist breeding grounds. New generations form groups that forage in unison. Once their wings grow, the herds become the flying swarms in search of more food and habitat. Although uncommon, a major plague begins, when swarms travelling at 30–60 miles a day (Maggiacomo and Kaya 2019)

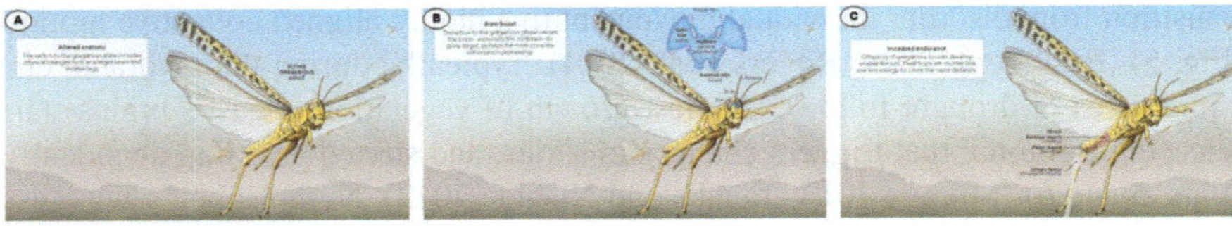

Fig. 6 Physiological changes associated with transformation of solitarious phase locust into the gregarious phase. The switch to gregarious phase includes physical changes such as a larger brain and shorter legs (**a**). Transformation to the gregarious phase causes the brain especially the mid-brain to grow larger, perhaps for more complex information processing (**b**). Offsprings of gregarious locusts develop smaller femurs. Their hops are smaller but use less energy to cover the same distance (**c**) (Maggiacomo and Kaya 2019)

Fig. 7 Global map showing the distribution of ranges invaded by the different phases of desert locusts (Maggiacomo and Kaya 2019)

4 Life Cycle of a Desert Locust

4.1 Eggs

Females lay their eggs in bare sandy soil areas (Harry 1970) with moist soil from a depth of 0–15 cm (Fig. 8) (Cressman 2013). This depth of soil provides optimal moisture for locusts breeding and development when there has been more than 25 mm of rainfall over the past two consecutive months (Maeno et al. 2020). Eggs dry out if laid in dry soil, except it rains immediately after the eggs are laid. However, excessive rainfall can be detrimental, as the eggs can be destroyed if heavy rains or floods occur after spawning (Piou et al. 2019). Furthermore, the desert locust eggs development also depends on soil temperature at depth where eggs are laid (Hurst 1965). Optimal temperature of soil for viability of locusts eggs vary from 15–35 °C, while temperature above 35 °C potentially kills the eggs (Cressman 2013; Symmons and Cressman 2001).

4.2 Hoppers

The wingless locusts at the juvenile stage are called "Hoppers" and on average take 36 days to develop from eggs. However, hoppers could develop quickly in 24 days or slowly in 95 (Cressman 2016). Typically, hoppers moult five or six times earlier ultimately becoming adult in around 4–6 weeks (Fig. 8). Transformation of the wingless hoppers of final moult to the winged adults is called "fledging". The new adult in the fledgling state has soft wings that need out to dry and harden after a few days before it can fly (Cressman 2013). Development from eggs into hoppers is also a function of temperature which decreases as everyday temperature rises from 24–32 °C (Wang and Kang 2005). Warmer temperature, faster the hoppers will reach adulthood. Importantly, correlation between air temperature and hoppers is a bit less clear than relationship between air temperature and eggs; because hoppers can largely control body temperature by basking in sun or seeking shade whenever needed (Symmons and Cressman 2001). At midday when temperature is warmest,

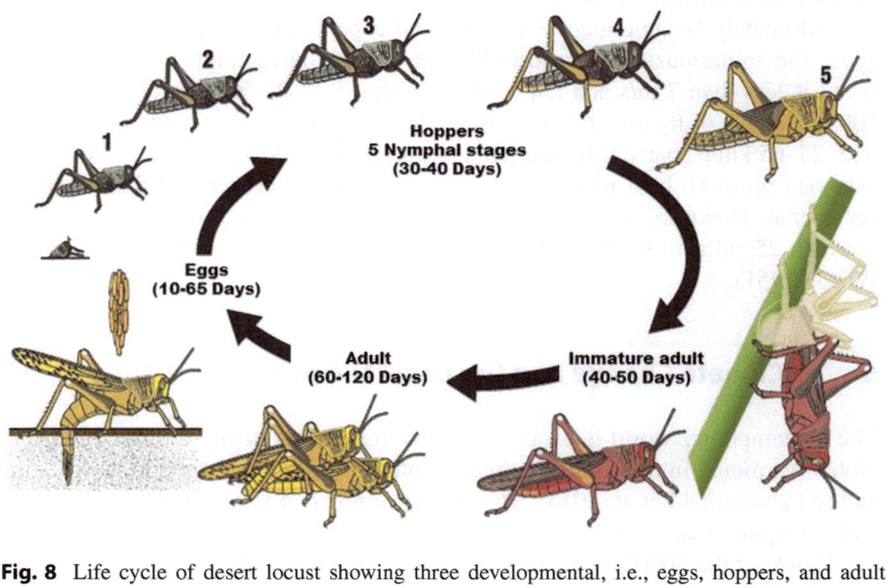

Fig. 8 Life cycle of desert locust showing three developmental, i.e., eggs, hoppers, and adult (Onyango et al. 2020)

hoppers seek shelter inside plants. Hoppers indirectly need rainy circumstances when require edible vegetation to consume during their growth (Cressman 2013).

Under favorable conditions hoppers can congregate to form bands that will transform into adults and swarm. The grouping of hoppers (bands) normally occurs in less uniform and open terrains where there are patches of dense vegetation interspersed with large areas of bare soil (Cressman 2013). These bands will stride downwind for searching vegetation to feed during warm, sunny days, however, do not move on cloudy days (Buhl et al. 2006). In warm and sunny days, bands will move in an alternate pattern of marching along with roosting in vegetation whole day. When vegetation is too dry, bands may continue marching whole night for searching green vegetation (Ellis 1963). They need to feed on a large number of plants to provide nutrition for the growth, mating, and reproduction period. On an average, the small locust can consume about 100 grams of plants in its life time, while large locusts can eat hundreds of grams of plants. The scientists have estimated that in a day 30–50 million locusts can travel 150 km and eat 200 tons of crops (Peng et al. 2020).

4.3 Adults

On an average adults take 2-4 months to mature but may mature as rapidly as in 3 weeks or to 9 months (Fig. 8) (Cui et al. 2019). Adults inhabiting in areas receiving "significant rains" in the recent past matured more rapidly than those occupying the dry habitats or areas with low temperatures (Branson 2017; Veran et al. 2015). The adult stage is of significant value because they are able to fly and reproduce (Munro and Saugstad 1938). Adults are sexually immature when they first fly, however, ultimately became mature sexually and capable of mating as well as laying eggs. The solitarious adults tend to fly and migrate during night time when wind speed is less than 7 m/s and temperature is above 20–22 °C (Taylor and Thomas 2003). Heights of fly might be to 1800 m, however, usually stay underneath 400 m with 25–65 km/h. speed. Flying period is usually limited to a few hours but can be extended up to 10 h. If temperature falls under 20 °C, continuous flights will be occasional. However, it is observed that the solitarious locusts are most active between 25 °C and 30 °C of the soil temperature (Neville and Weis-Fogh 1963; Wilson 1961).

5 Climate Change and the Desert Locust

The contemporary world is experiencing the repercussions of "climate change or global warming" in the form of unprecedented rise in in atmospheric temperature and sea level (Field et al. 2014). The Intergovernmental Panel on Climate Change (IPCC) reports that it has been warming each year for the past three decades, with the decade of the 2000's being the warmest (Field et al. 2014) and similar to other living organisms, insects are also sensitive to warming climate (Guo et al. 2011).

The rising temperature is likely to affect the insect population by causing an expansion in ranges, shift in phenology, and acceleration of developmental rates (Lehmann et al. 2020; Skendžić et al. 2021). Each of these three responses is due to different grasshopper species that have been shown to be expanding their distribution range northwards (Olfert et al. 2011), exhibiting earlier phenology compared to previous decades (Guo et al. 2009), and indicating faster development both at embryonic (Fielding and Defoliart 2010) and postembryonic stages (Guo et al. 2009; Wu et al. 2012). For locusts, however, there is scarcity of information on effects of global climate change, but only limited work is available about the oriental migratory locust (*L. migratoria manilensis*) (Latchininsky 2013). In the past, a millennium (957–1956) analysis of China's locust outbreaks and weather data revealed unexpected results: the outbreaks appeared to be associated with cold, wet rather than hot, dry periods (Stige et al. 2007) and with drought/flood frequencies (Pörtner et al. 2022). When the analysis period was expanded to almost 2000 years, the negative relation with temperature persisted, but it became inconsistent with precipitation, with more locusts were found in dry years (Tian et al. 2011). Based on it, the researchers concluded that global warming will be useful in terms of decreasing the frequency of locusts outbreak (Zhang et al. 2019). However, given the highly inconsistent and even contradictory results of their studies, such prediction appears more a speculation (Latchininsky 2013). There is no doubt that changing ecological circumstances like emerging of lush green vegetation because of unusual precipitation can lead to a surge in desert locust numbers (Hielkema et al. 1986). For example, disaster that occurred due to locust in Sahel area in north-western Africa between July 2003 and April 2004 was caused by above-average rainfall (Ceccato et al. 2007). In the foregoing paragraph, an analysis of the role of climate change in the irruption of latest locust crisis (2019–2020) is being presented.

6 Climate Change and the Current Locust Crisis

Since biblical times, the desert locust has been found in the Horn of Africa and its intense outbreak in the recent past (2019–2020) is being associated with anthropogenic climate change and augmented occurrence of extreme events of weather (Roussi 2020). Although, other factors like poor governance and the political uncertainties of these countries also exasperated the plague and placed particular pressure on the already devastated nations (Salih et al. 2020; Schlosser and Saulnier 2000). The exceptional wet-weather in the East Africa is being connected with wider climate system of Indian Ocean Dipole (IOD) (Marchant et al. 2007) which exerts influence on the weather of ocean on both sides, from East Africa and Arabian Peninsula to Indonesia, Papua New Guinea, and Australia (Saji et al. 1999). The oceans absorb about 90% of anthropogenic heat (Barnett et al. 2005; Zanna et al. 2019), and western portion of Indian Ocean most rapidly warms over the tropical Ocean system, with an average increase of 1.2 °C in the summer (Roxy et al. 2014). This warming has increased the frequency and intensity of extreme climate events of the neighborhood areas (Murakami et al. 2017). Historically, the strongest on record to affecting Arabian Peninsula took place during past 15 years (Lecoq and Cease 2022a). The dipole has three phases, i.e., positive, negative, and neutral (Fig. 9) (Saji and Yamagata 2003). The positive phase is characterized to have high temperatures in western Indian Ocean than in eastern and produces a positive index (Vinayachandran et al. 2009). Subsequently, it weakens the westerly and occasionally easterly winds, drawing warm waters towards Arab Peninsula and Horn of Africa. It plays a role in producing cyclones and heavy rainfall in the region as additional warmth along with moisture brought by climate system acts as fuel for budding storms (Fig. 9) (Ashok et al. 2001; Huang et al. 2005). Importantly, the IOD has been in its positive phase since June 2018 and reaching its most extreme positive level during 40 years in 2019. The frequency of occurrence of positive dipole phase measured in 30 years time-period increased from roughly four times during first half of twentieth century to ten times in the latter half, between 1979 and 2009 (Cai et al. 2009). Uncontrolled future climate change can pave to a three times increase in positive dipole events by 2099 compared to the period between 1900 and 1999 (Cai et al. 2014). If world temperature upsurge is restricted to 1.5 °C, quantity of positive dipole events could still be double numbers in the pre-industrial time (Cai et al. 2018). Furthermore, the climate experts forecast additional extreme events like droughts, floods, and cyclones. The drought conditions tend to cause a decrease in the locust population while floods and cyclones often results in their outbreaks (Salih et al. 2020). Increased local rainfall might favor breeding circumstances for locusts by defining extent of feeding zones which can lead to changes in the plague development (Ceccato et al. 2007; Peng et al. 2020).

The latest locust plague (2019–2020) happened in a series of events. It all started during May 2018, when an exceptionally dominant tropical cyclone "Mekunu" made landfall over Arabian Peninsula (Fig. 10). The moisture laden cyclone allowed the growth of rich vegetation to grow in the otherwise barren environment, making it conducive for many desert locusts hunting for food (Stone 2020), to move into the area and transform themselves into the gregarious phase once they reached a certain density (Dominy and Fannin 2021). Tropical cyclones usually weaken when they make landfall; however, Mekunu crossed over Oman, causing excessive rains which developed the desert lakes over "Empty Quarter" in Kingdom of Saudi Arabia.

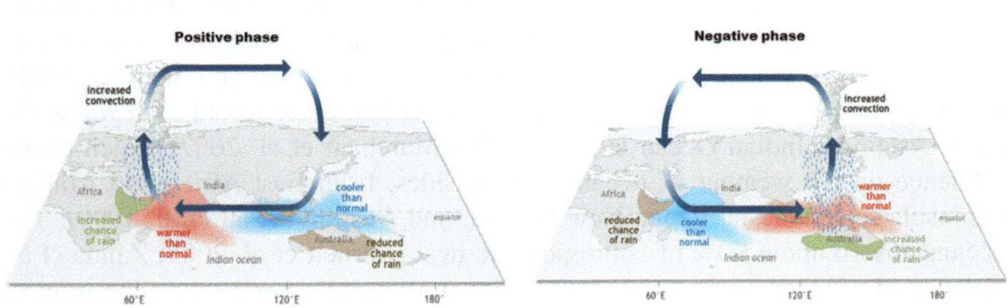

Fig. 9 Diagrammatic presentation of the positive and negative phases of the Indian Ocean Dipole (Johnson 2020)

Warm, sandy, and moist soil provided an ideal environment for locusts to hatch from eggs, develop, and breed (Cressman 2013; Salih et al. 2020). Normally, dry circumstances in this area would kill the breeds, however, alternative tropical cyclone "Luban" followed during October 2018, providing a lifeline for continuance of initial outbreak (Dominy and Fannin 2021). Outbreak range across Yemen was out of control because of political instability and lack of coherent government reply (Meynard et al. 2020). This period was tracked with a specifically mild-winter, allowing locusts to survive in greater quantities. During late 2019, winds of yet another tropical cyclone "Pawan" helped movement of locusts to East Africa (Fig. 10) (FAO 2020a). The time, when the insects moved through East Africa, area was hit by extraordinarily wet circumstances and more cyclones—allowing swarms to grow more larger (Salih et al. 2020). It is worthwhile to mention here that Horn of Africa was hit by eight cyclones, the largest number in any year, since 1976 (FAO 2020a; Salih et al. 2020). Overall, all this escalated frequency of cyclones along with extreme climate variability paved to upsurge likelihood of desert locust outbreaks and spread (Fig. 10) (Meynard et al. 2020; Roussi 2020).

Climate experts are predicting that temperature will continue to increase (Engelbrecht and Monteiro 2021; WMO 2022). Temperature governs rapidity of locust development along with swarm movements. Increased temperature under climate change scenario could potentially shorten long maturation as well as incubation periods during spring and allow an extra breeding generation to happen in North-West Africa, Arabian Peninsula, and South-West Asia (Cressman 2013). In nutshell, this could increase the number of locusts that occur annually in these regions and enhance the risk of locusts plague. The climate change scenario is likely to influence the El Niño and La Niña events (Rosenzweig et al. 2001), which might affect the breeding of desert locusts in winter in Horn of Africa as well as in summer in West African Sahel (Cressman 2013). Any change in wind speed, direction, and

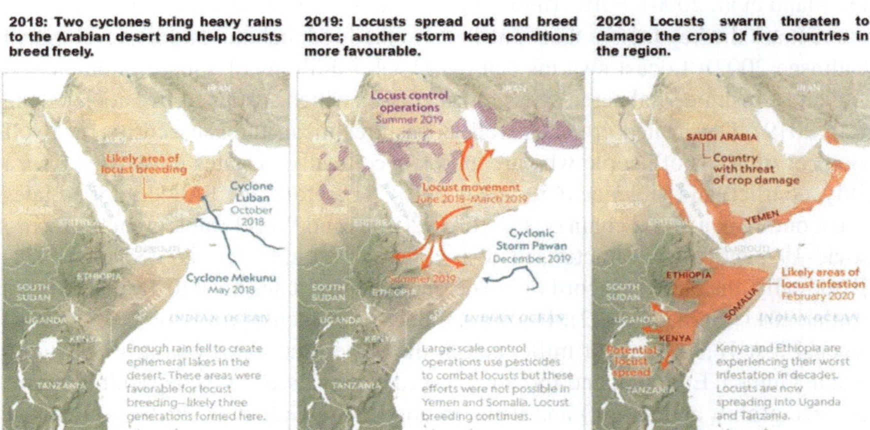

Fig. 10 Global map showing how the climate change and the occurrence of rains lead to the locust crisis of 2019–2020 (Stone 2020)

circulation flows is likely to alter migration of locust, allowing adults along with swarms to reach in new zones at different times of year. Their ability to establish, survive, and breed in new zones depends on ecological, habitat, along with weather conditions. This is how unexpected weather conditions and cyclones fueled by climate change played a role in driving locust outbreaks (Cressman 2013).

7 Effects of Desert Locust Invasion

According to "Global Report on Food Crises-2020" approximately 135 million people in 55 countries, or 16% of the total population analyzed, fell into the category of 'crisis condition I or worse' in 2019 and unfortunately four (Ethiopia, South Sudan, Sudan, Yemen) out of the ten most affected countries experienced the locust outbreak of 2019–2020 (Crises 2020; Kalakkal and Singh 2021). Food and Agriculture Organization (FAO) of United Nations (UN) stated that this locust outbreak (2019–2020) was the largest seen in the past 25, 30, 50, and 70 years in Somalia and Ethiopia, Pakistan, Iran, and Kenya, respectively (Kalakkal and Singh 2021). Desert locust, being a global threat presents serious economic, social, and environmental challenges (Fig. 11) (Lecoq 2004). Therefore, a precise outline of the losses incurred due to the desert locust is being presented.

7.1 Food Security and Nutrition

The polyphagous pest due to its voracious feeding not only destroys the agricultural landscape of the invaded areas but also affects the livestock and wild animals of that particular territory (Brader et al. 2006; Escorihuela et al. 2018). Therefore, the movement of this deadly pest towards the rural areas is always problematic (Despland et al. 2000). Furthermore, the availability of the favorable host and environment are the key factors that contribute to the food, habitat, and its development (Culmsee 2002). Locust swarms can overwhelm millions of square kilometers of field and consume vital cereals as well as grasses which sustained livestock (Zhang et al. 2019). Devastated crops as well as pastures mean food insecurity for livelihoods of millions of people who are at risk specifically in states devastated by fragility, conflict, and violence (Cressman 2016). Therefore, the FAO described latest locust outbreak a "food chain crisis" (FAO 2020c) and World Bank in the context of Covid-19 pandemic characterized it as a "crisis within a crisis" (Sparks 2021). Historically, the earliest record of havoc due to locust dates back to 1866 in Algeria (Cutler 2011; Spinage 2012). During the decade of locusts invasion (1952–1962), Senegal lost 16,000 tons of millet and at the same time many countries including Sudan, Guinea, Ethiopia, and India witnessed heavy losses in the agricultural crops (cereals, grains, and cotton) and fruits (oranges) (Le Gall et al. 2019). The locusts invasion of 1987–1989 in Mali and Mauritania destroyed most of the grazing lands and agricultural crops (De Vreyer et al. 2012). Both the countries again encountered the locust attack (2003–2005) that depleted 90% of their cereal crops (Pandey et al.

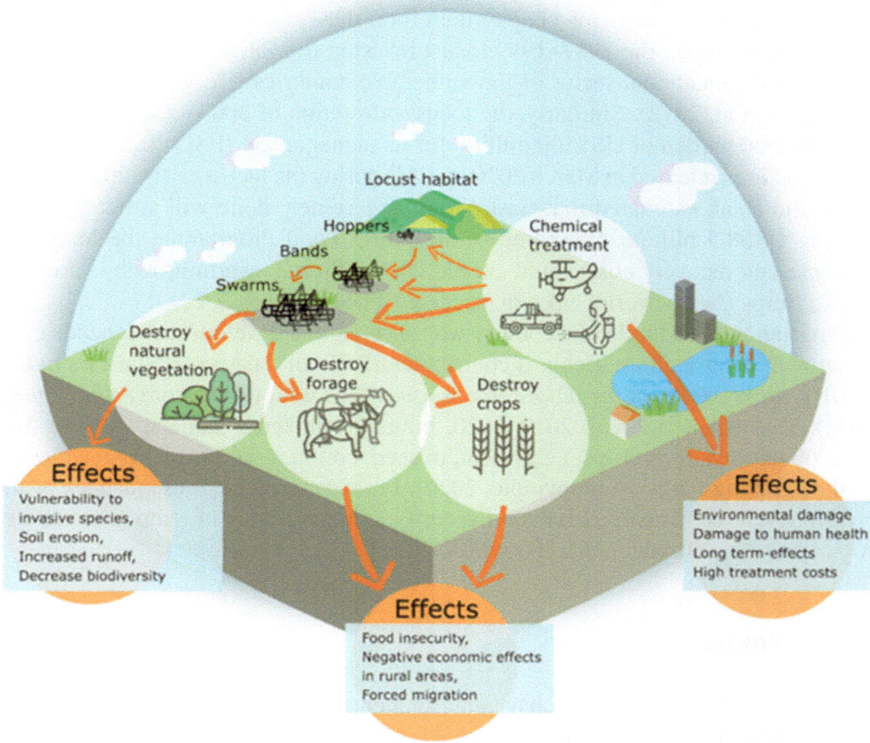

Fig. 11 Schematic effects of the locusts invasion during the gregarious phase on the agriculture, livestock, natural environment, and human settlements (Klein et al. 2021)

2021). In the recent past, the huge locust attack in Ethiopia questioned its food security (Nandelenga and Legesse 2020). The United Nations (UN) reported that the 2019–2020 locust outbreak affected 20 million persons in six (Ethiopia, Kenya, Somalia, South Sudan, Uganda, Tanzania) of eight East African countries and put them at danger of acute food insecurity (FAO 2020a). Situation was so distressing that the Somalia has to declare a state of national emergency in February, 2020, and FAO's Director-General Qu Dongyu has to appeal for swift action to avoid a humanitarian catastrophe (FAO 2020a). Unfortunately, current locust plague proved more horrific in the context of global pandemic of Covid-19, as it affected the already stressed food demand and supply chain by putting the food security of the affected areas in doldrums (Aday and Aday 2020; Barrett 2020).

7.2 Economic Development

Unfortunately, the costs associated with the desert locusts are not just restricted to crop losses (UN 2021). The control program involves surveillance and different operations, both air- and ground-based, large volume of pesticides and other material, as well as large staff of people, therefore, they can be very expensive (FAO 2021). For example, the 1987–1989 locust invasion in Mali massively affected the income and educational sector of the vulnerable countries (De Vreyer et al. 2015; De Vreyer et al. 2012). Similarly, the locust infestation of 2003–2005 in Sahel put an additional burden of US$500 million for its management (FAO 2021). Based on the projections released in March 2020 by FAO during the locust crisis (2019–2020), the World Bank estimated that East Africa and Yemen alone will suffer damages totaling US$8.5 billion. According to the World Food Programme the long-term response and recovery costs could exceed US$1billion if swarm growth is not controlled (https://www.worldbank.org, accessed 11 September 2022).

Pakistan is an example of a country whose economic development depends upon the growth in agriculture sector. It contributes 22.7% of the gross domestic product (GDP), while 37.9% of the labor force is associated with this sector (Annonymous 2022). The locust outbreak 2019–2020, initially inflicted a damage to more than 115,000 ha of various crops, i.e., wheat, oilseed, cotton, gram, fruits, vegetables, as well as grazing fields (Annonymous 2020). Furthermore, the consequences of the invasions can be disastrous for food security of many states in many countries of Africa and Asia (Brader et al. 2006; FAO 2021).

7.3 Environment

Locust control operations may have environmental side effects due to the high use of pesticides (Lazar et al. 2016). Inevitably, there is also collateral damage to the ecosystems, as it is not easy to discriminate between the locust swarms and other non-harmful organisms and vegetation. In addition to the direct impacts on the local ecosystem, there is also the possibility of pollinators, natural enemies, microbial fauna, humans, and livestock intoxication by the chemical pesticides used (Brader et al. 2006; Everts and Ba 1997).

8 Management Strategies

All over the world, the interventions against the locusts plague include huge aerial sprays of broad-spectrum pesticides (Roussi 2020) and other sustainable locust management techniques including monitoring, global positioning systems (GPS) and geo-referencing, insect growth regulators (IGRs), biopesticides, and semiochemical traps (Roussi 2020). Although, many are experiencing development still and not readily accessible in markets (Egonyu et al. 2021). The following management strategies are being used across the world for this notorious pest.

Fig. 12 Different cultural practices including hand picking (**a**), burning of locust eggs infested fields (**b**), and digging of pits to impede their movement (**c**) are commonly used to manage this notorious pest

8.1 Cultural Control

Cultural methods of pest control consist of destroying pest or regulating farm operations in ways that prevent them from causing economic loss (Hill 1987; Metcalf and Luckmann 1994). Desert locust is a rich protein source and may be used for food for people, livestock, duck, and fish (Egonyu et al. 2021). Hand picking of locusts (Figure 12a) for food and feed purposes decreases the population of the pest (Katel et al. 2021). Furthermore, killing or trampling bands, plowing or burning egg-infested fields (Figure 12b), and trapping hoppers in pits are also in practice (Figure 12c) (Sharma 2014).

8.2 Mechanical Control

Traditional method of mechanical control is effective when locusts infestation is low and labor is cheap (Hill 1987; Metcalf and Luckmann 1994). Making loud noises with acoustic (Fig. 13) and electronic equipment increases the randomness of a swarm which helps to decrease its population by breaking it apart (FAO 2020b; Ibrahim et al. 2013). The desert locusts are healthy during the day but at night they congregate on trees and open land without dense vegetation. They are dormant until the sun shines so that a mosquito net can catch the desert locusts (Wiktelius et al. 2003).

8.3 Chemical Control

Globally, the main locust management techniques being used over decades depended upon the conventional insecticides which are mainly neurotoxic to locusts (Dobson 2001; Mullié et al. 2021). Dieldrin, an insecticide belonging to the "organochlorine group" due to its persistence, was extensively used before 1980's to control the locust plague (Fig. 14). Although these insecticides proved more efficacious in managing the locust populations due to the prolonged activity of insecticide residues but concerns were raised about the increased harmful effects to the human and

Fig. 13 The utensils were beaten in the locust infested areas to thwart away the population of desert locusts

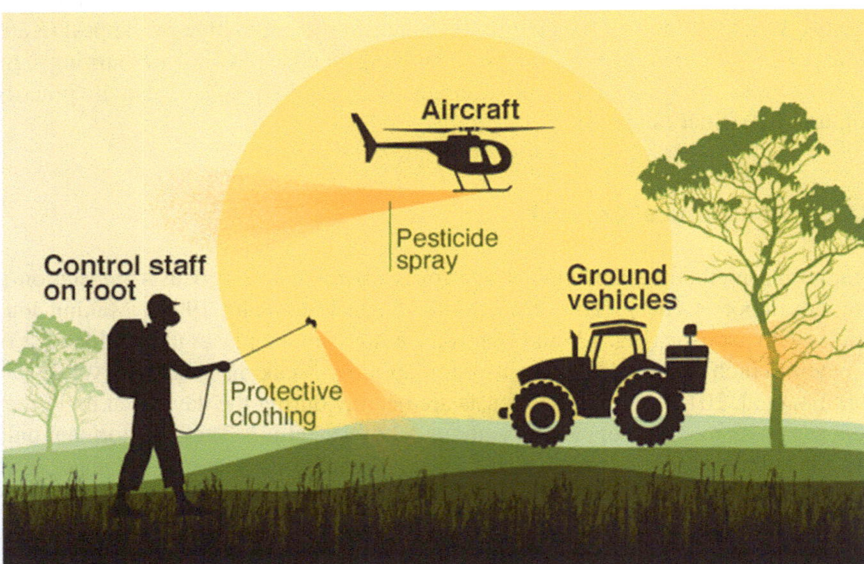

Fig. 14 Schematic of different strategies for the application of chemicals to manage the desert locust (UN 2021)

animals safety and the environment (Brader et al. 2006; Kimathi et al. 2020; Nicolopoulou-Stamati et al. 2016). Therefore, the use of organochlorine-based insecticides was banned and at the moment locust control mainly relies on organophosphate insecticides (fenitrothion and malathion), carbamates (bendiocarb), pyrethroids (deltamethrin and lambda-cyhalothrin), and the more recent products such as insect growth regulators (dimilin, triflumuron, fipronil, imidacloprid) (Abdelatti and Hartbauer 2020; Lecoq 2001) which have low persistence, making less effective as compared to dieldrin (Arthurs 2008; Lecoq 2005). After treatment with toxic chemicals, mortality of insects occur within hours, making it difficult to monitor the lethal effects of applied chemicals in case of highly mobile species (Langewald et al. 1997). As the upsurge begins, only some locusts aggregate into treatable targets thus making several individuals escape from toxic effects of chemicals which continue to increase and aggregate at higher populations (Huis 2007). Few insecticides are very effective for controlling locust outbreaks at earlier invasion stages in Saharan conditions; however, at the same time, the non-target insects are not immune to their toxicity (Balança and De Visscher 1997). In addition, outbreaks occurred in ecologically sensitive areas, particularly in wetlands close to human settlements and sometimes in sheltered areas with large numbers of migratory birds (Wiktelius et al. 2003). According to the FAO, between 2018 and June 2020, the locusts swarm invaded 22 countries and to suppress its population 2,792,840 ha were sprayed (Showler et al. 2021). For the management of desert locust, insecticides are usually applied using the ultra low volume (ULV) treatment technique (Matthews 1992). The greatest advantage associated with the ULV application is the management of the droplet size, as there is less waste from very large or very small droplets (Matthews 1992). Based on the hazardous effects of conventional insecticides, there is an urge to invest for alternate ways which are ecofriendly and sustainable like microbial insecticides for controlling the pest populations (Van der Valk et al. 2006; Wakil et al. 2022). Microbial biopesticides may provide an alternate and effective tool for locust's control as compared to synthetic chemicals (Zhang et al. 2019).

8.4 Microbial Control

8.4.1 Bacteria

Numerous entomopathogenic bacterial species have been identified as bio-controlling agent for locust. The locust infected by Gram-negative bacterium, *Serratia marcescens*, displays the symptoms of "behavioral fever" that greatly delays the progress of mycosis (Bundey et al. 2003). In the past different bacteria including *Bacillus weihenstephanensis* (Mashtoly et al. 2019), *Pseudomonas aeruginosa (*Ashrafi et al. 1965*)*, and *Bacillus cereus* (Reda et al. 2018) showing the insecticidal activities against the locust nymphs have been isolated from the gut microbiota of desert locust (Githae and Kuria 2021). These strains could be used for the development of potential biopesticides in controlling the nymphs of desert locust (Githae and Kuria 2021). Although, the Gram-positive bacterium, *Bacillus thuringiensis* (*Bt*) has been extensively explored for the development of microbial-based insecticides against diversified insect pest, yet there is paucity of knowledge for its activity against this deadly pest (Zelazny et al. 1997).

8.4.2 Fungi

Entomopathogenic fungi are a distinctive cluster of soil-dwelling microbes which infects and kills insects along with other arthropods by cuticle penetration (Mantzoukas et al. 2022). The fungi of genera *Beauveria*, *Metarhizium*, *Isaria*, *Hirsutella*, *Lecanicillium*, *Aspergillus,* etc. have been sporadically found in locust populations (Dakhel et al. 2019; Wood and Thomas 1996). Currently, experts have drawn their attention for the development of anti-locust mycoinsecticides especially from the *Beauveria* and *Metarhizium* genera (Lecoq and Cease 2022b; Lednev et al. 2020).

Due to their slow action than chemicals these are most suitable during early infestations (Vega et al. 2009). In the past, the research efforts were focused to use the fungi, *Metarhizium flavoviride*, as a biocontrol agent against the desert locust (Bateman et al. 1993; Lomer et al. 2001) due to its significant results under the laboratory as well as the field conditions (Langewald et al. 1997; Moore et al. 1992). Similarly, another entomopathogenic fungus, *Metarhizium anisopliae*, demonstrated its effectiveness against many species (Zimmermann 1993). *M. anisopliae* var. *acridum*, usually recognized as *M. acridum* in contemporary science (Bischoff et al. 2009), is an ecofriendly commercial biopesticide which has been established as ultra-low volume (ULV) spraying (Huis 2007). It kills about 70–90% of treated locusts within 14–20 days with no measurable impact on non-target organisms (Lomer et al. 2001). Over the past two decades, FAO has been working in different countries to develop biopesticides from *M. acridum* to manage grasshoppers and locusts population (Lecoq 2010). Importantly, in a first field study to evaluate the bio-effectiveness of four different entomopathogenic fungal formulations together—*M. acridum* (Green Muscle® and Green Guard®), *M. anisopliae*, and *Beauveria bassiana* against the nymphs and adults of *S. gregaria*, the authors have found the bio-control potential of all the formulations (Wakil et al. 2022). The obnoxious pest is likely to be exposed to *M. acridum* conidia by three means: (1) direct contact to toxic chemical, (2) pick-up of conidia from treated vegetation, or (3) horizontal transmission by infection during course of copulation and aggregation phenomena (Atheimine et al. 2014). The utilization of insect–pathogenic fungi has an additional benefits as they attack insects host by direct contact to insect's integuments in its place of ingestion by pathogen (Goettel and Glare 2010; Lacey and Shapiro-Ilan 2008). Furthermore, the ability of the fungi especially, *B. bassiana*, to opt different routes for entrance to its hosts makes it an ideal candidate to optimize sustainable and ecofriendly alternatives to chemicals (Fig. 15) (Mannino et al. 2019).

8.4.3 Virus

Insect pathogenic RNA viruses belonging to the family Reoviridae (Cologan 1986) were first found in the locust population, *Caledia captiva* (Fabr.), which increased the mortality of migratory locust due to bacteriosis caused by *Enterobacter cloacae* (Jord.) (Cologan 1986). While for DNA-containing viruses, an entomopoxvirus infecting adipose cells of *Anacridium aegiptium* L. (Lipa et al. 1994) and the iridescent virus that was isolated from crickets exhibited its faculty to infect desert (*S. gregaria*) and migratory (*L. migratoria*) locusts (Kleespies et al. 1999). However,

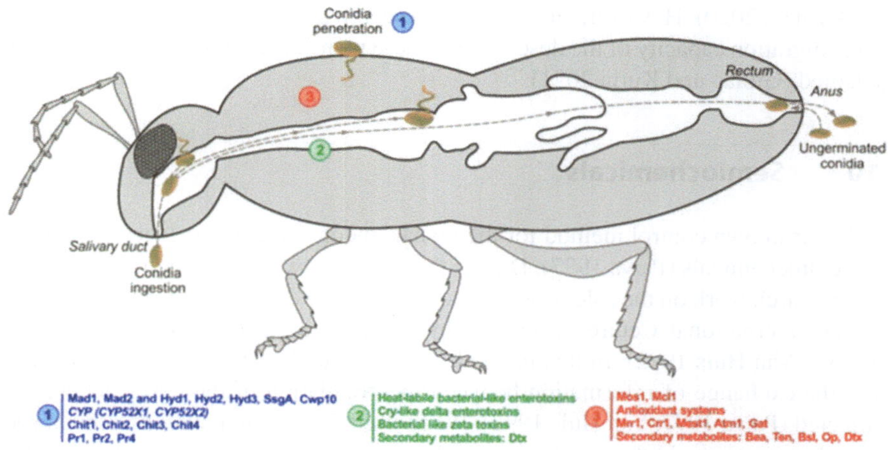

Fig. 15 Alternative routes of entry to a host for an entomopathogenic fungus. The fungal genes involved in conidial penetration through the cuticle are shown in blue (1); the fungal genes proposed to participate in oral infection are shown in green (2); and the fungal genes expressed into the hemolymph are shown in red (3) (Mannino et al. 2019)

due to their sensitivity to the ultra violet light, slow killing mechanism, and the limited research work about the use of entomopathogenic viruses against the threat of desert locust, these potential biocontrol agents are seldom used (Lomer et al. 2001; Van Huis 1992).

8.4.4 Microsporidia

Obligate intracellular, spore-forming protists, lacking metabolic activity outside the host cell are termed "Microsporidia" (Garcia 2002). Eighteen species of microsporidia have been described from different phylogenetic lines of Orthopteran insects (Ignatieva et al. 2019; Sokolova et al. 2006). Microsporidia can hijack the immunity system of insects hosts (Issi et al. 2005; Tokarev et al. 2007) and increases susceptibility of the infected insects to chemical (Johnson and Henry 1987) and biological plant protection products (Bauer et al. 1998). Since 1980s, microsporidia *Paranosema locustae*, after its registration in USA for biocontrol of grasshopper has been marketed under different trade names such as Nolo Bait™, Semaspore™, and Grasshopper Attack™ (Sukhoruchenko et al. 2020).

9 Botanicals

Botanicals are groups of biodegradable pesticides with greater selectivity and low mammalian toxicity (Mamadou and Sarr 2009). The vegetal extracts of neem (*Azadirachta indica*) and melia (*Melia volkensii*) in the past have demonstrated the effectiveness against the desert locust (Krall and Wilps 1994). Both the extracts are known to deter feeding and molting activity of the desert locust (Azhari et al. 2019;

Jaoko et al. 2020). However, due to limited research on the botanicals and the extensive migration capacity of the desert locust these plant-based pesticides could not be adapted (Githae and Kuria 2021).

10 Semiochemicals

A lesser-known control method for the management of the desert locusts is the use of semiochemicals (Paiva 1997). During the period between 1955 and1975, most of the research work on the role of semiochemicals in the desert locust was undertaken at the International Centre of Insect Physiology and Ecology (ICIPE), Nairobi, Kenya (Van Huis 1992). In this method, the compounds (semiochemicals) mediating the exchange of information between the population of the desert locusts are affected (Paiva 1997; Van Huis 1992). When used for communication within a species they are termed 'pheromones' (Seidelmann and Ferenz 2002). The desert locust emits pheromones that bring about the behavioral changes while interacting with one another. These pheromones can elicit multiple physiological changes that may range from mating to marching, swarming to scattering (Byers 1991). In addition to these, the semiochemicals can even cause effects such as increase in cannibalism and predation, and increased susceptibility to other forms of pesticides, pathogens, and natural diseases (Rai et al. 1997). Phenylacetonitrile pheromone induces stress by inhibiting the pheromonal communication among gregarious hoppers which eventually results in their mortality. This pheromone has the potential to become an alternative control agent to conventional pesticides. The male locust secretes a volatile substance to promote faster-maturation of young locusts, allowing swarms to last longer, and form faster (Loher 1990). The use of pheromone-based traps holds a promising future, as it can detect population growth at the transient stage, which is very difficult to study, and thus improves the preventative strategy by better forecasting the outbreak of the pest (Van Huis 1992). Furthermore, it is economical due to low cost. However, due to limited field experiments, it appears to be far from the practical use and operational strategies based on the use of such products have not yet been defined (Ferenz and Seidelmann 2003; Hassanali and Bashir 2011).

11 Remote Sensing (RS)

FAO has been actively involved in monitoring the movement of desert locusts around the world to provide an early warning of its imminent disaster for the timely implementation of control measures (Al-Ajlan 2007; Cressman 2012). For this purpose, it initiated Desert Locust Information Service (DLIS) to provide early warnings of potential locust outbreaks (Waldner et al. 2015). RS-based locust disaster surveillance typically employs indirect techniques for tracking outbreak areas through observing locust habitats (Lazar et al. 2016). During initial times of RS, common monitoring technique for predicting outbreak was based on analysis of variation vegetation which provided habitats for locusts (Bryceson and Wright

1986; Hielkema et al. 1986). Using this tool, quick decisions for its effective management can be made against the initial congregations of locusts (Ebbe 2010; Latchininsky 2013). Due to the availability of satellite imagery, the locust managers can target particular and highly vulnerable areas which reduces the management costs and contributes toward changing paradigm of locust control from curative to preventive (Magor et al. 2008). Currently, this tool is being used for monitoring and management of two species, *S. gregaria* and *C. terminifera*, and to a very limited extent for *L. migratoria* (Latchininsky et al. 2007; Propastin 2012). However, despite important advances in this direction, remote sensing alone cannot solve all locust problems, as 2003–2005 upsurge showed that the pest can get out of control (Lecoq 2005; Symmons 2009).

12 Conclusion

In a nutshell, this chapter provided an insight into the biology of the world's ancient and most ferocious pest which has always been an enigma for the people of all the times. Even, the contemporary world with massive flow of technological advancements is still at the mercy of this deadly pest. To add misery, the looming scenario of climate change is also playing a critical role in its irruption. Historically, all the locust crises affected the socioeconomic fabric of the societies and even put the food security of the invaded regions under question. Unfortunately, the last locust crisis (2019–2020) happened at a time when the whole world was in the grip of COVID-19. Since time immemorial, people have opted different means like cultural control (hand picking, digging pits) and mechanical control (using utensils to make noise) to manage the population of this horrific pest. In addition, conventional insecticides with long-term residual effects were intensively used, but unfortunately, these chemicals posed serious threats to human health, environment, and biodiversity. Therefore, there has been a realization to look for alternative control measures that could address all these issues. Fortunately, microbial pesticides derived from bacteria, fungi, virus, and microsporidia could be a potential solution. Moreover, with the passage of time new tools like georeferencing, global positioning system and remote sensing evolved to aid in the control of this notorious pest. The greatest advantage of using these novel interventions is to monitor and early predict the threat of an approaching locust swarm for its timely control. Therefore, it is need of hour that sanity should prevail and human beings should rationalize their activities that are contributing toward climate change and efforts should be geared to search for eco-friendly and sustainable desert locust management tools, before it is too late.

References

Abdelatti ZAS, Hartbauer M (2020) Plant oil mixtures as a novel botanical pesticide to control gregarious locusts. J Pest Sci 93:341–353

Aday S, Aday MS (2020) Impact of COVID-19 on the food supply chain. Food Qual Saf 4:167–180

Al-Ajlan AM (2007) Relationship between desert locust, *Schistocerca gregaria* (Forskål), infestation, environmental factors and control measures in Gazan and Makkah regions, Saudi Arabia. Pak J Biol Sci 10:3507–3515

Alexander D (1993) Natural disasters. Routledge, London

Annonymous (2020) Economic survey of Pakistan 2019–2020. Ministry of Finance, Govt. of Pakistan

Annonymous (2022) Economic survey of Pakistan 2021–2022. Ministery of Finance, Govt. of Pakistan

Anstey ML, Rogers SM, Ott SR, Burrows M, Simpson SJ (2009) Serotonin mediates behavioral gregarization underlying swarm formation in desert locusts. Science 323:627–630

Arthurs S (2008) Grasshoppers and locusts as agricultural pests. Springer, Dordrecht

Ashok K, Guan Z, Yamagata T (2001) Impact of the Indian Ocean dipole on the relationship between the Indian monsoon rainfall and ENSO. Geophys Res Lett 28:4499–4502

Ashrafi SH, Zuberi RI, Hafiz S (1965) Occurrence of *Pseudomonas aeruginosa* (Schroeter) Migula as a pathogenic bacterium of the desert locust, *Schistocerca gregaria* (Forskål). J Invertebr Pathol 7:189–191

Atheimine MO, Bashir MO, Ely SO, Kane CMH, Babah MAO, Benchekroun M (2014) Efficacy and persistence of *Metarhizium acridum* (Hypocreales: Clavicipitaceae) used against desert locust larvae, *Schistocerca gregaria* (Orthoptera: Acrididae), under different vegetation cover types. Int J Trop Insect Sci 34:106–114

Azhari OA, Magzoub NEHEA, Abd ESAI, Ahmed MAH (2019) Evaluation of the systemic action of neem (*Azadirachta indica* a. Juss) seed products against the desert locust immature *Schistocerca gregaria* (Forskal)(Orthoptera: Acrididae). Afr J Agric Res 14:1472–1486

Balança G, De Visscher M-N (1997) Impacts on nontarget insects of a new insecticide compound used against the desert locust [*Schistocerca gregaria* (Forskål 1775)]. Arch Environ Contam Toxicol 32:58–62

Barnett TP, Pierce DW, AchutaRao KM, Gleckler PJ, Santer BD, Gregory JM, Washington WM (2005) Penetration of human-induced warming into the world's oceans. Science 309:284–287

Barrett CB (2020) Actions now can curb food systems fallout from COVID-19. Nat Food 1:319–320

Barton GA (1913) A text from the oldest period of Babylonian writing. Orient Lit 16:3–8

Bateman R, Carey M, Moore DE, Prior C (1993) The enhanced infectivity of *Metarhizium flavoviride* in oil formulations to desert locusts at low humidities. Ann Appl Biol 122:145–152

Bauer MDO, Nascimento Júnior DD, Regazzi AJ, Silva EAMD (1998) Composiçao botânica da dieta de bovinos nos relevos côncavo e convexo, em pastagem natural de Viçosa, MG. Rev Bras Zootec 1:1–8

Bennett LV (1975) Development of a desert locust plague. Nature 256:486–487

Bischoff JF, Rehner SA, Humber RA (2009) A multilocus phylogeny of the *Metarhizium anisopliae* lineage. Mycologia 101:512–530

Brader L, Djibo H, Faye F, Ghaout S, Lazar M, Luzietoso P, Babah MO (2006) Towards a more effective response to desert locusts and their impacts on food security, livelihoods and poverty. In: Multilateral evaluation of the 2003–05 desert locust campaign. Food and Agriculture Organisation, Rome

Branson DH (2017) Effects of altered seasonality of precipitation on grass production and grasshopper performance in a northern mixed prairie. Environ Entomol 46:589–594

Bryceson KP, Wright D (1986) An analysis of the 1984 locust plague in Australia using multitemporal landsat multispectral data and a simulation model of locust development. Agric Ecosyst Environ 16:87–102

Buhl J, Sumpter DJT, Couzin ID, Hale JJ, Despland E, Miller ER, Simpson SJ (2006) From disorder to order in marching locusts. Science 312:1402–1406

Bundey S, Raymond S, Dean P, Roberts S, Dillon R, Charnley A (2003) Eicosanoid involvement in the regulation of behavioral fever in the desert locust, *Schistocerca gregaria*. Arch Insect Biochem Physiol 52:183–192

Byers JA (1991) Pheromones and chemical ecology of locusts. Biol Rev 66:347–378

Cai W, Cowan T, Sullivan A (2009) Recent unprecedented skewness towards positive Indian Ocean dipole occurrences and its impact on Australian rainfall. Geophys Res Lett 36

Cai W, Santoso A, Wang G, Weller E, Wu L, Ashok K, Masumoto Y, Yamagata T (2014) Increased frequency of extreme Indian ocean dipole events due to greenhouse warming. Nature 510:254–258

Cai W, Wang G, Gan B, Wu L, Santoso A, Lin X, Chen Z, Jia F, Yamagata T (2018) Stabilised frequency of extreme positive Indian ocean dipole under 1.5 °C warming. Nat Commun 9:1–8

Cease AJ, Elser JJ, Fenichel EP, Hadrich JC, Harrison JF, Robinson BE (2015) Living with locusts: connecting soil nitrogen, locust outbreaks, livelihoods, and livestock markets. Bioscience 65:551–558

Ceccato P, Cressman K, Giannini A, Trzaska S (2007) The desert locust upsurge in West Africa (2003–2005): information on the desert locust early warning system and the prospects for seasonal climate forecasting. Int J Pest Manag 53:7–13

Chen C, Qian J, Chen X, Hu Z, Sun J, Wei S, Xu K (2020) Geographic distribution of desert locusts in Africa, Asia and Europe using multiple sources of remote-sensing data. Remote Sens 12:3593

Çiplak B (2021) Locust and grasshopper outbreaks in the near east: review under global warming context. Agronomy 11:111

Cologan D (1986) Studies of the mortality of *Locusta migratoria* (L.) treated with a polyhedrosis virus from the grasshopper *Caledia captiva* (F.)(Orthoptera: Acrididae). Bull Entomol Res 76:539–544

Cressman K (2012) Satellites and GIS in desert locust monitoring worldwide: lessons learned. Proceedings 7th International IPM symposium March 27–29

Cressman K (2013) Climate change and locusts in the WANA region. In: Sivakumar MVK, Lal R, Selvaraju R, Hamdan I (eds) Climate change and food security in West Asia and North Africa. Springer, Dordrecht, pp 131–143

Cressman K (2016) Desert locust. In: Biological & environmental hazards, risks, and disasters. Elsevier, Amsterdam, pp 87–105

Crises GNAF (2020) Global report on food Crises: joint analysis for better decisions. IFPRI, Rome

Cui DN, Tu XB, Kun H, Raza A, Jun C, McNeill M, Zhang ZH (2019) Identification of diapause-associated proteins in migratory locust, *Locusta migratoria* L.(Orthoptera: Acridoidea) by label-free quantification analysis. J Integr Agric 18:2579–2588

Culmsee H (2002) The habitat functions of vegetation in relation to the behaviour of the desert locust *Schistocerca gregaria* (Forskal)(Acrididae: Orthoptera)-a study in Mauritania (West Africa). Phytocoenologia 32:645–664

Cutler BW (2011) Evoking the state: environmental disaster and colonial policy in Algeria, 1840–1870. University of California, Irvine

Dakhel WH, Latchininsky AV, Jaronski ST (2019) Efficacy of two entomopathogenic fungi, *Metarhizium brunneum,* strain F52 alone and combined with *Paranosema locustae* against the migratory grasshopper, *Melanoplus sanguinipes*, under laboratory and greenhouse conditions. Insects 10:94

De Vreyer P, Guilbert N, Mesplé-Somps S (2012) The 1987–1989 locust plague in mali: evidences of the heterogeneous impact of income shocks on education outcomes. Dauphine University, Paris

De Vreyer P, Guilbert N, Mesple-Somps S (2015) Impact of natural disasters on education outcomes: evidence from the 1987–89 locust plague in Mali. J Afr Econ 24:57–100

Despland E, Collett M, Simpson SJ (2000) Small-scale processes in desert locust swarm formation: how vegetation patterns influence gregarization. Oikos 88:652–662

Dobson HM (2001) Desert locust guidelines: control. FAO, Rome

Dominy NJ, Fannin LD (2021) The sluggard has no locusts: from persistent pest to irresistible icon. People and Nature 3:542–549

Ebbe B (2010) Biogéographie du Criquet pèlerin en Mauritanie. Ouvrage publié chez Hermann 286

Egonyu JP, Subramanian S, Tanga CM, Dubois T, Ekesi S, Kelemu S (2021) Global overview of locusts as food, feed and other uses. Glob Food Sec 31:100574

Ellis PE (1950) Marching in locust hoppers of the solitary phase. Nature 166:151–151
Ellis PE (1963) Changes in the social aggregation of locust hoppers with changes in rearing conditions. Anim Behav 11:152–160
Engelbrecht F, Monteiro P (2021) Climate change: the IPCC's latest assessment report. Quest 17:34–35
Escorihuela MJ, Merlin O, Stefan V, Moyano G, Eweys OA, Zribi M, Kamara S, Benahi AS, Ebbe MAB, Chihrane J (2018) SMOS based high resolution soil moisture estimates for desert locust preventive management. Remote Sens Appl Soc Environ 11:140–150
Everts J, Ba L (1997) Environmental effects of locust control: state of the art and perspectives. New strategies in locust control. Birkhauser, Basel, pp 331–336
FAO (2020a) Desert locust crisis: appeal for rapid response and anticipatory action in the greater horn of Africa. FAO, Rome
FAO (2020b) FAO Desert locust guidelines-latest edition (2001–2003). FAO, Rome
FAO (2020c) Five things you should know about an age-old pest: the desert locust. FAO, Rome
FAO (2021) The impact of disasters and crises on agriculture and food security. Report. Food and Agriculture Organization, Rome
Ferenz H-J, Seidelmann K (2003) Pheromones in relation to aggregation and reproduction in desert locusts. Physiol Entomol 28:11–18
Field CB, Barros VR, Dokken DJ, Mach KJ, Mastrandrea MD, Bilir TE, Chatterjee M, Ebi KL, Estrada YO, Genova RCEA (2014) Part A global and sectoral aspects; Contribution of working group II to the fifth assessment report of the intergovernmental panel on climate change (IPCC). In: Field CB, Barros VR (eds) . Cambridge University Press, New York, pp 1–32
Fielding DJ, Defoliart LS (2010) Embryonic developmental rates of northern grasshoppers (Orthoptera: Acrididae): implications for climate change and habitat management. Environ Entomol 39:1643–1651
Fontana P, Marino-Perez R, Sanabria-Urban S, Woller DA (2017) Studies in Mexican grasshoppers: three new species of *Dactylotini* (Acrididae: Melanoplinae) from Mexico and a review of existing conspecifics with comments on their geographical distributions. Zootaxa 4337:301–343
Garcia LS (2002) Laboratory identification of the microsporidia. J Clin Microbiol 40:1892–1901
Githae EW, Kuria EK (2021) Biological control of desert locust (*Schistocerca gregaria* Forskål). CAB Rev 16:13
Goettel M, Glare T (2010) Entomopathogenic fungi and their role in regulation of insect populations. In: Gilbert L, Latrou K, Gill S (eds) Comprehensive molecular insect science. Elsevier, Amsterdam, pp 361–406
Guo K, Hao S-G, Sun OJ, Kang L (2009) Differential responses to warming and increased precipitation among three contrasting grasshopper species. Glob Chang Biol 15:2539–2548
Guo K, Sun OJ, Kang L (2011) The responses of insects to global warming. In: Liu T, Kang L (eds) Recent advances in entomological research. Springer, Berlin, pp 201–212
Harmon K (2009) When grasshoppers go biblical: serotonin causes locusts to swarm. Sci Am 30
Harry OG (1970) Gregarines: their effect on the growth of the desert locust (*Schistocerca gregaria*). Nature 225:964–966
Hassanali A, Bashir MO (2011) Insights for the management of different locust species from new findings on the chemical ecology of the desert locust. Int J Trop Insect Sci 19:369–376
Hielkema J, Roffey J, Tucker C (1986) Assessment of ecological conditions associated with the 1980/81 desert locust plague upsurge in West Africa using environmental satellite data. Int J Remote Sens 7:1609–1622
Hill DS (1987) Agricultural insect pests of the tropics and their control. Cambridge University Press, Cambrige
Horne CF (1917) The sacred books and early literature of the east. Babylonia and Assyria. Parke, Austin, and Lipscomb, New York, NY
Huang X, Tian B, Niu Q, Yang J, Zhang L, Zhang K (2005) An extracellular protease from Brevibacillus laterosporus G4 without parasporal crystals can serve as a pathogenic factor in infection of nematodes. Res Microbiol 156:719–727

Huis AV (2007) Strategies to control the desert locust *Schistocerca gregaria*. Area-wide control of insect pests. Springer, Cham, pp 285–296

Hurst GW (1965) Meteorology and locust migrations. Nature 205:661–662

Ibrahim AG, Oyedum OD, Awojoyogbe O, Okeke S (2013) Electronic pest control devices: a review of their necessity, controversies and a submission of design considerations. Int J Eng Sci 2:26–30

Ignatieva AN, Gerus AV, Senderskiy IV, Malysh SM, Dolzhenko VI, Tokarev YS (2019) Infection of *Chorthippus loratus* (Orthoptera: Acrididae) with *Liebermannia* sp.(microsporidia) in South-Western Russia. J Eukaryot Microbiol 66:680–683

Issi I, Dolgikh V, Sokolova YY, Tokarev YS (2005) Factors of the pathogenicity of microsporidia, intracellular parasites of insects. Plant Protect News 3:16–25

Jaoko V, Nji Tizi Taning C, Backx S, Mulatya J, Van den Abeele J, Magomere T, Olubayo F, Mangelinckx S, Werbrouck SPO, Smagghe G (2020) The phytochemical composition of *Melia volkensii* and its potential for insect pest management. Plants (Basel) 9:143

Johnson N (2020) Meet ENSO's neighbor, the Indian Ocean Dipole. NOAA Climate.gov. https://www.climate.gov/news (Accessed on August 02, 2022)

Johnson DL, Henry JE (1987) Low rates of insecticides and *Nosema locustae* (microsporidia: Nosematidae) on baits applied to roadsides for grasshopper (Orthoptera: Acrididae) control. J Econ Entomol 80:685–689

Kalakkal J, Singh A (2021) Desert locusts' upsurges: a harbinger of emerging climate change-induced crises. The United Nations Environment Programme (UNEP), Nairobi

Katel S, Mandal HR, Neupane P, Timsina S, Pokhrel P, Katuwal A, Subedi S, Shrestha J, Shah KK (2021) Desert locust (*Schistocerca gregaria* Forskal) and its management: a review. J Agric Appl Biol 2:61–69

Kevan DKME (1978) Land of the locusts; parts I-before 450 A.D. Memoir n6. Lyman Entomological Museum and Reserch Laboratory, McGill University, Quebec, QC

Kimathi E, Tonnang HE, Subramanian S, Cressman K, Abdel-Rahman EM, Tesfayohannes M, Niassy S, Torto B, Dubois T, Tanga CM (2020) Prediction of breeding regions for the desert locust *Schistocerca gregaria* in East Africa. Sci Rep 10:1–10

Kleespies R, Tidona C, Darai G (1999) Characterisation of a new iridovirus isolated from crickets and investigations on the host range. J Invertebr Pathol 73:84–90

Klein I, Oppelt N, Kuenzer C (2021) Application of remote sensing data for locust research and management- a review. Insects 12:233

Krall S, Wilps H (1994) New trends in locust control. Ecotoxicology, botanicals, pathogenes, attractants, hormones, pheromones, remote sensing. CAB International, Wallingford

Lacey LA, Shapiro-Ilan DI (2008) Microbial control of insect pests in temperate orchard systems: potential for incorporation into IPM. Annu Rev Entomol 53:121–144

Langewald J, Kooyman C, Douro-Kpindou O, Lomer C, Dahmoud A, Mohamed H (1997) Field treatment of desert locust (*Schistocerca gregaria* Forskal) hoppers in Mauritania using an oil formulation of the entomopathogenic fungus *Metarhizium flavoviride*. Biocontrol Sci Tech 7:603–612

Latchininsky AV (2013) Locusts and remote sensing: a review. J Appl Remote Sens 7:075099

Latchininsky AV, Sivanpillai R, Driese KL, Wilps H (2007) Can early season landsat images improve locust habitat monitoring in the Amudarya river delta of Uzbekistan? J Orthoptera Res 16:167–173

Latchininsky A, Sword G, Sergeev M, Cigliano MM, Lecoq M (2011) Locusts and grasshoppers: behavior, ecology, and biogeography. Hindawi, London

Latchininsky A, Piou C, Franc A, Soti V (2016) Applications of remote sensing to locust management. Land surface remote sensing. Elsevier, Amsterdam, pp 263–293

Lazar M, Piou C, Doumandji-Mitiche B, Lecoq M (2016) Importance of solitarious desert locust population dynamics: lessons from historical survey data in a lgeria. Entomol Exp Appl 161:168–180

Le Gall M, Overson R, Cease A (2019) A global review on locusts (Orthoptera: Acrididae) and their interactions with livestock grazing practices. Front Ecol Evol 7:263

Lecoq M (2001) Recent progress in desert and migratory locust management in Africa. Are preventative actions possible? J Orthoptera Res 10:277-291, 215
Lecoq M (2003) Desert locust threat to agricultural development and food security and FAO/international role in its control. Arab Society Pl Prot 21:188–193
Lecoq M (2004) Vers une solution durable au problème du criquet pèlerin? Science et Changements Planétaires/Sécheresse 15:217–224
Lecoq M (2005) Desert locust management: from ecology to anthropology. J Orthoptera Res 14(179–186):178
Lecoq M (2010) Integrated pest management for locusts and grasshoppers: are alternatives to chemical pesticides credible? J Orthoptera Res 19:131–132
Lecoq M, Cease A (2022a) What have we learned after millennia of locust invasions? Agronomy 12:472
Lecoq M, Cease AJ (2022b) Are mycopesticides the future of locust control? vol 12. Multidisciplinary Digital Publishing Institute, Basel, p 2344
Lecoq M, Pierozzi I (1995) *Rhammatocerus schistocercoides* locust outbreaks in Mato Grosso (Brazil): a long-standing phenomenon. Int J Sustain Dev World Ecol 2:45–53
Lednev G, Levchenko M, Kazartsev I (2020) Entomopathogenic microorganisms in locusts and grasshoppers populations and prospects for their use for control of this pest group. BIO web of conferences. EDP Sciences, Les Ulis, p 00025
Lehmann P, Ammunét T, Barton M, Battisti A, Eigenbrode SD, Jepsen JU, Kalinkat G, Neuvonen S, Niemelä P, Terblanche JS (2020) Complex responses of global insect pests to climate warming. Front Ecol Environ 18:141–150
Lewis WJ, Van Lenteren J, Phatak SC, Tumlinson Iii J (1997) A total system approach to sustainable pest management. Proc Natl Acad Sci 94:12243–12248
Lipa JJ, Hernandez-Crespo P, Gonzalez-Reyes JA, Santiago-Alvarez C (1994) A newly recorded entomopoxvirus B in *Anacridium aegyptium* (Orthoptera: Acrididae). Biocontrol Sci Tech 4:343–345
Loher W (1990) Pheromones and phase transformation in locusts. In: Biology of grasshoppers. Wiley, Hoboken, NJ, pp 337–355
Lomer C, Bateman R, Johnson DL, Langewald J, Thomas M (2001) Biological control of locusts and grasshoppers. Annu Rev Entomol 46:667–702
Maeno KO, Piou C, Ghaout S (2020) The desert locust, Schistocerca gregaria, plastically manipulates egg size by regulating both egg numbers and production rate according to population density. J Insect Physiol 122:104020
Maggiacomo T, Kaya LB (2019) How to set off a plague of locusts. National Geographic. National Geographic Society, Washington. https://www.nationalgeographic.com/magzine/graphics/how-to-setoff-a-plague-of-locusts-interactive (Accessed on August 03, 2022)
Magor JI, Lecoq M, Hunter DM (2008) Preventive control and desert locust plagues. Crop Prot 27:1527–1533
Marino-Perez R (2022) Locusts and other pest grasshoppers (Orthoptera: Acrididae) on stamps. Dugesiana 29:171–179
Mamadou A, Sarr M (2009) Impact of two insecticides used in the control of the desert locust on *Psammotermes hybostoma* Desneux (Isoptera: Rhinotermitidae) in Niger. Afr Entomol 17:147–153
Mannino MC, Huarte-Bonnet C, Davyt-Colo B, Pedrini N (2019) Is the insect cuticle the only entry gate for fungal infection? Insights into alternative modes of action of entomopathogenic fungi. J Fungi 5:33
Mantzoukas S, Kitsiou F, Natsiopoulos D, Eliopoulos PA (2022) Entomopathogenic fungi: Interactions and applications. Encyclopedia 2:646–656
Marchant R, Mumbi C, Behera S, Yamagata T (2007) The Indian Ocean dipole–the unsung driver of climatic variability in East Africa. Afr J Ecol 45:4–16
Mashtoly TA, El-Zemaity MS, Abolmaaty A, Abdelatef GM, Marzouk AA, Reda M (2019) Phylogenetic characteristics of novel *bacillus weihenstephanensis* and *pseudomonas* sp.

to desert locust, *Schistocerca gregaria* Forskål (Orthoptera: Acrididae). Egypt J Biolo Pest Control 29:1–9

Matthews G, Lomer CJ, Prior C (1992) Biological control of locusts and grasshoppers. The principles of ultra-low volume spraying in relation to the application of microbial insecticides for locust control. C.A.B. International Oxon, Oxon

Metcalf RL, Luckmann WH (1994) Introduction to insect pest management. Wiley, Hoboken, NJ

Meynard CN, Lecoq M, Chapuis M-P, Piou C (2020) On the relative role of climate change and management in the current desert locust outbreak in East Africa. Glob Chang Biol 26:3753–3755

Mohamed EK, Zyaan OH, Ahmed NH, Guneidy NA, Khaled AS, Badawy NS (2018) Biochemical and molecular studies related to phase change in gregarious and solitarious desert locust, *Schistocerca gregaria* (Forskal)(Orthoptera: Acrididae). Egyptian academic journal of biological sciences. Egypt Acad J biol Sci Entomol 11:149–161

Moore D, Reed M, Le Patourel G, Abraham Y, Prior C (1992) Reduction of feeding by the desert locust, *Schistocerca gregaria*, after infection with *Metarhizium flavoviride*. J Invertebr Pathol 60:304–307

Morgan J (2014) Locust swarms 'high'on serotonin. BBC News

Mullié WC, Cheke RA, Young S, Ibrahim AB, Murk AJ (2021) Increased and sex-selective avian predation of desert locusts *Schistocerca gregaria* treated with *Metarhizium acridum*. PLoS One 16:e0244733

Munro JA, Saugstad S (1938) A measure of the flight capacity of grasshoppers. Science 88:473–474

Murakami H, Vecchi GA, Underwood S (2017) Increasing frequency of extremely severe cyclonic storms over the Arabian Sea. Nat Clim Chang 7:885–889

Nandelenga R, Legesse T (2020) Impact of desert locust infestation on household livelihoods and food security in Ethiopia. Joint Assessment Findings, Geneva

Neville A, Weis-Fogh T (1963) The effect of temperature on locust flight muscle. J Exp Biol 40:111–121

Nicolopoulou-Stamati P, Maipas S, Kotampasi C, Stamatis P, Hens L (2016) Chemical pesticides and human health: the urgent need for a new concept in agriculture. Front Public Health 4:148

Olfert O, Weiss RM, Kriticos D (2011) Application of general circulation models to assess the potential impact of climate change on potential distribution and relative abundance of *Melanoplus sanguinipes* (Fabricius) (Orthoptera: Acrididae) in North America. Psyche 2011:980372

Onyango D, Mehari G, Saliou N, Collin M, George O, John N, Nicholas M, Robert K, Ivan R, Willis O, Daniel K, Mibei H (2020) Desert locust field handbook: identification and management. CABI, Connecticut. https://www.cabi.org/what-we-do/locusts (Accessed on August 02, 2022

Paiva MR (1997) Potential for the use of semiochemicals against *Locusta migratoria* migratorioides (R. & F.). New strategies in locust control. Springer, Cham, pp 293–303

Pandey M, Suwal B, Kayastha P, Suwal G, Khanal D (2021) Desert locust invasion in Nepal and possible management strategies: a review. J Agric Food Res 5:100166

Peloquin C (2013) Locust swarms and the spatial techno-politics of the French resistance in world war II. Geoforum 49:103–113

Peng W, Ma NL, Zhang D, Zhou Q, Yue X, Khoo SC, Yang H, Guan R, Chen H, Zhang X, Wang Y, Wei Z, Suo C, Peng Y, Yang Y, Lam SS, Sonne C (2020) A review of historical and recent locust outbreaks: links to global warming, food security and mitigation strategies. Environ Res 191:110046

Pflüger HJ, Bräunig P (2021) One hundred years of phase polymorphism research in locusts. J Comp Physiol A Neuroethol Sens Neural Behav Physiol 207:321–326

Piou C, Gay PE, Benahi AS, Babah Ebbe MAO, Chihrane J, Ghaout S, Cisse S, Diakite F, Lazar M, Cressman K (2019) Soil moisture from remote sensing to forecast desert locust presence. J Appl Ecol 56:966–975

Pörtner HO, Roberts DC, Adams H, Adler C, Aldunce P, Ali E, Ara Begum R, Betts R, Bezner Kerr R, Biesbroek R, Birkmann J, Bowen K, Castellanos E, Cissé G, Constable A, Cramer W, Dodman D, Eriksen SH, Fischlin A, Zaiton Ibrahim Z (2022) Climate change 2022: impacts, adaptation and vulnerability. IPCC, Geneva, Switzerland

Propastin P (2012) Multisensor monitoring system for assessment of locust hazard risk in the lake Balkhash drainage basin. Environ Manag 50:1234–1246
Rai M, Hassanali A, Saini R, Odongo H, Kahoro H (1997) Identification of components of the oviposition aggregation pheromone of the gregarious desert locust, *Schistocerca gregaria* (Forskal). J Insect Physiol 43:83–87
Reda M, Mashtoly T, El-Zemaity M, Abolmaaty A, Abdelatef G, Marzouk A (2018) Susceptibility of desert locust, *Schistocerca gregaria* (Orthoptera: Acrididae) to *Bacillus cereus* isolated from Egypt Arab Universities. J Agric Sci 26:725–734
Rogers SM, Cullen DA, Anstey ML, Burrows M, Despland E, Dodgson T, Matheson T, Ott SR, Stettin K, Sword GA, Simpson SJ (2014) Rapid behavioural gregarization in the desert locust, *Schistocerca gregaria* entails synchronous changes in both activity and attraction to conspecifics. J Insect Physiol 65:9–26
Rosenzweig C, Iglesius A, Yang X-B, Epstein PR, Chivian E (2001) Climate change and extreme weather events-implications for food production, plant diseases, and pests. NASA Publications, Washington, DC
Roussi A (2020) Why gigantic locust swarms are challenging governments and researchers. Nature 579:330–331
Roxy MK, Ritika K, Terray P, Masson S (2014) The curious case of Indian Ocean warming. J Clim 27:8501–8509
Saji N, Yamagata T (2003) Possible impacts of Indian Ocean dipole mode events on global climate. Clim Res 25:151–169
Saji NH, Goswami BN, Vinayachandran PN, Yamagata T (1999) A dipole mode in the tropical Indian Ocean. Nature 401:360–363
Salih AAM, Baraibar M, Mwangi KK, Artan G (2020) Climate change and locust outbreak in East Africa. Nat Clim Chang 10:584–585
Schlosser S, Saulnier V (2000) Mozambique: a country ravaged by civil war ana nature. J Conv Weapons Destr 4:22
Seidelmann K, Ferenz HJ (2002) Courtship inhibition pheromone in desert locusts, *Schistocerca gregaria*. J Insect Physiol 48:991–996
Sharma A (2014) Locust control management: moving from traditional to new technologies–an empirical analysis. Entomol Ornithol Herpetol 4:2161–0983
Showler AT, Ould Babah Ebbe MA, Lecoq M, Maeno KO (2021) Early intervention against desert locusts: current proactive approach and the prospect of sustainable outbreak prevention. Agronomy 11:312
Simpson SJ, McCaffery AR, Hägele BF (1999) A behavioural analysis of phase change in the desert locust. Biol Rev 74:461–480
Skendžić S, Zovko M, Živković IP, Lešić V, Lemić D (2021) The impact of climate change on agricultural insect pests. Insects 12:440
Sokolova YY, Lange CE, Fuxa JR (2006) Development, ultrastructure, natural occurrence, and molecular characterisation of *Liebermannia patagonica* ng, n. sp., a microsporidian parasite of the grasshopper Tristira magellanica (Orthoptera: Tristiridae). J Invertebr Pathol 91:168–182
Sorokin PA, Horowitz IL (2017) Man and society in calamity. Routledge, New York, NY
Sparks DL (2021) Economic transformation in sub-Saharan Africa-the way forward. Routledge, New York, NY
Spinage CA (2012) Locusts the forgotten plague part I: locusts and their ecology. In: African ecology. Springer, Cham, pp 481–532
Steedman A (1988) Locust handbook. Overseas Development, London
Stige LC, Chan K-S, Zhang Z, Frank D, Stenseth NC (2007) Thousand-year-long Chinese time series reveals climatic forcing of decadal locust dynamics. Proc Natl Acad Sci 104:16188–16193
Stone M (2020) A plague of locusts has descended on East Africa. Climate change may be to blame. National Geographic. National Geographic Society, Washington. https://www.nationalgeographic.com/science/articles/locust-plague-climate-science-east-africa (Accessed on August 03, 2022)

Sukhoruchenko G, Burkova L, Ivanova G, Vasilyeva T, Dolzhenko O, Ivanov S, Dolzhenko V (2020) The assortment formation of chemical means of pest control in the XX century. Plant Protect News 1:5–24
Sword GA, Simpson SJ, El Hadi OT, Wilps H (2000) Density-dependent aposematism in the desert locust. Proc Biol Sci 267(1438):63–68
Symmons P (2009) A critique of "preventive control and desert locust plagues". Crop Prot 28:905–907
Symmons P, Cressman K (2001) Desert locust guidelines: biology and behaviour. FAO, Rome
Tanaka S, Nishide Y (2013) Behavioral phase shift in nymphs of the desert locust, *Schistocerca gregaria*: special attention to attraction/avoidance behaviors and the role of serotonin. J Insect Physiol 59:101–112
Taylor GK, Thomas AL (2003) Dynamic flight stability in the desert locust *Schistocerca gregaria.* J Exp Biol 206:2803–2829
Tian H, Stige LC, Cazelles B, Kausrud KL, Svarverud R, Stenseth NC, Zhang Z (2011) Reconstruction of a 1,910-Y-long locust series reveals consistent associations with climate fluctuations in China. Proc Natl Acad Sci U S A 108:14521–14526
Tokarev YS, Sokolova YY, Entzeroth R (2007) Microsporidia–insect host interactions: teratoid sporogony at the sites of host tissue melanization. J Invertebr Pathol 94:70–73
UN (2021) Desert locusts' upsurges: a harbinger of emerging climate change induced crises? United Nations Environment Programme: foresight brief early warning, emerging issues, and futures? UN Science Division, Nairobi
Usmani K, Usmani S (2018) Locusts. In: Pests and their management. Springer, Cham, pp 825–869
Uvarov BP (1921) A revision of the genus Locusta, L.(= *Pachytylus*, Fieb.), with a new theory as to the periodicity and migrations of locusts. Bull Entomol Res 12:135–163
Uvarov BP (1944) The locust plague. J Econ Entomol 37:93–99
Uvarov B (1977) Grasshoppers and locusts. In: A handbook of general acridology vol 2 behaviour, ecology, biogeography, population dynamics, vol 2. Centre for Overseas Pest Research, London
Van der Valk H, Del Castello R, Cressman K, Monard A, Eriksson H, Ammati M, Bartoleschi P, De Brouwer B, Gazza S, Everts J (2006) Fighting the locusts... safely. Pesticides in desert locust control: balancing risks against benefits. FAO, Rome
Van Huis A (1992) New developments in desert locust management and control. Entomol Exp Appl 3:2–18
Vega FE, Goettel MS, Blackwell M, Chandler D, Jackson MA, Keller S, Koike M, Maniania NK, Monzon A, Ownley BH (2009) Fungal entomopathogens: new insights on their ecology. Fungal Ecol 2:149–159
Veran S, Simpson SJ, Sword GA, Deveson E, Piry S, Hines JE, Berthier K (2015) Modeling spatiotemporal dynamics of outbreaking species: influence of environment and migration in a locust. Ecology 96:737–748
Vinayachandran P, Francis P, Rao S (2009) Indian Ocean dipole: processes and impacts. Curr Trends Sci 46:569–589
Wakil W, Ghazanfar MU, Usman M, Hunter D, Shi W (2022) Fungal-based biopesticide formulations to control nymphs and adults of the desert locust, *Schistocerca gregaria* Forskål (Orthoptera: Acrididae): a laboratory and field cage study. Agronomy 12:1160
Waldner F, Babah Ebbe MA, Cressman K, Defourny P (2015) Operational monitoring of the desert locust habitat with earth observation: an assessment. ISPRS Int J Geo Inf 4:2379–2400
Wang X-H, Kang L (2005) Differences in egg thermotolerance between tropical and temperate populations of the migratory locust Locusta migratoria (Orthoptera: Acridiidae). J Insect Physiol 51:1277–1285
Wewetzer A, Krall S, Schulz F (1993) Methods for the assessment of crop losses due to grasshoppers and locusts. Deutsche Gesellschaft fur Technische Zusammenarbeit (GTZ) GmbH, Bonn
Wiktelius S, Ardö J, Fransson T (2003) Desert locust control in ecologically sensitive areas: need for guidelines. Ambio 32:463–468

Wilson DM (1961) The central nervous control of flight in a locust. J Exp Biol 38:471–490
WMO (2022) Climate change 2022: mitigation of climate change. IPCC Sixth assessment report. WMO, Geneva
Wood S, Thomas MB (1996) Space, time and persistence of virulent pathogens. Proc R Soc London Ser B Biol Sci 263:673–680
Wu T, Hao S, Sun OJ, Kang L (2012) Specificity responses of grasshoppers in temperate grasslands to diel asymmetric warming. PLoS One 7:e41764
Xu Z, Elomri A, El Omri A, Kerbache L, Liu H (2021) The compounded effects of COVID-19 pandemic and desert locust outbreak on food security and food supply chain. Sustainability 13:1063
Yuga M, Wani P (2022) Assessing the impact of desert locust infestation on crops, pasture and livestock health in eastern equatoria state, South Sudan. Eur J Appl Sci 10:332–341
Zanna L, Khatiwala S, Gregory JM, Ison J, Heimbach P (2019) Global reconstruction of historical ocean heat storage and transport. Proc Natl Acad Sci 116:1126–1131
Zelazny B, Goettel M, Keller B (1997) The potential of bacteria for the microbial control of grasshoppers and locusts. In: The memoirs of the Entomological Society of Canada, vol 129. Cambridge University Press, Cambridge, pp 147–156
Zhang L, Lecoq M, Latchininsky A, Hunter D (2019) Locust and grasshopper management. Annu Rev Entomol 64:15–34
Zimmermann G (1993) The entomopathogenic fungus *Metarhizium anisopliae* and its potential as a biocontrol agent. Pestic Sci 37:375–379

Role of Horticulture in Disaster Risk Management

Adnan Sami, Muhammad Saeed, Muhammad Shafiq, Syed Mohsin Abbas, Alishpa Anum, Hamza Haider, Muhammad Hamza Tariq Bhatti, Muhammad Arham Raza, Narmeen Khan, and Nuhammad Adnan Shahid

Abstract

Role of horticulture in disaster management refers to the implementation of systematic approaches including strategies, operational skills, and administrative decisions to reduce the impact of natural hazards caused by environmental disaster. Sufficient food production and consistent food supply become under pressure during disaster caused by environmental stress or climate change. During crisis, sustainable food supply is disrupted and prices become uncontrollable, whereas substantial food demand is increased in such areas. The steady supply of horticultural crops in areas prone to disaster can fulfil the basic need of the helpless people. Fruits and vegetables are vital source of food supply to support human health during disaster conditions. However, like agronomic crops, horticultural crops are also seriously threatened by natural disasters. The physiological functioning of crops has been disturbed during droughts, floods, and other extreme weather conditions, which causes stunted growth of the plants and leads to great yield reduction leads to food security. This book chapter discusses various disas-

Adnan Sami and Muhammad Saeed contributed equally with all other contributors.

A. Sami · A. Anum · H. Haider · M. H. T. Bhatti · M. A. Raza · N. Khan
Faculty of Agricultural Sciences, University of the Punjab, Lahore, Pakistan

M. Saeed · M. Shafiq (✉) · S. M. Abbas
Department of Horticulture, Faculty of Agricultural Sciences, University of the Punjab, Lahore, Pakistan
e-mail: shafiq.iags@pu.edu.pk

N. A. Shahid
Horticultural Science Department, University of Florida, Institute of Food and Agricultural Sciences, North Florida Research and Education Center, Quincy, FL, USA

M. Ahmed, S. Ahmad (eds.), *Disaster Risk Reduction in Agriculture*, Disaster Resilience and Green Growth, https://doi.org/10.1007/978-981-99-1763-1_18

trous conditions caused by climatic changes and technological interventions that protect horticultural crops from natural disasters to supply food in disastrous areas. Remote sensing uses spatial data of crops and unmanned aircraft systems (UAS) like drones and cameras to estimate the irrigation needs of the plants in order to conserve water and prevent drought stress in plants. Vegetable grafting plays an important role in boosting plants' resistance against biotic and abiotic stresses. Use of some valuable rootstocks in tomato, grapevines, apples, and cucumbers during drought stress results in enhancing significant drought tolerance in these crops. Plant diseases and disorders are prevented by effective management practises. The most successful method for reducing the negative effects of climate change is to discover new plant genetic resources that are tolerant to extreme weather conditions. Numerous microbes which function as biostimulants have been identified that help plants in a variety of physiological cycles that promote plant growth and development. In hypoxic conditions, soil loses oxygen, causing stresses. A study found that silicon in nano form generates antioxidant species and scavenges free radicals, making plants more resilient to stress. The measures discussed have the potential to support crops during unfavourable conditions as well as can contribute to provide food and nutritional supply to the population dwelling in disaster-prone area.

Keywords

Horticulture crops · Disaster risk management · Remote sensing · Water use efficiency · Rootstock · Biostimulants · Bioherbicide

1 Introduction

The availability of favourable climatic conditions is crucial to agricultural productivity and sustainable food supply. Natural disasters pose a risk to economic development, crop production, livestock, aquaponics, fisheries, and forestry. These are all negatively affected by weather-related events, which have increased over the past few decades. As a result of fuel consumption, deforestation, urbanization, and industrialization, there is an increase in greenhouse gases like carbon dioxide, nitrous oxide, ozone, and methane. These gases may have an impact on global temperatures, water availability, and biotic and abiotic stresses. It is a serious threat that directly affects agricultural and water supplies. Vegetables and fruits are extremely beneficial to human health and are responsible to fulfil the nutritional needs (Li et al. 2014). Droughts, floods, and forest fires caused by an increase in global temperature and changes in the water cycle are examples of disasters that are affecting agricultural productivity due to synchronization between the environment and climate change (Godfray et al. 2010). Fruit and vegetable crops require a significant amount of water. The consumption of fruits and vegetables rises with population growth. According to Kumar et al. (2017), growing of fuit and vegetables has been increased in arid and semi-arid regions where there is a lack of water. When there is a drought, there is less water available, so irrigation water is not enough for plants, and eventually plants are under water stress, which causes them to change their normal growth and development (Singh et al. 2017). A new method of grafting may help plants to survive under biotic and abiotic stressful conditions (Rheinbay et al. 2020). At first, grafting was used to improve the efficiency of horticultural crops. In China, Japan, and Korea, the use of vegetative propagation to control soil-borne diseases emerged. However, grafting is now utilized as a method for developing resistance to drought, salinity, nutritional, and chilling stresses (Sánchez-Rodríguez et al. 2013). Deficit irrigation is a method for maximizing water efficiency in which plants are given irrigational water that is less than the total amount of water required for plant growth and development throughout the entire crop cycle (Hatfield and Prueger 2015).

2 Drought Management in Horticulture

2.1 Remote Sensing for Irrigation Optimization

Evaluation and observation of crops prompted various earth observation satellite systems which operate at an altitude of 180–2000 km (Belward and Skøien 2015). 2015). Satellite and manned aircraft systems are useful in regional-scale characterization; however, unmanned aircraft systems are more advanced in mapping intra-field variability. A major issue in the execution of remote sensing of horticultural crops is the proportion of inter-row ground/vegetation cover that leads to mixed pixels that increase with the decrease in spatial resolution of image. By using thermal spectrum, canopy temperature and water status were characterized by manned aircraft in horticulture (Bellvert et al. 2021) (Table 1).

Table 1 Adverse climatic effects on vegetables

Incident	Source	Adverse effects	Affected group	Reference
Lack of chilling	Less cold temperature in winter	Delayed spears reduced spear growth	Asparagus	Burghardt (2002)
Lack of vernalization	Less cold temperature in winter	Delayed head formation	Cauliflower	Van Zyl (2006)
Warm and dry period	Pressure to complete life cycle	Bolting	Lettuce	Paraskevopoulou (2021)
Warm and dry periods	Lack of CA transport	Tip burn and blossom end rot	Lettuce, tomato	De Bon et al. (2015)

2.2 Crop Water Status Indicators

Various indices and quantitative products, such as the crop water status indicator, can be used to estimate the biochemical and biophysical characteristics of crops. Some thermal and spectral indices can be used to assess water status in horticultural crops. Proximal and remote sensing methods are used for estimating grapevine water stress. Estimation of instantaneous and seasonal variability of plant water status in vineyard is done by using high-resolution UAV-based thermal imaging. Plant's temperature rises when it is in stress, which is a sign of water stress. In this way, water stress and canopy temperature in a thermal image are measured.

Most common spectral index in horticulture is the NDVI:

$$\mathrm{NDVI} = \mathrm{Rnir} - \mathrm{Rr} / \mathrm{Rnir} + \mathrm{Rr}$$

where Rnir and Rr represent the spectral reflectance at the NIR and red spectral regions. In horticulture, DNVI is used to assess vigour, biomass, and water status. A plant shows vigorous canopy when more leaves regulate more water when irrigated and shows water stress when unirrigated (Goetz 1997). The ratio of canopy temperature to NDVI is termed as temperature vegetation dryness index (TVDI) which is found to be useful in the assessment of water status in horticulture. TVDI depicts that a vegetation with high NDVI will have lower surface temperature unless the vegetation is under stress. Usually, vegetation remains green in initial stress (Goetz 1997).

A Case Study in Almond

Almonds need a lot of water to grow in semi-arid areas. Almond's growth, productivity, and yield all depend on how well the water it needs is managed. Multi-sensor NDVI, surface temperature, and biophysical variables at a mixed grassland site were analysed by unmanned aircraft systems that have been used to analyse spatial patterns using remote sensing to determine the almond tree's water status. The crown temperature of an almond tree in California was measured with a thermal camera, and the temperature of the crown and the air (Tc – Ta) was compared to the water potential. This case study demonstrates that there is a negative correlation between the crown temperature and the water potential. It varies depending on the time, with a weak morning measurement and a strong afternoon measurement (Mirfenderesgi et al. 2016).

2.3 Vegetable Grafting to Control Drought Stress

It has been shown that the grafting procedure makes it possible for plants to withstand drought stress, although the main mechanism by which this occurs is unknown. It was found that tomato plants' biomass increased and they maintained a healthy water status when they were grafted onto drought-tolerant seedlings. In 2016, Liu made the claim that a rootstock Luffa had strong shoot growth when grown on a cucumber plant scion. Under water stress, this was observed as delay in leaf withering. Therefore, under well-watered conditions, there was no difference in the leaf area of grafted and non-grafted plants. However, rootstock was involved in adaptation against drought response. Grapevine's ability to tolerate drought is significantly influenced by rootstock.

The drought stress caused by photooxidation results in a decrease in chlorophyll concentration. According to a study, the level of chlorophyll is maintained in the cambium of drought-tolerant grafted plants while it drops in water-stressed grafted plants. It was found that tolerant rootstock can increase the blocking of Pn and Tr brought on by drought stress, but susceptible plants were found to be unable to maintain water under drought stress. Grafting increases tomato photosynthetic capacity and lowers ROS accumulation, which increases the plant's ability to withstand drought (Zhang et al. 2019).

A Case Study in Tomato Plant

Two tomato genotypes, the tolerant "Serina" and the sensitive "Jesifina," were discovered to have enhanced tolerance to water deficiency. This improved tolerance was brought about by rootstocks. The rootstock "Beaufort" from another tomato hybrid reduced the growth inhibition in tomatoes exacerbated by drought stress. To test the resilience of tomato rootstock lines to drought, about 144 lines were examined. Of these, 38% demonstrated good shoot growth when compared to self-grafted plants in dry conditions (Kumar et al. 2017).

2.4 Deficit Irrigation

Recent years have seen a lot of research on water use efficiency (WUE). It is characterized in terms of the plants' carbon gain in relation to their water loss (Franks et al. 2013). The recent idea of deficit irrigation has the potential to address the global water issues, especially because agriculture is the major consumer of fresh water. With the use of less irrigation water, this technique enables plants to maintain low yields with less water. There are two types of deficit irrigation: partial rootzone drying (PRD) and regulated deficit irrigation (RDI). For the RDI technique, irrigation is kept up with a meagre supply of water, typically when the plant is less vulnerable to drought stress. Another method is partial root drying, which tells the plant to close the stomata to lessen water evapotranspiration by alternately drying and watering the roots. Abscisic acid (ABA) hormone controls the opening and closure of stomata to control this physiological change in plants (Costa et al. 2007).

Effect of Deficit Irrigation on Citrus

Vegetative growth of fruit plants is slowed down by DI. The plant exploits this reduction in foliage as a defence strategy because it results in less plant interception, which reduces water loss through reduced transpiration (Tejero et al. 2011).

Table 2 Adverse climatic effects on fruits

Incident	Source	Adverse effects	Affected group	References
Lack of chilling	Lack of cold temperature during winter	Delayed flowering and increase in the risk of frost	Apple and other fruit crops	Bigey (2000)
Frost	Slight increase in the risk of frost	Damage to flowers and fruitlessness	Cherry, apple, apricot and others	Müller and Spreitzenbarth (2013)
Sunny and hot periods	Rise of fruit temperature >50 °C	Smaller and softer fruit Less fruit colour	Apple and other fruit crops	Dannemand et al. (2019)
Warm and dry periods	Insufficient water supply to fruit	Tip burn and blossom end rot	Apple and other fruit crops	Floriancic et al. (2020)

2.5 Mitigation of Climate Variability on Horticulture (Table 2)

3 Finding New Genetic Resources

The first priority to make climate-resilient agriculture is to develop new cultivars which are resistant to extreme weather events such as heat and drought. Plants which are resistant to bolting and tip burn give high yields despite of high temperature. In plants like lettuce, plant architecture has high influence on the occurrence of tip burn; however, it is not mandatory for all lettuce cultivars. For selection of fruits and vegetables low chilling and frost-tolerant cultivars should be included in breeding goals. Knowing the upcoming threats of climate change new plant genetic resources should be found from natural habitats of in situ, on farms, and in gene banks. In case of self-production by small farmers, vegetable seeds should be produced in gardens because seeds develop there under specific environment conditions and are tolerant to extreme weather events (Bisbis et al. 2019).

4 Management of Production Technology

In greenhouses, vegetables can be grown using a variety of techniques, including direct planting in the soil and the most recent form of hydroponic growth, which involves either growing in various soilless combinations utilizing containers or using liquid nutrient media.

5 Soil and Its Management

Vegetables are typically grown in greenhouses by planting directly into the existing, well-drained soil. The soil structure in the greenhouse may be completely destroyed if the same crop is continuously planted over an extended period of time. Initial procedures like sanitation, solarization, mulching, or fumigation are all part of soil preparation. In addition to eliminating pathogens, soil solarization boosts growth responses and functions as a bioremediation method for soils contaminated with pesticides. To determine soil fertility, soil testing must be done, and any nutrient deficiencies must be remedied as needed. The use of bioagents not only aids in the control of several soil-borne infections, such as nematodes, but also encourages plant development.

6 Irrigation and Fertigation

Protected cultivation can result in more effective water use than open field settings. In greenhouse production systems, drip irrigation technology has become a mainstay that aids in water conservation but may also be responsible for lowering the incidence of illnesses that thrive in humid environments. Fertigation, or the application of fertilizer along with irrigation, necessitates taking into account plant needs, soil characteristics, and technological requirements. It enables the targeted and uniform delivery of nutrients to the region where the active roots are gathered. Crop foliage can be maintained dry with fertigation through a drip irrigation system, preventing leaf burn and postponing the growth of plant infections. Depending on the variety, season, and crop stage, the timing of fertilization differs for various vegetables.

7 Good Agricultural Practices (GAPs)

GAPs are characterized as "on-farm practices that address environmental, economic, and social sustainability, resulting in safe and high-quality agricultural products, both food and non-food." Due to the high input costs and strict quality standards, greenhouse production systems must follow GAP guidelines. Adopting GAP is essential because the perishable nature of fresh fruits and vegetables poses a challenge to the food chain (Sabir and Singh 2013).

7.1 Role of Biostimulants

Food security can be secured if resilience and resistance would be induced in plants to mitigate climate-induced stresses (Rouphael and Colla 2018). Biostimulants are being effectively used against stresses (Calvo et al. 2014). These biostimulants are less toxic, do not store in plants, and less liable to choose resistant pathogens;

therefore, they are not hazardous for environment. In Europe biostimulants have EUR 578 million market value (Ricci et al. 2019).

Inorganic compounds like humic substance, amino acids, seawood extracts, protein hydrolates and some microbial biostimulant such as yeast, bacteria, and fungi have potential to promote plant growth and development, to make them stress tolerant. Microbes are utilized as biofertilizers to provide necessary nutrients to plants through the mechanisms of nitrogen fixation (e.g., *Rhizobium*), cellulolytic activity (e.g., *Aspergillus*, *Penicillium* spp.), and acidification of soil (e.g., *Bacillus subtilis*). In fungi group, mainly trichoderma is applied because it increases plant growth and promotes antimicrobial substances and plant defences. For biostimulant role, *Burkholderia*, *Bacillus*, and *Pseudomonas* are the most studied bacteria.

7.2 Effect of Biostimulant in Heat Tolerance in Potato

Microbial stimulants induce defences in plants by producing enzymes for ROS degradation and by reducing H_2O_2 level. When microbial stimulants are applied, they reduce ethylene production during stress, to save plants from negative effects of stress. A bacteria containing 1-aminocyclopropane-1-carboxylic acid (ACC) deaminase was found helpful in degradation of ethylene precursor to avoid its negative effects in heat stress.

When plant faces stress, it induces the synthesis of ethylene and indole acetic acid (IAA). The enzyme ACC synthase converts *S*-adenosyl methionine (SAM) to 1-aminocyclopropane-1-carboxylic acid (ACC) that ultimately converts into ethylene that restricts plant growth. By inoculation of ACC deaminase, ACC converts into ammonium (NH_4^+) and α-ketobutyrate (αKB) rather than ethylene. Additionally, nutrition stress (shortage of nitrogen) stimulates bacteria to produce IAA (Table 3). Hence, plant growth is promoted with decreased ethylene and increased IAA synthesis (Sangiorgio et al. 2020).

8 Organic Horticulture

Horticulture crops are highly sensitive to changing climate. Plant stages like fruiting, vegetative growth, and flowering are susceptible to extreme weather events. The relationship between crops, pollinators, diseases, insect pests, and weeds is

Table 3 Microorganism role in plants against stress caused by thermal events

Microorganism	Mode of action	Crop plant	Reference
Mychorrhizae	H_2O_2, reduction of ROS in roots, leaves	Tomato	Dighton (2009)
Ps. fluorescens A506	Competes with INA + bacteria	Apple and pear	Dighton (2009)
Paraburkholderia phytofirmans	Production of ACC deaminase	Grapevine	Donoso et al. (2017)

adversely being affected by changing climate. In advanced agriculture some technologies have been developed to counter the negative effects of climate; however, modification in agronomic practices, use of efficient inputs, and soil and crop management can be useful to minimize the adverse effects of climate change.

Organic farming has huge potential to mitigate climate catastrophes in various ways: (1) it enhances soil biological activity; (2) it sustains soil fertility for long duration; (3) plant waste can be recycled as green manuring; and (4) it encourages use of soil, air, and water that are environment friendly to reduce all forms of pollution. Diversifications and soil organic matter make soil more resilient to extreme climatic events. Crop production under organic farming is less susceptible to weather events such as flooding, drought, and water lodging (Kolambe et al. 2015).

8.1 Use of Bioherbicide in Organic Horticulture

Weeds in agriculture compete with crops for nutrients and water causing adverse effects on the growth and quality of plants. Annual weeds spread by seeds while the perennials are reproduced by stolons, roots, rhizomes, etc. In organic horticulture integrated use of hand weeding and cultural control should be encouraged. As synthetic weedicides are prohibited in organic production, bioherbicides taken from extracts, natural products, and microbes are being used successfully.

8.2 Bioherbicide Approach

Generally two approaches exist for biological control of weeds. In classical approach a natural enemy is introduced in the area to kill the weeds. However, there is a risk of attack on non-target plant species. In bioherbicide, control agents are applied within specific range of target weeds to kill them and to avoid their negative effects on crop yield. Bioherbicide is preferred over classical approach because it has more diversity to be used in gardens, agriculture, and lawns (Cai and Gu 2016).

8.2.1 Sources of Bioherbicides

Pathogens Several microbes have been identified as bioherbicides for trees, horticultural crops, and turf. First bioherbicide DeVine containing *Phytophthora palmivora* was used on citrus in Florida to suppress strangler vine (*Morrenia odorata*). Using biological control approach can cause damage to target weed. For suitability these pathogens must be produced in mass, having field efficiency and host range. Under greenhouse and field conditions of tomatoes, when a fungal bioherbicide *Myrothecium verrucaria* was used, it showed no effect on tomato growth; however, it killed 90%–95% purslane species and 85%–95% spurge species.

Natural Products Some natural products have been developed as bioherbicide to kill weeds of crops. Mustard seed meal (SAM) a byproduct of mustard oil has the potential to kill targeted weeds. Study showed that application of SAM to soil surface 113, 225, and 450 g.m^2 decreased bluegrass numbers by 60%, 86%, and 98%, respectively. In organic production, the rate, time, and type of seed meal should be planned to manage the weeds successfully.

Extracts Extracts from various substances have shown potential role as bioherbicides. Extracts from rice hull have significantly reduced germination, growth, and weight of barnyard grass. Extracts from walnuts are commercially used as bioherbicides. A product (NatureCur, Redoxchemical, USA) extracted from black walnut at a concentration of 33.8% has inhibited horseweed (*Conyza canadensis*) and hairy fleabane (*Conyza bonariensis*) and potentially used as bioherbicide for pre and post emergence.

9 Impact of Floods on Horticultural Crops

Horticultural crops are highly vulnerable to be adversely affected by disasters. Fruits, vegetables, ornamental plants, medicinal, and aromatic plants all are included in horticultural crops. Disasters disrupt the cycle of crop growth, development, and yield (Reddy 2012). Soil is the plant-growing medium which is adversely affected by natural disasters (Fig. 1). Percentage of water in root section of soil controls plant growth; the change in water percentage changes the behaviour of soil (Tampubolon et al. 2018). In flood conditions, due to speed and pressure of water

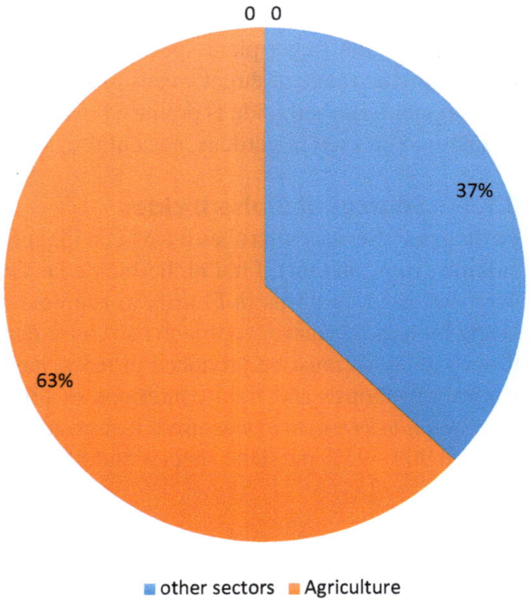

Fig. 1 Damage and loss in agriculture relative to combined industry, commerce, and tourism, 2008–2018 (Scawthorn et al. 2006)

causing soil runoff the root zone of soil is disturbed which the plant cannot tolerate and ultimately dies (Fig. 2). Due to floods plant water increases causing reduction in contact between air and soil. Under such conditions oxygen depletes in the soil and disrupts the redox potential of soil which induces the production of toxins, metals, and fatty acids. These abnormalities in plants lead to reduction in photosynthesis (Malhotra 2017).

9.1 Flood Control in Horticultural Crops

In subtropical areas heavy rainfall, storms, hurricanes, and irregular irrigation cause flooding conditions. In flooding under hypoxia conditions reactive oxygen species (ROS) starts accumulating causing degradation of proteins, lipids, and DNA which eventually leads to plant death.

9.2 Role of Silicon to Control Flooding

Silicon is abundantly present in soil. It rehabilitates plants from stress to non-stress conditions such as salinity, drought, and flooding. As Si solubility is dependent on surface area, so when Si is available to plants in nano size its efficiency increases, which decreases the effects of abiotic stresses on plants. Silicon plays significant role in plants protection; epidermal thickening protects cellular membrane, decreases toxic substances taken through roots, and sustains photosynthate. Si nano particles increase antioxidant activity of enzymes which reduces abiotic stress. This

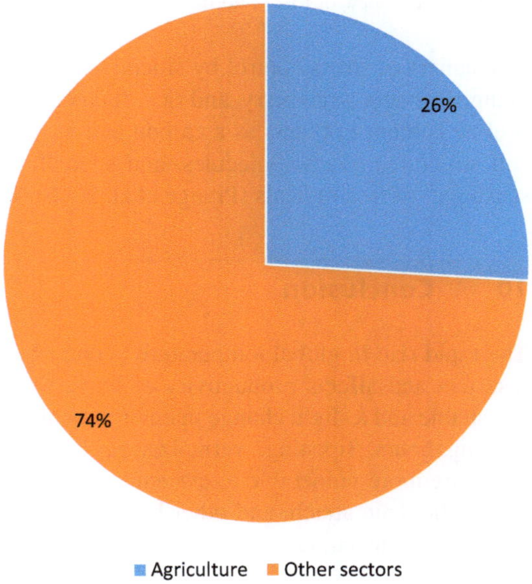

Fig. 2 Damage and loss in agriculture as a share to total damage and loss in all sectors, 2008–2018 (Wyer et al. 2022)

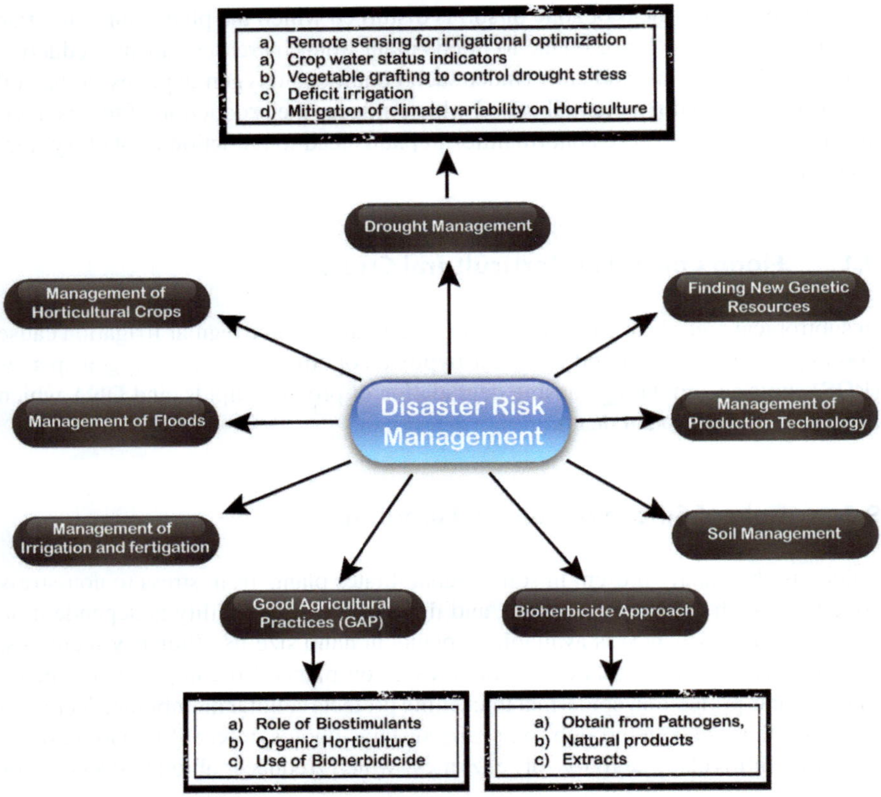

Fig. 3 Overall view of role of horticulture in disaster risk management

mechanism of stress control by silicon has been reported in many crops including, mango, tomato, strawberry, and rice. It is reported that silicon application increases proline content in plants as a carbon and nitrogen source, helps in stabilization of cell structures, macromolecules, and subcellular substances. Further, it acts as a scavenger of free radicals (Prange et al. 2006) (Fig. 3).

10 Conclusion

The rapid rise in global temperature is contributing to an increase in other natural disasters and affects productivity of horticultural crops. Because they are highly perishable and a slight change in environmental conditions can affect processes like chilling hours, ripening, vernalization, and many other processes, horticultural crops are more vulnerable to extreme weather conditions. Drought, salinity, flooding, pathogenic attach, and other biotic and abiotic stresses are some examples of climate-driven natural disasters that can be lessened with the help of advanced technological approaches and adaptative research. It was successful to graft scion onto rootstock that has stress-resistant genetic ability. Similarly how biostimulants help plants grow better and how bioherbicides effectively kill weeds can contribute to resilient crop production. In organic horticulture, soil conservation and the use of naturally derived fertilizer are integral parts of a holistic approach to sustainable production. The approaches pointed out in this chapter are practical, reliable, effective, and sustainable having the potential of productivity enhancement of horticultural crops to develop resilient food system in the areas prone to be sensitive to onset of natural disaster and climate change.

References

Bellvert J, Nieto H, Pelechá A, Jofre-Čekalović C, Zazurca L, Miarnau X (2021) Remote sensing energy balance model for the assessment of crop evapotranspiration and water status in an almond rootstock collection. Front Plant Sci 12:608967

Belward AS, Skøien JO (2015) Who launched what, when and why; trends in global land-cover observation capacity from civilian earth observation satellites. ISPRS J Photogramm Remote Sens 103:115–128

Bigey J (2000) Chilling requirements and compensation for the lack of chilling in strawberry. IV Int Strawber Sympos 567:269–272

Bisbis MB, Gruda NS, Blanke MM (2019) Securing horticulture in a changing climate—a mini review. Horticulturae 5:56

Burghardt J (2002) Natural hazard perceptions, natural disaster experiences and recovery at American public horticulture institutions: University of Delaware

Cai X, Gu M (2016) Bioherbicides in organic horticulture. Horticulturae 2:3

Calvo P, Nelson L, Kloepper JW (2014) Agricultural uses of plant biostimulants. Plant Soil 383:3–41

Costa JM, Ortuño MF, Chaves MM (2007) Deficit irrigation as a strategy to save water: physiology and potential application to horticulture. J Integr Plant Biol 49:1421–1434

Dannemand M, Perers B, Furbo S (2019) Performance of a demonstration solar PVT assisted heat pump system with cold buffer storage and domestic hot water storage tanks. Energ Buildings 188:46–57

De Bon H, Holmer RJ, Aubry C (2015) Urban horticulture. In: Cities and agriculture. Routledge, pp 236–272

Dighton J (2009) Mycorrhizae. In: Encyclopedia of microbiology. Elsevier, pp 153–162

Donoso R, Leiva-Novoa P, Zúñiga A, Timmermann T, Recabarren-Gajardo G, González B (2017) Biochemical and genetic bases of indole-3-acetic acid (auxin phytohormone) degradation by the plant-growth-promoting rhizobacterium Paraburkholderia phytofirmans PsJN. Appl Environ Microbiol 83:e01991–e01916

Floriancic MG, Berghuijs WR, Jonas T, Kirchner JW, Molnar P (2020) Effects of climate anomalies on warm-season low flows in Switzerland. Hydrol Earth Syst Sci 24:5423–5438

Franks PJ, Adams MA, Amthor JS, Barbour MM, Berry JA, Ellsworth DS, Farquhar GD, Ghannoum O, Lloyd J, McDowell N (2013) Sensitivity of plants to changing atmospheric CO_2 concentration: from the geological past to the next century. New Phytol 197:1077–1094

Godfray HCJ, Beddington JR, Crute IR, Haddad L, Lawrence D, Muir JF, Pretty J, Robinson S, Thomas SM, Toulmin C (2010) Food security: the challenge of feeding 9 billion people. Science 327:812–818

Goetz S (1997) Multi-sensor analysis of NDVI, surface temperature and biophysical variables at a mixed grassland site. Int J Remote Sens 18:71–94

Hatfield JL, Prueger JH (2015) Temperature extremes: effect on plant growth and development. Weather Clim Extrem 10:4–10
Kolambe B, Patel K, Kaswala A, Dubey P (2015) Organic farming for climate smart horticulture
Kumar P, Rouphael Y, Cardarelli M, Colla G (2017) Vegetable grafting as a tool to improve drought resistance and water use efficiency. Front Plant Sci 8:1130
Li B, Yu Z, Liang Z, Song K, Li H, Wang Y, Zhang W, Acharya K (2014) Effects of climate variations and human activities on runoff in the Zoige alpine wetland in the eastern edge of the Tibetan plateau. J Hydrol Eng 19:1026–1035
Malhotra S (2017) Horticultural crops and climate change: a review. Indian J Agric Sci 87:12–22
Mirfenderesgi G, Bohrer G, Matheny AM, Fatichi S, de Moraes Frasson RP, Schäfer KV (2016) Tree level hydrodynamic approach for resolving aboveground water storage and stomatal conductance and modeling the effects of tree hydraulic strategy. J Geophys Res Biogeo 121:1792–1813
Müller T, Spreitzenbarth M (2013) Frost. In: International conference on applied cryptography and network security. Springer, pp 373–388
Paraskevopoulou A (2021) Horticulture, design, and ecology: how to deal with the urban environment? In: VIII international conference on landscape and urban horticulture, vol 1345, pp 1–12
Prange RK, Ramin AA, Daniels-Lake BJ, DeLong JM, Braun PG (2006) Perspectives on postharvest biopesticides and storage technologies for organic produce. HortScience 41:301–303
Reddy PP (2012) Organic farming for sustainable horticulture. Scientific Publishers
Rheinbay E, Nielsen MM, Abascal F, Wala JA, Shapira O, Tiao G, Hornshøj H, Hess JM, Juul RI, Lin Z (2020) Analyses of non-coding somatic drivers in 2,658 cancer whole genomes. Nature 578:102–111
Ricci M, Tilbury L, Daridon B, Sukalac K (2019) General principles to justify plant biostimulant claims. Front Plant Sci 10:494
Rouphael Y, Colla G (2018) Synergistic biostimulatory action: designing the next generation of plant biostimulants for sustainable agriculture. Front Plant Sci 9:1655
Sabir N, Singh B (2013) Protected cultivation of vegetables in global arena: a review. Indian J Agric Sci 83:123–135
Sánchez-Rodríguez E, Romero L, Ruiz JM (2013) Role of grafting in resistance to water stress in tomato plants: ammonia production and assimilation. J Plant Growth Regul 32:831–842
Sangiorgio D, Cellini A, Donati I, Pastore C, Onofrietti C, Spinelli F (2020) Facing climate change: application of microbial biostimulants to mitigate stress in horticultural crops. Agronomy 10:794
Scawthorn C, Flores P, Blais N, Seligson H, Tate E, Chang S, Mifflin E, Thomas W, Murphy J, Jones C (2006) HAZUS-MH flood loss estimation methodology. II. Damage and loss assessment. Nat Hazards Rev 7:72–81
Singh H, Kumar P, Chaudhari S, Edelstein M (2017) Tomato grafting: a global perspective. HortScience 52:1328–1336
Tampubolon J, Nainggolan H, Ginting A, Aritonang J (2018) Mount Sinabung eruption: impact on local economy and smallholder farming in Karo Regency, North Sumatra. In: IOP conference series: earth and environmental science. IOP Publishing, p 012039
Tejero IG, Zuazo VHD, Bocanegra JAJ, Fernández JLM (2011) Improved water-use efficiency by deficit-irrigation programmes: implications for saving water in citrus orchards. Sci Hortic 128:274–282
Van Zyl K (2006) Reducing disaster risk through vulnerability assessment: an agricultural perspective. Jàmbá J Disast Risk Stud 1:19–24
Wyer KE, Kelleghan DB, Blanes-Vidal V, Schauberger G, Curran TP (2022) Ammonia emissions from agriculture and their contribution to fine particulate matter: a review of implications for human health. J Environ Manag 323:116285
Zhang Z, Cao B, Gao S, Xu K (2019) Grafting improves tomato drought tolerance through enhancing photosynthetic capacity and reducing ROS accumulation. Protoplasma 256:1013–1024

Disaster Hazards and Vulnerabilities in Agriculture: Role of Food Technologist

Ehsan Ul Haque, Akbar Hayat, Sohaib Afzaal, Muhammad Asim, Shakeel Hanif, and Tahir Awan

Abstract

Numerous external phenomena, including climate change elements, frequently cause disasters that have an impact on entire communities. Individuals in developing countries, particularly vulnerable groups like children, the elderly, and people with disabilities, as well as communities living in challenging environments, frequently fall victim to crime. Extreme natural calamities like earthquakes, wind cyclones, and floods frequently affect populations worldwide, but the higher vulnerability of developing nations makes those calamities disasters. Humans require three essential things to survive: food, shelter, and clothing. Food is the most important basic requirement for the communities affected by disasters and emergencies to survive. In typical circumstances, 10% of the world's population becomes ill after consuming ill-prepared food. For different towns and areas, maintaining food safety as well as a sufficient supply in catastrophe and emergency scenarios becomes more difficult. Food aid rations should be planned with cultural preferences, the environment, and general needs of population for energy and essential nutrients in mind so that a high standard of living may be sustained.

Keywords

Climate change · Vulnerable groups · Food · Shelter · Clothing · Food aid rations

---

E. U. Haque · A. Hayat · S. Afzaal · M. Asim
Citrus Research Institute, Sargodha, Pakistan

S. Hanif
Fodder Research Institute, Sargodha, Pakistan

T. Awan (✉)
Punjab Agriculture Research Board, Lahore, Pakistan

M. Ahmed, S. Ahmad (eds.), *Disaster Risk Reduction in Agriculture*, Disaster Resilience and Green Growth, https://doi.org/10.1007/978-981-99-1763-1_19

1 Introduction

A wide range of disaster threats includes droughts, earthquakes, storms, floods, cyclones, epidemics, insect infestations, and landslides. Avalanches and river erosion are global problems that affect many nations and areas. Human-induced risks like civil conflicts, community displacement, transportation accidents, industrial disasters, and urban and forest fires have also had a significant negative impact on Pakistan and other developing nations throughout the world. High-priority risks on the list of catastrophes offered may include windstorms, earthquakes, floods, and landslides that result in widespread damage and the loss of people, animals, and livelihoods. A few of the causes contributing to community vulnerabilities include poor management of wildlife, agriculture, construction methods, a fragile natural environment, a lackluster early warning system, inadequate training and awareness, and a lack of essential amenities and infrastructure (Quarantelli 1998). The increasing frequency and severity of losses to lives, livelihoods, and communities as a result of natural disasters, climate change, and hazards have drawn international organizations from a wide range of nations to develop policy guidelines for technological, legal, financial, and institutional steps to reduce vulnerabilities and associated destructions. A common disaster risk management idea integrates the concepts of prevention, mitigation, and readiness with reaction, while related organizations advise proactive disaster risk reduction policies at pre-disaster phases. Due to catastrophic weather events and difficult climatic circumstances brought on by climate change, vulnerable groups of people, particularly those living in mountainous areas, have become even more vulnerable. As certain risks become catastrophes for certain communities and regions in the future, there may be greater personal, economic, social, and environmental losses as a result of their increased intensity, frequency, and effect.

Many international organizations and parties have defined the importance of food supply and availability in relation to "food security," particularly in emergency or disaster scenarios. In accordance with the Universal Declaration of Human Rights, "everyone has a right to the health and welfare of himself and his family for a level of living assuring health and nourishment." Similarly, "Humanitarian Charter and the Minimum Standards 1998" emphasizes the significance of calculating each person's needs for food, nutrition, water, and sanitation in times of tragedy and emergency. When a calamity strikes and the community is dependent on outside aid, extra care should be taken to prepare rations with the quantity, quality, and safety of the general public in mind. In order to provide the least amount of energy, fat, vitamins, protein, and micronutrients, the food aid rations should be well balanced. For distinct populations that demand varied considerations, the minimum energy requirement varies in nature. Numerous international agencies, including the World Health Organization (WHO), the World Food Program (WFP), UNICEF, and UNHCR, propose 2100 kcal per person per day as the minimum energy intake. Priority is given to subgroups including children, infants, women, and older people to meet the fundamental energy needs of the general population during emergencies, while subsequent consideration is given to these subgroups to address food and nutrition deficits. Food products are chosen to offer appropriate, balanced nutrition while taking into account local management practices, public acceptance, physiological requirements, and environmental factors (Shaluf 2007).

During a disaster when a territory is adversely affected, no industry remains unaffected including food-processing facilities. The apparent destruction is recognized as occurring to the agricultural farms and livestock working as the base line for the food production. But actual deep losses occurring to the local food industry and entire supply chain stretch much deeper and have adverse effects for longer period of time disrupting various businesses and production facilities. Food supply chains are diversified and complex in nature in which each step is integrated and dependent upon various other stakeholders and processes. Breakdown in any step may stop or adversely affect the entire production and supply process in the chain. The affected block in the supply chain may have a domino effect of the natural disaster disrupting the food safety and quality systems in place. Industrialists associated with food production and food supply chains should recognize vulnerabilities of their processes through risk estimations. As a preventive step, prevailing hazards and upcoming disasters and risks should be considered and their possible impacts to food safety should be minimized through better planning and execution.

2 Hazards, Disasters, and Risks

United Nations International Strategy for Disaster Reduction (UNISDR) defines hazard as a natural phenomenon or process posing negative impact to society, economy, and ecology. Hazards are hindering human development and global sustainability giving rise to serious disasters. Hazards may be natural, human originated, or complex in nature associated with detrimental effects to human society. There are various classifications of hazards available in literature depending upon nature, intensity, causes, and environment. Common groups are geophysical, hydrological, meteorological, climatological, biological, and extraterrestrial hazards as detailed in Fig. 1. The main events included in hazards are flood, earthquake, volcanic activity, storms, extreme temperature, fog, mass movement, cyclones, drought, wildfire, disease insect infestation, and many others.

Disasters are originated from hazards causing greater and widespread losses to human, environment, economic, and associated structure. Usually affected community is not able to manage disasters through its own resources especially in developing world. In recent history, natural disasters and human-triggered disasters are so complex in nature to identify them separately.

Risk is the potential loss involving uncertainty about the negative outcomes in the future due to prevailing hazards. UNISDR 2004 defines risk as a probability of negative consequences occurring due to natural or human-made hazards or disasters. Hazards intensity, social factors, and distribution of impact after occurrence of disaster usually define the risk management.

Fig. 1 Types of disasters and their management

Fig. 2 Link among hazard, disaster, and risk

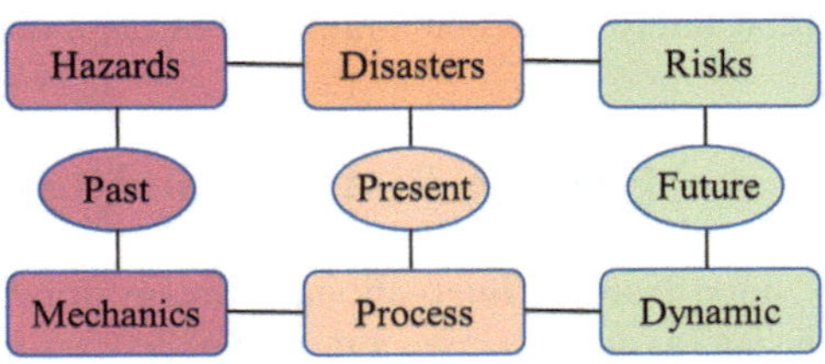

Hazards are natural or man-made negative events retarding human development as evident from the historical data. Disasters are the losses or widespread negative impact of hazards on human societies which is measured in death tolls, economic, and material losses. Risks are future threats likely to accrue through disasters taking place due to prevailing hazards in a region. Relationship among hazard, disaster, and risk is shown in Fig. 2.

When a hazard affects more than 100 people or causes more than 10 causalities it is declared as disaster. There is an estimate that United States, China, and India are considered top three countries suffering from disasters. Natural disasters are striking more intensively since 1970, as shown by the fourfold increase in events recorded till 2015. In only 2018 United States had experienced 14 such disasters each causing a loss of more than 1 billion USD. According to *The Economist* number of disasters is quadrupled since 1970 while hydrological events had increased six times.

As more than 57,000 food processors operate in the United States, there is greater risk prevailing for the food industrialists. All hazards, disasters, and extreme weather conditions pose different types of risks for the food processors. Food industries have to implement procedures and emergency preparedness to cope the situation ensuring the continuity of their operation to ensure the food supply in emergencies. Food processors must proactively work together to implement the strategies to ensure safe and quality food supply minimizing the risks and negative impacts of the disasters to the community.

3 Hazards

Hazards are posing serious threat to human development since ages with varying intensities to various regions of the world. Pakistan, like other nations around the world, is subject to a variety of natural and human-caused dangers and hazards. The degree of Pakistan's vulnerability to natural disasters and hazards can range from moderate to severe. Pakistani society is at risk from natural disasters such as avalanches, earthquakes, cyclones and storms, river erosion, droughts, epidemics, floods, glacial lake outbursts, pest assaults, and tsunami. The society, economy, and environment are all at risk due to a number of human-caused risks. They include those related to industry, transportation, oil disasters, forest and city fires, civil unrest, and community internal migration. High-priority hazards include earthquakes, droughts, flooding, windstorms, and landslides because of their frequency and scope of impact. These disasters have historically resulted in significant losses and property damage. An outline of the main threats to Pakistan is provided below.

Each type of hazard when intensified into a disaster adversely affects the food supply in the region. Water and food are the hardest but inevitable commodities to find after onset of a disaster in the region. Scarcity of food and water may grow up to the level that one cannot procure with plenty of cash and resources in hand. Management of food supply could depend upon the prevailing conditions in the region and better planning and management.

3.1 Earthquakes

The Indo-Australian plate, on which Pakistan, India, and Nepal are located, is continuously subducting beneath the Eurasian plate and migrating northward, causing earthquakes and the formation of the Himalayan Mountains. The Northern Areas and Chitral district in NWFP, Kashmir, including Muzaffarabad, Quetta, Chaman, Sibi, Zhob, Khuzdar, Dalbandin, the Makran coast, including Gwadar and Pasni in Balochistan, and the Suleiman, Hindu Kush, and Karakoram mountain ranges are all located in high- or very-risk regions. The cities of Peshawar, Karachi, and Islamabad are situated on the outskirts of high-risk regions. In the twentieth century, Pakistan was struck by four significant earthquakes, including Quetta earthquake in 1935, Makran coast earthquake in 1945, Northern Areas in 1976, and Kashmir/NWFP in 2005 (Fig. 3).

One or more significant earthquakes, according to seismologists like Dr Roger Bilham and colleagues, may be overdue in a significant portion of the Himalayas.

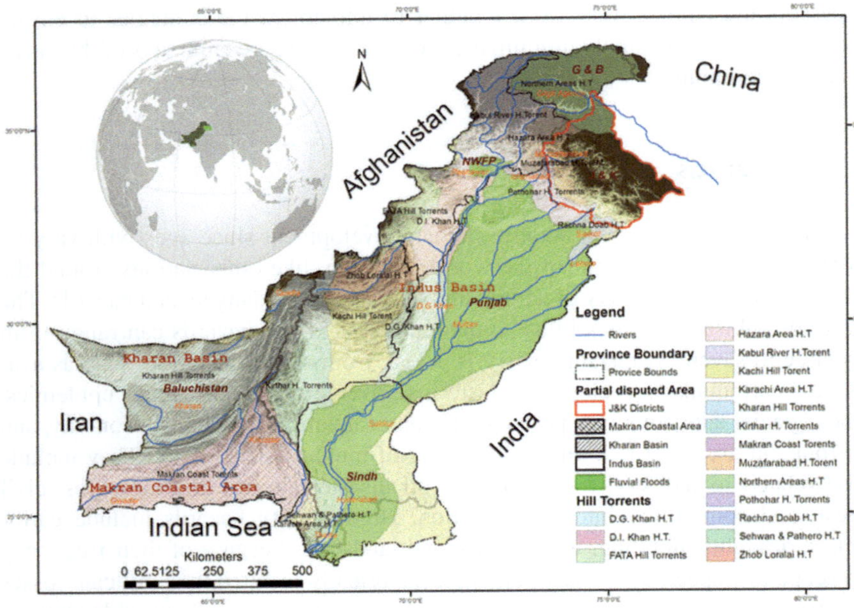

Fig. 3 Map of Pakistan showing main hydrological and geopolitical regions

Additionally, they cannot completely rule out the possibility of ruptures in the Balochistan region with magnitudes between 7.5 and 8.

3.2 Droughts

In Pakistan, droughts are occurring more frequently, which has serious repercussions for sustainable development in the areas of food security, agriculture, water resources, livestock, hydropower, and environment. The climate in Pakistan is characterized by dramatic temperature swings and little precipitation. About 60% of the total geographical area is categorized as arid, receiving less than 200 mm of rainfall yearly. Cholistan, D. I. Khan, Kohistan, Tharparkar, D. G. Khan, and Western Balochistan are among the principal desert rangelands. Balochistan and Sindh get about 160 mm of yearly precipitation on average, compared to 400 mm in Punjab and about 630 mm in NWFP. Additionally, there is a significant amount of seasonal variation in rainfall. Arid and hyper-arid conditions prevail throughout the lower southern part of the nation. In every region, there are some areas that continue to experience extreme dryness and are constantly susceptible to drought with a slight departure from the low mean rainfall. In some regions, there are 2–3 years of drought every 10 years.

3.3 Floods

Pakistan comprises over 70% of the land of the Indus River Basin, or 56.6% of its total area (ICSU 2008). Major floods typically happen in the Indus Basin in the late summer (July–September), when the South Asian region experiences strong monsoon rainfall.

Generally, tributaries like Jhelum and Chennab are the main cause of floods in the high to mid reaches of the basin. Major flooding is mostly related to the monsoon low depression, which forms in the Bay of Bengal and travels through India in a westward or northward trajectory before entering Pakistan. Punjab and Sindh are severely affected by river floods, while the hilly regions of NWFP, Balochistan, and northern portions are frequently affected by hill torrents. The NWFP region's Charsadda, Mardan, Peshawar, and Nowshera districts are under risk of flooding from the Kabul River. Dam failures in Pakistan can also result in flooding, as was the case in Pasni in February 2005 after the Shadi Kot Dam failed. Large cities are now more susceptible to flooding than they were previously. Flooding has occurred in cities including Lahore, Karachi, and Rawalpindi as a result of the sewerage system's failure to handle the severe rainfall. The recent year 2022 floods had affected Northern Areas, southern Punjab, and various regions of the Balochistan and Sindh provinces (Fig. 4). In Punjab, flash floods can also occur in mountainous and steep places, which can result in landslides and road erosion. Over Lahore (like it did in 1996), Rawalpindi, Islamabad, and Jhelum, cloud burst flash floods (CBFF) could potentially happen (Fig. 5).

3.4 Landslides

The province of NWFP, the Northern Areas, and the regions of Kashmir are particularly at risk from landslide hazards. Accelerated deforestation is a significant factor in the rise of the number of landslides, in addition to the young geology and delicate soil types of mountain ranges. In the aforementioned regions, there are typically small-scale, localized landslide dangers. Given that the forest cover is decreasing by 3.1%, the frequency of landslides could rise in the future (7000–9000 ha taken away annually).

3.5 Tsunami

Tsunami disasters have also occurred in Pakistan in the past. On November 28, 1945, an earthquake of 8.3 M off the coast of Makran Coast caused a significant tsunami. At least 4000 people were killed in Pasni and the other areas by sea waves of 12–15 m height that the tsunami created. Sea waves up to 6 feet high were experienced in Karachi, which is located roughly 450 km from the epicenter. Cities like Karachi should pay attention to improving local capacities for disaster risk

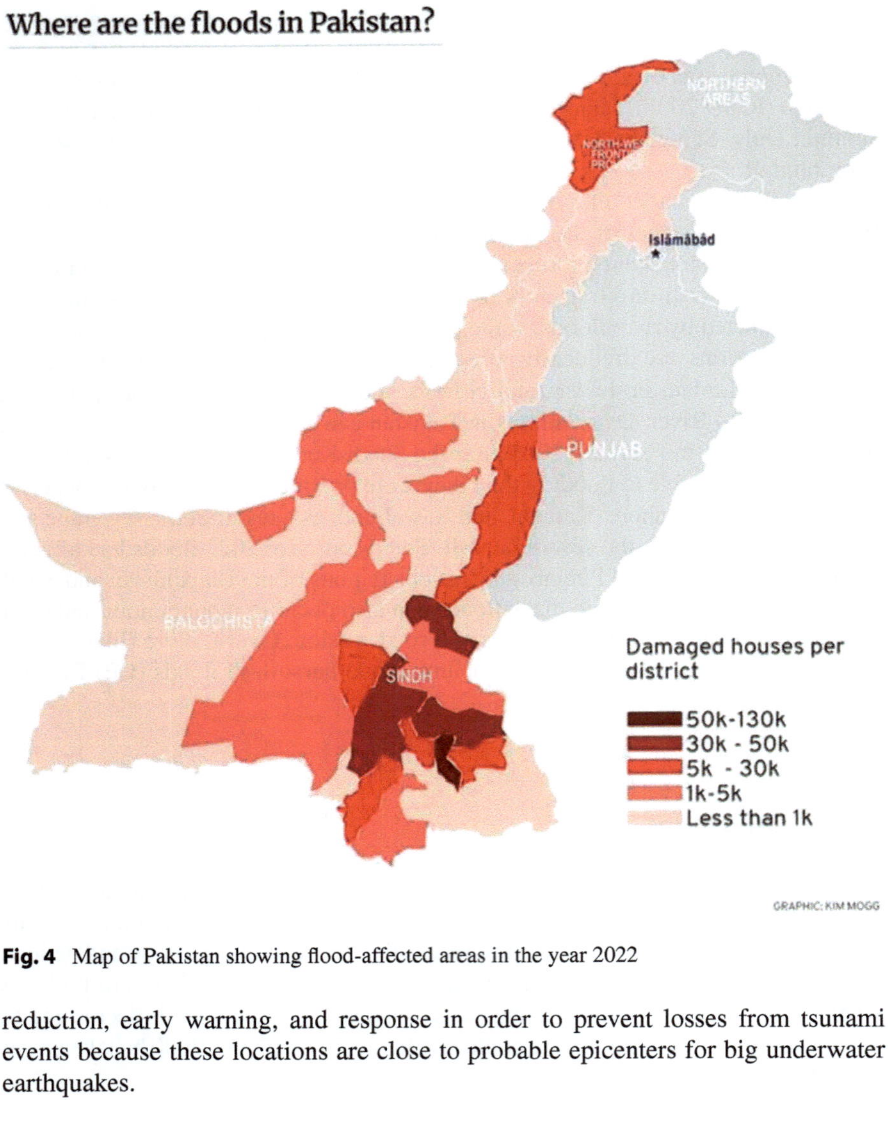

Fig. 4 Map of Pakistan showing flood-affected areas in the year 2022

reduction, early warning, and response in order to prevent losses from tsunami events because these locations are close to probable epicenters for big underwater earthquakes.

3.6 Cyclones/Storms

The coastal region of Pakistan, particularly Sindh, is extremely susceptible to cyclones and the ensuing storm surges. Between 1971 and 2001, 14 cyclones were reported. Cyclones have the potential to seriously harm Sindh's and Balochistan's coastal regions. In the Thatta and Badin districts, 73 towns were destroyed, 168 people perished, and 11,000 livestock perished in the storm of 1999. A little under

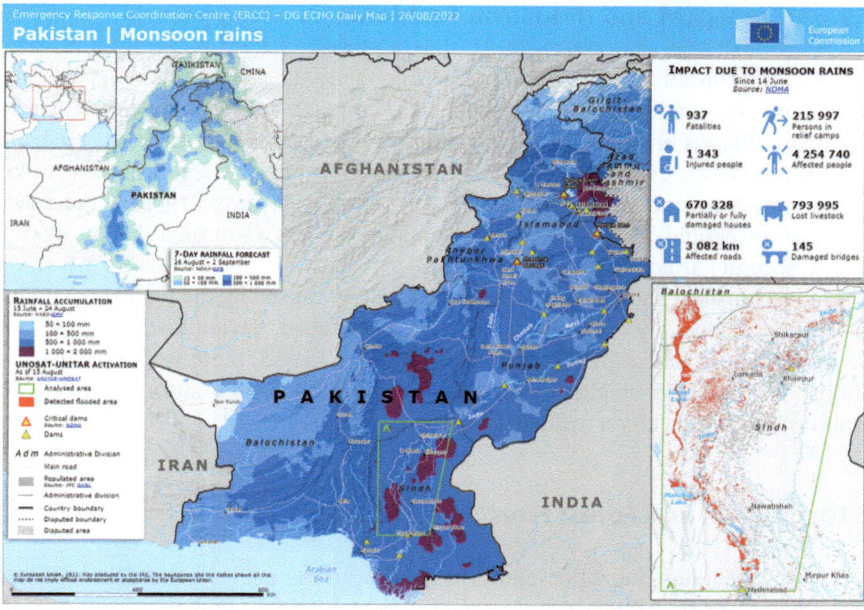

Fig. 5 Map of Pakistan showing monsoon rains

0.6 million people were impacted. It cost Rs. 380 million in losses as it destroyed 1800 small and large boats and partially damaged 642 others. The cost of the infrastructure losses was put at Rs. 750 million. The frequency and severity of storms may increase due to climate change, and their paths may also vary.

3.7 Glacial Lake Outburst Flood (GLOF)

Another potential scenario that might occur is the explosion of glacial lakes in the Indus Basin's upstream due to heat waves, a process known as the glacial lake outburst flood (GLOF). Of the 2420 glacial lakes in the Indus Basin 52 are potentially harmful, according to a new research, and might cause considerable harm to life and property in the event of a GLOF. The study also showed that future GLOF potential may rise due to global warming.

3.8 Avalanches

Avalanches occur frequently during specific seasons in Pakistan's northern regions and the Kashmir region. Tourists and residents in the dangerous area are at risk from this danger. The probable avalanche paths were identified by a study done by WAPDA between 1985 and 1989 as part of the Snow and Ice Hydrology Project.

3.9 Transport and Industrial Accidents

In Pakistan, transportation accidents occur frequently. The train system in particular is infamous for crashes. Road accidents and aeroplane crashes happen frequently. The ports of Karachi and Gwadar are particularly susceptible to maritime mishaps. The collapse of the Greek oil ship Tasman Spirit in August 2003 in Karachi resulted in enormous environmental losses and health risks for businesses, port workers, and neighboring people. A total of 28,000 tons of oil leaked throughout the harbor region, severely affecting marine life.

Major industrial and chemical disasters could be caused by the increasing industrialization, especially in urban areas like Faisalabad, Hyderabad, Gujranwala, Karachi, Quetta, Sialkot, Peshawar, and Lahore. In 1985, there was a gas leak in Bhopal that affected neighboring India, killing 5000 people and causing severe health issues for the city's residents.

3.10 Urban and Forest Fires

With growing industrialization and urbanization in the country, the risks of urban fires are on the rise. All urban areas have CNG gas stations constructed, and the gas is also offered for sale in small shops for use in homes. Cities also frequently see the sale of petroleum products within residential areas. Urban services are typically ill-equipped to combat these threats, and these activities significantly increase the risk of fire in urban environments. Forest fires are common in Pakistan's and AJK's northern regions.

3.11 Civil Conflicts

Pakistan is a society that is varied in terms of ethnicity, language, religion, and culture. This diversity has occasionally resulted in civil wars between various social classes, which mostly affect women, minorities, and children. For instance, throughout the 1980s and 1990s, sectarian violence erupted in Pakistan. In addition to causing property damage and fatalities, these clashes made many social groups in the impacted communities feel unsafe. The social fabric of Pakistan was also harmed by the hosting of roughly six million Afghan refugees.

4 Vulnerabilities

The main factors that make Pakistan vulnerable to hazards are the low quality of the housing stock, buildings, and infrastructure (especially in rural areas), the delicate natural environment, ineffective livestock and agricultural management techniques, the lack of early warning systems, a lack of awareness and education, and poverty.

Communities' vulnerabilities in the aftermath of disasters are further exacerbated by a lack of vital infrastructure and facilities.

In Pakistan, adobe makes up the majority of rural dwellings, making it particularly susceptible to natural disasters like earthquakes, floods, and landslides. Without any reinforcement, people construct homes by piling stones on top of one another in Kashmir, FATA, NA, and NWFP.

The traditional method of light-weight, timber-laced construction has been replaced by more substantial masonry and reinforced concrete structures, which offer excellent defense against inclement weather but are frequently inadequately built to resist powerful earthquakes. Housing and infrastructure in metropolitan areas suffer from a lack of application of building codes. Slum expansion and urban poverty have made dangerous building practices even worse. Even Quetta, which was completely destroyed by an earthquake in 1935, does not employ safer building techniques. The causes are a lack of political will, commercial interests, corruption, ignorance of the situation, and a shortage of skilled labor.

Conditions of vulnerability have also been made more vulnerable by the fragility of the natural environment in locations upstream of the Indus River Basin. The northern region has a high rate of soil erosion as a result of extensive deforestation. Only 4% of Pakistan's land is covered with trees and other vegetation, compared to a required 25%, resulting in a strong and continuous flow of water, particularly during monsoon seasons. Flood and landslide risks have increased as a result of this and increasing melting in the Himalayan glaciers. In order to lessen vulnerabilities, pressures on forests and other natural resources must be released.

Drought conditions have gotten worse due to overgrazing on marginal lands in Balochistan and Tharparkar as well as the growing of crops that require a lot of water, like rice and sugar cane. In dry regions, the number of animals has multiplied, which has resulted in overuse of rangelands without giving them time to recover. Parallel to this, widespread tube-well construction in Balochistan has intensified groundwater extraction, rapidly lowering water tables. Modifications to agricultural and livestock management practices are necessary to address the problems of drought and water scarcity in arid zones. For example, the size of the livestock population must be reduced to be compatible with the carrying capacity of rangelands, and water-intensive crop varieties must be replaced with drought-resistant ones.

A little under one-third of Pakistanis, many of whom reside in hazardous areas, live below the poverty line. This social group suffers greatly from disasters since it is unrealistic to expect them to prioritize reducing disaster risk given how difficult it is for them to manage everyday risks.

Physical isolation, dispersed habitation patterns, and harsh weather conditions are to blame for the specific vulnerability of mountain communities in northern Pakistan. Due to high building costs and the terrain, infrastructure development for health, safe drinking water, education, and sanitation is frequently neglected. Physical remoteness is complicated further by fragile ecosystems that are vulnerable to landslides, erosion, and biodiversity loss.

5 Dynamic Pressures

The increasing susceptibility of Pakistani society and economy to catastrophes is being driven by a number of main stressors, including population expansion, urbanization, industrialization, the resulting environmental degradation, and climate change/variability.

6 Population Growth and Size

Population growth in Pakistan has become a significant dynamic pressure, adversely affecting many aspects of the social, economic, and environmental life. Since independence in 1947, the population has increased by 35%. With a contribution of 133 million people through 2025, Pakistan would be the second largest donor to the global population, after China. Multiple mechanisms contribute to population growth's creation and escalation of vulnerabilities.

Population density in areas at risk for hazards also contributes to increased loss of life and property during disasters. In the upcoming years, a much higher proportion of people would reside in regions vulnerable to droughts, floods, and earthquakes if population growth patterns keep up at their current rates. Pakistan must stop the escalating patterns of vulnerability by controlling population increase.

There has also been a significant increase in the animal population in Pakistan's dry and drought-prone regions. In Tharparkar and Balochistan regions, the population of livestock has increased five to nine times faster than the capacity of the nearby rangelands. This has an impact on the local climate and weather, which in turn exacerbates the dry and drought conditions.

Livestock size needs to be reduced, to lessen pressure on the local carrying capacity in these areas. Creating channels for livestock export from dry areas to other areas of the nation can help with this. To reduce extreme reliance on livestock-based livelihoods and to diversify the sources of income in dry regions alternative ways of subsistence should be encouraged.

7 Urbanization, Industrialization, and Environmental Degradation

Pakistan is in the process of changing from an agricultural and rural to a contemporary industrial economy. Rapid urbanization, infrastructural growth, environmental deterioration, soil erosion, water and air pollution, etc., are all effects of this progressive transformation. Due to the high rate of rural—urban migration, urban expansion is occurring more quickly. The preference for building infrastructure and services in urban areas, together with the availability of jobs and higher salaries, has served as pull factors in drawing both educated and uneducated rural populations to metropolitan areas. Consumption patterns dramatically increase as cities grow. Better services and greater natural resources (land, water, and forests) are needed in cities to support current lifestyles. More water, timber, and mineral resources are also needed as industrialization expands. As a result, natural resources in rural areas and upstream are used more quickly, causing environmental degradation such as the clearing of land for development, the chopping of trees, and the depletion of ground and surface water resources. According to studies, environmental degradation in Pakistan may cause soil erosion and degradation, which could exacerbate landslides in the Muree Hill tracts, Kashmir, and Northern Areas. Additionally, it can cause more flooding and intense rain. Sea intrusion has occurred in coastal Sindh as a result of the clearance of mangroves and a decrease in the amount of water discharged into the ocean. If this natural barrier disappears, coastal infrastructure and residents may be more frequently subjected to storms and flooding.

8 Climate Change and Variability

According to observations of WWF Pakistan, climate change is harming Pakistan's environment. Loss of biodiversity, modifications to weather patterns, and adjustments to the provision of fresh water are some of the effects felt and observed. According to a research by GTZ for WAPDA that examined temperature and precipitation patterns over the previous century in the Northern Areas, Skardu has experienced an increase in seasonal and annual temperatures. The mean annual maximum temperature climbed by more than 2.35 °C, while the mean annual temperature increased by 1.4 °C overall. Since 1961, the maximum winter temperatures have increased by up to 0.51 °C every decade, significantly greater than the maximum summer temperatures.

Mountain glaciers in the Karakorams have been shrinking for the past 30 years, according to observations made by the World Glacier Monitoring Service, which is based in Switzerland. According to experts, compared to the decade of 1975–1990, river flow rose throughout the 1990s, increasing the amount of ice that melted upstream. Additionally, studies show that some of Pakistan's glaciers have recently undergone major retreat. This, according to scientists, is a sign of climate change because it causes more snow to melt. The likelihood of flash flooding and catastrophic flooding may rise over the next few decades due to climatic changes. According to research done by SDPI, the amount of rain that falls on average in South Asia would rise by 17–59% with a twofold of CO^2. The frequency of high rainfall occurrences will double as a result of this. Additionally projected variable monsoons may result in increased droughts. A greater desiccation of arid regions brought on by global warming, according to experts, may threaten food production on the plains unless a significant number of trees are planted there.

9 Gender Power Imbalances

Countries that have gone through significant disasters show how costly it is to ignore gender in disaster preparedness, response, and recovery. This deepens already-existing poverty and inequality by ignoring the harm, requirements, and priorities of the most vulnerable people during disasters. The current political, social, and economic inequalities are made worse by a lack of gender-responsive assessments and programming. Despite the destruction they inflict, natural disasters offer chances for social and economic reform. Women should be given the opportunity to participate in decision-making on an equal footing before, during, and after disasters in order to minimize casualties, protect the home economy, and prevent the collapse of social safety nets.

10 Future Disaster Trends in Pakistan

The examination of hazard risks, vulnerabilities, and dynamic pressures highlights the possibility of an increase in the number of people residing in and around hazard-prone locations. With a growing population in risky locations, new communities would continue to appear. Given that Pakistan's population is anticipated to double within the next 25–30 years, this trend may get worse over time. On the other hand, as a result of environmental deterioration and climate change, certain risks, such as droughts, flooding, soil erosion, and landslides, are becoming more frequent, severe, and intense. These scenarios suggest that disasters will occur more frequently in the future and will have greater social, economic, and environmental effects than they did in the past. Previously unaffected areas may now face certain threats (such as droughts and flooding) in the future (Lindell 2013).

11 Urbanization in Pakistan

Cities in Pakistan are growing significantly more quickly than the nation's population as a whole. Many Indian immigrants settled in metropolitan areas when their country gained independence in 1947. The urban population doubled from 1951 to 1981. In the 1950s and 1960s, the yearly pace of urban growth was greater than 5%. It averaged around 4.6% from 1980 to the beginning of 1994; 32% of Pakistanis were residing in urban areas in 1994. In 2001, the number of people living in urban areas was estimated at 47.7 million using a growth rate of 3.5%. According to estimates, during the period 1951–1998, when the population as a whole increased by almost four times, the urban population increased by seven times.

12 Livestock and Droughts in Balochistan

Approximately 20% of the nation's livestock is kept in Balochistan. Balochistan has an estimated livestock population of 1,402,000 cattle, 213,000 buffaloes, 10,761,000 sheep, and 10,098,000 goats in 1999. Since 1947, the number of cattle in Balochistan has increased by 90%, from 2,327,000 to 22,483,000 heads. In typical years, there are 3.3 million tons of dry edible matter produced annually, compared to the 6.3 million tons of feed needed by the animal population. As a result, the carrying capacity of the range lands has been exceeded over time by an increasing animal population.

Balochistan's livestock population suffered significantly from the drought that lasted from 1998 to 2002. According to estimates from an FAO/WFP joint mission study, farmers may have lost up to 50% of their sheep and up to 40% of their goats due to greater mortality and forced culling during the drought. According to estimates, 2.18 million out of 10.65 million heads of livestock died. Five million animals in Sindh were impacted, and 30,000 of them perished.

13 Role of Food Scientists in Natural Disasters

Natural disasters like flood, earthquake, tsunami, and hurricanes drastically affect food security by affecting the agricultural production, food availability, and stability of long-term food security. The nature of food and nutrition problems depends on the type of disaster, its duration, and the size of area affected. Earthquakes usually have little effect on long-term food supplies while hurricanes, floods, and tsunami directly affect the availability and safety of food by destroying the crops and stored food.

Food scientists and nutritionists can play an important role in food supplying and prevention of malnutrition by developing value-added and nutritive food products which can be consumed for a long-term period. It is hard to prepare food for consumption during an emergency situation. In this condition, food technologists play a key role in developing instant and RTS/RTD food products, canned juices, powdered milk, non-perishable pasteurized milk, high-energy foods like peanut butter, dry cereals, and food products with better shelf life. These foods are not only suitable for consumption in disaster situation but also provide required dose of calories and ease of transportation.

There is a high chance of infection during disasters due to food and waterborne diseases and lack of hygienic foods. But all these problems can be tackled by producing biotechnologically processed food products with high nutritional value. Food professionals choose the foods having longer shelf life and foods with high energy which remain stable during ambient temperature, keeping in mind the power outages (Fig. 6). Food technologists preserve the food by adding different kinds of preservatives which enhance their shelf life and can be used for a long-term period.

In a natural disaster, best foods to Stockpile

Prefer these nutrient-dense foods to leave you scavenging through more stable shelf life foods

Food stuff	Food Stuff
Bottled water	Canned soups
Canned fish like Tuna	Canned or jarred tomatoes
Nuts, dried fruits	Bread, crackers, candies
Apples	Plant milk (Almond, coconut etc)
Dehydrated meat	Dry Oats, wheat, rice or cereals
Protein/ fruit/ nut bars	Non-perishable milk/ powders
Peanut butter	Infant formula foods
Boxed pasta	Canned fruit juices
Canned beans	Dried herbs and spices
Canned vegetables	Staples like sugar, salt, pepper

Fig. 6 Best food to choose during disasters

Food technologists also play a role in maintaining and enhancing the quality of food products and moreover they have a major role in supervising food and nutritional need, storage, awareness, data collection and analysis, and local food utilization.

14 Role of Food Scientist in Emergency Food Supply Chain

Food is one of the most important parts of life and it plays a vital role in the survival of the organisms. It is so difficult to live without food even in the emergency condition. Food technology helps human beings for their survival by designing food products using different techniques. Food supply chain is a regular chain in the environment or the society in which food can be run from the glut (high concentration or high production point) to the area of low production or high demand with low production (mostly occurs in unfavorable climatic zones, starvation area, or the disaster effected/emergency zone).

- In the emergency conditions, mostly high-caloric foods and food with long shelflife are designed for the disaster-suffering public.
- In this regard food technologists design the products which have low perishability and low water activity.
- Mostly dry products are preferred in this regard by the food technologists.

- Food technologists make instant foods for the sufferers of the emergency area as they are small in quantity but rich in energy.
- They make energy-rich products and also some type of RTDs and RTS for the sufferers.
- Food technologists also design some special type of baked products and cookies for the emergency food supply chain. Food technologists make the handling of foods easy.
- They also try to make the food pass through high preservation techniques to avoid food losses and prevent the sufferer from starvation and malnutrition.
- Food technologists emphasize on the quality and the microbial attack control during food production and sometimes prefer the production of those food products which act as food and nutraceuticals at the same time.

For example, in the World War 2 noodles were made first time for the army soldiers to manage their feed and prevent them from hunger.

15 Role of Food Science and Technology in Humanitarian Response

Food science has exceptional ability to change lives, societies, and nation as well as contribute professionally in humanitarian aid and relief in case of emergencies and disasters (Fig. 7).

- Food science can contribute heavily to humanitarian aid and relief.
- The principles of food science help to deal with humanitarian crisis.
- The methodology of food science includes these characteristics:

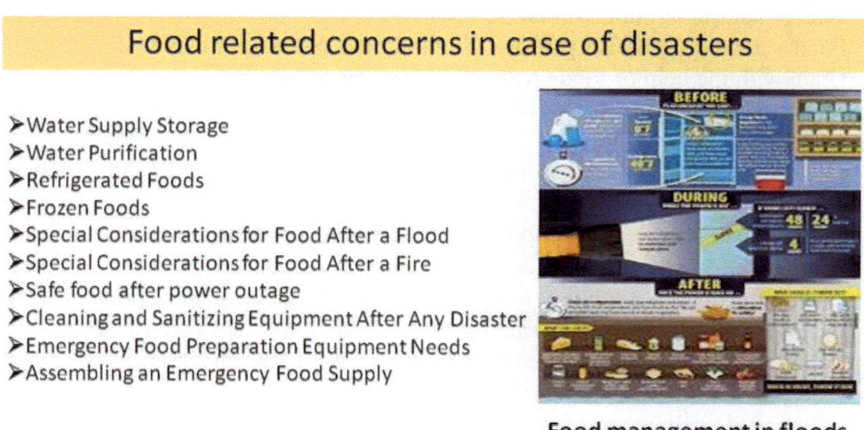

Fig. 7 Food-related concerns during disasters

  - It utilizes various resources for different materials.
  - It utilizes creative food solutions.
  - It is beneficial for long-term emergency situations.
-
- Food science helps in humanitarian response in a sense that it provides assistance for the purpose of saving lives.
- In case of an emergency, food science can respond in a way of providing a sustainable approach.
- Therefore, all the food-related problems including humanitarian response can be solved by food science.
- A committee of food scientists is designed that provides recommendations and experience to provide assistance for people suffering from humanitarian crisis.
- The basic goal of food scientists is to provide necessities to people during the time of crisis.
- Food science can raise awareness among people to be prepared for crisis.
- Food science can respond immediately to such crisis.
- Food science contributes heavily to reduce the chances of such calamities.
- On the other hand, food scientists can play roles as the following:
  - Food scientists can enhance production of food and nutrition.
  - They can contribute to eco-friendly rehabilitation and can decrease risk of these tragedies.
  - They can raise awareness among people regarding food security.
-
- Food science plays a role in strengthening the food security phase classification.
- Food science plays a leading role in global, national, and local analysis.
- Food science can stabilize food security and can improve livelihood opportunities.
- Main goal of food scientists is to ensure that people can easily defend themselves against natural or human-induced disasters.
- Food scientists can also provide information regarding threats to human food chain with the help of food chain crisis management framework.
- Food science ensures the livelihoods of farmers as well as their better living and also food safety.
- Food science contributes to food processing and providing food in areas that are crisis prone.
- Food scientists can enhance food supply chain in private sector and can produce innovative food services in case of humanitarian response.
- The aim of food scientists here is to enhance food security, to improve the productivity of primary production, and to make full use of agricultural sector.
- Therefore, food scientists make sure that high-quality and nutritionally good food is available to people during the time of crisis.
- Better food chain is must for attaining developmental goals.
- Food security response emergency systems can be developed.

- Food science can boost agricultural productivity to provide long-term support for the purpose of better livelihood.
- Food scientists can design such products that can be used in time of emergency because disasters can occur anytime; so being a food scientist, one must make use of his/her knowledge to provide basic necessity like food to disaster-prone people.
- Food scientists should design safe, affordable food that is relevant to local regulations.
- Such food that is utilized in acute and chronic aid situations must be available to the person who needs it.
- Food scientist can make use of food principles to overcome food insecurity during the time of crisis.
- An emerging field called Food Science for Relief and Development is providing creative food solutions to reduce malnutrition and food insecurity.
- Food Science for Relief and Development improves health, economic, and food security for humanitarian purposes.
- Therefore, it provides food that is available long term and has higher sustainability and is best for the time of crisis.
- This field improves the value of food products and creates innovative options that are beneficial to many.
- Food scientists can make use of process engineering and food development programs to produce efficient food under critical conditions.
- Food scientists can design protein-enhanced foods with the help of new functional ingredients so that they can be easily available.
- Food scientists should also raise awareness among people to get ready if any calamity arises.
- Food scientists can make solutions for the affected by provision of relief food products and to strengthen the food system.
- Agriculture-based food directly benefits the poor people to make their livelihood better.
- In the end, food science and technology can act as an accelerator to enhance the development of basic food chain that provides food to crisis-prone people.
- Food science and technology can address food security issues during and after humanitarian response.
- Food science can also improve the local resilience for the sufferers to ensure better livelihood for them.

16 Role of Food Scientist and Food Technologist in Global Food Security

According to the United Nations' Committee on World Food Security, food security is defined when "all the people, at all the time, have physical, social, and economical access to sufficient, safe and nutritious food that meet their food preferences and dietary needs for an active and healthy life."

Over the coming decades, changing climate, growing global population, rising food prices, and environmental stressor will have sufficient yet certain impact on global food security especially in developing and under-developed nations. By 2050, the world population will surpass 9 billion people, so there will be higher demand for available food, water, and arable land and greater environmental impacts. Food safety issues, nutrition deficiencies, postharvest losses, regulation inconsistencies, and consumer attitudes are all striking challenges which must be met to maintain food security and sustainability. Possible solutions include advancements in food-processing technologies, nanotechnology, innovation in food formulations, and the use of genomic approaches in examples such as alternative protein sources, insect flour, nutrigenomics, 3D food printing, biomimicry, food engineering and merging technology. International organizations such as International Union of Food Science and Technology (IUFST) also play important roles in securing the world's food supplies by providing expertise through their respective country memberships and sharing the experience with other developing and under-developed nations. The present review addresses the role of food science and technology in meeting current challenges and investigates possible solutions to feed the world in the near future.

Food scientists research and analyze the chemical, physical, and nutritional aspects of food while food technologists are mainly concerned with the processes of making, improving, preserving, and storing food and drinkable products but these two jobs overlap in many areas. Now we will discuss the role of each one in global food security separately.

First of all, food scientists analyze the food at the level of testing, consumption, nutrition, and production which may help to develop more efficient and safe packaging and production systems so that food retain its nutritional value for longer period of time. Food scientists work in quality assurance and will closely examine the food for its nutrition and safety standards. Food scientists who work in research spend most of their time in labs and execute the scientific experiments to improve the food products by increasing and enhancing their nutritional values and improving the shelf life of that specific product so that a quality food product can be made available for vast majority of people for longer period of time. Moreover, food scientists ensure that the government standards for safety and hygiene are being met and keep the records of nutritional and safety requirements. Last but not the least, food scientists contribute to global food security by suggesting the most affordable foods which are equally nutritious, shelf stable, and as safe as compared to the expensive food commodities.

On the other hand food technologists work to make food accessible and safe for the people of the globe without the loss of quality of the food products. One way forward is through the continued development of preservation and stabilization technologies such as high-pressure processing, drying, fermentation, canning, and lyophilization. Furthermore, food processors have developed new extraction methods for the biomass recovery and separation technologies to limit the food loss and utilize the different components of food as much as possible to develop new products for the consumption of people. These latest technologies may create the new value-added food ingredients to further reduce food waste and make it available in the reach of every average individual globally. The strategies to increase food preservation are vital because with the help of preservation techniques one can preserve the food products for a longer period of time and consume them without compromising on nutritional and quality standards. Another supporting approach is to reduce consumption in the regions where overconsumption is common because at present more than two billion people are currently obese or overweight and reducing the consumption in this population can improve the global food security. Food technologists work on increasing the nutritional value and availability of higher quality packaged, frozen, and shelf-stable food products to meet the goal of global food security. Food processing and technology are essential in transforming food into different products that have elongated shelf life, improved functional properties, desired nutritional properties, and higher quality. Often in the developing nations, there is a shortage of food supplies due to lack of storage facilities until the next harvest season. Thus, dependence on importing food greatly affects a country's economy. That is why emerging technologies applied to food are critical in reversing food insecurity and maintaining a constant food supply. Different technologies such as microwave vacuum drying do not cause thermal degradation like traditional drying and also protect the nutritional value of the products. Nano sensors can be installed near fresh food commodities that detect bacteria such as Salmonella and other toxic substances and can communicate the information with product manufacturers and monitor the food freshness during storage and retail. Similarly other technologies include DNA barcoding (to minimize food fraud, misleading, and adulteration especially in sea foods by comparing the short genetic markups in the product with reference DNA), protein utilization (use of plant protein as an alternate of animal protein), and multicomponent solutions (these are nutrient-dense, stable, and portable food products that can be developed in fighting world hunger and nutrient imbalance).

Future work can include development of cultured beef (in vitro meat), alternative protein resources such as edible insects (which are dense in nutrient and have high-quality proteins, amino acids, and vitamins and they have higher food conversion rates than livestock), insect-fortified flour, and 3D printing to create "Smoothfood" for the people suffering from dysphagia to make food more appetizing.

17 Conclusion

During disaster there are high chances of infection due to food and waterborne diseases, lack of hygienic food, etc. But all these problems can be reduced with the help of biotechnologically processed food and processed water, which eventually lower the health hazard, thereby lowering the economic expenses of government in health and medicine sectors. Another thing is that the government should send food technologists along with food aid so that they could preserve food for a long time, thus helping in preventing malnutrition and upliftment of the country. The main role performed by food technologist may include the production of prolonged shelf life-containing foods having high nutritional contents with small servings.

References

ICSU (2008) A science plan for integrated research on disaster risk: addressing the challenge of natural and human-induced environmental hazards. ICSU
Lindell MK (2013) Disaster studies. Curr Sociol 61(5–6):797–825
Quarantelli EL (ed) (1998) What is a disaster? Perspectives on the question. Psychology Press
Shaluf IM (2007) Disaster types. Disaster Prev Manag 16(5):704

Role of Livestock for Disaster Risk Reduction

Asim Faraz, Nasir Ali Tauqir, Hafiz Muhammad Ishaq, Syeda Maryam Hussain, Amir Ismail, Abdul Waheed, Aashir Sameen, and Muhammad Arslan Akbar

Abstract

Animals are factories for converting roughage into quality products being consumed by mankind as food like milk and meat. It is an established fact that animal proteins are better than vegetable ones for supplying vital amino acids and are considered a paramount food source. Livestock products are not only industrial raw materials for inland and export products but also creating markets and the basis of movement of capital from cities to villages and periphery. For the landless poor community, it is considered as societal security and need-based

A. Faraz (✉) · H. M. Ishaq · A. Waheed
Department of Livestock and Poultry Production, Bahauddin Zakariya University, Multan, Pakistan
e-mail: drasimfaraz@bzu.edu.pk

N. A. Tauqir
Department of Animal Nutrition, The Islamia University of Bahawalpur, Bahawalpur, Pakistan

S. M. Hussain
Department of Livestock Production and Management, Pir Mehr Ali Shah Arid Agriculture University,
Rawalpindi, Pakistan

A. Ismail
Institute of Food Science and Nutrition, Bahauddin Zakariya University, Multan, Pakistan

A. Sameen
Department of Agronomy, Pir Mehr Ali Shah Arid Agriculture University, Rawalpindi, Pakistan

M. A. Akbar
Department of Breeding and Genetics, Cholistan University of Veterinary and Animal Sciences, Bahawalpur, Pakistan

M. Ahmed, S. Ahmad (eds.), *Disaster Risk Reduction in Agriculture*, Disaster Resilience and Green Growth, https://doi.org/10.1007/978-981-99-1763-1_20

ready cash; hence it acts as a mobile bank account that can fulfill their needs at a time when no other resources are available or ruined. Different livestock species are involved in entertainment such as sports activities and have attained a decent status in many areas of Pakistan as well as the world, so in this way, they contribute to a major part of the socioeconomic system. Livestock is very imperative and plays a principal role in mitigating the disastrous effects of floods, storms, pandemics, and other socio-unbalancing situations. Their importance in the national economy of the country is well extolled, and mainly the rural poor rely on the livestock sector for their customary income and household chores. This chapter provides a brief look at the role of the livestock sector in disastrous conditions.

Keywords

Livestock · Economy · Rural · Disaster · Risk · Pakistan

1 Introduction

Human population and food security are interconnected, and both have significant implications for global sustainability and well-being. The global population has been steadily increasing over the years. The world's population is now around 7.9 billion. According to the food and agriculture organization (FAO), the number of undernourished persons and population who cannot afford a healthy diet is increasing day by day (Fig. 1a, b). So malnutrition is increasing due to the present food systems. Also the cost of healthy diet around the globe has increased that results in hunger, food security, and malnutrition. Recent inflation rise in Pakistan (the highest rate in South Asia) is a clear example which led to the issue of food security and malnutrition in the year 2023 as most of the families are unable to buy quality foods filled with fruits and meat.

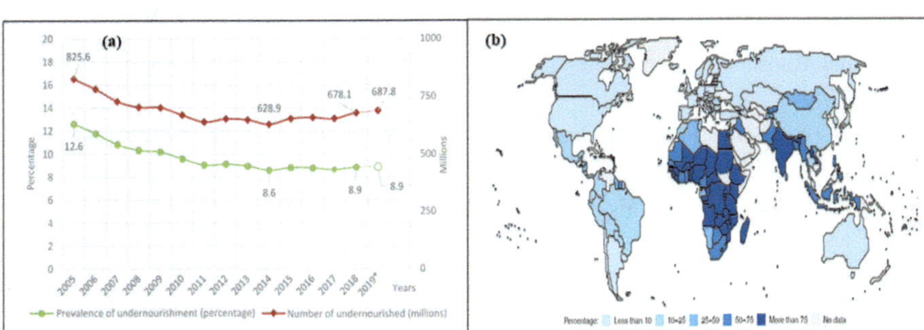

Fig. 1 (**a**) Global number and percentage of undernourished persons; (**b**) Percentage of population that cannot afford a healthy diet (Source: FAO)

Food security exists when all people, at all times, have access to sufficient, safe, and nutritious food to meet their dietary needs for an active and healthy life. Achieving and maintaining food security is crucial for individual and societal well-being. Food security has multiple challenges that include:

1. Limited Resources: As the population grows, there is an increasing strain on natural resources such as arable land, water, and energy. Meeting the food demands of a growing population becomes more challenging without sustainable resource management.
2. Climate Change: Changing weather patterns, extreme events, and shifts in temperature and precipitation patterns due to climate change pose risks to agricultural production. These changes can lead to reduced crop yields, increased pest and disease outbreaks, and altered growing seasons, affecting food production and availability.
3. Food Systems: Food production occupies 50% of earth habitat and it uses 70% of fresh water. Furthermore, food production produces around a quarter of global greenhouse gases (GHG), and it is a major driver of biodiversity loss, air and water pollution, soil degradation, deforestation, antibiotic resistant bacteria, and water scarcity. Certain food and food production choices have higher carbon footprint (Fig. 2).
4. Poverty and Inequality: Food security is closely linked to poverty and inequality. Many individuals and communities lack the financial means to access an adequate and nutritious diet. Addressing poverty and reducing income disparities are crucial for improving food security.

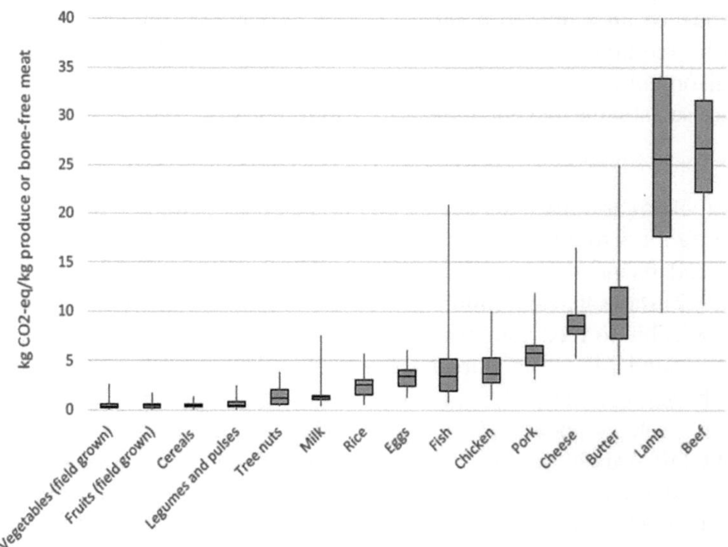

Fig. 2 CO_2 emissions from selected broad food categories (Source: FAO)

5. Food Waste and Loss: A significant proportion of the world's food is lost or wasted along the supply chain, from production to consumption. Reducing food waste and improving post-harvest handling and storage practices can enhance food security without increasing production.

Solutions for food security include:

1. Sustainable Agriculture: Promoting sustainable agricultural practices that minimize environmental impacts, conserve resources, and enhance productivity is essential. This includes practices like agroecology, precision agriculture, and organic farming.
2. Technology and Innovation: Harnessing advancements in agricultural technology, such as precision farming, genetically modified crops, and improved irrigation techniques, can increase agricultural productivity and resilience. Improved infrastructure and enhancing rural infrastructure, including transportation, storage facilities, and irrigation systems, can reduce post-harvest losses and improve access to markets.
3. Policy Interventions: Governments and international organizations play a vital role in developing policies that support smallholder farmers, improve land tenure systems, ensure fair trade, and provide safety nets for vulnerable populations.
4. Education and Empowerment: Promoting education and providing farmers, especially smallholders, with the necessary knowledge and skills to adopt sustainable agricultural practices can enhance food security.
5. Diversification of Diets: Encouraging diverse and balanced diets, including traditional and locally available food sources, can improve nutrition and reduce dependence on a limited number of crops. Addressing the challenges of food security requires a multidimensional approach involving governments, international organizations, farmers, scientists, and individuals working together to ensure sustainable and equitable access to food for all (FAO et al. 2021).

Livestock is carefully associated with the social and cultural lives of several million poor farmers for whom animal possession ensures several levels of sustainable farming and financial balance (Faraz et al. 2019). Economic Survey of Pakistan (2021–2022) states that agriculture is the most important gadget in Pakistan's economy responsible to feed the complete rural along with the urban community contributing around 24% of the Gross Domestic Product (GDP) along with debts for half of the labor force and is the vibrant source for foreign exchange earnings. Livestock's role is about 61% in value-added agriculture along with 14.0% in national GDP. Animal farming is the distinguished monetary hobby of the inhabitants of rural regions of Pakistan. More than eight million rural households are engaged in animal production and are deriving around 35–40% of their subsistence from livestock (Faraz and Waheed 2017). The gross value addition of farm animals has advanced from Rs 5269 billion (2020–2021) to Rs 5441 billion (2021–2022), displaying a boom of 3.26%. More than ten million heads of draught animals are engaged in agricultural activities. Their substitute with mechanization needs heavy monetary inputs approximately equal to 5.12 billion rupees (Sarwar et al. 2002).

The geographical and climatic situations of Pakistan have large variability; it enjoys the benefits of each irrigated, rain fed, and a variety of vicinities that extend from the coastal levels in the south to the alpine pastures in the north. Due to horizontal growth of farm animals, poor germplasm, immoderate sickness risks, bad control practices, scarce feed supply, and inflated prices of quality feeds, per animal production is very low, which reflects a horizontal linear graph (Waheed et al. 2017).

2 Livestock Sector as Socioeconomic Device

Livestock husbandry is a key indicator of economic development, environmental protection, and food security in the world, which is vulnerable to environmental changes and economic shocks (Yang et al. 2022; Han et al. 2020). Almost eight million families are doing livestock rearing business with greater than 35–40% earnings originating from farm animals production activities. Livestock could play a vital role in poverty relief in the state. Livestock falls in a subsistence region; as it is ruled via small households through meeting their demands like milk, meat, and egg production along with daily cash earnings. Livestock keeping is a secure supply of revenues for landless, awful, and small farmers and a supply of employment for agricultural communities. It can play an essential role in poverty reduction through uplifting of the socioeconomic status of the rural agricultural masses. Livestock husbandry can help to achieve Sustainable Development Goals (SDGs) as animal husbandry production system can help to provide nutritional food for residents, which could contribute to SDG1 (no poverty), SDG2 (zero hunger), SDG3 (good health and well-being), SDG12 (responsible consumption and production), and SDG17 (partnerships for the goals) (Mehrabi et al. 2020). Similarly, the livestock sector plays a significant role as a socioeconomic device in many countries around the world. It encompasses various activities related to the rearing, breeding, and processing of livestock, including cattle, sheep, goats, poultry, and pigs. Here are some ways in which the livestock sector functions as a socioeconomic device:

1. Employment Generation: The livestock sector provides employment opportunities for millions of people, particularly in rural areas. It involves a wide range of jobs, including animal husbandry, veterinary services, meat processing, transportation, and marketing. This sector helps alleviate rural poverty by creating income-generating activities and reducing unemployment.
2. Livelihood Support: Livestock rearing often serves as a primary source of livelihood for small-scale farmers and pastoralists. It allows them to diversify their income sources and reduce dependence on crop production, which can be vulnerable to environmental and market uncertainties. Livestock-based activities provide a safety net during times of crop failure and contribute to food security.

3. Food Security and Nutrition: Livestock products, such as meat, milk, eggs, and honey, are important sources of nutrition and protein for human consumption. The livestock sector contributes to food security by ensuring a regular supply of animal-based food products. Additionally, livestock can efficiently convert low-quality feed resources, such as grass and agricultural by-products, into high-quality protein sources.
4. Economic Growth and Trade: The livestock sector contributes to economic growth through the production, processing, and marketing of livestock and related products. It stimulates trade both domestically and internationally, as countries import and export livestock and livestock products to meet consumer demand. The sector's growth can lead to increased revenue, foreign exchange earnings, and investment opportunities.
5. Rural Development: Livestock activities are often concentrated in rural areas, where they contribute to the overall development of communities. The sector creates opportunities for infrastructure development, such as roads, veterinary clinics, and processing facilities. It also encourages the growth of ancillary industries, such as feed manufacturing, equipment suppliers, and cold storage facilities, thus promoting rural entrepreneurship.
6. Social Cohesion and Cultural Significance: Livestock rearing is deeply ingrained in the cultural fabric of many societies. It plays a role in traditional ceremonies, social events, and cultural practices. Livestock ownership can confer social status and facilitate social interactions within communities. In this way, the livestock sector fosters social cohesion and helps preserve cultural heritage.
7. Environmental Sustainability: While the livestock sector can have environmental challenges, such as greenhouse gas emissions and land degradation, sustainable livestock practices can contribute to environmental conservation. Proper management of grazing lands, efficient feed utilization, and waste management systems can minimize negative impacts and promote sustainable land use practices.

Overall, the livestock sector serves as a socioeconomic device by providing employment, supporting livelihoods, ensuring food security, promoting economic growth, contributing to rural development, preserving cultural traditions, and offering opportunities for sustainable practices. However, it is important to address the environmental and social challenges associated with the sector to ensure its long-term viability and sustainability.

Pakistan's agriculture community consists of small farmers having numerous obstacles in their everyday farming practices, graded in the lower to middle ranged economy fulfilling the propensity to cater to the food requirements of its growing population. Presently, 90% of overall farmers (7.4 million) are categorized as smallholder farmers in Pakistan as they own less than 12.5 acres of land (5 Ha). Smallholder farmers continue to be the backbone of the agriculture sector, contributing to national food security as well as export earnings. However, populace boom, urbanization, modified eating styles, increase consistent with capita incomes, export opportunities and education, cost of transportation, distance from farm to market, and access to market information are fueling the choice of farm animals products within the country (Ahmed et al. 2016). Climate change is also causing a big trouble to livestock farmer; thus farmers are taking some adaptation measures to cope with issue of climate change. The measures include mix farming, reduction in animals, provision of more drinking water, use of tree shades, livestock diversification, use of muddy roof and floor in order to cope with climate changes (Shahbaz et al. 2020). Furthermore, some other regulatory measures taken are focused on some point of the stated insurance of husbandry practices. They may be geared to enhancing per unit animal productivity via improving health insurance, management practices, animal breeding practices, artificial insemination practices, use of balanced ration, and controlling diseases of economic significance. The goal is to make the maximum livestock production and its functionality for monetary growth, food safety, and rural socioeconomic upliftment (Varijakshapanicker et al. 2019). The poultry sector is also one of the vibrant segments of the livestock section in Pakistan. This area gives employment (direct/oblique) to over 1.5 million population. More than 90% of rural families involved in cattle farming have indigenous poor performing breeds that need to be genetically improved to raise their profits and ultimately enhance their socioeconomic status (Hussain et al. 2015).

3 Livestock Strength

Pakistan's principal natural resources are cultivable land and water. The main agricultural province is Punjab where wheat, rice, sugar cane, and cotton are the favorite crops. Table 1 shows the land utilization statistics in Pakistan while Fig. 3 shows the area distribution for cultivation in the country. Pakistan is endowed with diverse farm animals' genetic resources. In reality, it is hypothesized that facilities of animal domestication lay in this fragment of location. Pakistan has a massive cattle population, fairly designed to diverse ecological zones (Shahbaz et al. 2020). The current animal population in Pakistan represents 43.7 million buffaloes, 53.4 million cattle, 31.9 million sheep, 82.5 million goats, and 1.1 million camels (Table 2). The buffaloes in Pakistan are riverine type, i.e., Nili-Ravi and Kundi. The Nili-Ravi is a remarkable dairy breed in this area while ten remarkable cattle breeds are located in Pakistan. But the ones (Sahiwal and Red Sindhi) that breed probably extremely well make up to 30% of the population while the remaining of the

Table 1 Land utilization statistics (geographical area of Pakistan is 79.6 million hectares)

Classification of area	Pakistan (area in million hectares)	Punjab (area in million hectares)
Cultivated area	22.02	12.47
Total reported area	57.99	17.50
Current fallow	6.69	1.89
Net area sown	15.38	10.63

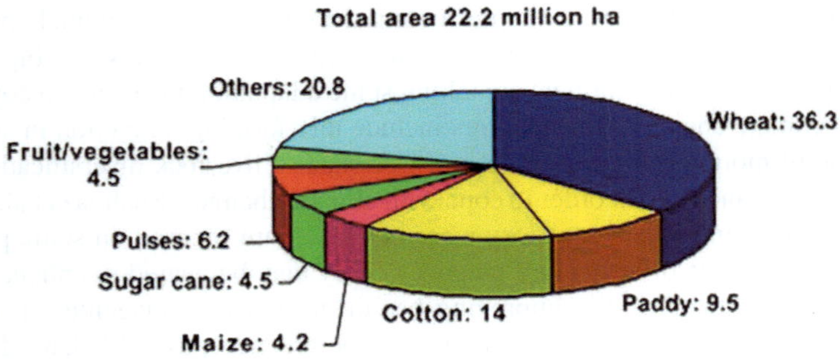

Fig. 3 Area distribution for cultivation in Pakistan

Table 2 Livestock strength of Pakistan (Economic Survey of Pakistan 2021–2022)

Animals	Livestock species population (Millions)
Cattle	53.4
Buffalo	43.7
Sheep	31.9
Goat	82.5
Camels	1.1
Horses	0.4
Asses	5.7
Mules	0.2

populace is usually categorized as non-descript. Tables 3 and 4 show agricultural growth and milk and meat production in the country.

Farm animals make up an essential part of agricultural systems, raising productivity and providing a continuous flow of food and income for communities (FAO et al. 2021. Within the rural zones of the globe, the ties between human beings and their animals are considered essential because they are a source of food, field work, and transportation. Seventy percent of the world's farm animal owners and poor landless people are most vulnerable to disaster. In Pakistan, livestock business is typically a crop-livestock business and production structures of farm animals are of subsistence type. Camels are of great importance in the country as they are the main pillars of pastoral economy, and the livelihood of marginal people mainly relies on these camels. They are of great socioeconomic importance and are the main source of milk, meat, and other by-products for the deep desert herders.

In the future, production will increasingly be affected due to climate change, competition for natural resources, especially land and water, and competition among food and forage through the carbon-limited monetary tools (Shahbaz et al. 2020).

Table 3 Agriculture growth percentage (Economic Survey of Pakistan 2021–2022)

Classes	2020–2021	2021–2022
Agriculture	3.48	4.4
Cash crop	5.83	7.24
Another crop	0.87	5.44
Cotton ginning	−13.08	9.19
Livestock	2.38	3.26
Forestry	−0.45	6.13

Table 4 Milk and meat production in the country (Economic Survey of Pakistan 2021–2022)

Species	2020–2021	2021–2022
Estimated milk production (000 tones)		
Cow	18,686	19,390
Buffalo	30,691	31,603
Sheep	41	42
Goat	991	1018
Camels	932	944
Estimated meat production (000 tones)		
Beef	2380	2461
Mutton	765	782
Poultry meat	1809	1977

Cattle production is probably increasing in number, but it may face constraints such as carbon emissions, environmental regulations, and animal welfare guidelines. Call for farm animals' merchandise in future may be carefully moderated by using socio-economic elements along with aspects of human health issues and converting socio-cultural values.

4 Source and Strength of Livestock Sector in Disaster

During disasters, the livestock sector can face significant challenges, but it also possesses certain strengths that contribute to disaster resilience and recovery. Following could be possible options:

1. Food Security and Emergency Response: Livestock can play a crucial role in ensuring food security during and after a disaster. They provide a source of immediate nutrition through the consumption of animal products such as milk, meat, and eggs. In emergency situations, the presence of livestock can help sustain affected communities and alleviate immediate food shortages.
2. Livelihood Support and Asset Protection: For many people, especially in rural areas, livestock ownership represents a significant part of their livelihood. During a disaster, protecting and preserving livestock becomes essential for preserving

livelihoods. Policies and interventions that support livestock protection, such as early warning systems, animal evacuation plans, and emergency veterinary services, can help safeguard the assets and income of vulnerable communities.

3. Resilience and Adaptation: Livestock rearing can contribute to community resilience and adaptation to disasters. Livestock possess certain characteristics that make them resilient in the face of adverse conditions. For example, certain breeds of livestock may be more tolerant to extreme temperatures, drought, or disease outbreaks. The knowledge and experience of livestock keepers regarding animal care, breeding, and management practices can also enhance their resilience in the face of disaster events.
4. Economic Recovery: The livestock sector can play a vital role in the economic recovery of disaster-affected areas. It offers opportunities for income generation and employment as communities rebuild their lives. Rebuilding livestock-related infrastructure, such as barns, sheds, and processing facilities, can create employment opportunities and stimulate local economies.
5. Social and Cultural Significance: Livestock holds social and cultural significance in many communities, and this can contribute to the resilience of affected populations. The presence of livestock can provide a sense of continuity, stability, and identity during times of crisis. Livestock-related practices and traditions may help maintain social cohesion, foster community support, and provide emotional support to individuals and families affected by disasters.

Fig. 4 Camels grazing at Thal Desert, Punjab, Pakistan

Fig. 5 Camel at Clifton Beach Karachi, Pakistan (a socioeconomic impact)

Fig. 6 Yak at Skardu, Gilgit-Baltistan, Pakistan (a lifeline of mountains)

It is important to note that the impact and effectiveness of the livestock sector in disasters can vary depending on factors such as the type of disaster, the local context, infrastructure availability, and pre-existing support systems. Local authorities, disaster management agencies, and agricultural organizations play a vital role in implementing appropriate strategies, policies, and interventions to harness the strength of the livestock sector in disaster situations (FAO 2017). Furthermore, there might be a huge number of nomads worldwide who rely upon animals for their livelihood. The animal has become a crucial cultural element, vital in preserving man's traditional existence mode. As an example, the Camel and Yak have made it potential for people staying in harsh regions wherein forage production is surely not feasible (Figs. 4, 5, and 6). Animal genetic biodiversity has helped human beings to stay in an extensive form of environment. Genetic diversity allows animals to adapt to contamination, adverse climates, inadequate feed, and specific conditions, thereby reducing stress on the environment.

There are about 60% of cattle farming labor consisted of ladies and greater than 90% of labors associated with animal care are rendered via womenfolk of the household. However, the reality is that chicken rearing has assumed the role of an employer in some sections; about 50% of birds are reared in backyards, where they produce eggs along with meat by foraging in surrounding habitats. During the course of flash floods people lose their belongings as well as their livelihoods. Animals and animal-driven carts have been used as ambulances to transport injured and diseased people in farther areas as no proper means of transportation is available. Animals are being utilized for clearing spots. Lack of plants, substructures, or devices takes time to recover. Animals, if saved, could provide the public with food, strength, delivery, and the possibility of utilities in a proper away. Unlike apparent vegetation or substructure, animals may be saved and carried to extra comfy locations, in which disaster could not be expected. In transit, animals can be utilized for milk, manure, shipping, and gasoline. Animals are utilized for drawing water and water transportation in desert areas, in which machinery or human strength is not feasible nor can be employed. In part, damaged vegetation and grains cannot be utilized for human consumption and may be advantageously used for animal feeding.

Fish farming, also known as aquaculture, can indeed play a role in disaster risk reduction. There are several ways in which fish farming can contribute to reducing the impact of disasters (Shava and Gunhidzirai 2017). It includes food security as fish farming can help to ensure a stable and consistent food supply, even during or after natural disasters. In regions prone to droughts, floods, or other weather-related events, traditional fishing may be severely affected. Fish farms provide an alternative source of protein and nutrients, reducing the risk of food shortages and malnutrition during times of crisis. Furthermore it is also a source of economic resilience as it can strengthen the economic resilience of communities, particularly in areas vulnerable to natural disasters. By diversifying income sources, communities can be less dependent on a single sector that may be adversely affected by disasters. Fish farming offers opportunities for livelihoods, employment, and income generation, contributing to the overall economic stability of the region. Ecosystem preservation is also possible through fish farming as sustainable fish farming practices can help to protect and restore ecosystems, reducing the vulnerability of natural resources to disasters. Well-managed fish farms can serve as sanctuaries for fish populations, preserving biodiversity and supporting the recovery of ecosystems impacted by disasters such as pollution, overfishing, or habitat destruction. Similarly fish farming can be a good option for climate change adaptation as climate change increases the frequency and intensity of natural disasters, including storms, heatwaves, and sea-level rise. Fish farming, especially in coastal areas, can provide a buffer against the impacts of climate change. Integrated systems such as aquaponics, which combine fish farming with hydroponics (soilless plant cultivation), can help mitigate the effects of extreme weather events by utilizing and recycling resources efficiently. Fish farming can enhance the resilience of communities by fostering local capacity building and knowledge transfer (Roy et al. 2022). Training programs in aquaculture techniques, resource management, and disaster preparedness can empower communities to respond effectively to disasters and recover more quickly. Additionally, community-based fish farming initiatives encourage social cohesion and cooperation, facilitating collective actions for disaster risk reduction. It is important to note that while fish farming can contribute to disaster risk reduction, proper planning and management are essential. Practices should be environmentally sustainable, ensuring that fish farming does not exacerbate existing vulnerabilities or harm natural ecosystems. Government regulations, research, and monitoring are crucial to support responsible aquaculture practices and maximize the benefits while minimizing potential risks.

5 Conclusion and Way Forward

Livestock plays a significant role in disaster risk reduction, particularly in rural and agricultural communities. The role of livestock to reduce disaster can be further enhanced by using livestock as an early warning system for impending disasters. Animals often exhibit behavioral changes or signs of distress before the onset of natural hazards like earthquakes or tsunamis. By observing unusual behaviors in livestock, farmers and communities can take early precautionary measures and evacuate if necessary. Implementing proper livestock management practices can mitigate the impact of disasters, for example, ensuring that animal shelters or barns are built in safe locations, away from flood-prone areas or landslide zones. Proper fencing and containment systems can help prevent livestock from straying or being swept away during floods. Additionally, maintaining adequate food and water supplies, including emergency reserves, is crucial for sustaining livestock during and after a disaster. Similarly, promoting the diversification of livestock species can enhance resilience to disasters. Different livestock species have varying tolerances to different environmental conditions and hazards. By diversifying the livestock portfolio, farmers can reduce the risk of losing their entire herd to a single disaster event. For example, integrating cattle, sheep, goats, and poultry can help spread the risk and ensure continuity of livestock-based livelihoods. Encouraging farmers to diversify their livelihoods beyond traditional livestock rearing can enhance their resilience to disasters. This can include activities such as beekeeping, fish farming, or crop cultivation. Diversifying income sources reduces dependence on a single sector and provides alternative sources of food and income during and after disasters. Similarly, providing training and capacity building programs to livestock owners and communities can enhance disaster preparedness and response. Training can include topics such as livestock first aid, emergency evacuation procedures, livestock health management during disasters, and early warning signs in animals. Additionally, knowledge sharing platforms and community-based networks can facilitate the exchange of information and best practices among livestock owners. Access to livestock insurance and financial support can aid in recovery and rebuilding efforts after disasters. Insurance schemes that cover livestock losses can provide farmers with a safety net and help them recover their losses more quickly. Government or nongovernmental organizations can also provide financial assistance, grants, or loans to farmers affected by disasters, enabling them to rebuild their livestock assets. Finally, effective collaboration and coordination among various stakeholders are essential for leveraging the role of livestock in disaster risk reduction. This includes government agencies, disaster management organizations, agricultural extension services, veterinarians, researchers, and local communities. By working together, these stakeholders can develop comprehensive disaster risk reduction strategies and implement timely response measures. By implementing these measures, livestock can contribute significantly to disaster risk reduction efforts, safeguarding the livelihoods and food security of communities in disaster-prone areas.

References

Ahmed U, Ying L, Bashir M, Abid M, Elahi E, Iqbal M (2016) Access to output market by small farmers: the case of Punjab, Pakistan. J Anim Plant Sci 26(3):787–793

Economic Survey of Pakistan (2021–2022) Pakistan Bureau of Statistics Statistics, Government of Pakistan, House, 21-Mauve Area, G-9/1, Islamabad

FAO (2017) The Food and Agriculture Organization of the United Nations. FAO, Rome

FAO, IFAD, UNICEF, WFP, WHO (2021) The State of Food Security and Nutrition in the World 2021. Transforming food systems for food security, improved nutrition, and affordable healthy diets for all. Rome, FAO https://doi.org/10.4060/cb4474en

Faraz A, Waheed A (2017) Livestock scenario in Pakistan. AVN Agro Veterin News 24(01):04

Faraz A, Waheed A, Ishaq HM, Mirza RH (2019) Rural development by livestock extension education in southern Punjab. J Fisher Livestock Prod 7(1):287. https://doi.org/10.4172/2332-2608.1000287

Han C, Wang G, Zhang Y, Song L, Zhu L (2020) Analysis of the temporal and spatial evolution characteristics and influencing factors of China's herbivorous animal husbandry industry. PLoS One 15(8):e0237827

Hussain J, Rabbani I, Aslam S, Ahmad HA (2015) An overview of poultry industry in Pakistan. Worlds Poult Sci J 71(4):689–700. https://doi.org/10.1017/S0043933915002366. Epub 2015 Dec 2. PMID: 26696690; PMCID: PMC4684835

Mehrabi Z, Gill M, van Wijk M, Herrero M, Ramankutty N (2020) Livestock policy for sustainable development. Nat Food 1(3):160–165

Roy A, Chatterjee P, Das BK (2022) Nexus of climate change with fish production and its implications on livelihood and nutritional security. In: Sinha A, Kumar S, Kumari K (eds) Outlook of climate change and fish nutrition. Springer Nature, Singapore, pp 85–96. https://doi.org/10.1007/978-981-19-5500-6_8

Sarwar M, Khan MA, Iqbal Z (2002) Feed resources for livestock in Pakistan, status paper. Int J Agric Biol 04(1):186–192

Shahbaz P, Boz I, ul Haq, S. (2020) Adaptation options for small livestock farmers having large ruminants (cattle and buffalo) against climate change in Central Punjab Pakistan. Environ Sci Pollut Res 27(15):17935–17948. https://doi.org/10.1007/s11356-020-08112-9

Shava E, Gunhidzirai C (2017) Fish farming as an innovative strategy for promoting food security in drought risk regions of Zimbabwe. Jamba 9(1):1–10. https://doi.org/10.4102/jamba.v9i1491

Varijakshapanicker P, Mckune S, Miller L, Hendrickx S, Balehegn M, Dahl GE, Adesogan AT (2019) Sustainable livestock systems to improve human health, nutrition, and economic status. Anim Front 9(4):39–50. https://doi.org/10.1093/af/vfz041. PMID: 32002273; PMCID: PMC6951866.

Waheed A, Faraz A, Yaqoob M, Tariq MM (2017) Livestock and national economy. Farm Reform 02(05):03

Yang J, Wang Y, Zhang H, Su Y, Wu X, Yan S, Yang S (2022) Impact of socio-economic and environmental factors on livestock production in Kyrgyzstan [Original Research]. Front Environ Sci 10:1049187. https://doi.org/10.3389/fenvs.2022.1049187

Role of Social Sciences in Reducing Disaster Risk in Agriculture

Javaria Nasir, Bukhtawar Nasir, and Muhammad Ashfaq

Abstract

Agriculture sector and farm livelihoods are facing serious threats from rising disasters and their damages. The geological, climatic, hydrological, and biophysical disaster hits the farming community through direct and indirect ways. The agricultural value chain as a whole is prone to disasters; however, farming communities face serious challenges to sustain in such alarming conditions. Disasters threaten all three pillars of sustainable development: social, environmental, and economic. Agriculture continues to bear the brunt of disaster impacts as new risks and correlations emerge. Urgent efforts are necessary to build disaster-, disease-, and climate-resilient agricultural systems that will be capable of improving the nutrition and food security of present and future generations, even in the face of mounting threats.

Keywords

Agricultural value chain · Social · Environmental · Economic · Climate-resilient agricultural systems

J. Nasir (✉) · M. Ashfaq
Institute of Agricultural and Resource Economics, University of Agriculture, Faisalabad, Pakistan
e-mail: Javaria.nasir@uaf.edu.pk

B. Nasir
Centre of Agricultural Biochemistry and Biotechnology, University of Agriculture, Faisalabad, Pakistan

M. Ahmed, S. Ahmad (eds.), *Disaster Risk Reduction in Agriculture*, Disaster Resilience and Green Growth, https://doi.org/10.1007/978-981-99-1763-1_21

1 Introduction

1.1 Disaster and Risk

Risk is "the probability of an outcome having a negative effect on people, systems or assets. It is depicted as the combined impacts of hazards, the exposure of communities along with vulnerability". Risk is the ultimate outcome of our own decisions, the decision about hazards and the exposure of our communities as a result of our decisions. The decision about construction, infrastructure, budget allocations, development, and resilience affected the exposure and surveillance of hazards and their risks. Risk assessment is determined by the awareness and resource allocations of communities and policymakers about social integration and care about vulnerable societies and assets (Munawar et al. 2022).

The research and understanding about risk assessment and its management usually depend on the traditional data analysis and forecasting approaches. And these approaches can define and formulate the historical traceability of risk exposure to provide some guidelines. However, because of the complexity of its full topography, risk assessment and management requires some holistic scientific analytics that provide a clear insight into disasters and hazards, instead of guidelines about past and future projections about them. The planet and its life-supporting systems are complex and interrelated, and risks towards these systems are complex (Perazzini et al. 2022).

The traditional cycle of disaster management comprised disaster–response–recovery–repeat. It is evident that emergencies are expensive and comprise of big social and economic cost but the risk and disaster management also need some budgets. After the happening of an event or disaster, the rebuilding and reconstruction must consider the concept of "building back better" to avoid the repeat cycle in traditional management system.

Floods, draughts, hurricanes, and other types of extreme weather are all climate-related calamities. Because management capabilities vary, there are differences in the exposure and susceptibility of the global repercussions.

Weather, climate, and water hazards caused 50% of all disasters, 45% of all recorded fatalities, and 74% of all reported economic losses between 1970 and 2019. More than 91% of these fatalities took place in underdeveloped nations (using the United Nations Country Classification). Droughts (650,000 deaths), storms (577,232 deaths), floods (58,700 deaths), and severe temperatures were among the top 10 disasters that caused the greatest number of fatalities (55,736 deaths).

From 1970 to 2019, there was a nearly threefold decline in deaths. Less than 20,000 people died in the 2010s, compared to nearly 50,000 in the 1970s. There were 170 associated deaths each day on average in the 1970s and 1980s. This average decreased by a third in the 1990s to 90 linked fatalities per day, and it continued to decrease through the 2010s to 40 related deaths per day.

The top 10 occurrences in terms of economic damages are storms (US$ 521 billion) and floods (US$ 115 billion). Damage totaling US$ 202 million occurred on average per day during the course of the 50-year period. The number of economic losses has multiplied sevenfold since the 1970s. The recorded losses from 2010 to 2019 were seven times greater than those reported from 1970 to 1979 (US$ 49 million), averaging US$ 383 million per day throughout the ten-year period. The most frequent cause of destruction and highest economic losses worldwide came from storms. It is the only risk for which the portion that is ascribed keeps rising.

The hurricanes Harvey (US$ 96.9 billion), Maria (US$ 69.4 billion), and Irma (US$ 58.2 billion) were three of the ten disasters that cost the most in 2017. From 1970 to 2019, these three hurricanes alone caused 35% of combined economic losses of the top 10 global disasters. From 1970 to 2019, there were 3454 disasters in Asia, resulting in 975,622 fatalities and US$ 1.2 trillion in reported economic losses. Nearly half (47%) of all weather, climate, and water-related disasters reported globally occur in Asia, where it also accounts for a third (31%) of the associated economic losses. Most of these disasters were caused by flooding (45%) and storms (36%).

Floods caused the most economic losses (57%), while storms had the largest effects on life, taking 72% of the lives lost. The top 10 natural disasters in Asia are responsible for 70% (680,837 deaths) of all casualties and 22% (US$ 266.62 billion) of the region's monetary losses.

The existing risk management strategies must consider the adaptive behavior and mitigation strategies in the mitigation framework for better resilience and making risk informed decision-making in the triangle of environment, social, and economic conditions, to break the norms. It pertains that resilience of societies is not just depending on bouncing back and getting the socioeconomic momentum but building back better with radical transformation of societies, a political momentum and commitment are needed to achieve the resilience considering all the segments of societies. The public sector alone is not able to achieve the desired targets, so the help from global organizations and private sector are needed. Business communities and residents of disaster-prone areas could drive change with more vigor instead of just being recipients of the suggested changes.

2 Identification of Disaster Risks

Disaster risk management is a multifaceted strategy that includes identifying hazards' threats, analyzing and processing those threats, understanding people's insecurity, evaluating communities' resilience, developing strategies for preventing future risk reduction, and developing the operational capabilities to implement the proposed measures (Agrawal 2018). Figure 1 explains the risk function in comprehensive way.

Fig. 1 Disaster risk management. (Source: Agrawal 2018)

3 Disaster Risk Management and Agriculture

Agriculture sector is complex, is interrelated and has wide socioeconomic, environmental, and ecosystem linkages. Agricultural progressive achievements ensure food and nutritional security but also threaten natural capital and ecosystem. Agricultural production systems also face transitions and sustainability issues. The external and internal stress factors make it more prone to natural calamities and water deficiencies (Nasir et al. 2021a, 2021b). The impact of shocks and disasters is evident on poor farm families in developing and underdeveloped nations (Borowski and Patuk 2021). Agriculture sector faces numerous challenges in terms of biophysical, institutional, market, socioeconomic, and policy indicators. Global food security is not achievable with rising population and resource degradation, frequent extreme events, economic, and markets stress (Mishenin et al. 2021).

Population growth, urbanization, and advent of modern civilization play important roles in economic growth, investment, and trade; likewise the resource competition also increased manifolds. Climate change is also an evident phenomenon and affected the communities largely. Climate change affects are vital, intense, and unavoidable for subsistence farming communities. Poverty, unemployment, and lack of opportunities formulate the inequitable social setup with lots of social disparities and unrest (Buttel 2013).

The main disaster that hit agriculture includes crop pests and animal diseases infestation, storms, earthquake, land sliding, mass movements, floods, extreme

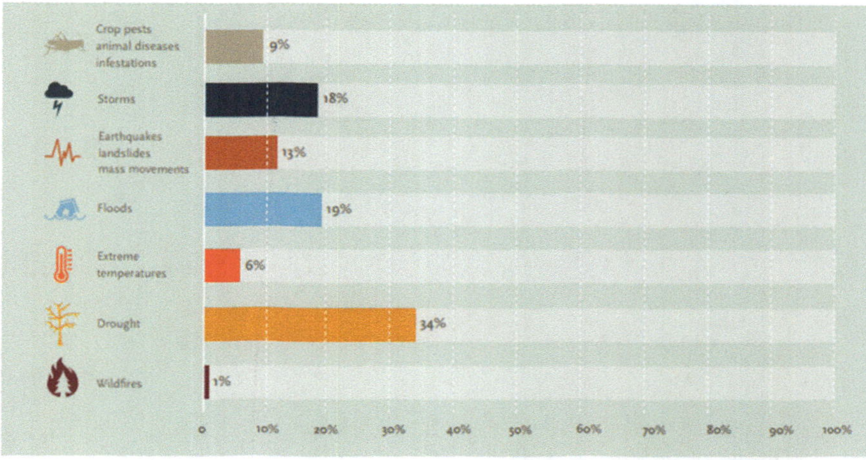

Fig. 2 Total production of crop and livestockloss per disaster, LDCs and LMICs, 2008–2018. (Source: FAO, EM-DAT CRED)

temperatures, droughts, and wildfires. The FAO data showed that highest crop and livestock losses observed by droughts are 34% of total losses followed by flood and storms. The data showed that 71% of losses observed were due to drought, floods, and storms (Fig. 2).

The impacts of natural hazards and disasters on agriculture and its subsectors are vital, and alarming for developing countries. The 22% of economic losses due to natural hazards and disasters pertains to the agriculture sector that comprises crop, livestock, fisheries and forestry (FAO 2015). The impacts assessment of these disasters is not estimated exactly and underweighs the losses due to lack of expertise, resources, data, estimation, and analytical tools in developing nations. The data gaps on impacts of disaster also hinder the efforts for its education, management, and resilience. The sectoral data sets must be formulated systematically and included in the national and international disaster risk management and resilience framework to combat the challenges. The financial instruments must be utilized to find out the risk reduction strategies, investments, and policies. The portfolio of investments and informed policy-making could assist the agricultural system sustainable development and resilience in food systems.

The agriculture sector must be prioritized and designed as effective systems that require for the delivery of the post-2015 framework on disaster risk reduction so as to improve local action and build resilience in the most vulnerable, which are often also the most food insecure (Fig. 3).

Agriculture is the main source of livelihood in developing nations and provides food, clothing, and shelter to millions but lacks the support system from policymakers and other sectors of economies and communities for resilience and sustainable development. However, the livelihoods of farm families are not protected by financial institutions and regulatory frameworks.

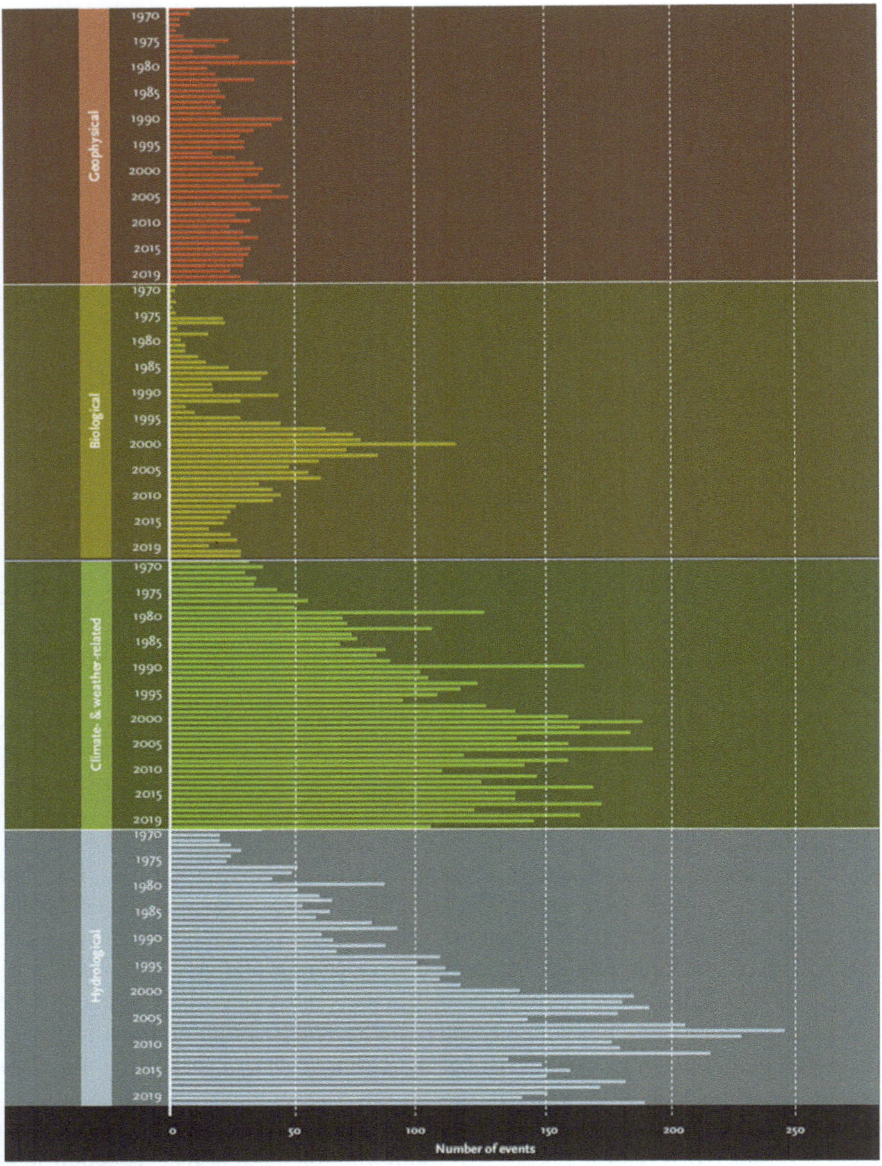

Fig. 3 Global disaster occurrence by type, 1970–2019. (Source: FAO, EM-DAT CRED)

The global data showed that the disasters get a peak value in terms of number of events in 2000 and the number of events started to decline after that but we can divide the events into geophysical, biological, climate, and weather-related hazards, and hydrological disasters. The historical data showed that climate-related disaster and hydrological events increased substantially.

Table 1 Direct and indirect impacts of disaster on agriculture

Direct impact of disaster in agriculture	Indirect impact of disaster
• Loss of crops, livestock, fisheries, and other agri-business • Loss of agri development infrastructure • Loss of fertile soil and its fertility • Loss of production and productivity	• Weakens the farming employment • Lowers food saving leading to food insecurity • Lowers farmer's income • Weakens the multiplier effect to the other sectors of economy

Conflicts and disasters are major hurdles in achieving food security and sustainable development. These disasters and hazards also affected the economies and communities in many ways. The trade policies also change after a disaster along with other socioeconomic policies. The disaster risk management is not that much easier for developing nations and they are just repeating the cycle of disaster–recover–respond–repeat due to lack of resources and better planning efforts for mitigation.

A study was conducted by FAO for the needs assessment undertaken for the medium- to large-scale disasters in 48 developing countries in Africa, Asia, and Latin America for the decade of 2003–2013. The study found that the reported damage and losses were US$ 140 billion in all sectors of economy while they were US$ 30 billion in agriculture and allied subsectors alone. Overall, the share of agriculture in economies of developing nation is meaningful and has important place in terms of food and nutritional security and livelihood of households. The agriculture sector absorbs 22% of losses due to the disasters. The largest share was reported in crop subsector (42%) followed by livestock (36%). Almost 60% of these damages and losses were caused by floods, followed by storms with 23%.

Out of the 78 catastrophes indicated in the post-disaster needs assessments, 45 of them resulted in damage and losses to the fisheries subsector totaling US$ 1.7 billion, or around 6% of all damage and losses within the agricultural sector. Tsunamis, a rare occurrence, were responsible for over 70% of this economic impact, but storms like hurricanes and typhoons only account for about 16% of the economic damage to fisheries and floods only account for 10%.

The most severely affected region in terms of geographical distribution of production losses is Asia, which accounts for 40% of all agricultural and livestock production losses (US$ 28 billion), followed by Africa (US$ 26 billion). The most severely impacted region in terms of predicted agricultural and livestock production is Africa, which lost 3.9% of that total. Central Asia came in second with 3.8% (Table 1).

Disaster Risk Reduction Progress Score (1–5 Scale; 5 = Best)

The Hyogo Framework Priority 1 National Progress Reports self-assessment scores, which range from 1 to 5, are averaged to provide the disaster risk reduction progress score. In 2005, 168 nations adopted the Hyogo Framework, a global framework for disaster risk reduction. Four indicators had been used in assessments of "Priority 1" to show how much countries have given catastrophe risk reduction and the development of pertinent institutions a high priority (Table 2).

Table 2 Disaster risk reduction progress score

Country name	Score	Country name	Score	Country name	Score	Country name	Score
Ecuador	4.75	Nicaragua	3.75	El Salvador	3.25	Czech Republic	2.75
Switzerland	4.75	Norway	3.75	Ghana	3.25	Fiji	2.75
Brazil	4.5	Paraguay	3.75	Guatemala	3.25	Georgia	2.75
Costa Rica	4.5	Sweden	3.75	India	3.25	Mongolia	2.75
Cuba	4.5	Thailand	3.75	Indonesia	3.25	Nepal	2.75
Japan	4.5	Zambia	3.75	North Macedonia	3.25	Senegal	2.75
Canada	4.25	Algeria	3.5	Poland	3.25	Venezuela, RB	2.75
Germany	4.25	Bulgaria	3.5	Romania	3.25	Cote d'Ivoire	2.5
Mexico	4.25	Cabo Verde	3.5	St. Lucia	3.25	Lesotho	2.5
Australia	4	Finland	3.5	Armenia	3	Bolivia	2.25
Bangladesh	4	Italy	3.5	Botswana	3	Lao PDR	2.25
Barbados	4	Mauritius	3.5	Cayman Islands	3	Maldives	2.25
Kenya	4	Pakistan	3.5	Dominican Republic	3	Turks and Caicos Islands	2.25
Mozambique	4	Samoa	3.5	Lebanon	3	Yemen, Rep.	2.25
Nigeria	4	Sri Lanka	3.5	Morocco	3	Solomon Islands	2
Colombia	3.75	St. Kitts and Nevis	3.5	Panama	3	Vanuatu	2
Honduras	3.75	Syrian Arab Republic	3.5	Peru	3	Comoros	1.75
Jamaica	3.75	Tanzania	3.5	Sierra Leone	3	Malawi	1.75
Madagascar	3.75	United States	3.5	Antigua and Barbuda	2.75	Marshall Islands	1.75
Malaysia	3.75	Argentina	3.25	Brunei Darussalam	2.75	Guinea-Bissau	1
New Zealand	3.75	Burundi	3.25	Chile	2.75		

Source: UNISDR, 2009–2011 Progress Reports, World Development Indicator, 2011

4 Disasters and Sustainable Development Goals

The development undermines by hazards and disasters especially in developing nations and underdeveloped world that has low capacity to manage the risk and disasters. Nothing undermines development like a disaster and slows down the process of progress (Wisner et al. 2014). To achieve the sustainable development goals in the presence of rising population and natural disasters is a serious challenge. The farm societies are weaker parts of global population facing inequality and face serious risks and uncertainty. The natural disasters affected these communities' livelihood and assets drastically (Ghimire 2021). The zero-hunger goal, achievement of nutritional security, sustainable farm livelihood and sustainable land use pattern along with management of natural hazards require global help to create technologies and potential for adoption of measures that can enable the communities towards protecting themselves from risk and uncertainties (Nasir et al. 2020). Global cooperation for climate-related risk and capacity building in developing world regarding hydrological disasters (Cabello et al. 2021).

5 Disaster Risk Reduction at Global Level (The Sendai Framework and SDGs)

Risk in agriculture is systemic and if global organizations intended to achieve the sustainability and reduce the risk they must work together and join up to work cross-sectors, between institutional arrangements and harmony in policy actions. The Sendai framework was the best known global policy agenda that combines the efforts to achieve SDGs with risk reduction and management strategies. It is evident that risk must be addressed if countries want to achieve sustainable development as nothing curbs the developing process and infrastructure as disasters do. The economic losses from disasters increased by 150% over the last two decades, and the losses distribution is also disproportionate and mostly borne by poor countries. The Sustainable Development Goals (SDGs) outcomes and the Sendai Framework are both the results of interconnected social and economic processes. As a result, the two policy instruments have a lot of overlap (Koloffon and von Loeben 2019).

6 Socioeconomic Assessment of Disaster Risk Management in Agriculture

The poor, underdeveloped nations have severe damages due to natural disasters and hazards along with low capability to manage the risk and mitigate that in future. Mostly the conflict hits hardest the rural areas and causes hunger and food-deficient scenarios. The factors of vulnerability include gender, socioeconomic status, ethnicity, and cultural values in communities. The different social classes also experience different levels of exposure and have variation in coping strategies against disaster management (FAO 2017). In 2007 a study by the London School of Economics indicated that, on average, natural disasters and their aftereffects kill more women than men, and at an earlier age than men. The study used a sample of 141 nations between 1981 and 2002 (Neumayer and Plümper 2007).

The disaster events are vital and increased over a period of time and maximum losses for crop and livestock production in these events occurred in 2015, and as showed in graph the pattern of losses is not sequential. The losses in Asia in livestock sector are maximum for the production of milk, eggs, and honey (Fig. 4).

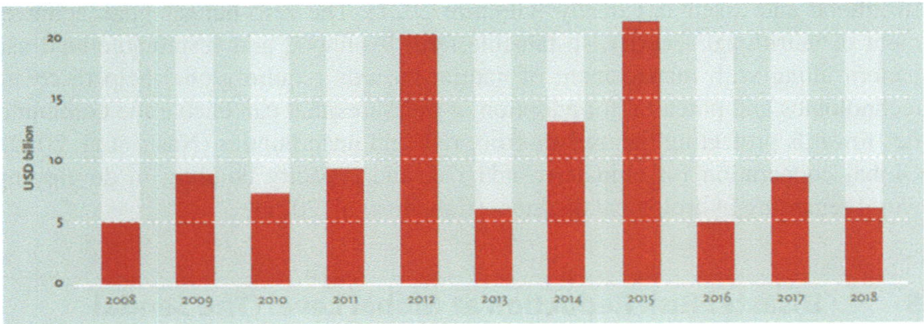

Fig. 4 Total losses in crop and livestock production of developing countries (three selected regions: Asia, Africa, and Latin America, 2008–2018, USD billion)

7 Disaster and Risk Management

The main elements of disaster and risk management include risk management and identification, recovery, preparedness, prevention, and mitigation. The risk identification and assessment comprise hazard analysis and monitoring, vulnerability analysis, and determination of risks. To recover from the disastrous event the process involves rehabilitation, reconstruction, and rescue services. Preparedness involves early warning, evacuation, and emergency planning. The policy makers and all stakeholders must intake the land use planning, land management and non-structural measures to combat the challenges (Fig. 5).

7.1 Risk Identification and Assessment

The data about disaster risk must be collected before and after the events (Table 3).

7.2 Disaster and Risk Management Frameworks in Agriculture

The disaster and risk management framework must be formulated utilizing integrated efforts to achieve global food and nutritional security along with market and business growth in sustainable manner. The disaster has multifaceted impacts on livelihoods and whole value chains of farm produce (Nasir et al. 2020). The following ways could help farming community to better cope with the disaster and risk management.

7.3 Crop and Livestock Insurance

The impacts of disaster, be it biological, climatic, hydrological or geophysical, on agriculture are vital. The farming community is facing severe risk under blue sky

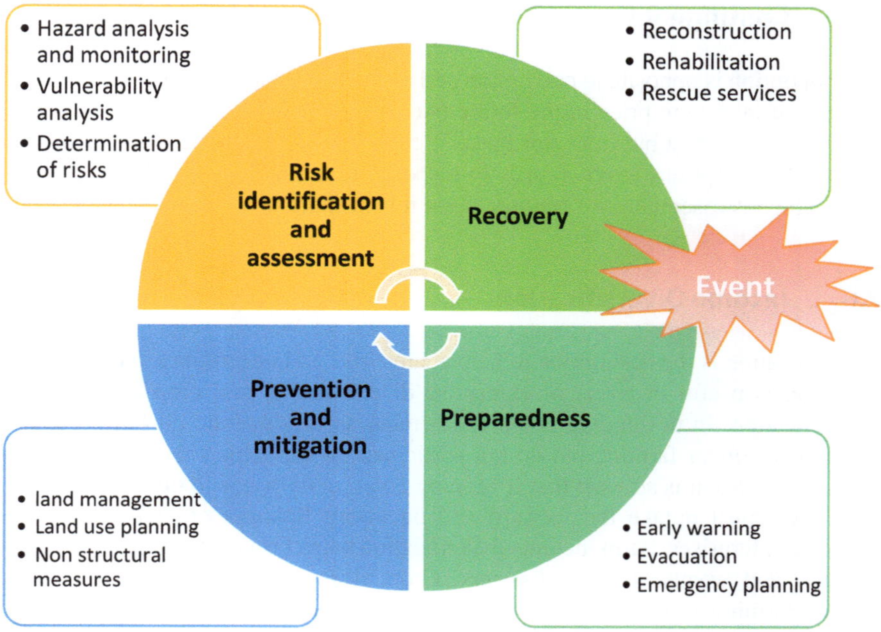

Fig. 5 Disaster risk management. (Source: Enemark 2009)

Table 3 Main disaster management system requirement for data

Status of disaster	Before disaster	After disaster
Requirements of disaster management strategies	Reliability, accessibility, precision, implementation, and usability	
Application of disaster management strategies	Pre-disaster forecast, early warning, and imitation application	Emigration, rescue assistance, monitoring, logistics management

with all their assets and investments. The developing world lacks financial instruments that could help and protect farming community under risk-prone areas. The financial support toward farming community needs public agencies' help at every level. The weather-based insurance and long-term contractual arrangement in terms of financial arrangement can sustain the farmers (Nasir et al. 2019).

7.4 Contingent Import Agreement (Call Option)

Future contracts and forward marketing can support farmers to better equip with financial capacities to face risk and uncertainty in better ways. Call option is the tool that can provide the buyer to buy a stock for delivery later in the season. This is predefined contract for acquisition if required at ceiling price at a pre-agreed time in the future.

7.5 Contingent Export Agreement (Put Option)

The put option is opposite to call option and a specified arrangement to sell the farm produce at minimum price in the future but at pre-agreed contract. The put option allows buyer to buy but does not make it compulsory to sell the delivery in later season. These options fetch security in response to falling prices and provide elasticity which is essential for a country when it is in short supply in its home market.

7.6 Income Diversification

Diversification is the best option to help against risk and uncertainty. Income diversification is useful in terms of enterprise diversification and crop and livestock diversification. In developing countries farmers usually operate all farm area and investments under farming but do not participate in the value chain and farm services sector. So, it is advised that if farming community wants to optimize the benefits, they must uptake the forward and backward linkages of their produce as alternate enterprises. Farm income diversification helps farmers to better combat the climate- and weather-related disasters. Crop, livestock, fisheries, agro forestry, orchard farming, olericulture and poultry are easy to adopt at farms with same set of resources and initial investments. However, farming community must invest in all enterprises in whole food value chains.

7.7 Disaster Payment and Relief Operations

The financial losses could be minimized if there exists a proper mitigation and support system. The payment toward affected communities would help them to cover their losses and decrease the severity of damage. The payment will also help farmers in their future investment portfolio. The flood usually causes the highest damages to farm income, which leaves the community helpless and unable to fulfill basic life amenities.

7.8 Disaster Risk Management by Big Data Analytics

The disaster management must utilize the latest technology and advancement in data sciences like big data analytics and the internet of things technologies that create opportunities for better management system and management. These allow the management to better coordinate and formulate better informed decision-making. The policy architect with better data and decision-making tools that include smart technologies and advanced projections allows accurate and timely decision-making (Shah et al. 2019) (Table 4; Fig. 6).

Table 4 Disaster data requirements and quality parameters for big data analytics

Data constraints of disaster management	
Accessibility	Is there a public or private data access? Does the data have any rights or legal restrictions on its use? Whether the data collection requires a unique information source?
Timeliness	Whether the historical database or run-time data collection is used to acquire the data Do periodic updates of the data need to be made?
Credibility	Is the data source identified? Any mechanism or organization able to validate the source?
Accuracy	In any sense data must be related to the disaster Is there no redundant data in the data?
Completeness	Data must be clear and understandable? Can the information be categorized to get the intended outcomes?

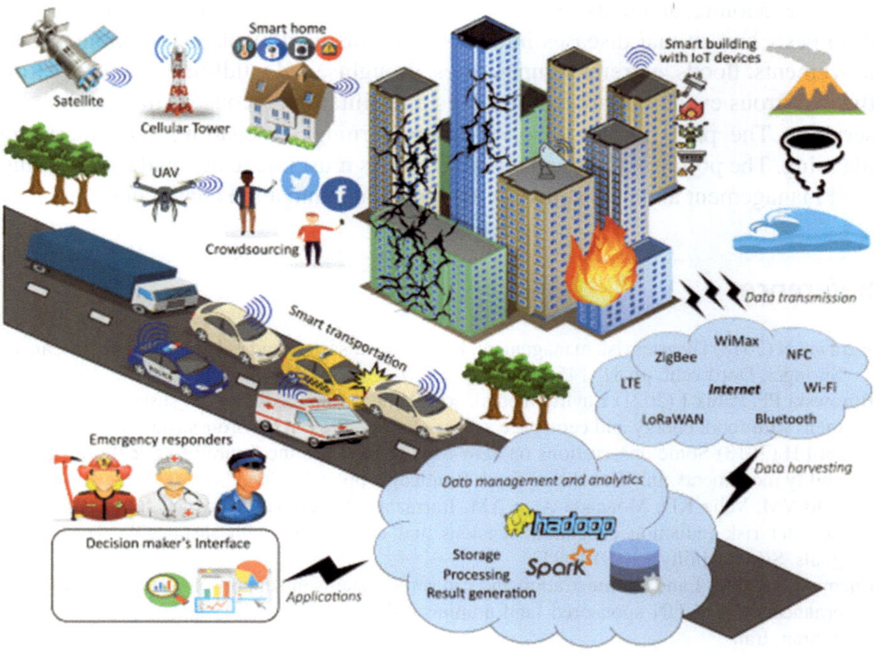

Fig. 6 Disaster risk management by big data analytics (BDA) and the internet of things (IoT). (Source: Shah et al. 2019)

7.9 Supplemental Agricultural Disaster Assistance

In this type of assistance financial support is provided for farm revenue losses due to natural disasters.

Other alternative strategies to combat the risk and disasters are soft loans, deficiency payments, future contracts, efficient disaster risk financing mechanism, effective global food security policy and plan, national policy improvement to assist farmers against challenges of disaster, and better social safety nets for farm families.

8 Conclusion

Risk is the ultimate outcome of our own decisions, the decision about hazards and the exposure of our communities as a result of our decisions. Risk assessment is determined by the awareness and resource allocations of communities and policymakers about social integration and care about vulnerable societies and assets. Agriculture sector is complex, is interrelated, and has wide socioeconomic, environmental, and ecosystem linkages. It faces numerous challenges in terms of biophysical, institutional, market, socioeconomic, and policy indicators. Global food security is not achievable with rising population and resource degradation, frequent extreme events, economic, and markets stress. The main disasters that hit agriculture include crop pests and animal diseases infestation, storms, earthquake, land sliding, mass movements, floods, extreme temperatures, droughts, and wildfires. To recover from the disastrous event, the process involves rehabilitation, reconstruction, and rescue services. The preparedness involves early warning, evacuation, and emergency planning. The policy makers and all stakeholders must intake the land use planning, land management and non-structural measures to combat the challenges.

References

Agrawal N (2018) Disaster risk management. In: Natural disasters and risk management in Canada. Springer, Dordrecht, pp 81–145

Borowski PF, Patuk I (2021) Environmental, social and economic factors in sustainable development with food, energy and eco-space aspect security. Present Environ Sustain Develop 15(1)

Buttel FH (2013) Some observations on agro-food change and the future of agricultural sustainability movements. In: Globalising food. Routledge, pp 264–286

Cabello VM, Véliz KD, Moncada-Arce AM, Irarrázaval García-Huidobro M, Juillerat F (2021) Disaster risk reduction education: tensions and connections with sustainable development goals. Sustainability 13(19):10933

Enemark S (2009) Land administration and cadastral systems in support of sustainable land governance. In: 3rd UN sponsored land administration forum for Asia and the Pacific region, Tehran, Iran

FAO (2015) The impact of natural hazards and disasters on agriculture. Food Agric Organ UN

FAO (2017) The future of food and agriculture – Trends and challenges. Rome

Ghimire KM (2021) Natural disasters and weak government institutions: creating a vicious cycle that ensnares developing countries. Law Develop Rev 14(1):59–104

Koloffon R, von Loeben S (2019) Disaster Risk Reduction and agriculture sector interrelated planning processes. Lessons learnt. Contributing Paper to GAR

Mishenin Y, Yarova I, Koblianska I (2021) Ecologically harmonized agricultural management for global food security. In: Ecological intensification of natural resources for sustainable agriculture. Springer, Singapore, pp 29–76

Munawar HS, Mojtahedi M, Hammad AW, Kouzani A, Mahmud MP (2022) Disruptive technologies as a solution for disaster risk management: a review. Sci Total Environ 806:151351

Nasir J, Ashfaq M, Adil SA, Hassan S (2019) Socioeconomic impact assessment of climate change in cotton wheat production system of Punjab, Pakistan. J Agric Res 57(3):199–206

Nasir J, Ashfaq M, Kousar R (2020) Climate policy. In: Global climate change: resilient and smart agriculture. Springer, Singapore, pp 337–358

Nasir J, Ashfaq M, Baig IA, Punthakey JF, Culas R, Ali A, Hassan FU (2021a) Socioeconomic impact assessment of water resources conservation and management to protect groundwater in Punjab, Pakistan. Water 13(19):2672

Nasir J, Ashfaq M, Baig I, Khair SM, Mangan T, Allan C, Punthakey J (2021b) Representative agricultural pathways and socioeconomic benefits of groundwater management interventions in Punjab, Sindh and Balochistan Provinces, Pakistan

Neumayer E, Plümper T (2007) The gendered nature of natural disasters: the impact of catastrophic events on the gender gap in life expectancy, 1981–2002. Ann Assoc Am Geograph 97(3):551–566

Perazzini S, Gnecco G, Pammolli F (2022) A public–private insurance model for disaster risk management: an application to Italy. Italian Econ J:1–43

Shah SA, Seker DZ, Hameed S, Draheim D (2019) The rising role of big data analytics and IoT in disaster management: recent advances, taxonomy and prospects. IEEE Access 7:54595–54614

Wisner B, Blaikie P, Cannon T, Davis I (2014) At risk: natural hazards, people's vulnerability and disasters. Routledge

Use of AI for Disaster Risk Reduction in Agriculture

Muhammad Hammad, Muhammad Shoaib,
Hamza Salahudin, Muhammad Azhar Inam Baig,
and Muhammad Usman Ali

Abstract

Agriculture, being significantly related to and dependent on natural factors, is most vulnerable to abnormal behaviors of nature. The trends, frequency, and intensity of natural disasters are becoming much more unpredictable with changing climatic and meteorological behaviors. The last decade has seen a significant boom in disastrous events; in turn, the socioeconomic loss incurred by different sectors of society also jumped all-time high. Agriculture tolerates the major parts of these damages and losses among all other sectors. Owing to such rising concerns, sophisticated technologies have made their way through to disaster risk reduction (DRR) and management in agriculture. Artificial intelligence (AI) and its domains, i.e., machine learning (ML) and deep learning (DL), are being increasingly used to predict, forecast, detect, monitor, and assess natural disasters in the context of agriculture. In this study, we have reviewed recently published literature relative to such application of AI in DRR in an agricultural context. An in-depth qualitative examination of recent studies reflected that AI itself is in the progression phase. Newer stand-alone and hybrid algorithms are being developed and applied to agriculture-related real-world problems. Similarly, the effectiveness and modeling accuracy of these models keep on getting better with every innovation and parametric requirement also decreases. We hope to see the pinnacle of AI application and minimal disaster-related losses as its outcomes in the prospect of AI integration in agricultural DRR.

M. Hammad · M. Shoaib (✉) · H. Salahudin · M. A. I. Baig · M. U. Ali
Department of Agricultural Engineering, Bahauddin Zakariya University, Multan, Pakistan
e-mail: msho127@aucklanduni.ac.nz

M. Ahmed, S. Ahmad (eds.), *Disaster Risk Reduction in Agriculture*, Disaster Resilience and Green Growth, https://doi.org/10.1007/978-981-99-1763-1_22

Keywords

Machine learning · Deep learning · Disaster management · Flood · Earthquake · Landslide

1 Introduction

Disaster, as defined by World Health Organization, is a sizable disorder within the usual course of operation of a society or a community at any level as a result of potentially dangerous natural or man-made developments which may result in physical, human, economic, environmental, and/or ecological losses (WHO 2002). Disasters can originate from solid earth (i.e., geophysical disasters) that include volcanic activities, earthquakes, and landslides, or can be related to climatic extremes (i.e., climatological disasters) including wildfires due to extreme temperatures and droughts. Similarly, disasters like avalanches and floods (i.e., hydrological disasters) mainly occur due to presence, motion, and segmentation of water on earth, and cyclones and storms (i.e., meteorological disasters) are caused due to temporary weather conditions. Lastly, insect attacks, animals plague, and epidemic outbreaks (i.e., biological disasters) are originated from exposure to toxic and viral constituents of living organisms. Man-made disasters include warfare, pollution, traffic, and industrial accidents. Such events usually upend the regular course of life and bring about a degree of loss that is more than what the afflicted community can bear. Moreover, disaster's effects may sometimes be instantaneous and confined, but they are often extensive and may persist for a relatively longer period. A society or community may not be able to deal with the repercussions by itself and may need help from other sources, such as those from nearby authorities, at the national or global level, or from other communities or societies.

A plethora of recent literature favors the evidence of increasingly transforming weathers and changing climatic conditions (Blanc and Reilly 2017; Blanc and Schlenker 2017; Corwin 2021; Malhi et al. 2021; Van Meijl et al. 2018), which in consequence, is also increasing the frequency and unpredictability of disaster events (Benevolenza and DeRigne 2019; Sahana et al. 2021) adding to the disaster-related risks and hazards. A recent Food and Agriculture Organization report demonstrates that a rise in disaster occurrences is in fact the new normal. Contrary to little above 100 in the 1980s and a tolerable 90 per annum in the 1970s, disastrous events averaged well over 360 separate incidents annually during the years 2010–2019 and 440 events annually during the first decade (FAO 2021). The trend of global economic losses and damages taken by all sectors due to geophysical, meteorological, climatological, and hydrological disasters from 1970 to 2019, as reported by Food and Agriculture Organization, is shown in Fig. 1. Globally, valuable infrastructure and individuals are seriously at danger from natural disasters (Ogie et al. 2018). A total of 3469 natural disasters (including landslides, storms, earthquakes, floods,

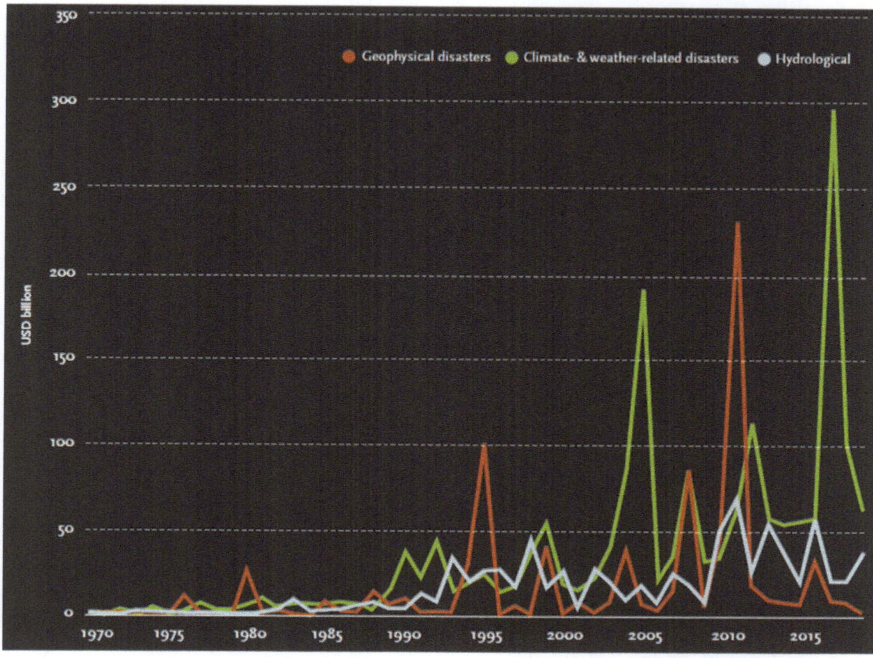

Fig. 1 Trend of economic losses taken globally due to different types of disasters during 1970–2019 (FAO 2021)

wildfires, earthquakes, and volcanic activities) have affected 1.4 billion people worldwide during 2007–2017 alone, resulting in 1.2 million injuries, 597,287 fatalities, and $1.6 billion in total damage (Guha-Sapir 2018). More recently, over 98.4 million people were impacted by 389 registered disastrous events in the year 2021 alone, resulting in over 15,000 fatalities and an additional $171.3 billion in terms of economic losses (CRED and UNDRR 2021).

Agriculture is particularly susceptible to the increasing recurrence and magnitude of severe weather-related and climate-induced events because of its intrinsic dependence on climate and the weather behaviors. An extravagantly interconnected global society and highly dynamic ground conditions provide agriculture with a range of known and unknown threats that have never been encountered in human history. Our whole food supply is in danger because disasters are occurring more often and more intensely, and because risk is pervasive. Disasters may drastically harm forests and other landscapes as well as agricultural production, cattle wellness, aquaculture, and fisheries output. The human food chain is also being severely impacted by an unprecedented rise in the frequency of incidences of transborder infestations and illnesses that endanger global fauna and flora. Moreover, agriculture sector tends to be a major consumer when it comes to bear the detrimental impacts of disasters. As shown in Fig. 2, crops, livestock, forests, aquaculture, and fisheries suffered 26% of the total damage brought on by a range of medium to

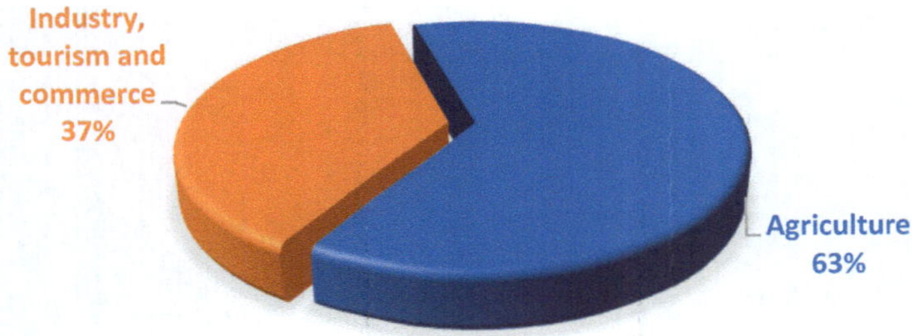

Fig. 2 Losses and damages incurred by agriculture sector compared to industry, tourism, and commerce sector combinedly during 2008–2018 (Jeggle and Boggero 2018)

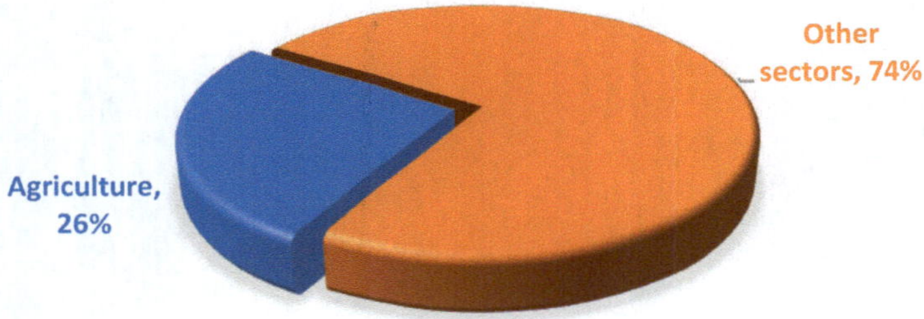

Fig. 3 Losses and damages incurred by agriculture sector compared to all other sectors during 2008–2018 (Jeggle and Boggero 2018)

large-category disasters from 2008 to 2018 (Jeggle and Boggero 2018). Furthermore, the sector alone carries a substantial part of the harm and losses caused by disasters, accounting for 63% of total losses and damages across agriculture, industries, businesses, and tourist industry, as shown in Fig. 3. Over 2.5 billion individuals depend on agriculture for their living globally. Building highly robust agricultural system requires immediate and aggressive effort due to the sector's inherent linkages with the nature, direct dependence on environmental factors for productivity, and importance for national economic growth. This degree of loss necessitates greater risk management of natural disaster concerns.

Disaster risk reduction is a theoretical blueprint of factors taken into account with the potential to reduce disaster hazards and susceptibility across a community, as well as to avert (prevent) or restrain (mitigate and prepare) the negative effects of calamities (UNISDR 2004). Authorities around the world have been trying to facilitate disaster risk management (DRR) with proper planning and strategizing at national and international levels, for instance, United Nations Office for Disaster Risk Reduction (UNDRR), Global Facility for Disaster Reduction and Recovery (GFDRR), and Asian Disaster Reduction Center (ADRC). These policies often provide a roadmap for developing and implementing cutting-edge effective DRR techniques globally by using modern and sophisticated technology. The Asia-Pacific regions have been making significant changes to disaster management policy under the Sendai Framework for DRR (UN 2015), including an increased use of artificial intelligence techniques in devising disaster response and risk reduction systems. Technologically self-sufficient countries like China (Renwick 2017), USA (Kuglitsch et al. 2022; Martire et al. 2021), Africa (UNESCO 2021) and Asian counties including Japan, India, and Indonesia (Pau et al. 2017) have been at the forefront of incorporating AI-powered solutions to bolster their disaster preparedness and response capabilities. For example, landslide vulnerability assessment, prediction, and forecasting by employing support vector machines (Lin et al. 2017; Marjanović et al. 2011; Zhou et al. 2017), multilayer perceptron neural networks (Yuan and Moayedi 2019; Zare and Pourghasemi 2013), logistic regression (Bai et al. 2010), genetic algorithms (Terranova et al. 2015), deep belief networks (Huang and Xiang 2018), and recurrent neural networks (Mutlu et al. 2019) have been done during the recent times. Similarly, recent risk reduction studies related to drought prediction, forecasting, detection, and monitoring are mainly based on MLP (Feng et al. 2019), ANN (Belayneh et al. 2014; Liu et al. 2020; Saha et al. 2021), SVM (Feng et al. 2019; Rahmati et al. 2020), hidden Markov model (Khadr 2016), and random forest (Rahmati et al. 2020) models. Likewise, detailed analysis of recent literature related to application of similar algorithms in disaster risk management of flood, avalanche, wildfires, and extreme rainfalls has been presented in the later subsections.

2 What Is Artificial Intelligence (AI)?

Artificial intelligence, from a wider perspective, is a quantitative replica of human intellect created by machines. While organic intelligence is best shown by people, AI is the level of intelligence expressed by robots, machines, and models. The development of the digitally configurable computer, a device solely dependent on mathematical logic, in 1940 prompted researchers to consider ways to create an artificial brain environment (Buchanan 2005). Turing (1950) put out strategies for teaching computing-based mechanical systems' intelligence and for testing that intelligence. Machine learning (ML), a branch of AI, involves processing raw data into models and algorithms that train from the provided input data to generate predictions and/or classifications. It can be further classified into main categories as supervised models, unsupervised models, deep learning, reinforcement learning, and deep reinforcement learning, as well as optimization.

As mentioned by Russell and Norvig (2016), unsupervised models uncover latent patterns from unstructured input using intrinsic properties without human intervention. Unsupervised models have several implications to data clustering and consolidation issues and are effective for identifying anomalous data and lowering the data dimensions. In order to recognize patterns, clustering techniques divide raw data into several categories depending on specific resemblance traits (Maulik and Bandyopadhyay 2002). Principal component analysis (PCA), for example, is a dimension reduction approach that helps minimize data intricacy and prevent overfitting.

Similarly, models that are developed on already existing data with user intervention are referred to as supervised models. Supervised models deduce a relationship between inputs and outputs by employing classification or regression techniques in order to estimate, predict, or forecast the magnitude or class of the output variable using labeled training data with predefined inputs and targets pairs (Russell and Norvig 2016). Generally, supervised algorithms have been utilized for voice recognition, pattern identification, and object detection in different contexts. Neural networks (including ANN, BPNN, FFNN), decision trees (DT), linear regression, logistic regression (LR), k-nearest neighbor (kNN), naive Bayes are some of the mostly used supervised learning algorithms.

A family of ML models known as "deep learning (DL)" uses numerous layers to gradually retrieve characteristics from input data, improving training efficiency and having a wide range of potential applicability (Pouyanfar et al. 2018). Deep learning methods are especially well suited to tackle issues of damage evaluation, motion sensing and monitoring, face identification, mobility predictions, and natural speech interpretation for helping disaster management, notwithstanding the disadvantage of needing a lengthy training period. Convolutional neural networks (CNN) are appropriate for image recognizing (Simonyan and Zisserman 2015) computer vision (Voulodimos et al. 2018), natural language processing (Otter et al. 2021), and speech processing (Song 2020), whereas recursive neural networks (RvNN) and recurrent neural networks (RNN) have already been effectively implemented to NLP (Xiao and Zhou 2020).

Reinforcement learning (RL) methods are modeled as Markov decision processes to handle task-oriented issues for taking choices in a successive way. RL programs learn through a succession of stimuli (applying punishments and reward as positive and negative stimuli) (Russell and Norvig 2016). RL is appropriate for tackling issues that require making a series of judgments in a complicated and unpredictable environment. Creating a learning condition that is appropriate and strongly connected to the activities that need to be done is the major problem in reinforcement learning. State-action-reward-state-action (SARSA) and Q-learning are examples of common RL algorithms (Sutton and Barto 2018).

Deep reinforcement learning (DRL) blends deep leaning models with reinforcement learning in order to create intelligent systems that can train on their own to design effective strategies for obtaining the greatest long-term benefits. DRL performs better at handling issues with complicated sequential processes. However, it can occasionally get quite computationally costly due to the massive quantity of training input data and training time required to achieve dependable effectiveness. Deep Q-network (DQN) and deep deterministic policy gradient (DDPG) are some of the DRL algorithms.

Although the use of AI techniques for disaster risk reduction is the main topic of this work, optimization is a crucial component of most AI approaches to identify the optimum model as determined by an objective function. Due to this, we clearly cite three particle swarm optimization and genetic algorithms as examples of effective procedures and explore how they might be used in disaster risk reduction.

3 Application of AI in Disaster Risk Reduction in Agriculture

Artificial intelligence (AI), in particular machine learning (ML), is playing an increasingly important role in disaster risk reduction (DRR). Its application ranges from the forecasting of extreme events (Hammad et al. 2021; Ke et al. 2020; Sachdeva et al. 2018) and the development of hazard maps (Chang et al. 2017; Doshi et al. 2018; Ngo et al. 2021) to the detection of events in real time (Lee et al. 2017a; Munawar et al. 2021), the provision of situational awareness and decision support, and beyond. The emergence of sophisticated high-accuracy AI and ML modeling algorithms and their successful applications in different research areas boosted the implication of these models in DRR. Moreover, Sendai framework (UN 2015) also emphasized on AI application to predict, forecast, early detect, and monitor natural disasters which urged the researchers to look for high-accuracy applications of AI in DRR.

According to a research work reported by Kuglitsch et al. (2022), it is evident that approaches connected to AI are being deployed to assist authorities in effectively managing the consequences of various natural hazards and calamities. However, further analysis of literature of application of AI in detection, forecasting, and monitoring of natural hazards published during 2018–2021 reflects clear overemphasizing of certain disasters, i.e., floods, earthquake, and landslides, as compared to others like avalanche, rainfall, and volcanic eruption. This is mainly due to the degree of risk and hazard posed by floods, earthquakes, and landslides, considering increased frequency, magnitude, and vulnerability of these disasters. Figure 4 shows the distribution of published articles based on their foci on risk reduction related to different disasters.

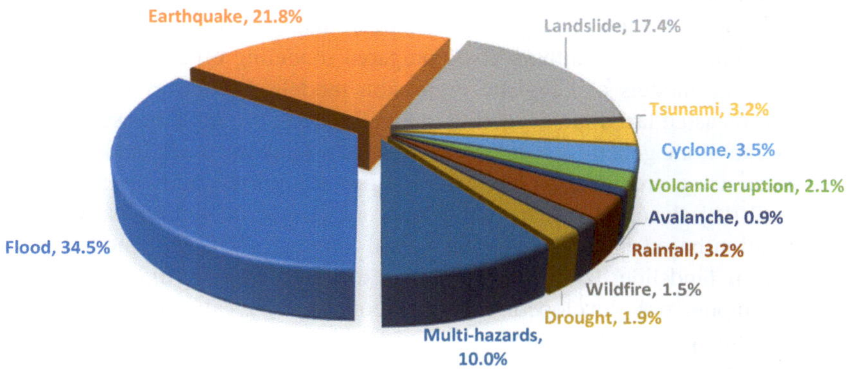

Fig. 4 Application of AI methods to detect and predict certain disasters obtained from research work published during 2018–2021 (Kuglitsch et al. 2022)

3.1 Natural Disaster Risk Reduction

Natural disasters are the cataclysmal events that originate from the natural forces like earth, water, climate, and that originating from any source of life like bacteria, virus, or pests. We have categorized natural disasters into five classes namely, geophysical disasters, climatological disasters, hydrological disasters, meteorological disasters, and biological disasters. In the following sections, we will look into how AI methods, in wider context, and ML algorithms, in specific, contribute toward risk reduction strategies of natural disasters.

3.1.1 Geophysical Disasters Risk Reduction

Geophysical DRR is related to those calamitous events that are caused mainly due to seismic and tectonic activities beneath the earth surface. Landslides, earthquakes, and volcanic eruptions are some major examples of geophysical disasters (Table 1).

Landslide is a general representation of downhill movement of earth mass, rocks, or debris due to earthquake, rainfall, volcanic eruption, or any other factors. Landslide can either start with a downslope rolling of single stone or detachment of bigger land mass, but debris, rocks, soil and earth mass continue to accumulate as it moves downward. This causes significant damage to in-path agriculture and infrastructure as well as endangers settlements and vegetation at the bottom of the slope. Land deterioration due to soil erosion, crop, forest, and livestock damage and failure due to landslide accumulation are some of the main risks that are associated with landslides from agricultural point of view.

AI algorithms, specifically machine learning methods, have been repeatedly applied in forecasting of landslide disasters, assessment of landslide vulnerability, and mapping susceptibility in a wider context. For example, Bai et al. (2010) used logistic regression-based GIS modeling to develop a map of landslide susceptibility in a section of Three Gorges Dam reservoir area in China. The authors used topographical data, geological maps, and different remote sensing products to effectively develop the vulnerability map. Likewise, Marjanović et al. (2011) deployed SVM models for the assessment of landslide vulnerability of Fruška Gora Mountain in Serbia by using morphological, geological, and environmental features of the area as input data. The authors also compared modeling efficiency of SVM with decision tree models and logistic regression models. Similarly, Terranova et al. (2015) forecasted landslide triggering in San Fili and Sorrento Peninsula of Italy by using stand-alone genetic algorithm (GA) models. High-accuracy landslide forecast was made using precipitation and ground slope data of the study area. Huang and Xiang (2018) developed deep belief network (DBN) including particle swarm optimization-based back-propagation neural network (PSO-BP) model for early warning of landslide disasters caused due to precipitation. High-accuracy early warning model was developed by using geomorphological information, soil and rainfall data, seismic data, and hydrogeological formation of earthquake affected area in Wenchuan, China. Furthermore, Yuan and Moayedi (2019) also predicted the occurrence of landslide events using stand-alone MLP-NN, hybrid MLP models (including genetic algorithm-MLP, particle swarm optimization-MLP, ant colony

Table 1 List of recent studies highlighting application of AI in geophysical disaster risk reduction in agriculture

Studies	Disaster category	Disaster type	AI model/ algorithm	Application purpose	Input data	Study area
Marjanović et al. (2011)	Geophysical	Landslide	LR, DTs, and SVM	Landslide susceptibility assessment	Geological, morphological, and environmental features	Fruska Gora Mountain, Serbia
Bai et al. (2010)	Geophysical	Landslide	LR	Landslide susceptibility map development	Topographical data, remote sensing data, thematic plots, geological plots	Three Gorges Reservoir region of China
Terranova et al. (2015)	Geophysical	Landslide	GA	Forecasting the triggering of landslides	Rainfall data and slope data	San Fili and Campania, Italy
Huang and Xiang (2018)	Geophysical	Landslide	PSO-BP and DBN	Precipitation-induced landslides climatic warning model	Geomorphological, topographical, and hydrogeological data	Wenchuan, China
Wang et al. (2013)	Geophysical	Landslide	LR-based FCM	Landslide susceptibility assessment		Mizunami City, Japan
Mutlu et al. (2019)	Geophysical	Landslide	RNN	Landslide susceptibility mapping	Geological, topographical, hydrogeological, and vegetation data	Buyukkoy Catchment, Turkey
Zhou et al. (2017)	Geophysical	Landslide	ANN and SVM	Landslide modeling	Topographical and satellite data	Three Gorges Dam, China
Pham et al. (2017)	Geophysical	Landslide	MLPNN	Landslide susceptibility assessment	Geographical and climate data	Himalayan area (India)
Saito et al. (2009)	Geophysical	Landslide	DTs	Landslide susceptibility analysis	Geographical and geological data	The Akaishi Mountains, Japan
Lin et al. (2017)	Geophysical	Landslide	SVM	Evaluation of vulnerability to rainfall-induced landslides	Geographical, topographical, and climate data	Kaoping River Basin, Taiwan

(continued)

Table 1 (continued)

Studies	Disaster category	Disaster type	AI model/ algorithm	Application purpose	Input data	Study area
Yuan and Moayedi (2019)	Geophysical	Landslide	Different types of MLPNN	Landslide prediction	Geographical and geological data	
Zare and Pourghasemi (2013)	Geophysical	Landslide	RBF and MLPNN	Landslide susceptibility mapping	Geographical, topographical, and climate data	Vaz Watershed (Iran)
Shirzadi et al. (2017)	Geophysical	Landslide	RS and NBT	Thin landslide susceptibility assessment	Geographical, topographical, and climate data	Bijar region, Kurdistan Province (Iran)
Xiong et al. (2021)	Geophysical	Earthquake	NB, IBPT, LR, FPL, CNN, GBM, RF, and DNN	Earthquake forecasting	Satellite data	
Wang et al. (2020)	Geophysical	Earthquake	LSTM	Earthquake prediction		
Last et al. (2016)	Geophysical	Earthquake	kNN, SVM, AdaBoost, IN, MIFN	Prediction of peak magnitude of future seismic activity	Historically recorded earthquake	
Narayanakumar and Raja (2016)	Geophysical	Earthquake	BPNN	Earthquake prediction	Event magnitude and time, location coordinates, and depth	
Saba et al. (2017)	Geophysical	Earthquake	BAT-NN	Earthquake prediction	Historically recorded earthquake	
Whitehead and Bebbington (2021)	Geophysical	Volcanic activity	BN, ET, and ANN	Short-term eruption forecasting	Eruption data	
Shirzaei and Walter (2010)	Geophysical	Volcanic activity	KF and GA	Time-dependent monitoring of volcano source		Centre of the caldera at Pozzuoli
Christophersen et al. (2022)	Geophysical	Volcanic activity	BN	Automated eruption forecasting	Lake temperature data, seismic data, gas flux data	Mt Ruapehu, Aotearoa, New Zealand

Studies	Disaster category	Disaster type	AI model/ algorithm	Application purpose	Input data	Study area
Nomura et al. (2020)	Geophysical	Volcanic activity	CNN	Volcanic eruption forecasting	Muograms data	Sakurajima Volcano in Kyushu, Japan
Sheldrake et al. (2017)	Geophysical	Volcanic activity	BNN	Evaluation of data regarding cessation of eruption		Soufriere Hills Volcano in Caribbean Islands

optimization-MLP, bio-geography optimization-MLP, probability-based incremental learning-MLP). In a similar manner, landslide susceptibility assessment has been done by using LR-coupled fuzzy c-mean (FCM) model in Mizunami city of Japan (Wang et al. 2013), stand-alone RNN model in Buyukkoy catchment, Turkey (Mutlu et al. 2019), SVM and ANN models in Longju area of Three Gorges Dam reservoir area in China (Zhou et al. 2017), MLP-NN model in Himalayan ranges of India (Pham et al. 2017), decision tree models in Akaishi mountain of Japan (Saito et al. 2009), stand-alone SVM model in Kaoping River, Vietnam (Lin et al. 2017), MLP-NN and radial basis function (RBF) models in Vaz watershed of Iran (Zare and Pourghasemi 2013), and naïve Bayes trees (NBT) models and random subspace (RN) assembling algorithm in Bijar region, Iran (Shirzadi et al. 2017). Input parameters like topographic details, hydrological and meteorological data, anthropogenic features, and vegetation information (all including precipitation, stream power and density, roads and infrastructure, lithology, plan and profile curvature, and slope degree) are mostly put in use to assess the landslide susceptibility by using AI and ML algorithms.

Another example of geophysical disaster is earthquake, which is the result of underground seismic or volcanic activity that causes the ground surface to vibrate or shake suddenly. This underground interaction of seismic waves with rocks or between two tectonic plates provides no physical warning or early sign, which makes earthquakes more hazardous and dangerous. Although earthquakes are regarded as relatively less risky for agriculture sector in terms of harvest and livestock loss, still early warning, prediction, and/or forecasting of earthquakes can be beneficial for risk reduction related to agriculture infrastructure including farm structures, water reservoirs, and canal systems. Earthquake disasters are responsible for only 0.7% of crop losses and 2.3% of livestock due to floods, storms, droughts, tsunamis, and earthquakes during 2003–2013 (FAO 2015).

Similar to other geophysical disasters, earthquakes have also been studied extensively to prevent and/or mitigate the disasters and reduce earthquake-related risks in agriculture and other dimensions of the society. Xiong et al. (2021) leveraged satellite data to train different machine learning algorithms including IBPT, FPL, GPM, DNN, RF, CNN, LR, and NB for earthquake forecasting. The anomalies in real-time seismic data were effectively captured by inverse boosting pruning trees (IBPT) model to detect and forecast earthquakes in short term. The IBPT model outperformed other baseline models used in the study. Similarly, deep learning algorithm LSTM was used by Wang et al. (2020) to forecast earthquakes using historical data of seismic signals. The LSTM successfully learned the spatio-temporal association between earthquake events worldwide and therefore used this information to forecast earthquake at a particular location with high accuracy. Last et al. (2016) trained kNN, AdaBoost, SVM, multi-objective info-fuzzy network (MIFN), and information network (IN), using historically recorded earthquake data as input, to predict the peak magnitude of future seismic events within the same region. In the comparative analysis of prediction accuracy, MIFN outperformed other models. In order to forecast earthquakes, Narayanakumar and Raja (2016) assessed how well BPNN models work. To create input data for the neural network, the authors collected data including event magnitude and time, location coordinates, and depth. The findings demonstrated that the BPNN algorithm yielded higher accuracy in the prediction of low-magnitude events as compared to that of higher than 5.0 earthquakes. Similarly, Saba et al. (2017) coupled bat algorithm with ANN to develop hybrid BAT-NN for predicting earthquake events in Pakistan. The developed model outperformed simple BPNN in terms of accuracy.

Similar to seismic activities, some researchers also focused their work on the disasters originating from volcanic activities. Shirzaei and Walter (2010) designed a hybrid GA-Kalman filter (GA-KF) model for time-dependent monitoring of the active volcanic sources. Whereas Nomura et al. (2020) pioneered the use of muographic data in CNN-based volcano eruption forecasting model in Sakurajima Volcano, Japan. Likewise, Whitehead and Bebbington (2021) developed a short-term forecasting model for volcanic eruption forecasts by using event tree models, ANN models, and belief network models. Input data included eruption onset time, phase duration, size, and vent locations. In a study, Christophersen et al. (2022) developed Bayesian network (BN) model for automated forecasting of volcanic activities in Mount Ruapehu, New Zealand. High-accuracy forecasts from BN data-driven model are based on the temperature data of adjacent lake, seismic data of the mountain, and gas flux data. BN has also been employed by Sheldrake et al. (2017) to study and monitor the volcanic activity of Soufriere Hills Volcano in Caribbean Islands.

3.1.2 Climatological Disasters Risk Reduction

Climatological DRR related to disastrous events that are characterized by the unusual behaviors of long-term weather mainly in the form of climatic extremes (Table 2). Droughts due to extremely low precipitation and wildfires due to extreme temperature are the main climatological disasters.

Drought is one of the most hazardous and impactful disaster types from agriculture point of view. The abnormally low rainfalls leading to increased moisture deficits have caused 85.8% of damage and losses, most among other disasters, to livestock and 15.1% share of losses to crops with respect to disaster types during 2003–2013 (FAO 2015). Due to highly dynamic weather behaviors and international campaigns on water scarcity, a lot of focus has been diverted toward disaster risk management of drought recently.

For instance, based on historical rainfall/precipitation data, Belayneh et al. (2014) developed long-term forecasting model of meteorological drought, in terms of standard precipitation index (SPI), in Awash River Basin, Ethiopia, using ANN, support vector regression (SVR). The authors also coupled ML models with wavelet transformation and formed hybrid WA-ANN and WA-SVR models. Feng et al. (2019) suggested bias-corrected random forest (BRF), MLP-NN, and SVM models to estimate agriculture drought on the basis of remotely sensed drought factors in New South Wales wheat belt, Australia. The authors observed that BRF produced drought distribution maps with highest accuracy as compared to other models used in the study. Similarly, Rahmati et al. (2020) used classification and regression tree (CART), flexible discriminant analysis (FDA), multivariate adaptive regression

Table 2 List of recent studies highlighting application of AI in climatological disaster risk reduction in agriculture

Studies	Disaster category	Disaster type	AI model/ algorithm	Application purpose	Input data	Study area
Feng et al. (2019)	Climatological	Drought	MLPNN, SVM, and BRF	Estimation of agricultural drought	Remotely sensed data and climate data	New South Wales, Australia
Belayneh et al. (2014)	Climatological	Drought	ANN and SVR	Long-term standard precipitation index (SPI) drought forecasting	Rainfall data	Catchment area of Awash River, Ethiopia
Rahmati et al. (2020)	Climatological	Drought	SVM, RF, BRT, MARS, FDA, and CART	Spatial modeling of agricultural droughts	Topographical, Geographical and Climate data, satellite data	South-east region of Queensland, Australia
Liu et al. (2020)	Climatological	Drought	ANN	Development of integrated agricultural drought index (IDI)	Geographical, climate, and satellite data	North China Plain (NCP), China
Zhang et al. (2021)	Climatological	Drought	Meta-Gaussian model	Prediction of agricultural drought (standardized soil moisture index, or SSI)	Monthly precipitation and soil moisture data	China
Khadr (2016)	Climatological	Drought	HMM	Forecasting of meteorological drought (standardized precipitation index)	Daily precipitation data, mean annual rainfall data	The upper Blue Nile River Basin, Ethiopia
Saha et al. (2021)	Climatological	Drought	ANN	Development of spatial drought vulnerability index	Meteorological and socio-economical parameters	Karnataka state of India
Lee et al. (2017b)	Climatological	Wild (forest) fires	CNN	Wildfire detection with unmanned vehicle (UAV)	Aerial images and videos	–
Muhammad et al. (2018)	Climatological	Wild (forest) fires	CNN	Fire detection in houses, forests	Aerial images and videos	
Sachdeva et al. (2018)	Climatological	Wild (forest) fires	ANN, SVM, RF, PSO-SVM, NB, DT, LR, EO-GBDT	Forest fire susceptibility mapping	Forest fire images, topographic factors, anthropogenic factors, soil texture, land cover, meteorological factors	Chamoli, Bageshwar, and Pithoragarh districts, India
Yu et al. (2005)	Climatological	Wild (forest) fires	WSN	Real-time forest fire detection	Temperature data, humidity data, aerial data	

splines (MARS), RF, and SVM methods to spatially model and monitor agricultural droughts in south-east regions of Queensland, Australia. Based on wetness indices, soil and vegetation data, and precipitation record, the RF model produced highest modeling accuracy as compared to others. In another study, Liu et al. (2020) developed integrated agricultural drought index (IDI) in North China Plain (NCP) using BPNN models. The authors used precipitation data, NDVI, land surface temperature (LST), and soil parameters as input data to highly accurate BPNN model. Zhang et al. (2021) predicted agricultural drought in China by modeling standardized soil moisture index (SSI) with meta-Gaussian model based on precipitation and soil moisture data. Similarly, Khadr (2016) used hidden Markov model (HMM) to forecast meteorological drought (in terms of SPI) in upper Blue Nile River in Ethiopia. Saha et al. (2021) used ANN to develop spatial drought vulnerability index for effective monitoring and risk reduction in Karnataka state of India.

Some serious episodes of forest or wildfires occurred worldwide in the recent years. For example, forest fires in Turkey, Greece, Algeria, California (USA), and Pakistan during the last 2 years have severely impacted millions of acres of forest land, natural habitat of wildlife, and hundreds of human lives. They resulted in thousands of wildlife losses. Following a deadly series of global wildfires, researchers and authorities have started focusing on early detection and effective monitoring of such disasters using unmanned aerial (UAV) vehicle and satellite imagery data. Lee et al. (2017a) and Muhammad et al. (2018) used CNN for early detection of wildfires using aerial and UAV images. Sachdeva et al. (2018) developed ML model for mapping of wildfire vulnerability based on data from wildfire image inventory, topographic profiles, soil and land cover information, and meteorological factors of three districts in India. The authors developed a hybrid decision tree model and compared performance accuracy with ANN, SVM, RF, PSO-SVM, DT, LR, and NB models. The developed hybrid model outperformed baseline models in the study.

3.1.3 Hydrological Disasters Risk Reduction

Hydrological DRR is the reduction of hazards associated with the disasters caused due to harmful and extreme changes in quality, distribution, and/or motion of water on earth surface (Table 3). Snow avalanche and floods are the two prime forms of hydrological disasters. Whether in solid form (snow) or in liquid (water), hydrological disasters are dangerous and hazardous from civil, economic, and agricultural point of view.

Snow avalanche can be described as a fast downhill movement of wet snow. It is mainly caused due to breaking and detachment of unstable mass of snow at the upper sections of the slope. The snow accumulates and forms a heavy mass as it sweeps down the hill or mountain. From agricultural point of view, snow avalanches might not be seriously harmful for crops and livestock, but sometimes agricultural farms and settlements get affected by these snowslides resulting in occasional crop and harvest failure as well as livestock losses. Similar to other disasters, AI techniques have been employed to forecast or assess the susceptibility of avalanches.

For example, Choubin et al. (2019) deployed SVM and multivariate discriminant analysis (MDA), in Karaj Watershed of Iran, to predict the disaster risk associated with snow avalanche. Meteorological information, locations of avalanche occurrences, and terrain characteristics were used as input to the models. Similarly, Bejiga et al. (2017) proposed an SVM- and CNN-based machine learning algorithm to assist in rescue and search operations during avalanches. The developed ML model was based on aerial and UAV imagery of different avalanche events. In a study, Dekanová et al. (2018) used snow cover data and weather conditions to develop high-accuracy ANN model for forecasting of avalanche events in Ziarska Valley in Slovakia. Heck et al. (2018) deployed continuous real-time seismic data-based HMM for automatic detection of avalanche generation in Davos, Switzerland. Chawla and Singh (2019, 2021) used RF and DT to forecast snow avalanche in upper mountain ranges of India.

Floods are known to cause highest share of damage to crops (i.e., 59.6%) among other disaster types (FAO 2015). The overflow of streams and rivers causes structural and economic damages in cities while submerges harvests and standing crops on the other side. Hill torrents and flash floods can also damage water-carrying bodies as well as canal system of the impacted area. Owing to the potential damage, increased severity and frequency of flooding events, prediction, forecasting, vulnerability assessment, detection, and monitoring of floods have always been the prime focus of DRR strategies. Flood simulation and prediction is one of the first applications of AI modeling in the field of hydrology and agriculture. The pioneer integration of machine learning algorithms into rainfall-runoff modeling is repeatedly credited to Daniell (1991).

In later studies, Pradhan (2009) used logistic regression-based GIS and RS models to develop map flood susceptibility and delineate the risk areas in Kelantan River Basin of Malaysia. The authors used precipitation data, topographic profiles, flow trends, land use classes, and historical record of flooding events as input data to the ML-collaborated qualitative models. Ren et al. (2010) employed BPNN and fuzzy clustering model (FCM) to develop real-time flood forecasting system in Liaoning province of China, using time series rainfall and flow data along with catchment parameters. Liu et al. (2016) used hybrid k-nearest neighbor model coupled with Kalman filter (kNN-KF) model to develop a real-time updating system for hydraulic model to assist in flood forecasting in Huai River Basin, China. The model uses historical time series data of flood events as input. Lee et al. (2017a) utilized DTF and TB algorithms in the geographic modeling of flood vulnerability of Seoul, South Korea. It was found that the machine learning approaches outperformed the study's other traditional methodologies. Wang et al. (2018) used FCM for flood risk assessment by flooding and non-flooding classification in Huai River Basin, China. Similarly, Wahab and Ludin (2018) employed MLP-NN to assess the flood disaster vulnerability of Muar region of Malaysia using soil data, precipitation data, topographic data, and land use classification. The authors emphasized on the high precision of ANN-based ML models in flood vulnerability assessment (FVA) problems. Lohumi and Roy (2019) developed an automatic system for the detection and monitoring of flood severity levels using deep learning model, i.e., CNN and gated recurrent units (GRU), based on aerial and UAV videos of previous flooding events. Ke et al. (2020) used decision trees, discriminant analysis, SVM, kNN, and ensemble models to predict flash pluvial floods in Shenzhen city of China. Authors observed

Table 3 List of recent studies highlighting application of AI in hydrological disaster risk reduction in agriculture

Studies	Disaster category	Disaster type	AI model/ algorithm	Application purpose	Input data	Study area
Choubin et al. (2019)	Hydrological	Avalanche	SVM and MDA	Snow avalanche hazard prediction	Meteorological data, and terrain characteristics	Karaj Watershed, Iran
Bejiga et al. (2017)	Hydrological	Avalanche	CNN and SVM	Development of algorithm for assisting avalanche search and rescue operations	Aerial and UAV images	
Dekanová et al. (2018)	Hydrological	Avalanche	ANN	Avalanche forecasting	Snow dept. and weather conditions	Ziarska Valley
Heck et al. (2018)	Hydrological	Avalanche	HMM	Automated detection of snow avalanches	Continuous seismic data	Davos, Switzerland
Chawla and Singh (2019)	Hydrological	Avalanche	DT and RF	Avalanche forecasting	Temperature data, snow cover data, wind speed data	Bandipore, India
Chawla and Singh (2021)	Hydrological	Avalanche	RF	Autonomous operational avalanche forecasting	Temperature data, snow cover data, sunshine hours	Bandipore, India
Ke et al. (2020)	Hydrological	Flood	SVM, DT, and KNN	Urban pluvial flooding prediction	Rainfall data, historical data of flood events	Shenzhen City, China
Wahab and Ludin (2018)	Hydrological	Flood	MLPNN	Flood vulnerability assessment	Rainfall data, topographic data, land use data, soil type	Muar region, Johor, Malaysia
Pradhan (2009)	Hydrological	Flood	LR and GIS	Flood susceptible mapping and risk area delineation	Rainfall data, topographic data, land use data, flow data, historical data of flood events	Kelantan River Basin, Malaysia
Lohumi and Roy (2019)	Hydrological	Flood	CNN and GRU	Recognition of flood severity level	Flood videotapes	
Liu et al. (2016)	Hydrological	Flood	KNN-KF	Real-time updating of hydraulic model in flood forecasting	Historical data of flood events	Huai River Basin, China

(continued)

Table 3 (continued)

Studies	Disaster category	Disaster type	AI model/ algorithm	Application purpose	Input data	Study area
Zhao et al. (2020)	Hydrological	Flood	RF, CNN, and SVM	Urban flood susceptibility assessment	Rainfall data, topographic data, land use data, anthropogenic data, historical data of flood events	Dahongmen Catchment, Beijing, China
Ngo et al. (2021)	Hydrological	Flood	PSO and DT	Flash flood susceptibility mapping with geospatial data	Historical data of flood events and satellite imagery	Tran Yen District, Vietnam
Pham et al. (2021)	Hydrological	Flood	AdaBoost-DT and Bagging-DT	Flood risk assessment	Physical, anthropogenic, and meteorological factors	Quang Nam Province, Vietnam
Ren et al. (2010)	Hydrological	Flood	ANN and FCM	Real-time flood forecasting	Historical data of flood events, rainfall data, and catchment data	Liaoning Province, China
Wang et al. (2018)	Hydrological	Flood	FCM	Regional flood risk assessment	Inundation data, flow data, land use data, anthropogenic data	Huai River Basin, China
Adikari et al. (2021)	Hydrological	Flood and drought	WANFIS, CNN, and LSTM	Forecasting floods and droughts in desert and tropical areas	Rainfall data and discharge data	Basin of Lower Darling River and Sekong River Basin, Australia
Munawar et al. (2021)	Hydrological	Flood	CNN	Real-time flood detection	UAV imagery	Indus River, Pakistan

higher accuracy of precipitation and flood event historic data-based ML models as compared to conventional mathematical models. Zhao et al. (2020) employed SVM, RF, and CNN to assess the susceptibility of urban flooding in Dahongmen catchment in China. Among the ML models based on time series data of flood events, anthropogenic data, land use and topographic classification, and precipitation data, CNN-based assessment was found to be highly accurate. Ngo et al. (2021) mapped flash flooding vulnerability in Tran Yen, Vietnam, using spatial data, satellite imagery, and time series flood data processed with a high-accuracy hybrid quantum-PSO-based credal decision tree ensemble (QPSO-CDT-ensemble) model. Similarly, Pham et al. (2021) also used hybrid decision tree models (AdaBoost-DT and Bagging-DT) for flood disaster risk assessment in Quang Nam province of Vietnam. Adikari et al. (2021) used hybrid machine learning model, wavelet-coupled adaptive neuro fuzzy inference system (WANFIS), and deep learning algorithms, CNN, and LSTM, to forecast floods and droughts in arid and tropical regions of Australia. Rainfall and discharge data of Darling River Basin and Sekong River Basin were used to forecast floods and droughts with CNN at higher accuracy as compared to WANFIS. Munawar et al. (2021) used unmanned aerial vehicle (UAV) imagery to develop and train CNN model for real-time flood detection in Indus River of Pakistan. Similarly, Khan et al. (2021) employed 5 ML techniques, i.e., MLP-NN, GEP, TB, SDT, and DTF models as stand-alone and also in combination with maximal overlap discrete wavelet transformation (MODWT), to simulate stream flows in four rivers against rainfall in Pothohar region of Pakistan. MODWT-DTF was found to outperform other stand-alone and hybrid models used in the study.

3.1.4 Meteorological Disasters Risk Reduction

Meteorological DRR relates to the disasters originating from short-term behavior of weathers (Table 4). For example, intense rainfall, hailstorms, typhoons, cyclones, etc., can be accurately estimated and forecasted using AI methods. Pioneer application of machine learning, more specifically ANN, in rainfall forecasting was introduced by French et al. (1992). The authors developed a stand-alone neural network model with back-propagation algorithm rainfall forecasting for the first time using ML model.

Similarly, Yu et al. (2017) compared the performance of RF and SVM models in satellite-based rainfall forecast in three catchments of Taiwan, by using rainfall data and elevation as inputs. The authors observed better accuracy of SVM-based forecasting model as compared to RF model. Likewise, Moon et al. (2018) developed an early warning system for short-term high-intensity rainfall event using stand-alone LR model built on meteorological data of South Korea. Pham et al. (2020) used hybrid PSO-ANFIS model and SVM and ANN models to predict daily rainfall in Hoa Binh, Vietnam. SVM model trained on temperature, humidity, wind, and solar radiation data outperformed the other two models. Hammad et al. (2021) developed novel wavelet-coupled multi-order time-lagged neural network (WMTLNN) to forecast daily rainfall in Upper Indus Basin, Pakistan. The developed model performed better than long short-term memory (LSTM) model, wavelet-coupled time-lagged neural network (WTLNN), and time-lagged neural network (TLNN) models. Adjei et al. (2021) forecasted daily rainfall of Axim, Ghana, by using different

Table 4 List of recent studies highlighting application of AI in meteorological disaster risk reduction in agriculture

Studies	Disaster category	Disaster type	AI model/ algorithm	Application purpose	Input data	Study area
Hammad et al. (2021)	Meteorological	Rainfall	WMTLNN, LSTM, WTLNN, TLNN	Forecasting of daily rainfall	Historical data of rainfall	Upper Indus Basin, Pakistan
Pham et al. (2020)	Meteorological	Rainfall	PSO-ANFIS, ANN, SVM	Prediction of daily rainfall	Temperature, humidity, wind, and solar radiation data	Hao Binh, Vietnam
Yu et al. (2017)	Meteorological	Rainfall	RF and SVM	Forecasting of daily rainfall	Satellite-based rainfall data and elevation	Taiwan
Moon et al. (2018)	Meteorological	Rainfall	LR	Development of early warning system for short-term high-intensity rainfall event	Meteorological data	South Korea
Adjei et al. (2021)	Meteorological	Rainfall	ANN and LSTM	Rainfall forecasting	Meteorological data	Axim, Ghana

combinations of meteorological data as inputs to ANN and LSTM models. The observations reflected that LSTM model well-suited the forecasting problem of the study. Similarly, multiple research works have evaluated the performance of AI algorithms in rainfall forecasting problems using single as well as multiple inputs (Abbot and Marohasy 2012; Gomes and Blanco 2021; Haidar and Verma 2018; Lee et al. 2018; Sharghi et al. 2019; Unnikrishnan and Jothiprakash 2017).

3.1.5 Biological Disasters Risk Reduction

Biological DRR is the risk reduction of disasters originating from any type of life (Table 5); for example, pests and insect infestation and epidemic outbreaks due to viral or fungal exposure. AI application in DRR has also made its way through to the detection and monitoring of agricultural viral epidemics and pest/insect infestation disasters. Some examples are detection and prediction of potato late blight disease in Sardinia using ANN (Fenu and Malloci 2019) and SVM models (Fenu and Malloci 2020). SVM model has also been applied to monitor and early forecast yellow rust in wheat in impacted parts of China (Dong et al. 2020) and in Great Britain (Skelsey 2021). Similarly, in the context of infestation, Gómez et al. (2018) conducted a study based on RF to locate breeding areas of desert locusts using satellite images in Africa. Likewise, Skawsang et al. (2019) also predicted the population of rice pest in Thailand by using ANN, RF, and MLR models. Among the adopted algorithms, ANN model yielded highest accuracy.

Table 5 List of recent studies highlighting application of AI in meteorological disaster risk reduction in agriculture

Studies	Disaster category	Disaster type	AI model/ algorithm	Application purpose	Input data	Study area
Fenu and Malloci (2019)	Biological	Diseases	ANN	Predict potato late blight disease	Temperature data, humidity data, rainfall data	Sardinia
Dong et al. (2020)	Biological	Diseases	SVM	Wheat yellow rust and oriental migratory locust monitoring and early forecasting	Spatial data, vegetative indices, disease index	Impact areas of China
Fenu and Malloci (2020)	Biological	Diseases	SVM	Potato late blight prediction Potato late blight prediction	Historical weather data such as temperature, humidity, rainfall, speed wind, and solar radiation	Sardinia
Skelsey (2021)	Biological	Diseases	SVM, one-class k-means, Gaussian mixture model	Potato late blight prediction Potato late blight prediction	Historical data of late blight epidemic outbreak	Great Britain
Gómez et al. (2018)	Biological	Diseases	RF	Locating breeding areas of locust's swarms	Satellite images	Africa
Skawsang et al. (2019)	Biological	Diseases	ANN, RF, and MLR	Prediction of rice pest population	Satellite data, meteorological data	Thailand

4 Current Challenges and Ways Forward

Complications might arise at any step of data collection, model construction, or operational execution when employing AI for DRR.

It is crucial to take into account the following while collecting and processing data: (a) distortions in training and testing data; (b) decentralized AI solutions inside the data domains; and (c) ethical concerns. It is crucial to check that data are accurately collected and that individual patterns are sufficiently represented for the given issue in order to avoid anomalies in training and testing data. Think about the difficulty of creating a dataset with instances of unusual catastrophic disasters. Consider the potential consequences of not providing enough data as well, such as inaccurate projections or distorted results. Once a dataset has been verified to be neutral, we must choose how to include novel distributed AI algorithms into the final dataset. Centralized information handling and administration might be challenging for AI since it depends on data transfer and the calculation of sophisticated machine learning models.

We also have to take constraints into account throughout the model construction phase once a dataset has been selected. Here, we emphasize transparency and the technological requirements. Because AI models often depend on intricate architecture, training them may be computationally costly. For instance, the image classification model VGG16 (Simonyan and Zisserman 2015) includes over 138 million trainable features. This magnitude of training networks needs costly and resource-intensive computational power, which is not usually available.

The output of an AI model must be comprehensible and agreeable to humans after it has been created. This may be difficult to achieve since there is not a standard immediate human–machine interaction that explains the reasoning behind the AI model's evaluations.

As a result, numerous academics are attempting to create reliable AI solutions. A detailed articulation of the issue, as well as the needs and objectives of the AI-based solutions, is crucial during modeling and model assessment. Only after that a problem-solving model and training method can be designed. Furthermore, selecting and creating appropriate assessment criteria are facilitated by being aware of the particular arrangement.

There is a significant demand in the area of DRR in investigating the implications of utilizing AI to support current approaches and tactics. This article offered a number of application examples illustrating how AI-based algorithms are improving DRR, but it also demonstrated that AI has drawbacks. Luckily, the potential of AI in DRR has prompted experiments to address these issues and influenced novel collaborations, bringing next to each other specialists from different scholarly disciplines (such as data science and earth sciences), from various industries (such as academia and government authorities), and from all over the world. Such collaborations are essential for advancing AI in DRR. We think that more work has to be done, namely, in the domain of developing learning media that will promote skill development, ensure the accessibility of computing capabilities and other technology, and close the technological gap. We can only ensure that everyone benefits as AI for DRR develops by doing this.

References

Abbot J, Marohasy J (2012) Application of artificial neural networks to rainfall forecasting in Queensland, Australia. Adv Atmos Sci 29(4):717–730. https://doi.org/10.1007/s00376-012-1259-9.1.1.1

Adikari KE, Shrestha S, Ratnayake DT, Budhathoki A, Mohanasundaram S, Dailey MN (2021) Evaluation of artificial intelligence models for flood and drought forecasting in arid and tropical regions. Environ Model Softw 144:105136. https://doi.org/10.1016/j.envsoft.2021.105136

Adjei C, Tian W, Onzo B-M, Chen S, Adu E, Darteh O (2021) Rainfall forecasting in sub-sahara Africa-Ghana using LSTM deep learning approach. Int J Eng Tech Res 10(3):464–470

Bai SB, Wang J, Lü GN, Zhou PG, Hou SS, Xu SN (2010) GIS-based logistic regression for landslide susceptibility mapping of the Zhongxian segment in the three gorges area, China. Geomorphology 115(1–2):23–31. https://doi.org/10.1016/j.geomorph.2009.09.025
Bejiga MB, Zeggada A, Melgani F (2017) A convolutional neural network approach for near real-time object detection from UAV imagery in avalanche search and rescue operations. In: The international geoscience and remote sensing symposium (IGARSS 2016). IEEE, Piscataway, NJ, p 9. https://doi.org/10.3390/rs9020100
Belayneh A, Adamowski J, Khalil B, Ozga-zielinski B (2014) Long-term SPI drought forecasting in the Awash river basin in Ethiopia using wavelet neural network and wavelet support vector regression models. J Hydrol 508:418–429. https://doi.org/10.1016/j.jhydrol.2013.10.052
Benevolenza MA, DeRigne LA (2019) The impact of climate change and natural disasters on vulnerable populations: a systematic review of literature. J Hum Behav Soc Environ 29(2):266–281. https://doi.org/10.1080/10911359.2018.1527739
Blanc E, Reilly J (2017) Approaches to assessing climate change impacts on agriculture: an overview of the debate. Rev Environ Econ Policy 11(2):247–257. https://doi.org/10.1093/reep/rex011
Blanc E, Schlenker W (2017) The use of panel models in assessments of climate impacts on agriculture. Rev Environ Econ Policy 11(2):258–279. https://doi.org/10.1093/reep/rex016
Buchanan BG (2005) A (very) brief history of artificial intelligence. AI Mag 26(4):53–60
Chang TK, Talei A, Alaghmand S, Ooi MPL (2017) Choice of rainfall inputs for event-based rainfall-runoff modeling in a catchment with multiple rainfall stations using data-driven techniques. J Hydrol 545(1):100–108. https://doi.org/10.1016/j.jhydrol.2016.12.024
Chawla M, Singh A (2019) Data efficient random forest model for avalanche forecasting. Nat Hazards Earth Syst Sci:1–33
Chawla M, Singh A (2021) A data efficient machine learning model for autonomous operational avalanche forecasting. Nat Hazards Earth Syst Sci, pp 1–18
Choubin B, Borji M, Mosavi A, Sajedi-hosseini F, Singh VP, Shamshirband S (2019) Snow avalanche hazard prediction using machine learning methods. J Hydrol 577:123929
Christophersen A, Behr Y, Miller CA (2022) Automated eruption forecasting at frequently active volcanoes using Bayesian networks learned from monitoring data and expert elicitation : application to Mt Ruapehu, Aotearoa, New Zealand. Front Earth Sci 10:905965. https://doi.org/10.3389/feart.2022.905965
Corwin DL (2021) Climate change impacts on soil salinity in agricultural areas. Eur J Soil Sci 72(2):842–862. https://doi.org/10.1111/ejss.13010
CRED and UNDRR (2021). Global trends and perspectives executive summary. https://cred.be/sites/default/files/2021_EMDAT_report.pdf
Daniell TM (1991) Neural networks—applications in hydrology and water resources engineering. Proc. inter. Hydrology and water symposium. Inst. Engrs. Australia, National Conf. Publ, 797–902
Dekanová M, Duchoň F, Dekan M, Kyzek F, Biskupic M (2018) Avalanche forecasting using neural network. In: Proceedings of the 2018 ELEKTRO. IEEE. 18: 383. https://doi.org/10.5194/nhess-18-383-2018
Dong Y, Xu F, Liu L, Du X, Ren B, Guo A, Geng Y, Ruan C, Ye H, Huang W, Zhu Y (2020) Automatic system for crop Pest and disease dynamic monitoring and early forecasting. IEEE J Sel Top Appl Earth Obs Remote Sens 13:4410–4418. https://doi.org/10.1109/JSTARS.2020.3013340
Doshi J, Basu S, Pang G (2018) From satellite imagery to disaster insights. In: 32nd conference on neural information processing systems (NIPS 2018), pp 1–6. http://arxiv.org/abs/1812.07033
FAO (2015) The impact of natural hazards and disasters on agriculture and food security and nutrition: a call for action to build resilient livelihoods. In: FAO report (issue May). http://www.fao.org/3/a-i4434e.pdf
FAO (2021) The impact of disasters and crises on agriculture and food security: 2021. Food and Agriculture Organization of the United Nations, Rome. https://doi.org/10.4060/cb3673en

Feng P, Wang B, Liu DL, Yu Q (2019) Machine learning-based integration of remotely-sensed drought factors can improve the estimation of agricultural drought in south-eastern Australia. Agr Syst 173:303–316. https://doi.org/10.1016/j.agsy.2019.03.015

Fenu G, Malloci FM (2019) An application of machine learning technique in forecasting crop disease. In: ACM international conference proceeding series. Association for Computing Machinery, New York, NY, pp 76–82. https://doi.org/10.1145/3372454.3372474

Fenu G, Malloci FM (2020) Artificial intelligence technique in crop disease forecasting: a case study on potato late blight prediction. In: Intelligent decision technologies, smart innovation, systems and technologies, vol 193. Springer Nature, Singapore, pp 79–89. https://doi.org/10.1007/978-981-15-5925-9_7

French MN, Krajewski WF, Cuykendall RR (1992) Rainfall forecasting in space and time using a neural network. J Hydrol 137:1–31

Gomes EP, Blanco CJC (2021) Daily rainfall estimates considering seasonality from a MODWT-ANN hybrid model. J Hydrol Hydromech 69(1):13–28. https://doi.org/10.2478/johh-2020-0043

Gómez D, Salvador P, Sanz J, Casanova C, Taratiel D, Casanova JL (2018) Machine learning approach to locate desert locust breeding areas based on ESA CCI soil moisture. J Appl Remote Sens 12(03):1. https://doi.org/10.1117/1.jrs.12.036011

Guha-Sapir D (2018) EM-DAT: the emergency events database. Center for research on the epidemiology of disasters. www.emdat.be

Haidar A, Verma B (2018) Monthly rainfall forecasting using one-dimensional deep convolutional neural network. IEEE Access 6:69053–69063. https://doi.org/10.1109/ACCESS.2018.2880044

Hammad M, Shoaib M, Salahudin H, Baig MAI, Khan MM, Ullah MK (2021) Rainfall forecasting in upper Indus basin using various artificial intelligence techniques. Stoch Env Res Risk A 35:2213–2235. https://doi.org/10.1007/s00477-021-02013-0

Heck M, Hammer C, Van Herwijnen A, Schweizer J, Fäh D (2018) Automatic detection of snow avalanches in continuous seismic data using hidden Markov models. Nat Hazards Earth Syst Sci 18:383–396. https://doi.org/10.5194/nhess-18-383-2018

Huang L, Xiang LY (2018) Method for meteorological early warning of precipitation-induced landslides based on deep neural network. Neural Process Lett 48(2):1243–1260. https://doi.org/10.1007/s11063-017-9778-0

Jeggle T, Boggero M (2018) Post-disaster needs assessment (PDNA): lessons from a decade of experience; https://www.gfdrr.org/sites/default/files/publication/Final_PDNA_Evaluation_Report.pdf

Ke Q, Tian X, Bricker J, Tian Z, Guan G, Cai H, Huang X, Yang H, Liu J (2020) Urban pluvial flooding prediction by machine learning approaches—a case study of Shenzhen City, China. Adv Water Resour 145:103719. https://doi.org/10.1016/j.advwatres.2020.103719

Khadr M (2016) Forecasting of meteorological drought using hidden Markov model (case study: the upper Blue Nile river basin, Ethiopia). Ain Shams Eng J 7(1):47–56. https://doi.org/10.1016/j.asej.2015.11.005

Khan MT, Shoaib M, Hammad M, Salahudin H, Ahmad F, Ahmad S (2021) Application of machine learning techniques in rainfall–runoff modelling of the Soan River Basin, Pakistan. Water 13:3528. https://doi.org/10.3390/w13243528

Kuglitsch M, Albayrak A, Aquino R, Craddock A, Edward-Gill J, Kanwar R, Koul A, Ma J, Marti A, Menon M, Pelivan I, Toreti A, Venguswamy R, Ward T, Xoplaki E, Rea A, Luterbacher J (2022) Artificial intelligence for disaster risk reduction: opportunities, challenges, and prospects. World Meteorological Organization. https://public.wmo.int/en/resources/bulletin/artificial-intelligence-disaster-risk-reduction-opportunities-challenges-and#:~:text=Artificialintelligence(AI)%2C.in,situationalawarenessanddecisionsupport%2C

Last M, Rabinowitz N, Leonard G (2016) Predicting the maximum earthquake magnitude from seismic data in Israel and its neighboring countries. PloS One 11(1):1–16. https://doi.org/10.1371/journal.pone.0146101

Lee S, Kim JC, Jung HS, Lee MJ, Lee S (2017a) Spatial prediction of flood susceptibility using random-forest and boosted-tree models in Seoul metropolitan city, Korea. Geomat Nat Haz Risk 8(2):1185–1203. https://doi.org/10.1080/19475705.2017.1308971

Lee W, Kim S, Lee YT, Lee HW, Choi M (2017b) Deep neural networks for wildfire detection with unmanned aerial vehicle. In: 2017 IEEE international conference on consumer electronics (ICCE). https://doi.org/10.1145/2647868.2654889

Lee J, Kim C, Lee JE, Kim NW, Kim H (2018) Application of artificial neural networks to rainfall forecasting in the Geum River basin, Korea. Water 10:1448. https://doi.org/10.3390/w10101448

Lin G, Chang M, Huang Y, Ho J (2017) Assessment of susceptibility to rainfall-induced landslides using improved self-organizing linear output map, support vector machine, and logistic regression. Eng Geol 224:62–74. https://doi.org/10.1016/j.enggeo.2017.05.009

Liu K, Li Z, Yao C, Chen J, Zhang K, Saifullah M (2016) Coupling the k-nearest neighbor procedure with the Kalman filter for real-time updating of the hydraulic model in flood forecasting. Int J Sediment Res 31(2):149–158. https://doi.org/10.1016/j.ijsrc.2016.02.002

Liu X, Zhu X, Zhang Q, Yang T, Pan Y, Sun P (2020) A remote sensing and artificial neural network-based integrated agricultural drought index: index development and applications. Catena 186:104394. https://doi.org/10.1016/j.catena.2019.104394

Lohumi K, Roy S (2019) Automatic detection of flood severity level from flood videos using deep learning models. In: 5th international conference on information and communication technologies for disaster management, ICT-DM 2018, pp 1–7. https://doi.org/10.1109/ICT-DM.2018.8636373

Malhi GS, Kaur M, Kaushik P (2021) Impact of climate change on agriculture and its mitigation strategies: a review. Sustainability 13(3):1–21. https://doi.org/10.3390/su13031318

Marjanović M, Kovačević M, Bajat B, Voženílek V (2011) Landslide susceptibility assessment using SVM machine learning algorithm. Eng Geol 123(3):225–234. https://doi.org/10.1016/j.enggeo.2011.09.006

Martire L, Constantinou V, Krishnamoorthy S, Komjathy A, Vergados P, Meng X, Bar-Sever Y, Craddock A, Wilson B (2021) Near real-time tsunami early warning system using GNSS ionospheric measurements. In: AGU fall meeting abstracts, p G45C

Maulik U, Bandyopadhyay S (2002) Performance evaluation of some clustering algorithms and validity indices. IEEE Trans Pattern Anal Mach Intell 24(12):1650–1654. https://doi.org/10.1109/TPAMI.2002.1114856

Moon SH, Kim YH, Lee YH, Moon BR (2018) Application of machine learning to an early warning system for very short-term heavy rainfall. J Hydrol 568:1042–1054. https://doi.org/10.1016/j.jhydrol.2018.11.060

Muhammad K, Ahmad J, Baik SW (2018) Early fire detection using convolutional neural networks during surveillance for effective disaster management. Neurocomputing 288:30–42. https://doi.org/10.1016/j.neucom.2017.04.083

Munawar HS, Ullah F, Qayyum S, Khan SI, Mojtahedi M (2021) Uavs in disaster management: application of integrated aerial imagery and convolutional neural network for flood detection. Sustainability 13:7545. https://doi.org/10.3390/su13147547

Mutlu B, Nefeslioglu HA, Sezer EA (2019) An experimental research on the use of recurrent neural networks in landslide susceptibility mapping. Int J Geoinform 8(578):1–21. https://doi.org/10.3390/ijgi8120578

Narayanakumar S, Raja K (2016) A BP artificial neural network model for earthquake magnitude prediction in Himalayas, India. Circuits Syst 07(11):3456–3468. https://doi.org/10.4236/cs.2016.711294

Ngo PTT, Pham TD, Nhu VH, Le TT, Tran DA, Phan DC, Hoa PV, Amaro-Mellado JL, Bui DT (2021) A novel hybrid quantum-PSO and credal decision tree ensemble for tropical cyclone induced flash flood susceptibility mapping with geospatial data. J Hydrol 596:125682. https://doi.org/10.1016/j.jhydrol.2020.125682

Nomura Y, Nemoto M, Hayashi N, Hanaoka S, Murata M, Yoshikawa T, Maeda E, Abe O, Tanaka MKM (2020) Pilot study of eruption forecasting with muography using convolutional neural network. Sci Rep 10(5):3–11. https://doi.org/10.1038/s41598-020-62342-y

Ogie RI, Forehead H, Clarke RJ, Perez P (2018) Participation patterns and reliability of human sensing in crowd-sourced disaster management. Inf Syst Front 20(4):713–728. https://doi.org/10.1007/s10796-017-9790-y

Otter DW, Medina JR, Kalita JK (2021) A survey of the usages of deep learning for natural language processing. IEEE Trans Neural Netw Learn Syst 32(2):604–624. https://doi.org/10.1016/B978-0-12-820273-9.00006-3

Pau J, Baker J, Houston N (2017) Artificial intelligence in Asia: preparedness and resilience; https://www.asiabusinesscouncil.org/docs/AI_briefing.pdf

Pham BT, Bui DT, Prakash I, Dholakia MB (2017) Catena hybrid integration of multilayer perceptron neural networks and machine learning ensembles for landslide susceptibility assessment at Himalayan area (India) using GIS. Catena 149:52–63. https://doi.org/10.1016/j.catena.2016.09.007

Pham BT, Le LM, Le TT, Bui KTT, Le VM, Ly HB, Prakash I (2020) Development of advanced artificial intelligence models for daily rainfall prediction. Atmos Res 237:104845. https://doi.org/10.1016/j.atmosres.2020.104845

Pham BT, Luu C, Van Phong T, Nguyen HD, Van Le H, Tran TQ, Ta HT, Prakash I (2021) Flood risk assessment using hybrid artificial intelligence models integrated with multi-criteria decision analysis in Quang Nam Province, Vietnam. J Hydrol 592:125815. https://doi.org/10.1016/j.jhydrol.2020.125815

Pouyanfar S, Sadiq S, Yan Y, Tian H, Tao Y, Reyes MP, Shyu ML, Chen SC, Iyengar SS (2018) A survey on deep learning: algorithms, techniques, and applications. ACM Comput Surv 51(5):1. https://doi.org/10.1145/3234150

Pradhan B (2009) Flood susceptible mapping and risk area delineation using logistic regression, GIS and remote sensing. J Spat Hydrol 9(2):1–18

Rahmati O, Falah F, Shaanu K, Deo RC, Mohammadi F, Biggs T, Moghaddam DD, Naghibi SA, Bui DT (2020) Science of the Total environment machine learning approaches for spatial modeling of agricultural droughts in the south-east region of Queensland Australia. Sci Total Environ 699:134230. https://doi.org/10.1016/j.scitotenv.2019.134230

Ren M, Wang B, Liang Q, Fu G (2010) Classified real-time flood forecasting by coupling fuzzy clustering and neural network. Int J Sediment Res 25(2):134–148. https://doi.org/10.1016/S1001-6279(10)60033-9

Renwick N (2017) China's approach to disaster risk reduction: human security challenges in a time of climate change. J Asian Secur Int Aff 4(1):26–49. https://doi.org/10.1177/2347797016689207

Russell SJ, Norvig P (2016) Learning from examples. In: Artificial intelligence: a modern approach, 3rd edn. Pearson, Harlow, pp 693–767

Saba S, Ahsan F, Mohsin S (2017) BAT-NN based earthquake prediction for Pakistan region. Soft Comput 21:5805–5813. https://doi.org/10.1007/s00500-016-2158-2

Sachdeva S, Bhatia T, Verma AK (2018) GIS-based evolutionary optimized gradient boosted decision trees for forest fire susceptibility mapping. Nat Hazards 92(3):1399–1418. https://doi.org/10.1007/s11069-018-3256-5

Saha S, Gogoi P, Gayen A, Chandra G (2021) Constructing the machine learning techniques based spatial drought vulnerability index in Karnataka state of India. J Clean Prod 314:128073. https://doi.org/10.1016/j.jclepro.2021.128073

Sahana M, Rehman S, Paul AK, Sajjad H (2021) Assessing socio-economic vulnerability to climate change-induced disasters: evidence from Sundarban biosphere reserve, India. Geol Ecol Landsc 5(1):40–52. https://doi.org/10.1080/24749508.2019.1700670

Saito H, Nakayama D, Matsuyama H (2009) Geomorphology comparison of landslide susceptibility based on a decision-tree model and actual landslide occurrence: the Akaishi Mountains, Japan. Geomorphology 109(3–4):108–121. https://doi.org/10.1016/j.geomorph.2009.02.026

Sharghi E, Nourani V, Molajou A, Najafi H (2019) Conjunction of emotional ANN (EANN) and wavelet transform for rainfall-runoff modeling. J Hydroinf 21:136–152. https://doi.org/10.2166/hydro.2018.054

Sheldrake TE, Aspinall WP, Odbert HM, Wadge G, Sparks RSJ (2017) Understanding causality and uncertainty in volcanic observations: an example of forecasting eruptive activity on

Soufrière Hills volcano, Montserrat. J Volcanol Geotherm Res 341:287–300. https://doi.org/10.1016/j.jvolgeores.2017.06.007

Shirzadi A, Tien D, Binh B, Pham T, Solaimani K (2017) Shallow landslide susceptibility assessment using a novel hybrid intelligence approach. Environ Earth Sci 76:1–18. https://doi.org/10.1007/s12665-016-6374-y

Shirzaei M, Walter TR (2010) Time—dependent volcano source monitoring using interferometric synthetic aperture radar time series: a combined genetic algorithm and Kalman filter approach. J Geophys Res 115:B10421. https://doi.org/10.1029/2010JB007476

Simonyan K, Zisserman A (2015) Very deep convolutional networks for large-scale image recognition. In: 3rd international conference on learning representations, ICLR 2015- conference track proceedings, pp 1–14. https://arxiv.org/pdf/1409.1556.pdf

Skawsang S, Nagai M, Tripathi NK, Soni P (2019) Predicting rice pest population occurrence with satellite-derived crop phenology, ground meteorological observation, and machine learning: a case study for the central plain of Thailand. Appl Sci 9(22):1–19. https://doi.org/10.3390/app9224846

Skelsey P (2021) Forecasting risk of crop disease with anomaly detection algorithms. Phytopathology 111(2):321–332. https://doi.org/10.1094/PHYTO-05-20-0185-R

Song Z (2020) English speech recognition based on deep learning with multiple features. Comput Secur 102(3):663–682. https://doi.org/10.1007/s00607-019-00753-0

Sutton RS, Barto AG (2018) Temporal-difference learning. In: Sutton RS, Barto AG (eds) Reinforcement learning: an introduction, 2nd edn. The MIT Press, Cambridge, MA, pp 119–140

Terranova OG, Gariano SL, Laquinta P, Iovine GGR (2015) GASAKe: forecasting landslide activations by a genetic-algorithms based hydrological model. Geosci Model Dev Discuss 8:1225–1291. https://doi.org/10.5194/gmdd-8-1225-2015

Turing A (1950) Computing machinery and intelligence. Mind 59(236):433–460. https://doi.org/10.1093/mind/VIII.2.145

UN (2015) Sendai framework for disaster risk reduction 2015–2030. In: United Nations

UNESCO (2021) Use of artificial intelligence for disaster risk reduction in Africa. https://www.unesco.org/en/articles/use-artificial-intelligence-disaster-risk-reduction-africa

UNISDR (2004) International strategy for disaster reduction. In: Living with risk: a global review of disaster reduction initiatives. https://web.archive.org/web/20040803191318/http://www.unisdr.org/eng/about_isdr/bd-lwr-2004-eng.htm

Unnikrishnan P, Jothiprakash V (2017) Data-driven multi-time-step ahead daily rainfall forecasting using singular spectrum analysis-based data pre-processing. J Hydroinf 20:645. https://doi.org/10.2166/hydro.2017.029

Van Meijl H, Havlik P, Lotze-Campen H, Stehfest E, Witzke P, Domínguez IP, Bodirsky BL, Van Dijk M, Doelman J, Fellmann T, Humpenöder F, Koopman JFL, Müller C, Popp A, Tabeau A, Valin H, Van Zeist WJ (2018) Comparing impacts of climate change and mitigation on global agriculture by 2050. Environ Res Lett 13:6. https://doi.org/10.1088/1748-9326/aabdc4

Voulodimos A, Doulamis N, Doulamis A, Protopapadakis E (2018) Deep learning for computer vision: a brief review. Comput Intell Neurosci 2018:1. https://doi.org/10.1155/2018/7068349

Wahab AM, Ludin ANM (2018) Flood vulnerability assessment using artificial neural networks in Muar region, Johor Malaysia. In: IOP conference series: earth and environmental science, vol 169(1). IOP Publishing, Bristol, p 012056. https://doi.org/10.1088/1755-1315/169/1/012056

Wang L, Sawada K, Moriguchi S (2013) Computers & geosciences landslide susceptibility analysis with logistic regression model based on FCM sampling strategy. Comput Geosci 57:81–92. https://doi.org/10.1016/j.cageo.2013.04.006

Wang Z, Wu J, Cheng L, Liu K, Wei YM (2018) Regional flood risk assessment via coupled fuzzy c-means clustering methods: an empirical analysis from China's Huaihe River basin. Nat Hazards 93(2):803–822. https://doi.org/10.1007/s11069-018-3325-9

Wang Q, Guo Y, Yu L, Li P (2020) Earthquake prediction based on Spatio-temporal data mining: an LSTM network approach. IEEE Trans Emerg Top Comput 8(1):148–158. https://doi.org/10.1109/TETC.2017.2699169

Whitehead MG, Bebbington MS (2021) Method selection in short-term eruption forecasting. J Volcanol Geotherm Res 419:107386. https://doi.org/10.1016/j.jvolgeores.2021.107386
WHO (2002) Disasters and emergencies: training package
Xiao J, Zhou Z (2020) Research progress of RNN language model. In: Proceedings of 2020 IEEE international conference on artificial intelligence and computer applications, ICAICA 2020, pp 1285–1288. https://doi.org/10.1109/ICAICA50127.2020.9182390
Xiong P, Tong L, Zhang K, Shen X, Battiston R, Ouzounov D, Iuppa R, Crookes D, Long C, Zhou H (2021) Towards advancing the earthquake forecasting by machine learning of satellite data. Sci Total Environ 771:145256. https://doi.org/10.1016/j.scitotenv.2021.145256
Yu L, Wang N, Meng X (2005) Real-time forest fire detection with wireless sensor networks. In: Proceedings–2005 international conference on wireless communications, networking and mobile computing, WCNM 2005, vol 2, pp 1214–1217. https://doi.org/10.1109/wcnm.2005.1544272
Yu PS, Yang TC, Chen SY, Kuo CM, Tseng HW (2017) Comparison of random forests and support vector machine for real-time radar-derived rainfall forecasting. J Hydrol 552:92–104. https://doi.org/10.1016/j.jhydrol.2017.06.020
Yuan C, Moayedi H (2019) Evaluation and comparison of the advanced metaheuristic and conventional machine learning methods for the prediction of landslide occurrence. Eng Comput 36:1801. https://doi.org/10.1007/s00366-019-00798-x
Zare M, Pourghasemi HR (2013) Landslide susceptibility mapping at Vaz watershed (Iran) using an artificial neural network model: a comparison between multilayer perceptron (MLP) and radial basic function (RBF) algorithms. Arab J Geosci 6:2873–2888. https://doi.org/10.1007/s12517-012-0610-x
Zhang Y, Hao Z, Feng S, Zhang X, Xu Y, Hao F (2021) Agricultural drought prediction in China based on drought propagation and large-scale drivers. Agric Water Manag 255:107028. https://doi.org/10.1016/j.agwat.2021.107028
Zhao G, Pang B, Xu Z, Peng D, Zuo D (2020) Urban flood susceptibility assessment based on convolutional neural networks. J Hydrol 590:125235. https://doi.org/10.1016/j.jhydrol.2020.125235
Zhou C, Yin K, Cao Y, Ahmed B, Li Y, Catani F, Pourghasemi HR (2017) Landslide susceptibility modeling applying machine learning methods: a case study from Longju in the three gorges reservoir area, China. Comput Geosci 112:23. https://doi.org/10.1016/j.cageo.2017.11.019

Disaster Risk Reduction Through Agricultural Engineering Technologies

Muhammad Asif, Muhammad Sultan, Zahid M. Khan, Shakeel Ahmad, Muhammad U. Khan, Md Shamim Ahamed, and Redmond R. Shamshiri

Abstract

Disasters in agriculture are inevitable; therefore, a disaster risk reduction plan is essential to reduce the hazards. This study aimed to comprehend the fundamentals of agricultural engineering sciences and associated technologies for disaster risk reduction in agriculture. Therefore, the study explored the disasters in agriculture from the viewpoints of temperature, storms, floods and droughts, earthquakes, land sliding, insect infestation, etc. In addition, the study presents a brief discussion regarding the pandemic effect on agriculture-related activities and disasters. Consequently, disaster risk reduction emerging technologies are discussed from the aspects of agriculture engineering and associated subdisciplines. Furthermore, the impact of various disasters on different regions of the world is studied. The frequency of these disasters, i.e., geophysical, hydrological, clima-

M. Asif · M. Sultan (✉) · Z. M. Khan
Department of Agricultural Engineering, Bahauddin Zakariya University, Multan, Pakistan
e-mail: muhammadsultan@bzu.edu.pk

S. Ahmad
Department of Agronomy, Faculty of Agricultural Sciences and Technology, Bahauddin Zakariya University, Multan, Pakistan
e-mail: shakeelahmad@bzu.edu.pk

M. U. Khan
Department of Energy Systems Engineering, Faculty of Agricultural Engineering and Technology, University of Agriculture, Faisalabad, Pakistan

M. S. Ahamed
Department of Biological and Agricultural Engineering, University of California, Davis, USA

R. R. Shamshiri
Agromechatronics, Leibniz Institute for Agricultural Engineering and Bio-economy, Potsdam, Germany

M. Ahmed, S. Ahmad (eds.), *Disaster Risk Reduction in Agriculture*, Disaster Resilience and Green Growth, https://doi.org/10.1007/978-981-99-1763-1_23

tological, and biological, are also highlighted. Climate change has increased the frequency of the disaster, i.e., the temperature may increase up to 2.5 °C by 2040. Throughout the literature, it has been found that flood disasters are responsible for the maximum destruction in the agricultural sector. This study concluded that emerging agriculture technologies, i.e., zero-carbon fuel to mitigate the climatic effect and green air conditioning to reduce the thermal stress in farm animals, could reduce the disaster risks effectively.

Keywords

Disasters in agriculture · Agriculture productivity · Disaster risk reduction · Planning and management

1 Introduction

Disaster is a serious hazard over a short or long period responsible for economic and environmental loss and affects the particular society (WHO/EHA 2002; Alrahelati and Aljawad 2017). Disasters can be classified as natural disasters and manmade/anthropogenic disasters. However, now disasters are classified as natural, manmade, and man-accelerated disasters (Gould et al. 2016; Smith 2006). Anthropogenic disasters can be controlled but natural disasters cannot be controlled. Generally, classification of the natural disaster, i.e., flood, landslide, volcano, wildfire, tornado, Tsunami. Climate change has been a major concern in the past few decades and one of the major reasons for the disaster. CO_2 gas emission leads to major changes in the climate. Disaster events occur due to climate change and fluctuation in climate patterns; climate change refers to weather disasters. Approximately all disasters frequency as well as intensity has been increased. It includes extreme temperature, precipitation, wind patterns, etc. Globally, certain regions are more sensitive to disasters as per their geographical location. Figure 1 shows the disaster-prone regions across the world. The disaster-prone risk index varies from 0 to 100, and Oceania is the most disaster-prone continent, whereas Pakistan lies in the 5.8 world risk index (Aleksandrova et al. 2020; Florian Zandt 2021).

In the broader sense, natural disaster is classified into six groups geophysical, hydrological, metrological, climatological, biological, and extraterrestrial (FAO 2017, 2021; CRED and UNDRR 2020; CRED 2021). Figure 2 shows that the frequency of geophysical, hydrological, climate and weather, and biological disaster events is increasing due to anthropogenic activities. Hydrological disasters are the most frequent event in the last few decades, which majorly impact agriculture. The higher frequency of disasters leads to maximum predisaster risk management.

The agricultural sector is majorly affected by disasters that can be either direct or indirect. Financial uncertainties and social disorder are also associated with the disaster (Khan et al. 2021a). South Asia is one of the most vulnerable regions because of its dependencies on agriculture and natural resources (Field and Barros 2014; Eckstein et al. 2019). In South Asia, climate risks have threatened food security and livelihood due to its dependence on agriculturally based economics.

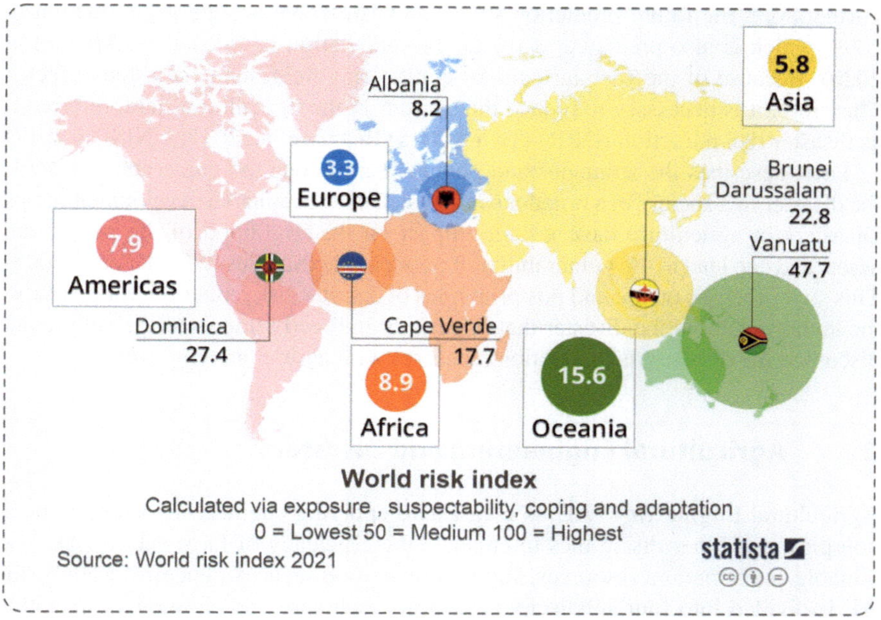

Fig. 1 Prone disaster places across the world (Florian Zandt 2021)

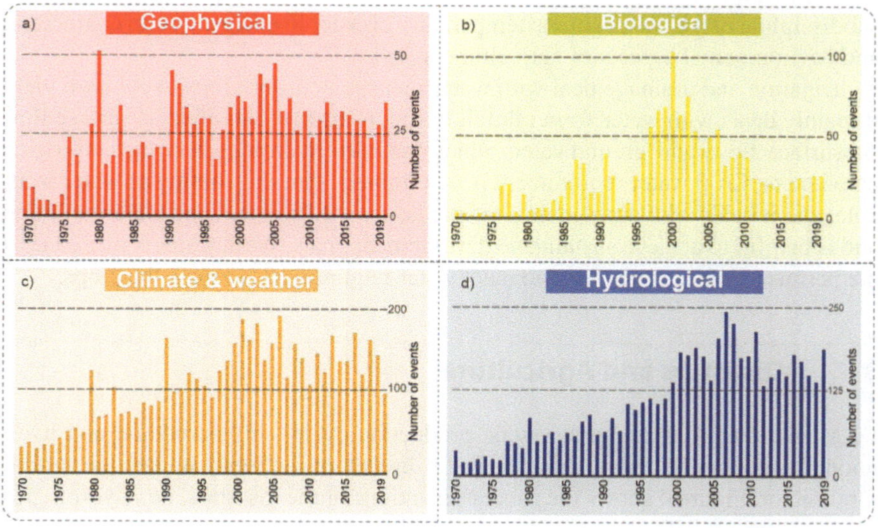

Fig. 2 Frequency of disaster events (**a**) geophysical, (**b**) biological, (**c**) climate and weather, and (**d**) hydrological, reproduced from (FAO 2021)

Furthermore, the future prediction shows a considerable decline in production by 2099, which is also predicted as the double population in the region (Aryal et al. 2020). Because of the vast agricultural engineering field, many disasters affect it. Therefore, it is necessary to reduce the disaster in the agricultural field, referred to as disaster risk reduction (DRR) (Twigg 2015; Řezník et al. 2017; UNISDR 2017).

DRR describes the strategies and policies that reduce the vulnerable impact of the disaster in a society in a broader sense to achieve sustainable development goals. Disasters in agriculture have a huge impact on the livelihood of the people and assets and can lead to the vulnerability of poor people (Sabates-Wheeler et al. 2008). This study focuses on pre and postprinciples of disaster risk reduction. Furthermore, the author explored the disaster that directly or indirectly impacted agriculture and discussed the disaster risk reduction measures with agriculture engineering.

2 Agricultural Engineering and Disasters

Agricultural Engineering (AE) is a vast field applying engineering sciences, technologies, and other disciplines to enhance the efficiency of farm enterprises, sustainable use of natural resources, and minimize food security concerns. Mainly, the AE is divided into four subdisciplines: farm machinery and equipment, irrigation and drainage systems, energy systems, and environment-related aspects of the agricultural systems. Discipline defining agricultural engineering varies from region to region, making it more complex. Figure 3 shows the disaster with the aspect of subdiscipline of AE, i.e., extreme temperature disaster in energy system engineering and environmental aspects of agricultural systems.

Irrigation and drainage deal with water management for crops to enhance yield. It mainly deals with water from rainfall to runoff, further classified as surface flow, subsurface flow, and groundwater, which includes hydrology and meteorological parameters. Crop water requirement is determined from the volume of water available, and it deals with agro-environmental issues, i.e., soil, air, and water pollution, and is helpful to achieve sustainable development goals. Wastewater is treated up to the permissible limits and used to meet water requirements to raise the crops.

3 Disasters and Agriculture

Disaster destroys agricultural assets, production cycles, infrastructure, and livelihood. As mentioned above, these disasters directly or indirectly impact the agricultural sector. Figure 4 shows the subclassifications of the disasters, i.e., hydrological disasters are classified as flood, landside and wave action; metrological disasters includes storm, extreme temperature, and fog; geophysical disaster categorized as earthquake, mass movement and volcanic activity; climatological disaster comprises on drought, glacial lake outburst, and wildfire; biological disasters incorporates animal accident, epidemic, and insect infestation; and extraterrestrial disaster includes space weather. Majorly the agricultural sector is affected by

Fig. 3 Disasters and subdisciplines of agricultural engineering

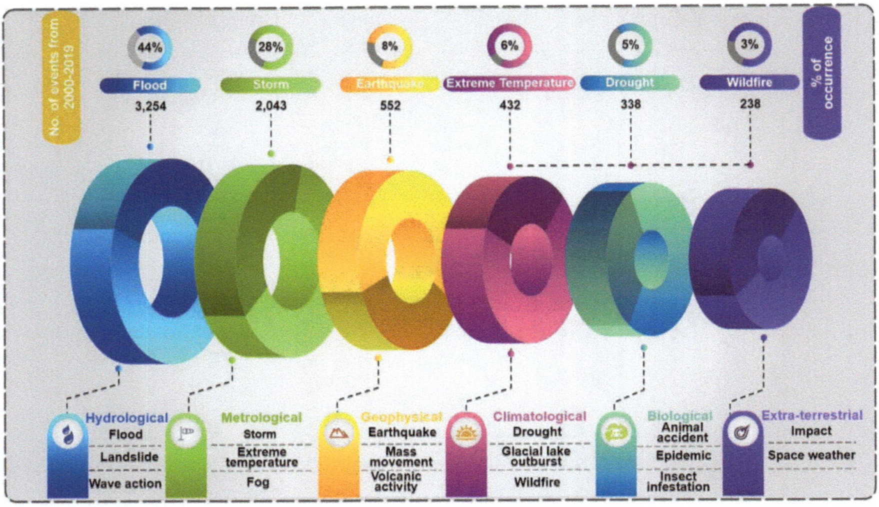

Fig. 4 Natural classification of disaster and percentage of disaster occurrence, this image is reproduced from the report on human cost of disasters (CRED and UNDRR 2020)

hydro-metrological disasters and climate change. The disaster directly impacts production. Furthermore, flood and hydrological and metrological disasters have the maximum chance of occurrence in the last few decades, i.e., flood occurrence frequency is 44%. Therefore, these disasters severely impacted the agricultural sector, as discussed below.

3.1 Temperature

Extreme temperature causes severe damage to agriculture. Extreme temperature refers to the hot temperature as well as the cold temperature. Extreme temperature also leads to global warming. As per the reports, it is stated that by 2050 the earth's temperature will be increased from 1.5 to 2.5 (Wigley et al. 1981; Krinner et al. 2013; Zhang et al. 2018; Bastin et al. 2019). Figure 5 represents the temperature prediction of the world up to 2050 caused by anthropogenic activities. Furthermore, it shows the fluctuations of the temperature from 1960 to 2050 and predicts the temperature for upcoming years (Wigley et al. 1981; Krinner et al. 2013; Zhang et al. 2018; Bastin et al. 2019). The subdisciplines of agricultural engineering are affected by the temperature. The major concerns of temperature in the agricultural engineering subdiscipline are energy consumption is increased as the temperature rises. As we move toward the serve weather condition, energy consumption increases. The main energy consumption in the agricultural sector is to enhance the shelf life of the products, achieve the thermal comfort of the poultry, livestock, and other farm animals using various air conditioning approaches. Because of the increase in temperature, it is difficult to achieve thermal comfort. It needs more

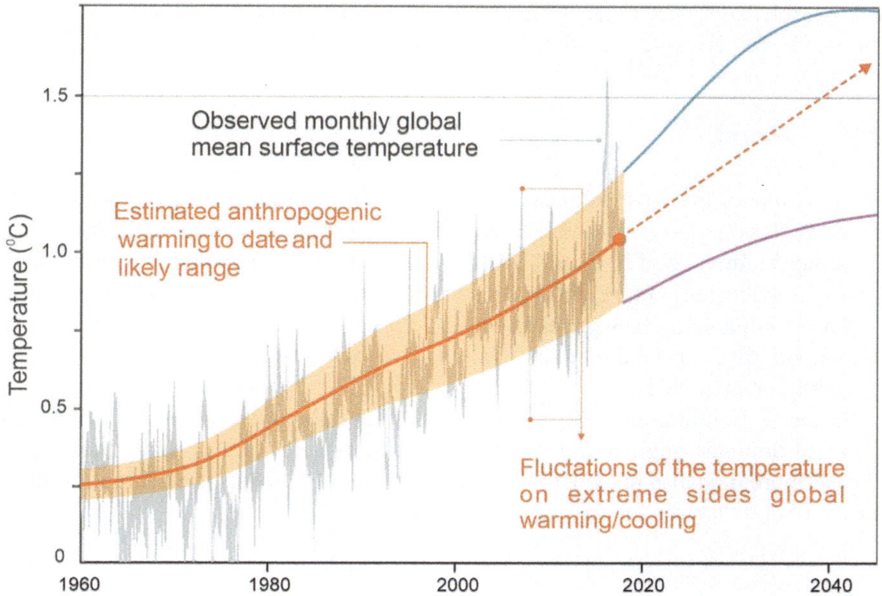

Fig. 5 Extreme temperature fluctuations and prediction of global temperature change (Wigley et al. 1981)

power as well as more resources and ultimately requires more capital cost. The temperature rises are also responsible for the glacier melts, changes in cyclone patterns, etc. The excessive glacier melts cause the flood. Somehow, it is also responsible for soil and water erosion.

3.2 Storm

A storm is an acute environmental disturbance affecting the earth's surface with heavy winds, thunderstorms, sandstorms, hail, ice, sleet, etc. It is the second most frequent disaster (NSDC 2020). There are different classifications of the storm based on the reason how they occur. Furthermore, these storms occur in every region with different climatic parameters. However, there are number of factors that play a significant role in storm occurrence such as heat bursts (sudden rise in temperature near the earth surface), moist air difference, and pressure difference. Storm severely impacts agriculture, energy, economics, and the environment.

Agriculture is one of the most affected sectors by the storms, severely impacting global food production. Sandstorm affects the photosynthesis process of the crops, and windstorm affects different fruits that fall and are wasted, i.e., mango (Stefanski and Sivakumar 2016). Storm also causes an imbalance in the spraying, referred to as spray drift, that is responsible for the health hazards (Lamela 2005; Nordgaard and Correll 2018; Xue et al. 2022). Storms are also responsible for soil erosion and reduce the fertility of agricultural land (Skidmore 1986).

3.3 Flood

Flood is one of the most frequent disasters with maximum chances of occurrence. Flood is classified as flash flood, coastal flood, storm surge, catastrophic and inland flooding. Mainly, flood is caused by heavy rainfall, rapid deglaciation, storm, etc. It deeply impacts agriculture, infrastructure, etc. Globally, flood is ranked as the second most influencing factor that affects crop production and livestock (FAO 2021). The flood disaster is the main challenge to achieving sustainable agriculture (Beddington et al. 2012).

In the agricultural engineering domain, which falls under the domain of irrigation and drainage engineering, water erosion ruins the swath of agricultural land. Furthermore, it is also responsible for waterlogging. It impacts the hydraulics infrastructure and disturbs the irrigation periods of crops.

3.4 Drought

A drought is a prolonged dry period in which a region or area receives below-normal precipitation. It is a slow-moving disaster marked by a lack of rainfall, resulting in water scarcity (Fathi-Taperasht et al. 2022). Moreover, droughts occur in almost all climatic zones, including both high and low-rainfall regions. There are various factors play an important role in the occurrence of droughts such as high temperature, low relative humidity, high winds, distribution of rainy days during crop growing season, duration and intensity of rainfall (Mishra and Singh 2010). Droughts seriously impact the environment, energy, economies, health, and agriculture.

Agriculture is always the first most affected sector by drought, one of the major natural disasters affecting global food production (Dilley 2005; Narasimhan and Srinivasan 2005; Helmer and Hilhorst 2006). In addition, the drought events have severely affected the crop growth process, livestock, and crop yield, creating an imbalance between grain production and consumption. This imbalance is the crucial factor that directly influences world food security (Tubiello et al. 2007). Moreover, droughts affect an estimated 55 million people worldwide each year, and they are the most serious threat to animals and crops in practically every corner of the world. Forty percent of the world's population has been affected by water scarcity, and up to 700 million people may be displaced because of drought by 2030. In the desert, many sources of fresh water such as lakes, rivers, and oases were reduced due to drought. Figure 6a shows perennial and seasonal water variations scare different countries' urban populations. Figure 6b shows the water scarcity issue in Pakistan, which is alarming and just a step up from the drought disaster. Figure 6c shows drought variation worldwide, and it is quite clear that the world is facing a

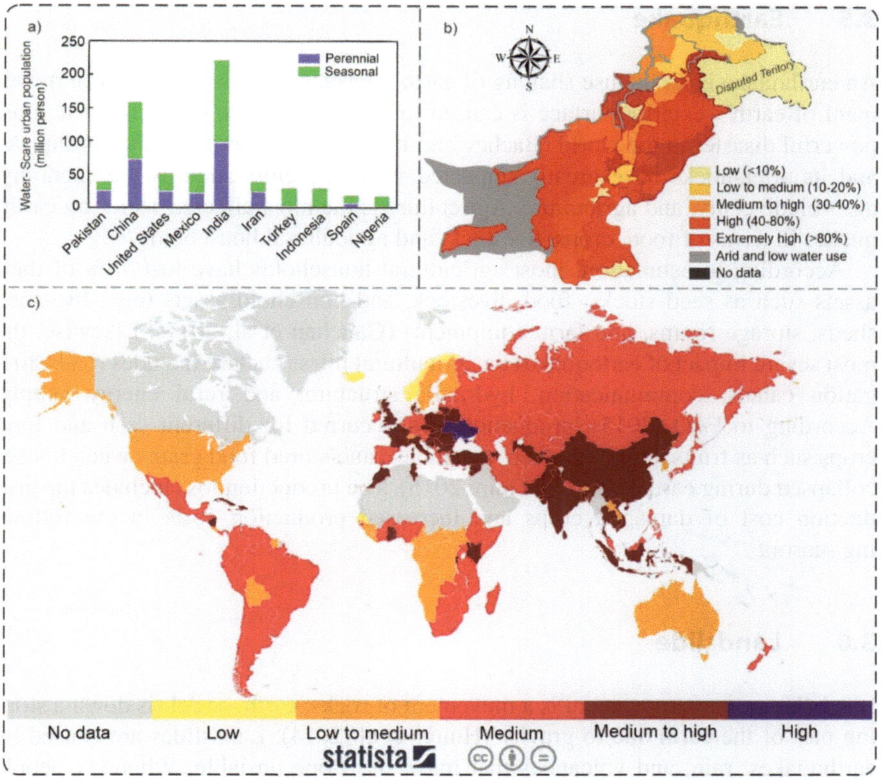

Fig. 6 (**a**) Seasonal and perennial water scare urban population (Bryan 2021); (**b**) water scarcity in Pakistan (Gassert et al. 2014); and (**c**) drought risk variation across the world (Katharina 2021)

severe drought situation (Gassert et al. 2014; Chakrabarti and Sen 2018; Wikipedia 2019; Bryan 2021; Katharina 2021).

An oasis is a tiny area in the desert where freshwater rises to the surface from deep underground (Li et al. 2016). It is a small area of greenery surrounded by desert. To keep the desert sands away from their sensitive crops and water, communities have traditionally planted strong trees, such as palms, along the edge of the oasis. Oases (more than one oasis) are irrigated by underground water resources and other natural channels (Cherif et al. 2015). Common oasis crops include dates, olives, cotton, citrus fruit, corn, and wheat. The oases in every desert of the world have become an important stop for humans, animals, and trades route. Traveling merchants and traders must stop in oases to refill their food and water supplies.

Many oases have decreased around the world due to an increase in the period of drought. The decline of oases has several causes, including the overexploitation of groundwater and neglect (Potchter et al. 2008). Furthermore, the high temperature, low humidity, and lack of precipitation throughout the year are common factors to reduce the number of oases.

3.5 Earthquake

An earthquake is an intense shaking of earth's surface (Shamsie 2012). The movement of earth's exterior surface is caused to shake the whole earth. It is the most powerful disaster in the world (Bachev and Ito 2017). Furthermore, the earthquake and its aftershocks have greatly impacted various sectors such as the economy, agro-biodiversity, and agriculture. Agriculture is the most affected sector by earthquakes in terms of food, crops, livestock, and agricultural households.

According to estimation, most agricultural households have lost 70% of their assets such as seed stocks, food, livestock, and household assets (e.g., livestock sheds, storage rooms, and farm equipment) (Gauchan et al. 2017). Likewise, the most severe impact of earthquakes on agricultural infrastructure includes roads, irrigation canals, communication, hydraulic structure, and rural energy supply. According to FAO (2015), production loss occurred for different cash and food crops such as fruits, animal fodder, vegetables, and stored food grains when houses collapsed during earthquake (McGuire 2015). The production loss includes the production cost of damaged crops and increased production costs in the following season.

3.6 Landslide

Landslide or mass movement is a movement of rocks, earth, or debris down a sloping part of the earth due to gravity (Hungr et al. 2014). Landslides are caused by earthquakes, rain, and volcanoes that make the slope unstable. Wherever people derive their livelihood from agriculture in steep lands and mountains, the soil's productivity is likely to be affected by erosion due to mass movements. Similarly, rock and soil properties are strongly affected by slope fall and are likely to increase with specific soil composition, geological origin, and disturb irrigation practices (Shanmugam and Wang 2015). As a result, the crop production yield is decreased due to changing all factors due to mass movement. The impact of landslides on agriculture farms has been studied in New Zealand, where the 20% production loss at the farm scale and 80% loss at the field scale have been measured (Blaschke et al. 2000). Another study in North Pacific America showed the 20–30% decline in forest wood volume due to mass movement (Garcia-Chevesich et al. 2020).

3.7 Insect Infestation

The presence of one or more pest species available in high numbers in an area or site, their impact on agriculture is potentially too great to tolerate (Rahaman et al. 2020). The insects were seen flying skies known as Tiddi Dal. They directly affect agricultural food production by boring through the stems, leaves, or roots, chewing crop plant leaves, spreading plant infections, and sucking out plant juices (Pener and Simpson 2009). They feed on natural plant fibers, damage wooden building materials, contaminate the stored grain, and hasten deterioration. This is the most dangerous disaster because when more insects go to the field, they do not save anything, not even a single leaf. Moreover, they consume or destroy around 10–25% gross national product in some developing countries (Rahaman et al. 2020). They can create food insecurity in the country if we do not control it properly.

3.8 Pandemics

The pandemic creates unusual conditions worldwide and affects the agricultural sector and food supply chain. The COVID-19 epidemic impacted agricultural labor. Labor mobility was restricted, which caused manpower scarcity, reduced mass production efficiency, the harvesting efficiency was 50%, and the demand gap increased (Pu and Zhong 2020; Tamru et al. 2020; Roubík et al. 2022). Recently, in Pakistan, lumpy skin disease has affected the livestock animals and caused a great loss in the economy (Tuppurainen and Oura 2012; El-Neweshy et al. 2013; Khan et al. 2021b). Approximately five million dairy animals are hit by lumpy skin disease. Burying the animals is also a laborious activity by doing excavation. So, the pandemic also affects livestock, poultry, and other farm animals. Pandemics create supply chain issues, economic loss, and agriculture loss.

4 Disaster Risk Reductions

Disasters are unavoidable; only their impact can be reduced to some extent, referred to as disaster risk reduction (WMO 2021). Disasters have also benefited and maintained the natural reserves and ecological balance, i.e., groundwater recharge during the flood (FAO 2018). As mentioned above, disasters are unavoidable, so it is necessary to take specific disaster risk reduction measures. Figure 7 shows the four principles to reduce the impact of the disaster. Postdisaster risk management includes the first three principles of the DRR cycle: response, rehabilitation and recovery, and prevention and mitigation. Predisaster risk management incorporates the fourth principle of the DRR cycle: preparedness for upcoming/future disasters.

Globally, some countries have already taken serious caution to meet the disaster risks. Referring Fig. 1 disaster-prone, it is quite clear and obvious that disaster risk management is only followed by few countries, however DRR remains an underappreciated field. Figure 8a shows the pictorial representation of the DRR progress across the world. Figure 8b shows the graphical representation of the DRR progress that varies from 0 to 5 where 0 for worst and 5 for best DRR management progress. Though different countries already have departments to tackle the hazardous disaster, it still needs serious attention to reduce its impact. Disaster risk reduction is required for agriculture because it is one of the most affected sectors. In AE, many emerging sciences and technologies could potentially reduce the disaster risk in the agricultural sector. As per the domain of agricultural engineering, the following sciences a technology will be able to reduce the disaster risk.

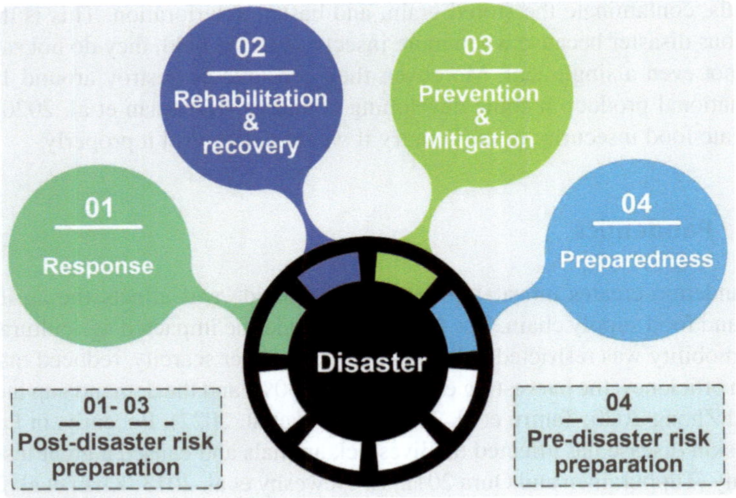

Fig. 7 Principal cycle of pre and postdisaster risk reduction

The main reason for the temperature disaster is climate change is CO_2 production influences the temperature rise; therefore, temperature could be reduced using net zero-carbon fuels that could be easily produced from biofuels, i.e., biogas, biodiesel, and bioethanol (Khan et al. 2022). However, extreme temperature causes more disastrous nuisance happening in the agricultural sector. It is hard to achieve thermal comfort for poultry, livestock, etc. Thermal comfort for poultry and livestock can be achieved using air conditioning methods. Green air conditioning systems are direct, indirect, Maisotsenko cycle, desiccant air conditioning that produces zero HFCs gases (Sultan et al. 2015a; Mahmood et al. 2016; Sajjad et al. 2021). The different integration of air-green conditioning systems is employed and tested for poultry (Aleem et al. 2020, 2022; Shahzad et al. 2021) and livestock (Ullah et al. 2022) under different climatic conditions. Extreme temperature affects the crops, so growing crops in a controlled environment is also a sustainable option. Greenhouses could be a substitute for growing crops. A controlled environment is provided in the greenhouse by green air conditioning systems to grow the crops and has better flourishment and product quality (Sultan et al. 2015b, 2016; Amani et al. 2020).

Flood disaster risk can be reduced by strengthening the existing hydraulic structure, i.e., dams and barrages, increasing their storage capacity, and constructing a new reservoir's structure. Furthermore, enhancing the number and capacity of hydraulic structures will further eradicate barren land. The other flood risk measure is to divert water in the canals and store water in barren lands by excavating ponds. Storm disaster risk can be reduced by planting trees on the sides of any orchard or crop to act as a windbreaker. The proper forecasting of the storm will lead the farmer to prepare for the disaster. Proper timing of harvesting, spraying, and intercultural operations will lead to proper management before the disaster.

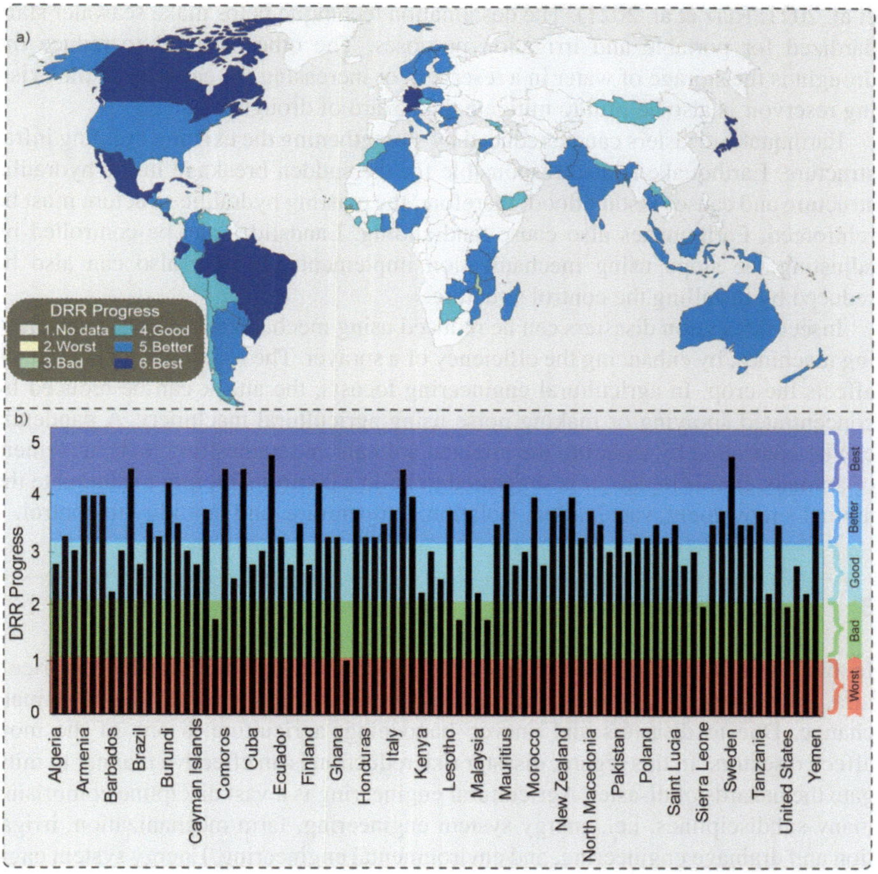

Fig. 8 Disaster risk reduction progress across the world (**a**) pictorial representation, and (**b**) graphical representation, reproduced from (DRR-Progress 2022)

Drought disaster risk can be reduced using emerging technologies that comprise atmospheric water harvesting and desalination. Atmosphere water harvesting collects the water from the atmosphere. Atmosphere water harvesting is classified as artificial rainwater collection, fog collection, and condensation collection (Klemm et al. 2012; Fessehaye et al. 2014; DeFelice and Axisa 2017; Wang et al. 2018; Xianming et al. 2022). Adsorption-based atmospheric water harvesting could be a sustainable option for harvesting the water using different sorbents (Sultan 2021; Bilal et al. 2022; Wasti et al. 2022). Atmospheric water harvesting is also a sustainable option to provide water in the oasis desert to provide the water. 97.5% of water is available in the sea, so in a drought disaster scenario, it can eradicate the severe effect of drought (Gleick et al. 2009). Desalination is the process of removing excess salt and other minerals from seawater. Adsorption-based desalination uses to provide water and colling as a byproduct (Ali et al. 2021; Ashraf et al. 2021; Mohammed et al. 2021; Riaz et al. 2021). The desalination technique helps make seawater standardized for portable and irrigation purposes. The other method to reduce the drought is the storage of water in a reservoir, or increasing the capacity of the existing reservoir is also helpful to mitigate the hazard of drought.

Earthquake disasters can be reduced by strengthening the existing building infrastructure. Earthquake is also responsible for the sudden breakage in the hydraulic structure and causes instant flood; therefore, the existing hydraulic structure must be reinforced. Earthquakes also cause land sliding. Landslides can be controlled by adjusting the slope using mechanization implements. Its risk also can also be reduced by installing the control structure.

Insect infestation disasters can be reduced using mechanization tools, i.e., spraying machines, by enhancing the efficiency of a sprayer. The locust (Teddi dal) attack affects the crop. In agricultural engineering locusts, the attack can be reduced by concentrated spraying or making noise using agricultural machinery. A pandemic can be controlled by isolating the affected animals and controlling their movement of animals. In this regard, it is suggested to build a control structure to automate the animal's movement, vaccination, isolation, temperature, and humidity to control.

5 Conclusions

In this study, the authors explored the different disasters and their impact on agriculture. In the past few decades, the disaster frequency has increased due to climate change. Due to disasters and uneven happening, agriculture is one of the most affected sectors. In this regard, disaster risk reduction is an effective manner to mitigate the hazards of disaster. Agricultural engineering is a vast discipline comprising many subdisciplines, i.e., energy system engineering, farm mechanization, irrigation and drainage engineering, and environmental engineering. Energy system engineering deals with on-farm and off-farm energy generation and consumption, farm mechanization deals with the on-farm and off-farm agricultural machinery, irrigation, and drainage engineering deals with the precision application of water to meet the crop water requirement, environmental engineering deals with the sustainable environment for the crop production. Disasters that directly or indirectly impact agriculture are, i.e. temperature, storm, drought, earthquake, land sliding, and insect infestation. Temperature disaster is responsible for thermal stress in farm animals, which increases energy consumption. It also boosts other disasters, i.e., sudden deglaciation due to high temperatures. Flood disaster ruins the swath/fertile soil, crops, and hydraulic structure. Drought disaster causes the crop growth process, livestock, and other farm animals. Earthquake is responsible for affecting the agricultural infrastructure and crops. Land sliding causes soil erosion, chokes the waterbodies, causes insect infestation triggers to decrease the crop yield, pandemics affect the farm animals, supply chain issues, etc. The authors concluded that emerging agriculture engineering sciences and technologies are capable and sustainable ways to reduce disaster risk. The climate change effect can be reduced using net zero-carbon fuel and precludes CO_2 generation. Temperature risk can be reduced by using green air conditioning to avoid the thermal stress in animal farming, and certain crops need to grow in control environment, i.e., a greenhouse minimizes the temperature risk. Drought risk could be reduced using atmosphere water harvesting, desalination technologies, and increasing the capacity of hydraulic structures. Insect infestation risk can be reduced by spraying with an efficient sprayer, whereas pandemic risk can be managed by building a controlled structure to isolate the affected farm animals.

References

Aleem M, Hussain G, Sultan M, Miyazaki T, Mahmood MH, Sabir MI, Nasir A, Shabir F, Khan ZM (2020) Experimental investigation of desiccant dehumidification cooling system for climatic conditions of Multan (Pakistan). Energies 13(21):5530. https://doi.org/10.3390/en13215530

Aleem M, Sultan M, Mahmood MH, Miyazaki T (2022) Desiccant dehumidification cooling system for poultry houses in Multan (Pakistan). In: Sultan M, Miyazaki T (eds) Energy-efficient systems for agricultural applications. Springer International, Cham, pp 19–42

Aleksandrova M, Balasko S, Kaltenborn M, Malerba D, Mucke P, Neuschafter O, Radtke K, Prutz R, Strupat C, Weller D, Wiebe N (2020) World risk report 2021 focus: social protection

Ali ES, Muhammad Asfahan H, Sultan M, Askalany AA (2021) A novel ejectors integration with two-stages adsorption desalination: away to scavenge the ambient energy. Sustain Energy Technol Assess 48:101658. https://doi.org/10.1016/j.seta.2021.101658

Alrahelati HA, Aljawad AH (2017) In the wake of the disaster-project of Iraqi scientific assets preservation (case study)

Amani M, Foroushani S, Sultan M, Bahrami M (2020) Comprehensive review on dehumidification strategies for agricultural greenhouse applications. Appl Therm Eng 181:115979. https://doi.org/10.1016/j.applthermaleng.2020.115979

Aryal JP, Sapkota TB, Khurana R, Khatri-Chhetri A, Rahut DB, Jat ML (2020) Climate change and agriculture in South Asia: adaptation options in smallholder production systems. Environ Dev Sustain 22(6):5045–5075

Ashraf S, Sultan M, Bahrami M, McCague C, Shahzad MW, Amani M, Shamshiri RR, Ali HM (2021) Recent progress on water vapor adsorption equilibrium by metal-organic frameworks for heat transformation applications. Int Commun Heat Mass Transf 124:105242. https://doi.org/10.1016/j.icheatmasstransfer.2021.105242

Bachev H, Ito F (2017) Agricultural impacts of the great east Japan earthquake–six years later. AVAILABLE SSRN 2977972

Bastin JF, Clark E, Elliott T, Hart S, Van Den Hoogen J, Hordijk I, Ma H, Majumder S, Manoli G, Maschler J, Mo L, Routh D, Yu K, Zohner CM, Crowther TW (2019) Correction: understanding climate change from a global analysis of city analogues (PLoS ONE (2019) 14:7 (e0217592) DOI: 10.1371/journal.Pone.0217592). PLoS One 14(10):1–13. https://doi.org/10.1371/journal.pone.0224120

Beddington JR, Asaduzzaman M, Bremauntz FA, Clark ME, Guillou M, Jahn MM, Erda L, Mamo T, Van Bo N, Nobre CA (2012) Achieving food security in the face of climate change: final report from the commission on sustainable agriculture and climate change

Bilal M, Sultan M, Morosuk T, Den W, Sajjad U, Aslam MMA, Shahzad MW, Farooq M (2022) Adsorption-based atmospheric water harvesting: a review of adsorbents and systems. Int Commun Heat Mass Transf 133:105961. https://doi.org/10.1016/j.icheatmasstransfer.2022.105961

Blaschke PM, Trustrum NA, Hicks DL (2000) Impacts of mass movement erosion on land productivity: a review. Prog Phys Geogr 24(1):21–52

Bryan BA (2021) Future global urban water scarcity and potential solutions. Nat Commun 12:4667. https://doi.org/10.1038/s41467-021-25026-3

Chakrabarti G, Sen C (2018) The globalization conundrum—dark clouds behind the silver lining: global issues and empirics. Springer, Singapore

Cherif H, Marasco R, Rolli E, Ferjani R, Fusi M, Soussi A, Mapelli F, Blilou I, Borin S, Boudabous A (2015) Oasis desert farming selects environment-specific date palm root endophytic communities and cultivable bacteria that promote resistance to drought. Environ Microbiol Rep 7(4):668–678

CRED (2021) Extreme events defining our lives executive summary. https://doi.org/10.1787/eee82e6e-en

CRED, UNDRR (2020) Human cost of disasters. https://doi.org/10.18356/79b92774-en

DeFelice TP, Axisa D (2017) Modern and prospective technologies for weather modification activities: developing a framework for integrating autonomous unmanned aircraft systems. Atmos Res 193:173–183. https://doi.org/10.1016/j.atmosres.2017.04.024

Dilley M (2005) Natural disaster hotspots: a global risk analysis. World Bank, Washington, DC

DRR-Progress (2022) Our world in data. In: Disaster risk reduct progress score. https://ourworldindata.org/grapher/disaster-risk-reduction-progress

Eckstein D, Künzel V, Schäfer L, Winges M (2019) Global climate risk index 2020. Germanwatch, Bonn

El-Neweshy MS, El-Shemey TM, Youssef SA (2013) Pathologic and immunohistochemical findings of natural lumpy skin disease in Egyptian cattle. Pakistan Vet J 33(1):60–64

FAO (2017) The impact of disasters and crises on agriculture and food security

FAO (2018) Multiple benefits, no regrets

FAO (2021) The impact of disasters and crises on agriculture and food security: 2021

Fathi-Taperasht A, Shafizadeh-Moghadam H, Minaei M, Xu T (2022) Influence of drought duration and severity on drought recovery period for different land cover types: evaluation using MODIS-based indices. Ecol Indic 141:109146

Fessehaye M, Abdul-Wahab SA, Savage MJ, Kohler T, Gherezghiher T, Hurni H (2014) Fog-water collection for community use. Renew Sustain Energy Rev 29:52–62. https://doi.org/10.1016/j.rser.2013.08.063

Field CB, Barros VR (2014) Climate change 2014–impacts, adaptation and vulnerability: regional aspects. Cambridge University Press, Cambridge

Garcia-Chevesich P, Wei X, Ticona J, Martínez G, Zea J, García V, Alejo F, Zhang Y, Flamme H, Graber A (2020) The impact of agricultural irrigation on landslide triggering: a review from Chinese, English, and Spanish literature. Water 13(1):10

Gassert F, Landis M, Luck M (2014) Baseline water stress map -Pakistan. https://geospatialwarehouse.wordpress.com/2016/02/05/baseline-water-stress-map-pakistan/

Gauchan D, Joshi BK, Ghimire KH (2017) Impact of 2015 earthquake on economy, agriculture and agrobiodiversity in Nepal. Rebuilding local seed system of native crops in earthquake affected areas in Nepal. Proceedings of National Sharingshop

Gleick PH, Cooley H, Morikawa M, Morrison J, Cohen MJ (2009) The world's water 2008–2009: the biennial report on freshwater resources. Island Press, Washington, DC

Gould KA, Garcia MM, Remes JAC (2016) Beyond "natural-disasters-are-not-natural": the work of state and nature after the 2010 earthquake in Chile. J Polit Ecol 23(1):93–114

Helmer M, Hilhorst D (2006) Natural disasters and climate change. Disasters 30(1):1–4

Hungr O, Leroueil S, Picarelli L (2014) The varnes classification of landslide types, an update. Landslides 11(2):167–194

Katharina Buchholz (2021) The world map of drought risk. https://www.statista.com/chart/25101/countries-by-drought-risk/

Khan NA, Gao Q, Abid M, Shah AA (2021a) Mapping farmers' vulnerability to climate change and its induced hazards: evidence from the rice-growing zones of Punjab, Pakistan. Environ Sci Pollut Res 28(4):4229–4244

Khan YR, Ali A, Hussain K, Ijaz M, Rabbani AH, Khan RL, Abbas SN, Aziz MU, Ghaffar A, Sajid HA (2021b) A review: surveillance of lumpy skin disease (LSD) a growing problem in Asia. Microb Pathog 158:105050

Khan MU, Rehman MAU, Sultan M, Rehman T, Sajjad U, Yousaf M, Ali HM, Bashir MA, Akram MW, Ahmad M, Asif M (2022) Key prospects and major development of hydrogen and bioethanol production. Int J Hydrogen Energy 47(62):26265–26283. https://doi.org/10.1016/j.ijhydene.2022.06.224

Klemm O, Schemenauer RS, Lummerich A, Cereceda P, Marzol V, Corell D, van Heerden J, Reinhard D, Gherezghiher T, Olivier J, Osses P, Sarsour J, Frost E, Estrela MJ, Valiente JA, Fessehaye GM (2012) Fog as a fresh-water resource: overview and perspectives. Ambio 41(3):221–234. https://doi.org/10.1007/s13280-012-0247-8

Krinner G, Germany F, Shongwe M, Africa S, France SB, Uk BBBB, Germany VB, Uk OB, France CB, Uk RC, Canada ME, Erich M, Uk RWL, Uk SL, Lucas C (2013) Long-term climate change: projections, commitments and irreversibility. Clim Chang 9781107057:1029–1136. https://doi.org/10.1017/CBO9781107415324.024. 2013 Phys Sci basis work gr I Contrib to fifth assess rep Intergov panel Clim Chang

Lamela N (2005) Pesticide drift. Tar Heel Nurse 67(1):23

Li X, Yang K, Zhou Y (2016) Progress in the study of oasis-desert interactions. Agric For Meteorol 230:1–7

Mahmood MH, Sultan M, Miyazaki T, Koyama S, Maisotsenko VS (2016) Overview of the maisotsenko cycle—a way towards dew point evaporative cooling. Renew Sustain Energy Rev 66:537–555. https://doi.org/10.1016/j.rser.2016.08.022

McGuire S (2015) FAO, IFAD, and WFP. The state of food insecurity in the world 2015: meeting the 2015 international hunger targets: taking stock of uneven progress. Rome: FAO, 2015. Adv Nutr 6(5):623–624

Mishra AK, Singh VP (2010) A review of drought concepts. J Hydrol 391(1–2):202–216

Mohammed RH, Rezk A, Askalany A, Ali ES, Zohir AE, Sultan M, Ghazy M, Abdelkareem MA, Olabi AG (2021) Metal-organic frameworks in cooling and water desalination: synthesis and application. Renew Sustain Energy Rev 149:111362. https://doi.org/10.1016/j.rser.2021.111362

Narasimhan B, Srinivasan R (2005) Development and evaluation of soil moisture deficit index (SMDI) and evapotranspiration deficit index (ETDI) for agricultural drought monitoring. Agric For Meteorol 133(1–4):69–88

Nordgaard A, Correll R (2018) Chapter 3—sampling strategies. In: Maestroni B, A C (eds) IAA for PM. Academic Press, Cambridge, MA, pp 31–46. https://doi.org/10.1016/C2017-0-00162-3

NSDC (2020) National storm damage center (types of storms). https://www.stormdamagecenter.org/common-storm-types/

Pener MP, Simpson SJ (2009) Locust phase polyphenism: an update. Adv In Insect Phys 36:1–272

Potchter O, Goldman D, Kadish D, Iluz D (2008) The oasis effect in an extremely hot and arid climate: the case of southern Israel. J Arid Environ 72(9):1721–1733

Pu M, Zhong Y (2020) Rising concerns over agricultural production as COVID-19 spreads: lessons from China. Glob Food Sec 26:100409. https://doi.org/10.1016/j.gfs.2020.100409

Rahaman M, Saha O, Rakhi NN, Chowdhury MK, Sammonds P, Kamal ASMM (2020) Overlapping of locust swarms with COVID-19 pandemic: a cascading disaster for Africa. Pathog Glob Health 114(6):285–286

Řezník T, Lukas V, Charvát K, Charvat K Jr, Křivánek Z, Kepka M, Herman L, Řezníková H (2017) Disaster risk reduction in agriculture through geospatial (big) data processing. ISPRS Int J Geo-Information 6(8):238

Riaz N, Sultan M, Miyazaki T, Shahzad MW, Farooq M, Sajjad U, Niaz Y (2021) A review of recent advances in adsorption desalination technologies. Int Commun Heat Mass Transf 128:105594. https://doi.org/10.1016/j.icheatmasstransfer.2021.105594

Roubík H, Lošťák M, Ketuama CT, Procházka P, Soukupová J, Hakl J, Karlík P, Hejcman M (2022) Current coronavirus crisis and past pandemics-what can happen in post-COVID-19 agriculture? Sustain Prod Consum 30:752–760

Sabates-Wheeler R, Devereux S, Mitchell T, Tanner T, Davies M, Leavy J (2008) Rural disaster risk–poverty interface

Sajjad U, Abbas N, Hamid K, Abbas S, Hussain I, Ammar SM, Sultan M, Ali HM, Hussain M, Tauseef-ur-Rehman, Wang CC (2021) A review of recent advances in indirect evaporative cooling technology. Int Commun Heat Mass Transf 122:105140. https://doi.org/10.1016/j.icheatmasstransfer.2021.105140

Shahzad K, Sultan M, Bilal M, Ashraf H, Farooq M, Miyazaki T, Sajjad U, Ali I, Hussain MI (2021) Experiments on energy-efficient evaporative cooling systems for poultry farm application in Multan (Pakistan). Sustain 13:2836

Shamsie Y (2012) Haiti's post-earthquake transformation: what of agriculture and rural development? Lat Am Polit Soc 54(2):133–152

Shanmugam G, Wang Y (2015) The landslide problem. J Palaeogeogr 4(2):109–166

Skidmore EL (1986) Soil erosion by wind: an overview. In: Physics of desertification, pp 261–273. https://doi.org/10.1007/978-94-009-4388-9_18

Smith N (2006) There's no such thing as a natural disaster. Underst Katrina Perspect from Soc Sci 11

Stefanski R, Sivakumar MVK (2016) Impacts of sand and dust storms on agriculture and potential agricultural applications of a SDSWS. In: IOP conference series: earth and environmental science 7(1):012016. Bristol, IOP. https://doi.org/10.1088/1755-1307/7/1/012016

Sultan M (2021) Adsorption-based atmospheric water harvesting: technology fundamentals and energy-efficient adsorbents. In: Bilal M (ed) Technology in agriculture. IntechOpen, Rijeka. p Ch. 19

Sultan M, El-Sharkawy II, Miyazaki T, Saha BB, Koyama S (2015a) An overview of solid desiccant dehumidification and air conditioning systems. Renew Sustain Energy Rev 46:16–29. https://doi.org/10.1016/j.rser.2015.02.038

Sultan M, Miyazaki T, Saha BB, Koyama S, Maisotsenko VS (2015b) Steady-state analysis on thermally driven adsorption air-conditioning system for agricultural greenhouses. Procedia Eng 118:185–192. https://doi.org/10.1016/j.proeng.2015.08.417

Sultan M, Miyazaki T, Saha BB, Koyama S (2016) Steady-state investigation of water vapor adsorption for thermally driven adsorption based greenhouse air-conditioning system. Renew Energy 86:785–795. https://doi.org/10.1016/j.renene.2015.09.015

Tamru S, Hirvonen K, Minten B (2020) Impacts of the COVID-19 crisis on coffee value chains in Ethiopia Seneshaw. COVID-19 Glob food Secur:81–83

Tubiello FN, Soussana J-F, Howden SM (2007) Crop and pasture response to climate change. Proc Natl Acad Sci 104(50):19686–19690

Tuppurainen ESM, Oura CAL (2012) Lumpy skin disease: an emerging threat to Europe, the middle east and Asia. Transbound Emerg Dis 59(1):40–48

Twigg J (2015) Disaster risk reduction- good practice review. Humanit Policy Netw 9:1–382

Ullah HS, Sultan M, Mahmood MH, Ashraf H, Ishaq M, Miyazaki T (2022) Evaporative cooling and desiccant dehumidification air conditioning options for livestock thermal comfort. In: Sultan M, Miyazaki T (eds) Energy-efficient systems for agricultural applications. Springer International, Cham, pp 43–63

UNISDR (2017) What is disaster risk reduction?

Wang G, Zhong D, Li T, Zhang Y et al (2018) Study on sky rivers: concept, theory, and implications. J Hydrol Res 21:109–117

Wasti TZ, Sultan M, Aleem M, Sajjad U, Farooq M, Raza HMU, Khan MU, Noor S (2022) An overview of solid and liquid materials for adsorption-based atmospheric water harvesting. Adv Mech Eng 14(3):1–27. https://doi.org/10.1177/16878132221082768

WHO/EHA (2002) Disasters & emergencies definitions. WHO/EHA train packag (March):1–26

Wigley TML, Jones PD, Kelly PM (1981) Global warming? Nature 291(5813):285. https://doi.org/10.1038/291285a0

Wikipedia (2019) Wikipedia. https://en.wikipedia.org/wiki/Water_scarcity

WMO (2021) Disaster risk reduction (DRR). In: WORLD metrol. Organ. https://community.wmo.int/activity-areas/drr

Xianming D, Nan S, Nielson OS, Boschitsch SB, Jing W, Shikuan Y, Tak-Sing W (2022) Hydrophilic directional slippery rough surfaces for water harvesting. Sci Adv 4(3):eaaq0919. https://doi.org/10.1126/sciadv.aaq0919

Xue S, Han J, Xi X, Zhao J, Lan Z, Wen R, Ma X (2022) Rapid velocity reduction and drift potential assessment of off-nozzle pesticide droplets. Chin J Chem Eng 46:243–254. https://doi.org/10.1016/j.cjche.2021.06.011

Florian Zandt (2021) The places most prone to disaster. https://www.statista.com/chart/25958/countries-most-at-risk-facing-natural-disasters-per-region/

Zhang K, Wang Z, Zhang P, Hua L, Fei M (2018) Research on accurate capture security strategy for high-speed visual inspection. Yi qi Yi Biao Xue Bao/Chinese J Sci Instrum 39(2):232–240. https://doi.org/10.19650/j.cnki.cjsi.j1702567

Integrated Approach for Disaster Risk Reduction in Agriculture Through Crop, Livestock, Forestry, Poultry, and Fish Farming

Muhammad Rizwan

Abstract

Climate change affects food security at local, regional and global levels. It disrupts food availability, affects food quality, and reduces access to food. Increasing temperatures, frequent extreme weather events, changes in precipitation patterns, floods, and unavailability of water may all result in declining agricultural productivity. Increases in the severity extreme weather events and frequency affect crops, livestock, forest, poultry and fish farming. Impacts on the global food supply concern the World because food shortage causes humanitarian crises and national security concerns. Moreover, global warming and more carbon dioxide in the atmosphere help some crops to grow faster. However, climate change, viz. floods, drought and warming of the atmosphere, reduces production. Livestock, poultry and fish farming could be at risk both from reduced quality of food supply indirectly and directly from heat stress. Fisheries are also affected by changes in water temperature and polluted temperature that make waters more hospitable to invasive species and shift ranges and lifecycles timing of fish species.

Keywords

Climate change · Livestock · Poultry · Fish farming · Fisheries

M. Rizwan (✉)
Veterinary Sciences, Bahauddin Zakariya University, Multan, Pakistan
e-mail: mrizwan@bzu.edu.pk

M. Ahmed, S. Ahmad (eds.), *Disaster Risk Reduction in Agriculture*, Disaster Resilience and Green Growth, https://doi.org/10.1007/978-981-99-1763-1_24

1 Introduction

1.1 Crops

Agriculture is a vital sector of the national economy. The crops, livestock, poultry and fish farming products contribute to the economy per year. When food and other agriculture-associated industries are included, it contributes more than doubled the GDP. All are highly dependent on climate. Increasing temperature and carbon dioxide can enhance crop yields in some regions, but other factors including soil moisture, fertiliser availability, and water availability must also be satisfied. Changes in the severity and frequency of floods and droughts could pose severe challenges for crops, livestock, poultry and fish farming, leading to threaten food security and safety. Meanwhile, increase in water temperature is likely to cause habitat ranges of many fish species to shift, which disrupt the ecosystem. Overall, the change in climate may cause difficulty in growing crops, raising domestic animals, poultry and fish framing. Therefore, the effect of climate change must be considered along with factors that affect agricultural and livestock, poultry and fish farming such as farming practices and technology (Backlund et al. 2008). Crops grown around the world are critical for food supply. In the US, farms supply about 25% of all grains, viz. wheat, rice, and corn, in the global market (USDA 2021). Changes in atmospheric CO_2 and temperature, and intensity and frequency of extreme weather have significant impacts on crop yield. The effect of increased temperature depends on the optimal temperature for growth and reproduction (USGCRP 2014). In some geographical location, atmospheric warming may benefit types of crops typically cultivated there or allow farmers to shift to crops currently grown in warmer areas. Conversely, if temperature becomes higher than crop optimum temperature it will lead to yield loss. Higher carbon dioxide affects crop production. Some research findings suggest that the increase in carbon dioxide can increase crop growth. While continuous changing temperature, water and nutrient constraints, change in ozone counteracts these potential increases in production. Moreover, increased carbon dioxide is associated with reduced nitrogen and protein contents in soybean and alfalfa crops, resulting in quantity and quality loss. The decline in forage and grain quality reduces the ability of rangeland and posture to support grazing livestock. More extreme temperature prevents crops from growing. For example, high night-time temperature affected the corn yield across US Corn Belt during 2010 and 2012. Dealing with drought could become a challenge in areas where the rising temperature in summer causes soils to become drier. Although increased irrigation might be possible in some places, water supplies may also be reduced, leaving less water available for irrigation when more is needed. Many pests, weeds and fungi thrive on warmer atmospheric temperature, increased carbon dioxide levels and wetter climates. The ranges and distribution of pests and weeds increase with climate change. This causes problems for crops already exposed to these species. Though the increase in carbon dioxide stimulates crops growth, it also reduces the nutritional value of most food crops. It also reduces protein concentration and essential minerals in most crops. This direct effect of the increase in carbon dioxide on the nutritional value of crops represents a serious threat to human and animal health. The health of animals and humans is also threatened by the increasing use of pesticides due to increased pest pressures and pesticide efficacy reduction.

1.2 Livestock

The animal agricultural subsector contributes about 40% of the agricultural gross domestic product (GDP) in developing countries. Livestock offers substantial opportunities for food, nutritional security (17% of the global calories and 33% of the global protein consumed) and sustainable development with appropriate adaptation to climate change. Worldly, livestock is raised by milk, meat, wool and employment. Climate change is a major threat to the sustainability of livestock production systems globally and, in particular, in developing countries. This threat will be both direct (through increased temperature, morbidity and mortality) and indirect (through reduced feed quality, quantity and increased animal diseases). Therefore, climate change has a considerable adverse impact on livestock growth rates, milk and egg production, and reproductive performance. Rural people are seriously affected by climate change because they highly dependent on livestock to survive. Meanwhile, livestock are a direct source of both nitrous oxide, methane, and an indirect source of those gases, carbon through land use and feed production. Globally, about 14.5% of greenhouse gas emissions come from livestock supply chains. Climate change affects livestock directly or indirectly. The change in climate increases the prevalence of diseases and parasites that affect livestock health and production. The earlier onset of warmer winters and spring allows pathogens and parasites to survive more easily. In flood-affected areas and in areas with increased rainfall, moisture-reliant pathogen thrive. Severe floods destroy feed supplies and pasture, resulting in the quantity and quality of forage available to the livestock (Fig. 1). Some areas experience intense and for a longer duration. Drought also threatens livestock directly or indirectly. It reduces the quality of fodder available for animals. Some geographical locations may experience longer and more

Fig. 1 Loss of livestock during flood in Pakistan

intense situation of drought, which results from the warmer summer atmospheric temperature and decline in precipitation. Therefore changes in the crops production due to severe drought become a problem for livestock. Under climate change, heat waves could threaten livestock directly. Heat stress affects livestock both indirectly and directly. Heat stress increases vulnerability to disease, reduces milk production and reduces fertility. An increase in the level of carbon dioxide may cause to increase the productivity of fodder but may also decrease their quality (Table 1). The quality of forages found in the pasturelands decreases with a high level of carbon dioxide. In consequence, the animal would need to eat more to maintain the same nutritional demands.

1.3 Fisheries

Nowadays, fish farming faces multiple problems due to water pollution that results due to climate change. In particular, change in water temperature significantly impacts on fish farming. The ranges of many fish species may change. In flood disasters, several economically important species have shifted. Some fish disease outbreaks are linked to climate change. Higher estuarine salinities and warmer temperatures have enabled an oyster parasite to spread farther north along the Atlantic coast. In the Arctic, winter warming contributes to salmon diseases in Bering Sea, reducting Yukon Chinook Salmon. Warmer temperature causes disease outbreaks. Water is gradually becoming more acidic due to the increase in atmospheric CO_2, and this increase in water acidity harms shellfish by weakening their shells, which are built by removing calcium from seawater. Moreover, acidification of water also

Table 1 Climate change impacts on livestock

Major influential factors	Observed impacts	Impact type
Increased temperature	Reduced feed intake	Direct impact
	Decline in milk and meat production	
	Decreased reproductive performances	
	Negatively affected immune system	
	Increased mortality	
Increased carbon dioxide level	Changes in feedstuff crops yields	
	Changes in pasture composition and forage production	
Increased temperature and carbon dioxide	Changes in forage quality	Indirect impact
Increased temperature	Shrinking water availability and increasing water use	
More frequent extreme climate events	Larger seasonal variation in resource availability	
Increased temperature and changes in precipitation pattern	Increased diseases, pests and parasites stress	

causes harm to structures of sensitive ecosystems. Climate changes also affect the timing of migration and reproduction. Many steps within an aquatic animal lifecycle are controlled by changing seasons and temperature. In the Northwest warmer water affects the lifecycle of salmon and increases in the occurrence of disease. Combined with other changes in climate impacts, these effects lead to a large reduction in the fish population. It is a source of employment, money and food for a huge number of human population, accounting for several million, but also an important protein source. About 50% of protein obtained from animal source is consumed in Ghana, Gambia and Sierra from fish. Over the years, small-scale fish farming has developed extraordinarily, approaching the current 90% of total fish production. For instance, in Mauritania, small-scale fish farming provides 80% of jobs in country. In recent years, fishing and aquaculture have been affected by demographic, natural and environmental changes. Moreover, this variability and climate change have changed fish supplies, while doubled population growth in the span of one generation has increased demand for fish and fish-associated products.

1.4 Forestry

Forestry, other than wood supply, also supplies many non-wood products, viz. fodder for animals, building materials and medicinal plants, food items, firewood, nutritional security and rural livelihoods to support national economies. Moreover, forest supply bush meat and consumption of bush meat is an important source of dietary protein for people living in fallow lands and nearby forests.

Trees are also an important resource, supplying food for animals during the dry season (e.g. acacia) and a source of income. According to the FAO report, Shea trees supply 600,000 tonnes of shea nuts per year, and about 4–5 M women are employed in treating Shea nuts and fat, collecting and marketing which accounts for about 80% of their income. Shea butter is used in cosmetic products and as fat in cooking. During the last three decades, forest cover was estimated to decrease by 1.2 M hectares. The forest land is receding owing to forest cover fragmentation in wetland, which result in not only extensive agriculture of cash crops but also Silviculturae, forest fires, mining and infrastructural developments. Moreover, harvesting timber for domestic purposes also has a growing impact on deforestation.

1.5 Poultry Farming

Poultry meat is a dietary source with good nutritional quality, essential micronutrients and macronutrients, and high vitamin bioavailability in poultry products, eggs and meat. Nutrients in eggs are well balanced to satisfy human nutritional needs. Poultry farming provides the cheapest source of meat, with high protein content to fulfil nutritional needs of protein, particularly in underdeveloped countries, and it is one of the most liked and utilized sources of meat: it is being eaten in every corner of the world in different forms. Instead of providing meat for human consumption, poultry farming also provides various byproducts along with meat, which is used in feed formulation and to fulfil the nutritional needs of different species of animals. Poultry is also the source of eggs. Production of poultry meat is easy and require a shorter time period as broiler gain 2 kg weight within 35 days, which is a very short period to obtain a protein meal as compared to protein sources of other species. Small-scale chicken farming is a reliable source of income for the family, particularly for impoverished landless people worldwide. Through transgenesis, alterations to the nutritional content, and processing, which need to be investigated, the value-added meat raises the return. Numerous aspects of rural life, including good nutrition, savings, food security, income and insurance, are influenced by small-scale chicken farming. In times of need, such as crop failure, drought, or illness, poultry farming offers a reliable source of income. The benefits of backyard poultry extend beyond money production and include providing nutritional security in many emerging and impoverished nations. Therefore, expanding poultry production and its value addition considerably increases the influence of poultry goods and transforms their function in ensuring the security of the world's food supply (Chatterjee et al. 2022).

2 Solutions to Disaster Reduction in Agriculture

2.1 Forests; as a Construction Material and Fire Wood

Forests are sources of high-quality timber, which can be used to construct houses for humans and shelters for animals. Portable and fixed animal feeding troughs can be made from wood, which is also useful for making different household items for human usage and different accessories used at animal farms. Dry wood, which is not valuable for construction and other important purposes, like can be used to ignite the fire. Fire from wood burning can be used to cook food and heat up rooms and animal sheds during cold weather conditions. In other terms helps in reducing disaster risk indirectly by providing food, and shelter.

2.2 Forests as Replacement to Pastures

The winter and spring seasons particularly present a challenging time period to maintain animal feeding as there is a scarcity of food due to floods and snow; this is the scenario in both eastern and western countries. This exigent epoch can be managed by shifting animals from pasture grazing to herbs grazing in forests. In many countries, animals are fed on deciduous forests canopy, which present a good approach in the period of scarcity. Animals feed well on forests canopy, so forests are helpful to raise adults and sub-adult livestock during disasters. The woodland's

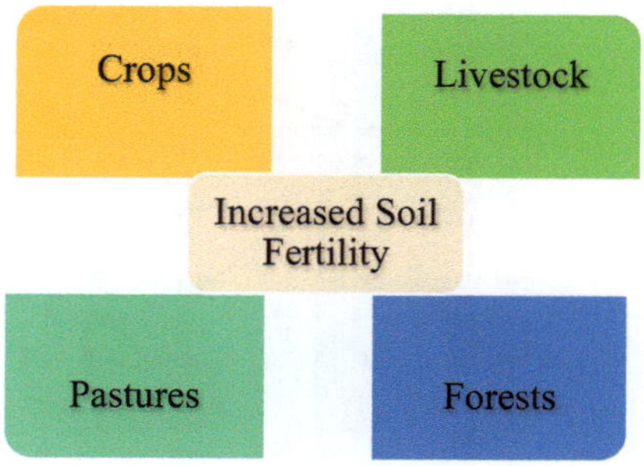

Fig. 2 Integration of forest with other production systems to improve soil fertility

leafy hay and shoots are the rich source of different nutrients, minerals and vitamins; therefore, they can be used to feed animal in order to fulfill their nutritional requirements (Gillis et al. 2022). Fodder trees can be used as a replacement for pastures and to compensate production meagerness, especially in periods of drought. They provide good nutrition to animals and are a predominant source of nutrition, peculiarly in countries like Africa and areas facing food shortage. These trees are uncomplicated to be grown and also become full of nutrition within small time-lapse. These are the source of different types of minerals and vitamins, so they are beneficial to improve livestock productivity (Figs. 2 and 3).

2.3 Fisheries; a Livelihood Asset to Manage Disaster Subversiveness

Disaster discourse usually emphasizes the significance of family assets to their pliability in disasters and rehabilitation. The cornerstone for rebuilding coastal livelihoods after catastrophes are human capital, particularly knowledge, skills and health. The availability of natural resources affects a household's capacity to survive a calamity. Their natural resources are productive agricultural land and fisheries, especially in the coastal area. For coastal areas, physical assets like boats, net and fishing hooks are principles to restore livelihood after disasters. Fish catching and then selling fish in the market can be used as sources of income and fish catcher can be used as a source of food (Islam and Walkerden 2022). Fish is a complete source of nutrition, provides high nutrition and is as a rich source of points, some oils, vitamins and minerals. It is a rich source of vitamin E, which acts as an antioxidant, helps to boost up the immune system, which is necessary to fight against diseases: as different contagious diseases emerge during and after disasters such as flood, earthquake and drought. It mainly comprises proteins that are beneficial to fulfil protein requirements of humans and poultry: as dead fishes are used as a fish meal in poultry feed.

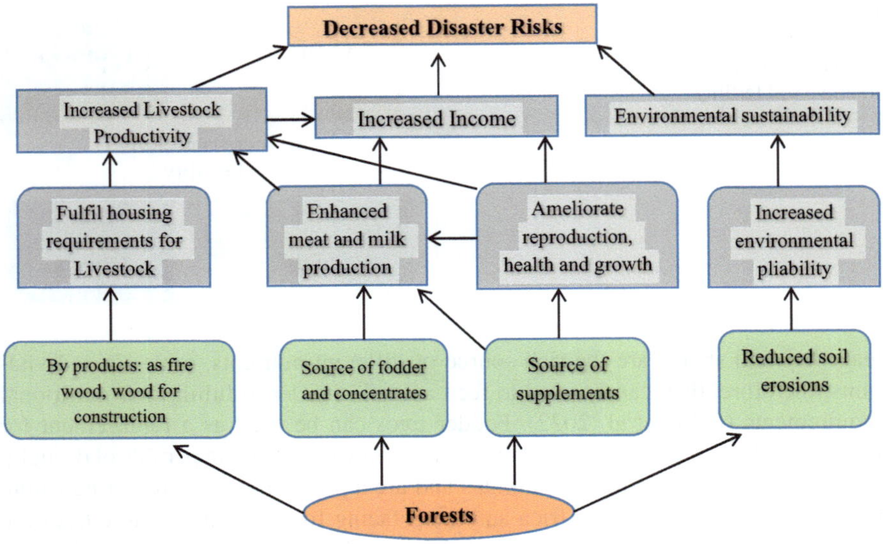

Fig. 3 Forests role to reduce catastrophes risks

2.4 Small Scale Fishing and Coastal Fishing Communities Development

The co-administration of sea resources, which comprises the community and people-centered perspective to manage decision-making, devolved upon the fishing communities. There is a need for sustainable resource management and rule enforcement to allow fishing only at small-scale communities. Fishers should be provided with facilities to take part in flood relief assistance. All fishing communities should be registered and easily identifiable (Alexander et al. 2006). Different policies should be developed to establish fish farm. Fish farming provides income source to a number of individuals and helps in raising their life style. Fish farming, if developed properly, will contribute a major proportion in raising GDP due to the agriculture sector.

1. Improving rangeland and pasture management, viz. restoring soil organic matter, reducing erosion and reducing biomass losses from fires and overgrazing, has positive environmental effects, viz. sequestration of soil carbon and biodiversity, and a positive impact on livestock productivity. Technical solutions should be available, for example, the beneficial use of forage trees, shrubs and cacti has been proven to be effective. Conservation through silage and use of forage cubes and pelleted feed gives greater efficiency in the use of a wide range of agricultural by-products. Supplementing animals with complete and balanced rations will decrease CH_4 and NO emissions. However, the adoption of these products has been slow and often due to a lack of knowledge on the part of farmers.

2. Traditionally, herd management includes strategies such as herd diversification, maintenance of high female herds, rearing of large herds and herd division. Many livestock keepers use herds of multiple species and breeds as a strategy to maintain a high degree of diversity in farm-level activities and as a buffer against climatic and economic adversity. Pastoralists keep all animal species that graze on the bottom layer (such as cows and sheep) or those that use upper grazing (such as camels and goats) in a conscious view of achieving the optimum use of pasture resources. In this way, pastoralists can harvest a wide variety of animal products, use more available fodder, different ecological activities and obtain animal products in different seasons. Increasing the proportion of females in the herd and increasing the size and division of the herd. Preserving the herds with a high percentage of females and herds with large numbers in order to ensure survival of herds for as long as possible, especially during crisis. Such traditional diversified practices are useful for adapting to climate change. Due to changes in climate and vegetation, the switching of animal types has already been observed in the Sahel region, where camels have replaced cows and goats have replaced sheep following the drought of the 1980s.
3. Travel is the most important means for pastoralists to adapt to spatial and temporal differences in rainfall. Climate change will bring more scarcity and greater variability in precipitation, so mobility will become more important than ever. As many pastoral societies become less mobile, they will become more vulnerable to the effects of climate change.
4. The movement of herds of livestock can lead to diseases and parasites due to the animals being grouped together or moved to unfamiliar areas where diseases are endemic, while blocking livestock routes will inevitably lead to overgrazing. Therefore, it is necessary to work on addressing the obstacles that lead to blocking the routes in order for the movement to be easy.
5. The indigenous breeds are part of the biodiversity and play a central role in the cultures of the people who keep them. Indigenous breeds are less susceptible to diseases and tolerant to drought and high temperatures. They are considered low risk for livestock keepers and can be used in response to disease outbreaks and climate change. Climate change will increase the need to conserve local animal genetic resources, particularly in the context of future food security concerns.
6. Water harvesting is practiced in both small and large areas, starting with the low rainfall savannah to the high rainy savannah for traditional agriculture and human and animal uses, as well as forest production. Adaptation activities for water harvesting include the construction of small dams, and water-harvesting projects for livestock drinking.
7. Protected areas can be used for promoting resistance to climate change, enhancing human health and wellbeing, and protecting biodiversity. Protected area consists of natural environment away from human activities and human population, in an area where human commercial activities are not possible to perform. These areas help in the restoration and conservation of the ecosystem. Freshwater water protected areas helps in the conservation of endangered fresh

water species like fishes and other invertebrates. These areas, if established in lowlands can help in both conservation and also helps to reduce disaster risks, as helps to promote biodiversity. Increasing freshwater biodiversity can be used as a food, so reduces disaster risk (Akasaka et al. 2022).

8. Zebu cattle are generally more heat resistant because they are characterized by maintaining lower rectal temperatures, lower respiratory rates and low water requirements. All countries must have conservation plans for their indigenous breeds.
9. The efficiency of animal breeds must be improved to be able to convert feed energy into production and reduce waste in production.
10. The greenhouse gases emitted by manure are mainly methane and nitrous oxide. Greenhouse gas emissions from manure make an important contribution to total emissions while providing mitigation opportunities.
11. Systems that amalgamate three elements; agriculture, forestry and livestock, increase species diversity and enable a variety of cropping methods (Fig. 4). Integration of animal, forest and grain production in the same location can reduce the environmental effect. This is especially helpful in the degraded pasture to improve soil quality by increasing mineral content in soil and by increasing soil fertility. The chemical, physical and biological properties of soil can be improved by using an integrating system of production of pastures, livestock

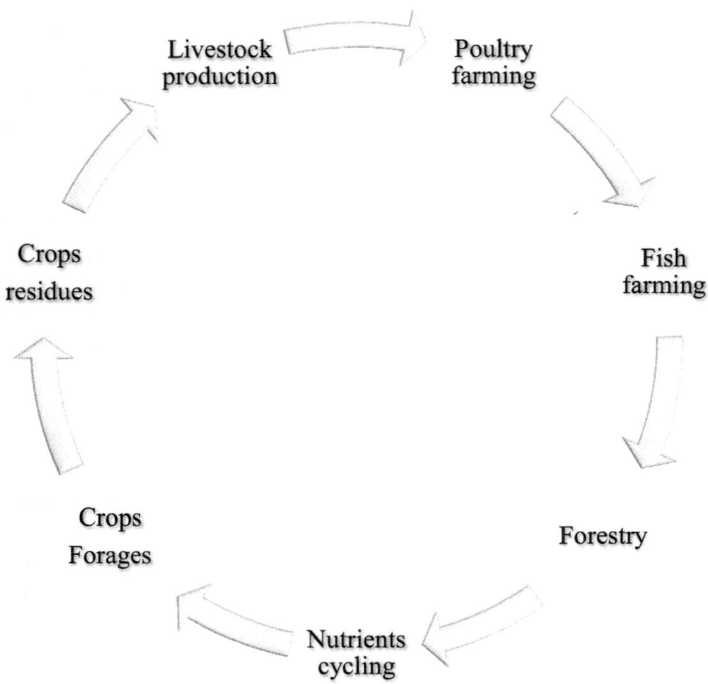

Fig. 4 Integrated farming system

and forest. As animals prefer pastures under trees shadow due to heat, feces and urine deposition by animals add nutrients to soil and help to improve soil quality (Borges et al. 2019).

3 Conclusion and Recommendation

Changes in the severity of extreme weather events, droughts, floods and changes in the precipitation pattern leads to a decline in agriculture productivity. This increase in severity and frequency affects crops, livestock, forestry, poultry and fish farming. These events cause humanitarian crises and national security concerns. Therefore, an integrated approach should be adopted among crops, livestock, forestry, poultry and fish farming to reduce losses in agriculture. The ulnerability of humans and ecosystems can be reduced by implementing adaptation options that lead to building resilience. The capacity of vulnerable populations for adaptation and mitigation measures must be strengthened.

References

Akasaka T, Mori T, Ishiyama N, Takekawa Y, Kawamoto T, Inoue M, Mitsuhashi H, Kawaguchi Y, Ichiyanagi H, Onikura N (2022) Reconciling biodiversity conservation and flood risk reduction: the new strategy for freshwater protected areas. Divers Distrib 28(6):1191–1201

Alexander B, Chan-Halbrendt C, Salim W (2006) Sustainable livelihood considerations for disaster risk management: implications for implementation of the government of Indonesia tsunami recovery plan. Disaster Prev Manag 3(6):45–49

Backlund P, Janetos A, Schimel D, Hatfield J, Boote K, Fay P, Hahn L, Izaurralde C, Kimball BA, Mader T, Morgan J, Ort D, Polley W, Thomson A, Wolfe D, Ryan M, Archer S, Birdsey R, Dahm C, Heath L, Hicke J, Hollinger D (2008) The effects of climate change on agriculture, land resources, water resources, and biodiversity in the United States. A Report by the U.S. Climate Change Science Program and the Subcommittee on Global Change Research U.S. Environmental Protection Agency, Washington, DC

Borges WL, Calonego JC, Rosolem CA (2019) Impact of crop-livestock-forest integration on soil quality. Agr Syst 93(6):2111–2119

Chatterjee R, Rajkumar U, Prince L (2022) Revolutionizing impacte of poultry resources in food security and rural economy. In: Agriculture livestock production and aquaculture: advances for smallholder farming systems. Springer International Publishing, Cham, pp 205–215

Gillis R, Kendall I, Roffet-Salque M, Zanon M, Anders A, Arbogast R-M, Bogucki P, Brychova V, Casanova E, Claßen E (2022) Forest ecosystems and evolution of cattle husbandry practices of the earliest central european farming societies, vol 45, pp 234–240

Islam R, Walkerden G (2022) Livelihood assets, mutual support and disaster resilience in coastal Bangladesh. Int J Disaster Risk Reduct 78:103148

USDA (2021) World agriculture supply and demand estimates. Department of Agriculture, Washington, DC

USGCRP (2014) Ch. 6: agriculture. In: Hatfield J, Takle G, Grotjahn R, Holden P, Izaurralde RC, Mader T, Marshall E, Liverman D (eds) Climate change impacts in the United States: the third national climate assessment. U.S. Global Change Research Program, Washington, DC, pp 150–174

Key Challenges and Financial Needs to Promote Climate-Smart Agriculture (CSA) in Pakistan: A Case Study of Hunza, Pakistan

Mehwish Aslam, Mujahid Rasool, Ikram Ullah Qadir, Uzair Zahid, Rifat Hayat, and Mukhtar Ahmed

Abstract

The undertaking of current research is to explore key challenges and financial needs to promote climate-smart agriculture (CSA) in Pakistan: a case study of Hunza (Latitude: 36° 18′60.00″N, Longitude: 74° 38′ 59.99″ E and altitude of 8200 ft.), Gilgit Baltistan, Pakistan. The goal of the climate-smart agriculture (CSA) notion is to better integrate agricultural development and meteorological responsiveness. It aims to ensure food security, reduce emissions of greenhouse gases, and strengthen resilience in the context of a changing climate. A field-based survey was conducted by visiting farmers in the Hunza region, i.e., Karimabad, Aliabad, Hyderabad. The collected data sample was heterogeneous as respondents were of different socio-economic and educational backgrounds. Results showed that climate change in the form of rise in temperature and intense rainfall resulted in the shift in the crop sowing and harvesting timings, decline in crop productivity, and occurrence of insect, pests, and disease attacks. Furthermore, farmers' knowledge about climate-smart agricultural practices and

M. Aslam (✉) · R. Hayat (✉)
Institute of Soil and Environmental Sciences, PMAS-Arid Agriculture University, Rawalpindi, Pakistan
e-mail: hayat@uaar.edu.pk

M. Rasool
Agriculture Credit and Microfinance State Bank of Pakistan, Karachi, Pakistan

I. U. Qadir · U. Zahid (✉)
State Bank of Pakistan, Karachi, Pakistan

M. Ahmed
Department of Agronomy, Pir Mehr Ali Shah Arid Agriculture University, Rawalpindi, Pakistan
e-mail: ahmadmukhtar@uaar.edu.pk

M. Ahmed, S. Ahmad (eds.), *Disaster Risk Reduction in Agriculture*, Disaster Resilience and Green Growth, https://doi.org/10.1007/978-981-99-1763-1_25

technologies was very limited. A survey also reports that farmers of this region are far behind than other farmers in the country. This could be due to poor connections between farmers, researchers, and extension workers, unavailability of modern machinery, lack of support from relevant government departments, and poor finance facilities. Hence it is important to provide all facilities to the farming community of this region so that farmers can have adequate adaptation and mitigation strategies against climate change.

Keywords

Hunza · Gilgit Baltistan · Climate-smart agriculture · Climate change · Adaptation and mitigation strategies

1 Introduction

Climate change is one of humanity's most critical challenges. The warming of the planet threatens food security, freshwater supply, and human health. The effects of climate change, including sea level rise, droughts, floods, and extreme weather, will be more severe if actions are not taken to dramatically reduce emissions of greenhouse gases into the atmosphere (Ahmed, 2020). While the link between human action and the planet's recent warming remains an almost unanimous scientific consensus, the links between population growth and climate change deserve further exploration (Fig. 1). The United Nations' most recent project indicates that by 2050,

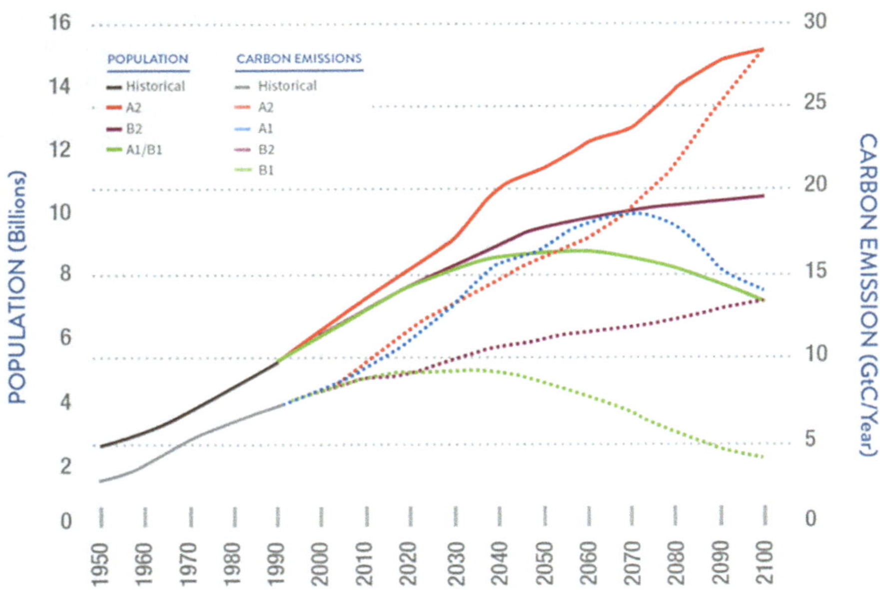

Fig. 1 Relationship between population changes and carbon emissions under Intergovernmental Panel on Climate Change (IPCC), special report on emission scenarios (SRES)

there will be 9.5 billion people on the earth. People who reside in underdeveloped nations are more susceptible to climate risks (Morton 2007). Hotspots areas across the globe have been shown in Fig. 2. Therefore, the poor in developing countries, who heavily rely on agriculture for their livelihood, are more rigorously exaggerated by the negative effects of climate change (Amole and Ayantunde 2019).

Agriculture uses around 38% of world land and around 1/3 of this area is used for crop production while the rest have been used for livestock grazing. Agriculture contributes 23% of greenhouse gas (GHG) emissions. Hence changes are needed to manage agricultural lands so that it releases less GHG. Climate-smart agriculture (CSA) is a novel, cleaner production alternative to conventional farming which can increase the productivity, resilience, and efficiency of the agricultural production system while reducing greenhouse gas (GHG) emissions. CSA was first introduced by the Food and Agriculture Organization (FAO) in 2010 (Imran et al. 2018). It is mainly a technique of addressing climate change and food security known and can be called as "Climate-Smart Agriculture" that involves continuous increase in productivity, enhancing climate resilience, reducing GHG emissions, and improving the achievement of national food security and sustainable development goals.

In particular, prior research demonstrates that farm households have implemented a variety of adaption approaches, for instance CSA practices, crop rotation changes, crop diversification, seed variety diversification, conservation of water and soil, the use of better-quality seed varieties, and income diversification (Issahaku and Abdulai 2020; Abdulai 2018). Furthermore, there is no one set of method for CSA that can be used everywhere. As a result, these procedures may differ from one country and one location to another (Food Agriculture Organization 2020).

According to FAO (2013), there are three key goals or pillars of the CSA where site-specific effort should be focused:

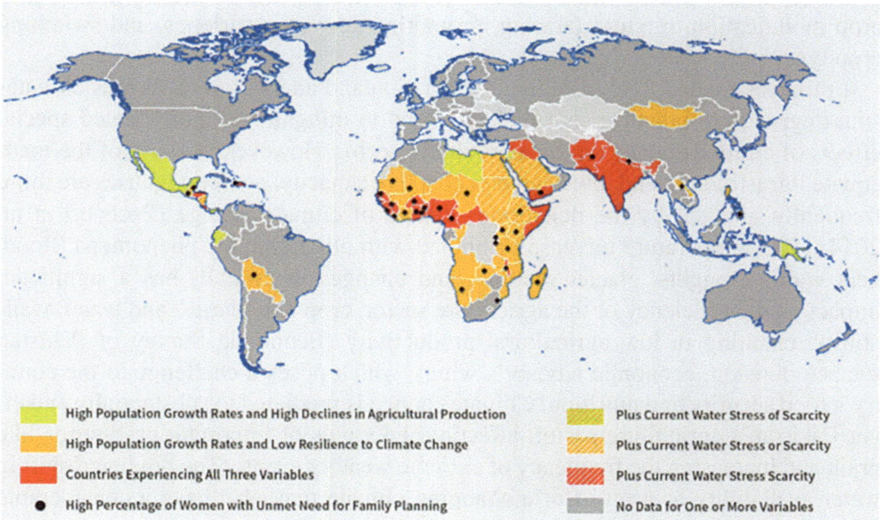

Fig. 2 Hotspots regions across the globe

- Better climate change adaptation and resilience.
- Increased agriculture revenue and output in a sustainable manner.
- Measures to reduce GHG.

Pakistan has become increasingly vulnerable to climate change (CC) over the past few decades that has had a huge negative impact on agriculture output, environment, food security situation, and rural resident's quality of life. Owing to the lack of mitigating strategies, the fast-changing climate, a lack of adaptive capacity, and poor infrastructure, Pakistan is placed seventh among highly susceptible nations. In 2010, the FAO unveiled CSA as a cutting-edge and sustainable farming method. Its objectives are to lessen the negative effects of CC, improve yield, raise production sustainably, minimize GHG emissions, and guarantee food security. This study was conducted with the objective to ascertain significant impediments that hinder the adoption of CSA practices and to investigate the role of microfinance in the agriculture sector from farmer's perspective.

2 Literature Review

Human-induced climate change is causing dangerous and widespread disruption in nature and affecting the lives of billions of people around the world, despite efforts to reduce the risks. Additionally, it has an adverse influence on the global economy (Intergovernmental Panel on Climate Change 2018). Other effects of climate change include significant weed and insect issues, ecological degradation, reduction of biodiversity, and insufficient water, all of which degrade the situation with regard to food security. Increased farm productivity for rural household is a result of the adoption of farm-level climate-smart agricultural techniques, including shifting planting dates, credit programmers, soil nutrient techniques, water saving strategies, crop modification, mixture farming, promoting new crop varieties, and switching crop type.

Initiatives related to climate, e.g., alleviation and adaptation to various climate-smart agriculture practices, are typically used to mitigate the multifaceted special effects of climate change on the agricultural sector. However, because of the inadequate infrastructure and low degree of adaptive capacity, poorer countries are more frequently affected by the deleterious effects of climate change (Pachauri et al. 2014). This temperature increase combined with other climatic phenomena (flood, heat wave, droughts, glacier melting, and changes in rainfall) has a significant impact on the efficiency of the agriculture sector, crop boundaries, and water availability, resulting in low agricultural productivity (Economic Survey of Pakistan finance division, economic advisor's wing), which poses a challenge to the country's food security and nutrition. Climate change is predicted to substantially endanger Pakistan's agriculture sector, affecting the yields of important cash crops like grain and increasing the frequency of extreme weather events. The predicted shift in water availability as a result of a changing climate may also have a considerable impact on the productivity of irrigated agriculture, which is solely dependent on the river system. A rise in temperature of 2–30 °C is predicted for Pakistan by 2050, as well as a substantial variation in rainfall patterns (Imran et al. 2018).

Moreover, the mainstream of farmers in Pakistan continue to use traditional farming methods. Therefore, a shortage of knowledge and innovation, market letdowns, weather shocks, and resources limitations are blamed for low staging per acre yield. In literature, including Imran et al. (2018, 2019), the lack of access is the primary cause of the poor acceptance frequency of CSA method and technologies, predominantly for small-scale and demoted farmers. Lack of the role of agricultural institution services and economic resources, such as agriculture extension servicear-eas, far fetching prices of agricultural commodities get a barrier to tackling the risks of climate change. Two thirds of Pakistan's people receive their living from the country's largest economic sectors, agriculture, which employs 42% of the labor force overall. Additionally, it makes up roughly 19% of the country's GDP. Firstly, the agriculture sector has not done well during the previous 20 years because agriculture growth is still weak and agricultural revenues are dropping. Parallel to the above-mentioned statement, the agriculture industry confronts significant difficulties in fulfilling the increasing food demands for a population that is expanding at a ratio of about 2% (Sardar et al. 2016).

The implementation of CSA like water and soil and crop controlling practices is assessed using cross-sectional figures from 350 cotton agriculturalists in Punjab, Pakistan's main cotton-growing districts. Factors influencing farmer adoption decisions and their effect on scarcity, profits, and yield are probable using logistic reversion and propensity score matching (PSM), respectively (Jamil et al. 2021). According to the finding, farmers' adoption behavior was favorably influenced by edification, access to acclaim, proprietorship of tube wells, agricultural knowledge, and availability of extension amenities. Additionally, PMS results showed that CSA practice adoption is economical, financially advantageous, environmentally preferable, and pro-poor (Jamil et al. 2021). However, in pakistan there is quite low implications of CSA practices it is mendatory to make people more aaware about it. Therefore, more data are needed to support the significance of CSA method in raising farmers' capacity for adoption to lessen the deleterious effect of climate modification (Sardar et al. 2021).

3 Materials and Methods

3.1 Study Area

This research is conducted in district Hunza. The Hunza valley is mountainous valley in the northern part of Gilgit-Baltistan located in the far north of Pakistan. It is located 100 kilometers from Gilgit, the main city, at 36.26^0 North latitude and 74.73^0 East latitude, at an elevation of 8200 ft. overhead to sea level (Fig. 3). The region has extremely harsh conditions, with summertime highs of over 40 °C and wintertime lows well below freezing. In Hunza, July and August are the warmest months of the year. Since there is only one farming season, from April to October, the local

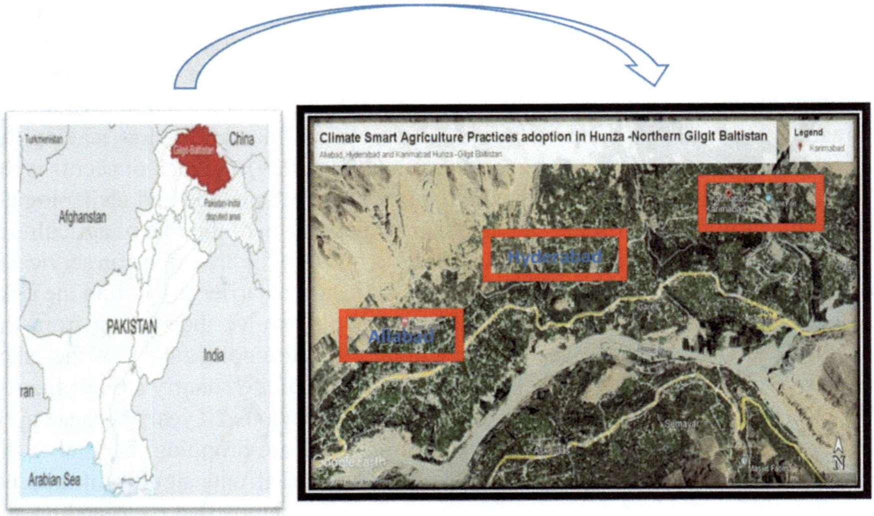

Fig. 3 Research data collected from areas of district Hunza

economy is dependent on subsistence farming (farmers' main aim is to produce enough to satisfy the household consumption demand). The major crops of Hunza are wheat and potato while grapes, apricot, apple, walnut, cherry, and almond are major fruits.

3.2 Data Collection and Sample Size

In order to scrutinize the research objectives I have worked on primary data. The survey was conducted in three villages of district Hunza: Aliabad, Karimabad, and Hyderabad. Information was collected through systematic random sampling technique (see fig Mention Sample size was 10 to 30 farmers of each village (Karimabad, Aliabad, Hyderabad)).

3.3 Questionnaire Designed

For data assemblage, a wide-ranging ingenious opinion poll was maneuvered to collect data (Fig. 4). The first part encompassed interrogations related to farmer personal information, for instance, age, edification, address, gender, etc. The second part of the opinion poll confined queries allied to farmland information: cultivable land, fallow land, animals they rear, etc. The third part of the questionnaire demonstrates adoption of CSA practices and climate change impacts. Last but not least, the fourth section reveals key challenges for the adoption of CSA practices and hindrance in getting bank loans.

Fig. 4 Questionnaire diagram

3.4 Conceptual Framework

The intangible framework of current study entails four components:

1. Climate change threats.
2. Adoption of CSA.
3. Key challenges.
4. Financial needs.

The climate transformation reduces farm efficiency and farm revenue. It is mendatory to opt CSM to enhance the farm productivity (Fig. 5).

4 Results and Discussions

4.1 Climate Change Impacts on the Agriculture Sector

The Karakoram-Hindu Kush Himalaya (HKH) Mountains are particularly susceptible to the effects of climate change (Fig. 6). The average temperature in the HKH section has altered over the past century at a pace of 0.10 °C each decade. According to climate predictions, the temperature will rise by 1–2 °C by 2050, and the pattern of precipitation will alter with a longer and more unpredictable monsoon and less frequent but more intense rainfall. Climate change will affect agricultural productivity and biodiversity including plant phenology.

People in all of the research regions had become aware of the shifting weather configurations and had discussed the deficiency of rain, fluctuations in temperatures, and the start of the seasons in their localities. Observation reveal an

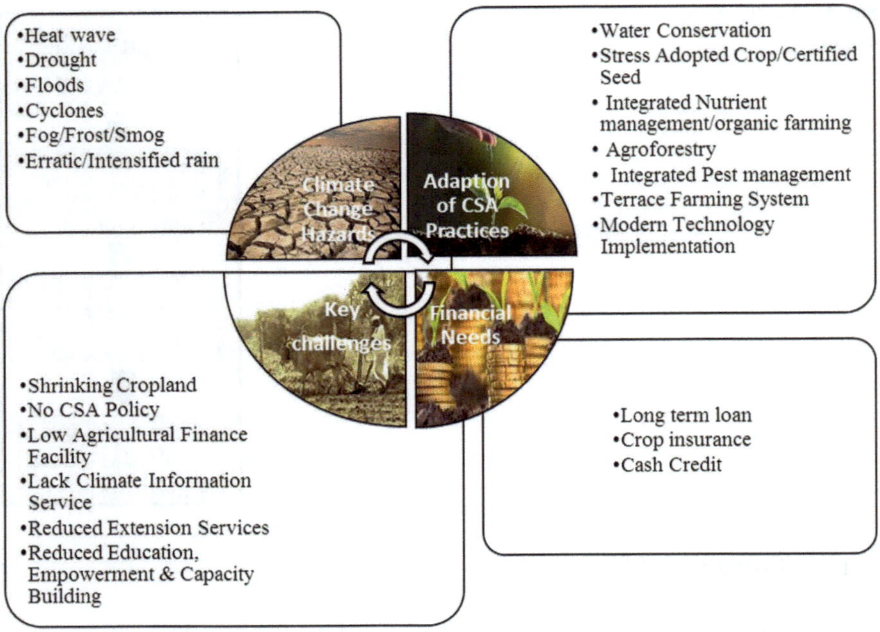

Fig. 5 Outline of research work

Fig. 6 Climate change impacts on the agricultural sector

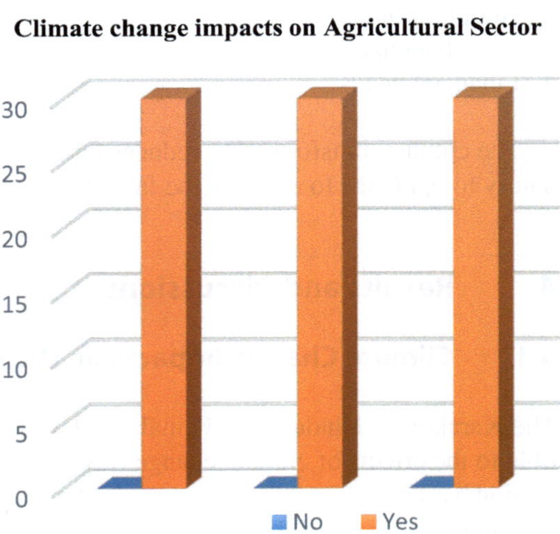

irregularity in the rainfall pattern, as reported by most of the farmers surveyed; they observed an asymmetry in rainfall distribution, accompanied by increasingly aggressive winds. These adverse weather conditions have had a detrimental impact on their crop yields, causing significant damage. Previously, they used to harvest crops 2 weeks prior, but due to fluctuation in rainfall pattern and high temperature it gets changed and the rains have really become unpredictable. Most of the farmers shared that all of the wheat was traditionally harvested in August and September, but the current delay is a result of the weather. Additionally, the produce doesn't adequately ripen. Wheat is chopped and gathered, but it's not of high quality as it formerly was. Furthermore, they claim that due to the heat and erratic rainfall crop yield is affected by a certain type of disease.

4.2 Adoption of Climate-Smart Agriculture

Climate change is a widespread matter in Hunza, and the people are well aware of its consequence on their farming community and revenue (Figs. 7 and 8). However, implementation of CSA practices are needed in the region so that farmers can have good returns from their lands. There are various CSA practices, but some selected measures have been used in this study to explore whether these measures have been adopted by farmers or not. With reference to the first question from farmers, i.e., Do they have acquaintances about CSA, only 30% of farmers responded positively that they are familiar with the term. Furthermore, we tried to evaluate what type of CSA practices have been used presently in Hunza (Table 1). Few of the CSA practices that were identified include soil conservation practices (agroforestry, intercropping, organic manure, mulching, green manure, crop diversification, and organic fertilizer), water conservation practices (water reservoirs, modern irrigation technology), and use of stress (heat and drought) tolerant seeds.

4.2.1 Soil Conservation CSA Practices Implication

Farmers frequently employ crop diversity as a technique to make the best use of their available land, water, and labor resources while avoiding the threats posed by climate alteration. It is the shift of single cereal-based agriculture to diversify cropping system. Only 30% of our farmers practice crop diversification. Likewise, crop rotation, green manure, mulching, intercropping, and agroforestry are the best

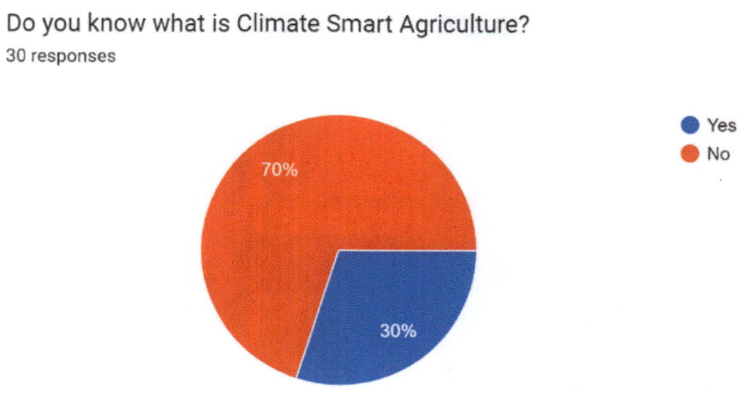

Fig. 7 Knowledge status of CSA

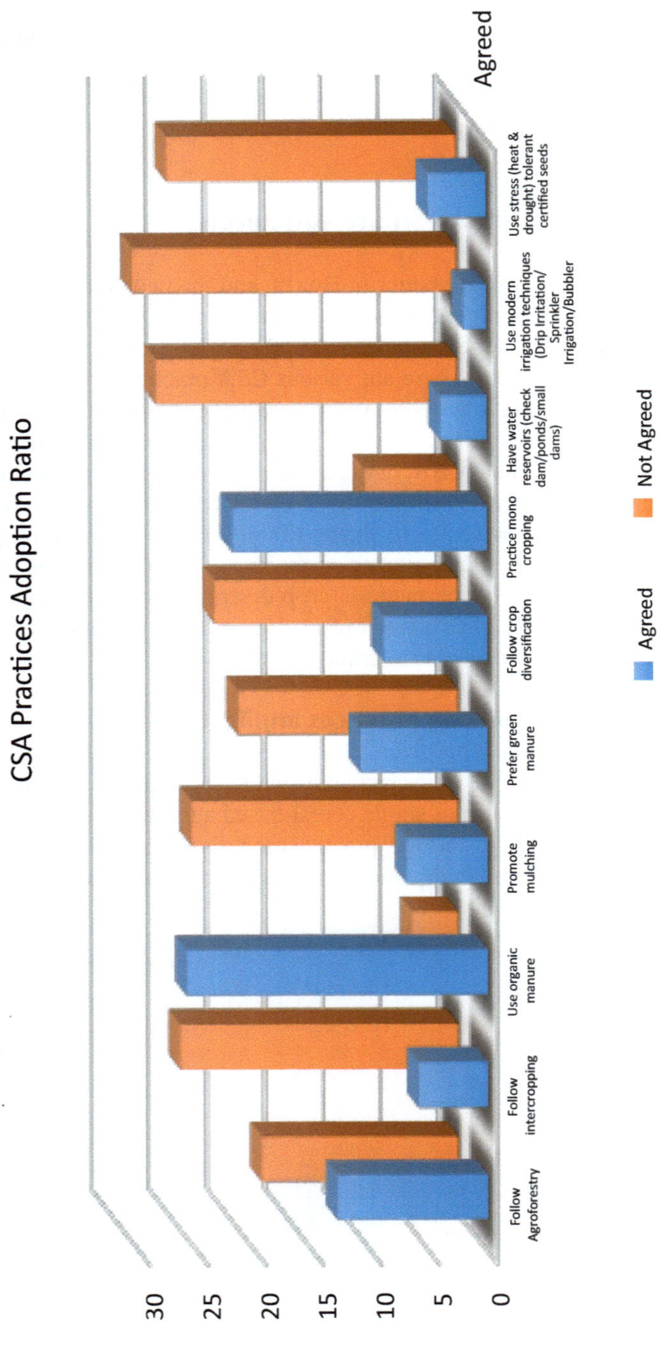

Fig. 8 CSA adoption ratio

Table 1 Climate-smart agricultural (CSA) practices

	Adoption of climate-smart agricultural (CSA) practices	Practice	Percentage (%)	Do not practice	Percentage (%)	Total no. of farmers
1.	Use stress (heat and drought) tolerant certified seeds	5	16.6%	25	83.3%	30
2.	Use modern irrigation techniques (drip irritation/ sprinkler irrigation/ bubbler)	2	6.6%	28	93.3%	30
3.	Have water reservoirs (check dam/ponds/small dams)	4	13.3%	26	86.6%	30
4.	Practice mono cropping	22	73.3%	8	26.6%	30
5.	Follow crop diversification	9	30%	21	70%	30
6.	Prefer green manure	11	36.6%	19	63.3%	30
7.	Promote mulching	7	23.3%	23	76.6%	30
8.	Use organic manure	26	86.6%	4	13.3%	30
9.	Follow intercropping	6	20%	24	80%	30
10	Follow agroforestry	13	43.3%	17	56.6%	30

measures that can be adopted for soil nutrient conservation, reduced erosion, and good crop production. There are fewer farmers who opt for CSA practices: 7% farmers promote mulching, 11% farmers prefer green manure, 6% farmers practice intercropping, 13% farmers practice agroforestry.

4.2.2 Water Conservation CSA Practices Implication

By enabling water to be used effectively, water conservation supports sustainable intensification by increasing agricultural productivity throughout the year and drought resistance, boosting farmer incomes, and enhancing food nutrition. In Hunza the water supply is limited and getting more precious every day. 93.3% of the farmers claim that they suffer from water scarcity issues for irrigation as mentioned in Fig. 9.

According to the farmers, they have no abundant water reservoirs. Only 13.3% have accommodation at their farm level due to which water is becoming scarcer with each passing days although this missed opportunity can be replaced by using irrigation techniques for water conservation. Water conservation practices include all modern irrigation techniques: drip irrigation, sprinkler irrigation, bubbler, etc. Unfortunately, only 6.6% of people have access to modern irrigation practices in Hunza Due to lack of modern Agriculture techniques information and low capital to buy agricultural commodities to work on farm.

4.2.3 Use of Stress-Tolerant (Heat and Drought) Seeds

Stress-tolerant certified seeds are the best option to mitigate the climate change hazard of productivity loss and reduced income. In this survey study it is anticipated

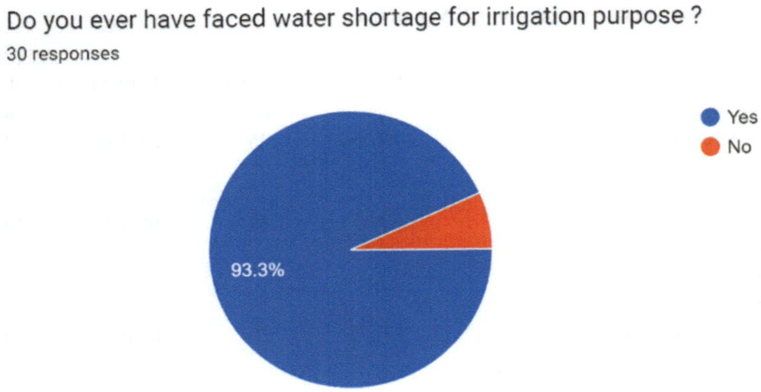

Fig. 9 Water scarcity

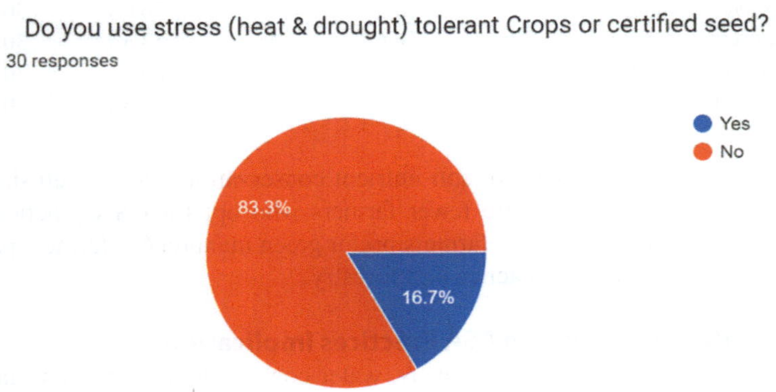

Fig. 10 Use of certified seed

that only 16.7% farmers have access to stress-tolerant seeds while rest of the farmers can not affort to buy stress tolerant seeds shown in Fig. 10.

4.3 Key Challenges

4.3.1 Access to Finance and Difficulties

About 90% of the respondents have bank accounts while 10% of farmers covered in this survey had no bank account (Figs. 11 and 12). Although the majority of farmers have bank accounts, 75.9% do not apply for agricultural loan. There are certain challenges, and that is why farmers do not have access to agricultural loan.

Data obtained through this survey have shown that about 40% of the farmers who did not or could not obtain bank loans claimed to be unaware of the financial

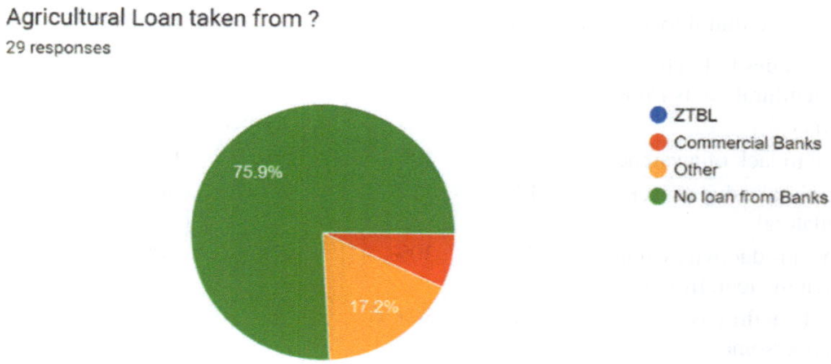

Fig. 11 Agricultural loan approaches

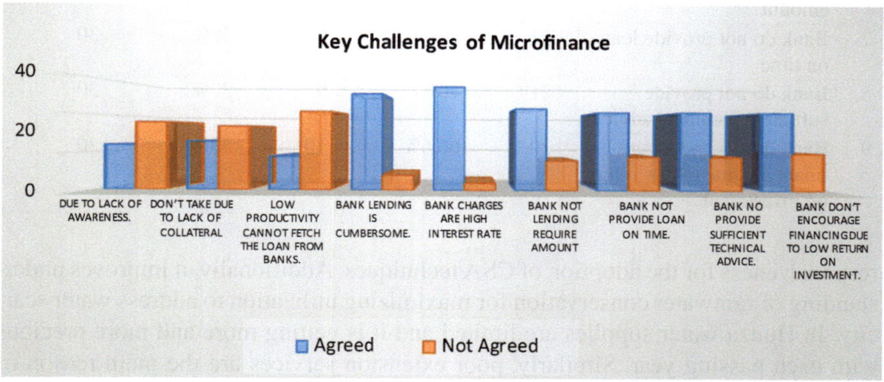

Fig. 12 Agricultural Loan Lending Barriers

services and products that have been provided by banks to the farming sector, while 60% people declared that though they are familiar about the terms and conditions, but unfortunately the interest rate of bank disappoints them. Furthermore, 43% of the growers could not afford bank advances. Similarly, some farmers state that loan procedure is bit problematic; those who apply for loan indicated that the bank does not deliver loan on time. Numerous records of farmers concerning financial barriers have been shown in Table 2 and Fig. 12. However, to encourage farmers to take out bank loans, procedures should be simpler with minimum documentation and interest rates.

4.3.2 Lack of Sufficient Information and Knowledge Transmission

The adoption of various sets of CSA practices is positively and significantly correlated with institutional indicators like access to extension institutions, credit services, weather forecasting information, market information, and aids on CSA skills. Given that institutions are crucial tools for supplying knowledge to farmers reduce the use of fertilizers, dry land soils, and soil preservation, it was found that institutional services might play a positive input by organizing training, workshops, and

Table 2 Agricultural loan lending barriers

	Key issues to fetch agricultural loans from banks	Agreed	Percentage (%)	Not agreed	Percentage (%)	Total no. of farmers
1.	Due to lack of awareness	12	40%	18	60%	30
2.	Don't take due to lack of collateral	13	43%	17	56%	30
3.	Low productivity cannot fetch the loan from banks	9	30%	21	70%	30
4.	Bank lending is cumbersome	26	86.6%	4	13.3%	30
5.	Bank charges are high interest rate	28	93.3%	2	6.6%	30
6.	Bank not lending required amount	22	73.3%	8	26.7%	30
7.	Bank do not provide loan on time	21	70%	9	30%	30
8.	Bank do not provide sufficient technical advice	21	70%	9	30%	30
9.	Bank do not encourage financing due to low return on investment	20	66.6%	10	33.4%	30

responsiveness for the adoption of CSA techniques. Additionally, it improves understanding of rainwater conservation for maximizing utilization to address water scarcity. In Hunza water supplies are limited and it is getting more and more precious with each passing year. Similarly, poor extension services are the main reason of low productivity, and data have shown that around 90% of the farmers have no access to extension services (Fig. 13). Another barrier is the gap between research organizations and farming communities. Smallholder farmers frequently hesitate to adopt new practices until they can clearly see the results of those activities and are given clear explanation of their advantages and disadvantages. Additionally, exposure to new technology is out of question as 86.7% farmers lack access of modern technology as mentioned in Fig. 14 (Table 3).

4.3.3 Poor Policy System and Deficiency of Government Investments

The prevalent key barrier for the implementation of CSA practices are worse agriculture policy system and lack of government intention towards the elevation of agriculture sector of Hunza. Furthermore the poor adaptation of CSA practices are because of lack of institutional assistance. Farmers are still using old traditional agricultural methods as adoption of CSA practices needs more money (e.g. Purchase of quality certified seeds etc.) and infrastructure. Hence, it is essential that policies that can assist farmers in overcoming the knowledge and financial obstacles should be promoted so that they can adopt CSA practices easily

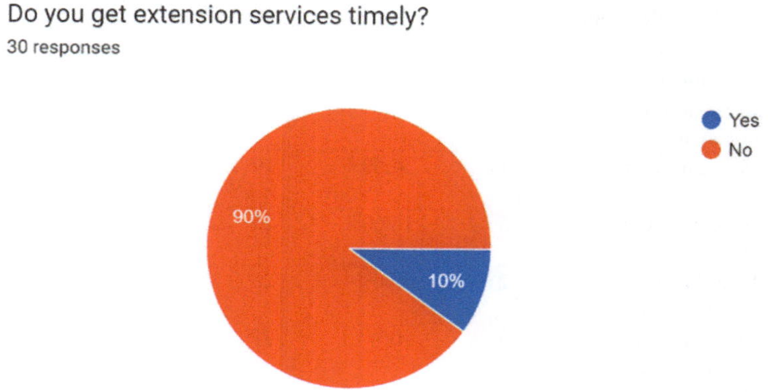

Fig. 13 Extension services

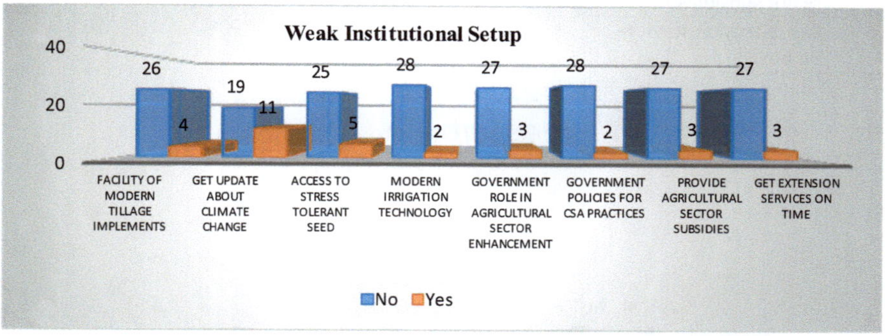

Fig. 14 Weak institutional setup

5 Recommendations

5.1 Utilization of Marginal and Fallow Land

Marginal and fallow lands are lands that are not actively used for agriculture or other human activities due to their poor soil quality, topography, or lack of water (Fig. 15). These lands can be used to minimize climate change in several ways:

1. **Reforestation**: Marginal and fallow lands can be planted with trees and shrubs, which can absorb and store carbon dioxide from the atmosphere through the process of photosynthesis. This can help to reduce greenhouse gas emissions and mitigate the effects of climate change. Sustainable agriculture: These lands can be used for sustainable agriculture practices such as agroforestry, which combines trees and crops to improve soil health and increase carbon sequestration.

Table 3 Institutional Setup

S.No	Institutional setup	Agreed	Percentage (%)	Not agreed	Percentage (%)	Total no. of farmers
1	Access of modern tillage implements	4	13%	26	86.60%	30
2	Get update about climate change	11	36.60%	19	63.30%	30
3	Access to stress-tolerant seed	5	16.60%	25	83.30%	30
4	Modern irrigation technology	2	6.60%	28	93.30%	30
5	Government role in agricultural sector enhancement	3	10%	27	90%	30
6	Government policies for CSA practices	2	6.60%	28	93.30%	30
7	Access of agricultural sector subsidies	3	10%	27	90%	30
8	Get extension services on time	3	10%	27	90%	30

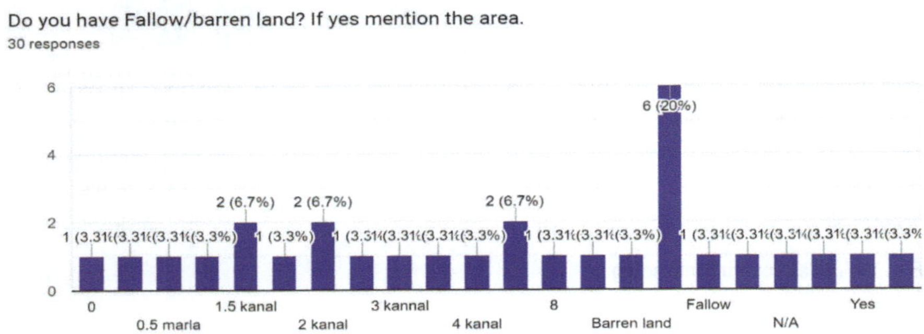

Fig. 15 Fallow cultivable land

Sustainable agriculture practices can also help to reduce greenhouse gas emissions from agricultural activities and improve food security.

2. **Renewable energy production**: Marginal and fallow lands can also be used for the production of renewable energy such as solar or wind power. This can reduce the use of fossil fuels and help to mitigate the effects of climate change.
3. **Carbon farming**: This is a method of land management that uses agricultural practices to increase carbon sequestration in the soil. Techniques such as no-till farming, cover cropping, and rotational grazing can help to build soil organic matter and reduce greenhouse gas emissions.
4. **Conservation**: Marginal and fallow lands can be used for conservation purposes such as creating wildlife habitats, preserving biodiversity, and protecting natural

resources. These efforts can help to maintain healthy ecosystems and reduce the impacts of climate change.

Furthermore this key challenging task can be accomplished through access to the following activities:

1. Great access to organic matter for soil management.
2. Energy source for land cleaning, degradation, and leveling.
3. Access of irrigation water.

Access to organic matter is a wise step to improve soil fertility and is a best CSA practice. The majority of farmers declare that they do have fallow land, which is cultivable, but due to water issues they cannot be brought into use. As Hunza is a mountainous area and most of the land is sloppy. This type of land can be brought under cultivation through a terrace farming system but farmers need huge agricultural machinery which they lack. Along that sufficient water is required for irrigation. While conducting the survey the majority of farmers complained about no access of irrigation water. If they accomplish to irrigate barren but cultivable lands by familiarizing about advanced technology, it can duple single cultivable landholdings, and enhance revenue and food safety.

5.2 Assistance of Microfinancing for the Implementation of Climate-Smart Agriculture

Microfinancing, which is the provision of small loans to individuals or groups who typically do not have access to traditional banking services, can be a useful tool for the implementation of climate-smart agriculture (CSA). Here are some ways in which microfinancing can help:

1. **Access to capital**: One of the primary benefits of microfinancing is that it provides access to capital for smallholder farmers and other rural communities. This can be crucial for implementing CSA practices, such as the adoption of climate-resilient crops or the use of sustainable farming techniques. By providing affordable credit, microfinancing can help farmers invest in the necessary resources and equipment to implement CSA practices.
2. **Support for risk management**: Climate change brings new and unpredictable risks to farming communities, such as extreme weather events, crop failures, and pest infestations. Microfinancing can help farmers manage these risks by providing them with the resources to implement risk-reducing practices, such as crop diversification or the use of drought-tolerant seeds.
3. **Improved access to information and training**: Microfinance institutions often work closely with their clients to provide them with training and technical assistance. This can help farmers adopt CSA practices and improve their productivity and profitability. By providing farmers with access to new knowledge and best

practices, microfinancing can help them adapt to the changing climate and become more resilient.

4. **Encouraging innovation**: By providing access to credit, microfinancing can encourage farmers to innovate and experiment with new CSA practices. This can lead to the development of new techniques or the adoption of more sustainable methods of production, which can have positive impacts on the environment and the local economy.

Smallholder farmers' decisions to use appropriate CSA methods are heavily influenced by their ability to access financial support services. Crop insurance scheme should be introduced for major crops in Hunza such as wheat and potato, 90% of the farmers grow wheat and potato, and these are the major crops of Hunza.

5.3 Education Empowerment and Capacity Building

Agri Extension staff should impart their knowledge to farmers through training programs. Training is a significant aptitude construction tool. To promote the implementation of CSA, it is also necessary to convey practical knowledge. For example, appropriate soil nutrient management, composting of locally sourced materials, conservation agriculture, and water collection techniques are all highly reliant on specific local circumstances, so practical training is frequently required. Therefore, the Extension workers should play a key role in the promotion of agricultural services and extension of knowledge or information.

5.4 Policy Support and Role of Government for the Implementation of Climate-Smart Agriculture

Research findings indicated that effective and increased adoption of CSA techniques may necessitate providing farmers with additional institutional assistance as well as financial resources. Farmers are still using traditional agricultural methods; subsidies for CSA-supporting technologies should also be offered, as well as the preferential availability of certified quality seeds. Policies should assist farmers in overcoming the knowledge and financial obstacles that are essential in promoting the adoption of CSA methods. Active support is needed to improve the agriculture sector of the study region. The main concern of the government should be to facilitate the farmers hailing from remote areas of Hunza. Through various ways to enhance the extension work services, broadcasting to spread accurate information on time in the right place. Additionally, in order to encourage widespread CSA adoption and the modernization of agriculture, the government should offer farm facilities to farmers. Thus, government interventions, such as increased approach to credit, extension services, meteorological information, and education, can minimize the obstacles to farmers in Hunza with relation to the implementation of CSA.

Table 4 Major climate smart agricutural (CSA) practices

	Major CSA practices	Adopting factors/ constrains	Potential solutions
1.	Water management, including irrigation	• High cost access to finance • Input accessibilities • Lack of awareness	• Education • Microfinancing • Deliver extension services
2	Improved modern tolerant varieties	• Lack of access • Costly to fetch • Poor policy system	• Microfinancing • Extension services • Imparting modern agricultural knowledge
3	Agroforestry	• Lack of skills/ experiences • Lack of awareness	• Microfinancing • Deliver extension services • Imparting modern agricultural knowledge
4	Intercropping and diversification	• Lack of awareness • Lack of experience	• Microfinancing • Extension service enhancement • Knowledge sharing
5	Climate information services	• Bad internet access • Lack of climate change perception	• Facility to weather forecast

Based upon my research work result I have computed foremost outcome based on the CSA implementation factors, constrains, and prospective resolutions for prospective solution to particular constrain summarized in Table 4.

6 Conclusion

The agricultural division, particularly in developing states, is more likely to be susceptible to the effect of climate modification due to poor implementation of Climate-Smart Agricultural (CSA) practices. The current research study in Hunza was intended to explore key issues faced by farmers to implement Climate-Smart Agriculture (CSA) practices, and financial prerequisite to encourage CSA acts in Northern Pakistan. Farmers shared that climate change has adversely affected the crop yield productivity, shifted the crop sowing and harvesting calendar, and led to irregular rainfall pattern; high temperatures have induced crop diseases. In order to mitigate this, farmers do not have much knowledge about the implementation of Climate-Smart Agriculture (CSA) practices and technologies. This research work made evident that the farmers of Hunza are deprived of modern agricultural knowledge, machinery, the poor role of extension department, no agricultural policy, lack of government attention towards agricultural sector enhancement, various challenges to access finance in order to mitigate the agricultural credit needs, and many more. Due to lack of resources, sufficient knowledge, and financial and policy backing, the acceptance of CSA is still constricted in Hunza. This research results thoroughly evaluated the narrow CSA techniques currently implemented by farmers, enabling to recommend opportunities and solutions for a long-term enhancement of agricultural systems. In this regard, I have given the following actions a particularly high priority actions:

1. Improving soil quality and restoring marginal and degraded land.
2. Endorsing training, and capacity-building programs for farmers and extension workers.
3. Developing specialized financing schemes and policies.
4. Facilitating with Agri Finance (crop insurance, livestock insurance, etc.).

The results would aid SBP in better planning and concentrating its policy, regulatory, and developmental measures for boosting the flow of funding to rural and agricultural communities of North Pakistan.

References

Abdulai A (2018) Simon brand memorial address-the challenges and adaptation to climate change by farmers in sub-Saharan Africa. Agrekon 57(1):28–39

Ahmed M (2020) Introduction to Modern Climate Change. Andrew E. Dessler: Cambridge University Press, 2011, 252 pp, ISBN-10: 0521173159. Sci Total Environ 734:139397. https://doi.org/10.1016/j.scitotenv.2020.139397

Amole TA, Ayantunde AA (2019) Options of making livestock production in West Africa "climate-smart". All Africa conference on Animal Agriculture, Accra, Ghana, 29 July-2 August 2019

FAO (2013) Climate-smart agriculture sourcebook; food and agriculture organisation of the United Nations. FAO, Rome

FAO (2020) Climate smart agriculture; Accessed 22 Dec 2020

Imran M, Ali A, Ashfaq M, Hassan S, Culas R, Ma C (2018) Impact of climate smart agriculture (CSA) practices on cotton production and livelihood of farmers in Punjab, Pakistan. Sustainability 10(6):2101

Imran MA, Ali A, Ashfaq M, Hassan S, Culas R, Ma C (2019) Impact of climate smart agriculture (CSA) through sustainable irrigation management on resource use efficiency: a sustainable production alternative for cotton. Land Use Policy 88:104113

IPCC (2018) Global warming of 1.5 °C. An IPCC special report on the impacts of global warming of 1.5 °C above pre-industrial levels and related global greenhouse gas emission pathways, in the context of strengthening the global response to the threat of climate change, sustainable development, and efforts to eradicate poverty. In: V Masson-Delmotte, P Zhai, HO Pörtner, D Roberts, J Skea, PR Shukla, A Pirani, W Moufouma-Okia, C Péan, R Pidcock, S Connors, JBR Matthews, Y Chen, X Zhou, MI Gomis, E Lonnoy, T Maycock, M Tignor, T Waterfeld

Issahaku G, Abdulai A (2020) Adoption of climate-smart practices and its impact on farm performance and risk exposure among smallholder farmers in Ghana. Aust J Agric Resour Econ 64(2):324

Jamil I, Jun W, Mughal B, Raza MH, Imran MA, Waheed A (2021) Does the adaptation of climate-smart agricultural practices increase farmers' resilience to climate change? Environ Sci Pollut Res 28(21):27238–27249

Morton JF (2007) The impact of climate change on smallholder and subsistence agriculture. Proc Natl Acad Sci 104(50):19680–19685

Pachauri RK, Allen MR, Barros VR, Broome J, Cramer W, Christ R, Dubash NK et al (2014) Climate change: synthesis report. Contribution of Working Groups I, II and III to the fifth assessment report of the Intergovernmental Panel on Climate Change, p 151

Sardar A, Javed SA, Amir-ud-Din R (2016) Natural disasters and economic growth in Pakistan: an enquiry into the foods related hazards' triad. Working paper series; Centre for Environmental Economics and Climate Change (CEECC), Pakistan Institute of Development
Sardar A, Kiani AK, Kuslu Y (2021) Does adoption of climate-smart agriculture (CSA) practices improve farmers' crop income? Assessing the determinants and its impacts in Punjab province, Pakistan. Environ Dev Sustain 23(7):10119–10140

Progressive Efforts in the Implementation of Integrated Water Resources Management (IWRM) in Uganda

Benson Turyasingura, Rogers Akatwijuka, Wycliffe Tumwesigye, Natal Ayiga, Tabukeli Musigi Ruhiiga, Abhishek Banerjee, Brahim Benzougagh, and Denis Frolov

B. Turyasingura (✉)
Institute of Tourism and Hospitality, Kabale University, Kabale, Uganda
e-mail: bturyasingura@kab.ac.ug

R. Akatwijuka
Department of Environmental Sciences, Kabale University, Faculty of Agriculture and Environmental Sciences, Kabale University, Kabale, Uganda
e-mail: rakatwijuka@kab.ac.ug

W. Tumwesigye
Department of Economics and Environmental Management, Bishop Stuart University, Mbarara, Uganda
e-mail: wtumwesigye@faest.bsu.ac.ug

N. Ayiga
Department of Social Work and Social Administration, Faculty of Arts and Social Sciences, Kabale University, Kabale, Uganda
e-mail: nayiga@kab.ac.ug

T. M. Ruhiiga
Department of Environmental Sciences Kabale University, Faculty of Agriculture and Environmental Sciences, Kabale University, Kabale, Uganda
e-mail: rmusigi@kab.ac.ug

A. Banerjee
State Key Laboratory of Cryospheric Sciences, Northwest Institute of Eco-Environment and Resources, Chinese Academy of Sciences, Lanzhou, China
e-mail: banerjee.abhishek@nieer.ac.cn

B. Benzougagh
Geophysics and Natural Hazards Laboratory, Department of Geomorphology and Geomatics (D2G), Scientific Institute, Mohammed V University in Rabat, Avenue Ibn Batouta, Agdal, Rabat, Morocco
e-mail: brahim.benzougagh@is.um5.ac.ma

D. Frolov
Lomonosov Moscow State University, Geographical Faculty, Moscow, Russia

M. Ahmed, S. Ahmad (eds.), *Disaster Risk Reduction in Agriculture*, Disaster Resilience and Green Growth, https://doi.org/10.1007/978-981-99-1763-1_26

Abstract

Worldwide, water is essential in creating the environment and human life. Uganda's Vision 2040 promises general use and safe access to drinking water for all Ugandans. Today, 14% of the population in Uganda have access to potable managed water, and 78% of rural communities have access to the sanitation system, and by 2040, the number is expected to have doubled." If the population is to exercise its constitutional right to receive a dependable, safe, and inexpensive water supply, it is imperative that infrastructure for water supply be provided. We assessed the progressive efforts in the implementation of integrated water resources management in Uganda, challenges, opportunities, and polies for water management." Explorations for the keyword's "water" AND "water resources", AND "water management" were used in this study. A total of 125 papers were selected for this study from 2013 to 2022, from Scopus, Google Scholar, Science Direct, and Web of Science, that were found relevant for this study and were selected and discussed. According to findings, the United States, the United Kingdom, China, and India were the leading in publications. On the analysis of sources of journal by publications, the Science of The Total Environment was the leading. According to the analysis by the organizations, the leading was Wageningen University and Research, followed by the University of Chinese Academy of Sciences. It was concluded that the problems of effectiveness in water resource management is still a global change and little effort has been made. To bridge the gap from science to policymaking in the water resources nexus, it is vital to enhance stakeholder collaboration, water resource tradeoffs and regional development and preservation by optimization of the ecological water resource pattern.

Keywords

Water resources · Integrated water resources management · Water management · CiteSpace · VOSViewer · Web of science · Elsevier · Scopus · Google scholar · Uganda

1 Introduction

Integrated water resources management has become a standard for the control of surface, coastal, and groundwater. Different international bodies, inclusive of the European Union (EU), the Global Water Partnership, and the United Nations (UN), are highly involved in ensuring Integrated Water Resource Management (IWRM) principles are well incorporated. Worldwide, information is needed about the IWRM as there is no known studies that have documented the progressive effort in water status. With their continued support, countries worldwide have seen the importance of this and are now taking on reforms to execute the principles and reorganize their domestic regional water governance arrangements (Benson et al. 2015). Basin or catchment management approaches, stakeholder participation, public involvement, the allocation of water resources, and an integrated approach to the management of water are traditionally regarded as the key elements in understanding the need for IWRM. IWRM focuses on conquering the outlines of fragmentation in regard to functions, societal interests, and political institutions, and this has led to water governance arrangements that previously were regarded as not worthy and irrelevant. In general, IWRM hypothetically answers wicked problems. In Africa, IWRM has existed since the adoption of the Africa Water Vision 2025 and the Johannesburg Declaration on Sustainable Development in 2002, which in turn led to the preparation and implementation of IWRM plans and building capability is required to support the development of solutions that take the water into consideration as in line with the study conducted by Turyasingura et al. (2023). Through IWRM, there was promotion of sustainable water resource management after the establishment of numerous entities at national, sub-regional, and regional levels in Africa (Kwesiga et al. 2023), which has more than 80 shared river basins and lakes and at most 60 transboundary systems (Merems et al. 2020; Goentzel et al. 2022).

According to O'Hanlon and Inouye (2020), 14% of the population has access to potable managed water despite the Ugandan government's great efforts since the 1990s to provide safe drinking water to the people across the nation (Alex and Johnson 2019). This is in line with Marks et al. 2020, who stated that 78% of rural communities had access to some sort of sanitation system.

Water is essential in both the creation of the environment and human life. The illusion of inexhaustibility is put to the test, according to Tränckner et al. (2012), because only 2.5% (Grover 2012) of the water is freshwater and the rest is saltwater (Katamba 2018).

Uganda is known for having abundant water resources and is located almost entirely inside the Nile Basin (Conway 1996). Receiving a lot of aid since the early 1990s (Leal Filho 2019), the Rio de Janeiro Earth Summit in 1992 allowed the nation to start an IWRM process, following the lead of Agenda 21's Chap. "Role of Horticulture in Disaster Risk Management" on Integrated Water Resources Management. Nevertheless, some two decades later, minimal IWRM interventions are being implemented despite this conference and government support.

Until in the 1980s when only a few persons in the water career began to realize that the overall state of the world's water resources was not as excellent as it seemed (Clarke 2013). This sentiment grew stronger in the 1990s, as more professionals realized that the only effective multi-disciplinary, multi-institutional approach to solving water problems had evolved into multi-dimensional, multi-sectoral, and multi-regional scaling (Pandey et al. 2021).

Water should be regarded as an economic good because it has a financial value in all of its competing uses. In order to fully implement this principle, it is crucial to first acknowledge that all people have a fundamental right to inexpensive access to sanitary facilities and clean water. Water has historically been used in wasteful and environmentally harmful ways due to a failure to realize its economic value. A key strategy for promoting effective and fair use, encouraging conservation, and safeguarding water resources is to manage water as an economic good (Koudstaal et al. 1992).

Integrated water resources management has been widely adopted by governments, international organizations, and several other parties. Although the idea is not new, it has been into existence for two generations. It was "rediscovered" by several water experts in the early 1990s, and numerous funders and international organizations then strongly promoted it. At the Dublin Conference in 1992, the Global Water Partnership was created to support the idea of integrated water resources management as a primary driver. This was aimed at guiding the development and protection of water resources through integrated approaches making it receive more attention (Chapman and Sullivan 2022; Donkor and Chitakira 2020; Sadeghi Pasvisheh et al. 2021).

According to Alaerts and Kaspersma (2022), IWRM should be seen as a method rather than a singular action, one that is cumulative rather than linear in character. IWRM has no set boundaries as it works towards changing the current unsustainable types of water development and management systems without just relying on one effective administrative paradigm. Gaudermann (2022) noted that the skill of IWRM lies in choosing, modifying, and using the ideal combination of these tools for a particular circumstance.

2 Literature Review

2.1 Urban Water Supply in Uganda

According to Nyenje et al. (2022), 68% of the population was anticipated to be using an upgraded water source in June 2020, below the 69% observed in June 2019 (Campbell et al. 2022). Only 47% of the data for new water sources implemented by local governments and recorded in the water supply database are used to support the 68% performance estimate as of June 2020 (Onohuean et al. 2021).

Amugsi et al. (2017) noted that "more than 10 million people (or 24% of the country's population) are thought to live in urban areas in Uganda, and by 2040 that number is expected to have doubled." If the population is to exercise its constitutional right to receive a dependable, safe, and inexpensive water supply, it is imperative that infrastructure for water supply be provided. "The rate of urban population increase in Uganda is faster than the pace of infrastructural growth (Cohen 2006)." Badmos (2019) noted that the movement of people from rural to urban regions and the development of new districts, cities, municipalities, and town councils are in line with Benson and Ayiga (2022), which has allowed for the designation of formerly rural areas as urban areas.

There are 498 urban centers in Uganda, including 15 cities (14 of which were recently approved), 55 municipalities, 47 newly approved counties, and 442 town councils and town boards (Maniple 2015). There are 68,731 villages in Uganda, of which 16,154 are located in large towns, 14,494 in small towns or rural growth centers, and 38,083 in rural areas.

Due to growing water scarcity, achieving SDG 6 and other water-related targets in other SDGs is a significant problem in Uganda (Vörösmarty et al. 2018). In order to achieve water-related sustainable development, a variety of integrated water resources management is essential for countries and river basins as the majority of nations are not on track to achieve SDG 6 by the deadline set for 2030 (Swamy et al. 2018). In order to fully utilize the potential of water resources at a time when most nations, including Uganda, are not on track to achieve SDG 6 by 2030, coordination across water-scarce areas is essential (Allen et al. 2018).

According to Benson et al. (2020), SDG 6.5: Integrated water resources management at all scales, particularly when necessary through transboundary collaboration. The implementation of integrated water resources management in water-scarce nations as well as transboundary basins can be aided by the efficient use of water resources. This also ensures transboundary planning and actions to create an environment supportive of new approaches to unlocking the potential of unconventional water resources (Bertule et al. 2018). If these goals are achieved by fostering international cooperation and offering developing countries assistance in building their capacity for projects and initiatives involving water and sanitation, including technologies for water recycling and reuse, International cooperation is necessary to support initiatives that enable professionals in poor nations to improve their ability to take advantage of the promise of water resources management (Ngene et al. 2021).

According to Katusiime and Schütt (2020), the construction of an enabling environment (Akello 2007; Griffiths and Lambert 2013), comprising proper policies, strategies, and legislation, an institutional framework, and management tools (Agarwal et al. 2000), is necessary for the execution of an integrated water resources management approach while applying the four water principles, i.e. (1) policy, (2) river basin management, (3) stakeholder participation, and (4) equity of access (Fritsch and Benson 2013).

Integrated water resources management is essential for the effective and efficient management of water resources and must be facilitated by appropriate policies and institutional frameworks (Al-Jawad et al. 2019). New and retrofitted water infrastructure, such as surface reservoirs, multipurpose dams, soil moisture conservation techniques, natural wetlands, rainwater harvesting for storage, rainwater harvesting for infiltration, urban green spaces, conjunctive use of surface and groundwater, managed aquifer charge, and source water. Hence, watersheds are natural environmental and land management units that influence a country's overall health. Poor ecosystem management in watersheds has resulted in and will continue to result in reduced watershed functioning, which in sensitive areas can lead to ecosystem collapse (Pourghasemi et al. 2020).

According to Ofumbi (2020), open freshwater sources cover 15.3% of Uganda's land area (rivers, lakes, streams, and swamps). The nation's 43.3 km^3 of total renewable water resources (TRWR) are available. The range of 19.1–39.9 mm represents the average annual groundwater recharge, which is quite substantial. Although there has been worrisome degradation, wetlands have continued to offer domestic water valued at USD 34 million annually (Aayog 2020). Uganda, therefore, does not have a water shortage. According to the national land cover mapping from 2019, wetlands make up about 13% of the country's overall land cover. Only 8.9% of this remains unchanged, with homes, farms leading to degradation (Turyasingura et al. 2022a), and tree plantations degrading the remaining 4.1% (Merems et al. 2020).

Despite its wealth of natural resources, Uganda still has problems with its water supply, including river and dam silting and water pollution (Mugume et al. 2021). The nation still faces problems with its water resources, such as encroachment on catchment areas and pollution from bacteria and chemicals that contaminate both surface and groundwater as a result of poor sanitation facilities, unsafe municipal and industrial waste disposal, and subpar farming practices combined with the deterioration of wetland and catchment areas (Qadir et al. 2022). Pollution has driven up the price of water treatment, making water provision currently among the most expensive in the area. Water stress in some locations is being exacerbated by the rate at which rivers and lakes are silting up.

Similar to this, the water crisis that is being increasingly reported around the globe is characterized by inefficient governance of water resources. As a result, they advocate integrating water resources management strategies as a fix (UNESCO 2006). Due to the apparent difficulties now attributed to the numerous components within the water resources management sector, Uganda's success in "the implementation efforts of Integrated Water Resources Management as a concept is discernible," albeit subjective (Nations), 2006). To track the progress so far made and over time, the cross-cutting criteria required when implementing the IWRM, the targets and aims the idea is aimed at attaining can be used to discuss the efforts made by the Government of Uganda to implement IWRM to ease diversity at the landscape level as in line with (Turyasingura et al. 2022b). However, their study never mentioned the relationship between water resources and diversity in landscape to ease conservation.

It's important to note that when the IWRM approach first gained popularity in the early 1990s, the nation was still experiencing political unrest, and the government was more concerned with restoring peace and democracy, guided by the NRA's 10-point program, which did not prioritize natural resource development and sustainability (Ebeku 2007; Turyasingura and Chavula 2022).

The government's desire to participate in and adapt to international policies that either directly or indirectly relate to integrated water resources management has further reinforced Uganda's legal and policy framework. The Ugandan government joined the Parish Agreement on climate action in 2015 (Ampaire et al. 2017). Parties to the agreement vowed to prioritize protecting the integrity of all ecosystems, including water bodies, and to recognize the importance of conservation (Act 1999).

According to Turyasingura and Chavula 2022, the nationwide efforts need to be made to protect and restore degraded wetlands, as well as to demarcate the boundaries of key wetlands to act as water sources. The pace of wetland degradation, however, is more than 70 times that of restoration (Wang et al. 2020). The National Development Plans (NDP II, III) made improving management of land and water resources, environmental protection, climate change, and household incomes a top priority. The NDP III places a high priority on the need to ensure that there are sufficient and dependable freshwater resources of high quality for all uses; to increase the area covered by trees, wetlands, and other vegetative cover; and to encourage inclusive, low-emission development at all scales in order to promote a resilient climate.

Government budgets and plans have evolved over time to more closely align with environmental and climate change concerns and incorporate these viewpoints into their sectoral development initiatives. To improve integrated water resources management, water and environmental conservation organizations have received greater financing and investment over time (Borchardt et al. 2016).

Although there are several studies undertaken on the integrated water resources, data on the progressive efforts on the integrated water resources management in Uganda is still missing. Still, there is vivid evidence on poor water resources, which can be attributed to limited knowledge and information on the inter-dependence between climate, human activities, and water resources hence, the need for the study.

3 Materials and Methods

3.1 Study Area

Uganda served as the site of the study to ease an understanding on water resources progress and its management practices. Uganda is located in southeast Africa between latitudes 1° N and 4° N and longitudes 30° E and 35° E (Bolwig 2002; Hernández et al. 2021) (Fig. 1). Lakes Albert, Kyoga, and Edward are among the several lakes found in Uganda. Uganda has an equatorial climate, which means it receives a lot of sunlight. In the southwest highlands, the average annual temperature is 16 °C, whereas it is 25 °C in the northwest and frequently exceeds 30 °C in the northeast (Tamale et al. 2022). While there are two rainy seasons in the south of Uganda, rain falls frequently in the northeast. In contrast, it rarely rains in the north from April to October, although it is dry from November to March.

3.1.1 Subsection Publication Per Year

Scientific mapping, a bibliometrics auxiliary technique, offers a spatial depiction of network structures. Scientific mapping involves the interdisciplinary subjects of applied science, including environmental sciences. In addition, with the recent rapid advancement of computer science, several academics have examined the probable dynamic mechanisms of discipline evolution using various science mapping methods (Shah et al. 2019). Two potent and useful tools for science map analysis are CiteSpace and VOS-viewer (Attar et al. 2022). These programs may generate and display a term cooccurrence network in a variety of areas where climate change has an impact on soil and water resources. This study chose VOSviewer software in comparison to other visualization tools (Nadi-Ravandi and Batooli 2022), which was crucial because it is used to handle massive data and create illustrations that more clearly illustrate the hotspots and research issues. This research analysed data from (2013–2022) to ease visualization and trend analysis.

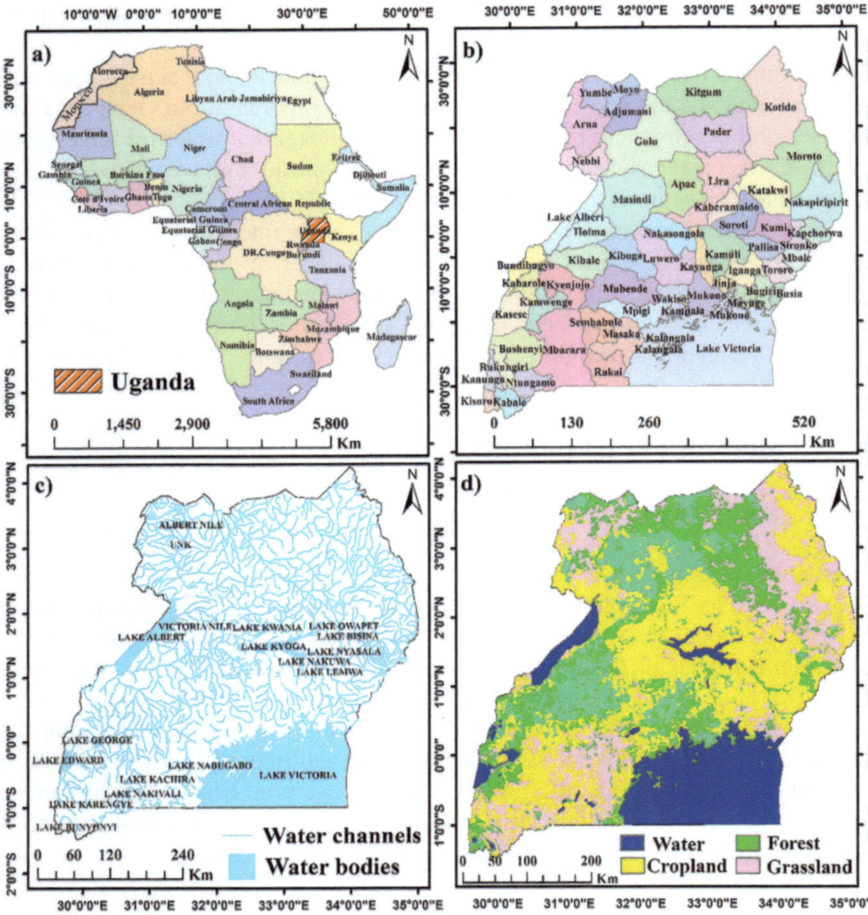

Fig. 1 (**a**) Location of Uganda, (**b**) administrative divisions of Uganda, (**c**) spatial distribution of major water bodies and channels, (**d**) major land use classification of Uganda (Hernández et al. 2021)

3.1.2 Literature Selection, Screening and Extraction

A total of 50 papers were selected for this study from 2010 to 2022, from Scopus, Google Scholar, Science Direct, and Web of Science, that were found relevant for this study and were selected and discussed. However, there is no single database which is sufficient to cover all relevant articles and information, and that's why different databases were used. Search terms included "Water", "integrated water management", AND "water resources", with a filter of mainly year of publication. Additional search terms were included to identify relevant papers that did not directly use the above terms. Selected documents were mainly peer-reviewed journal articles, a limited number of books, reports and online resources, including companies and organization reports and bulletins. From the retrieved papers' summaries,

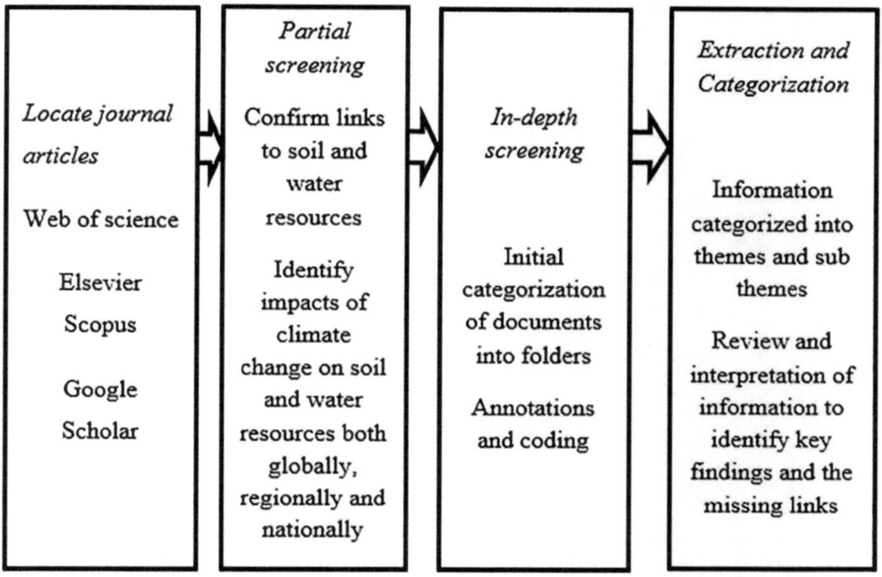

Fig. 2 Methodological approach

new graphics and tabulations were made to make the discussions and uncover gaps that need to be addressed. In this study, a methodological approach (Fig. 2) was used.

4 Results and Discussion

4.1 Author Trend Analysis in Uganda

The author captures the core idea of the article as it helps to quickly grasp the hot topics in the field. In this study, VOSviewer software was used to visualize authors (Huang et al. 2022). Nodes in the knowledge map represent keywords. The larger the node is, the higher the frequency is, and the lines between nodes represent the cooccurrence of particular authors (Nandiyanto and Al Husaeni 2021). In addition, in the VOSviewer knowledge map, different colors represent different clusters, and the same color represents the same cluster (Fig. 3).

4.2 Analysis of Country Network (ACN)

Figures 4 and 5 show that the number of articles published in United states, United Kingdom, China, and India is much more than in other nations. This is in line with Tang et al. (2021), who said that country network analysis helps to understand the most active countries in publishing. As a result, linkages between nodes are many

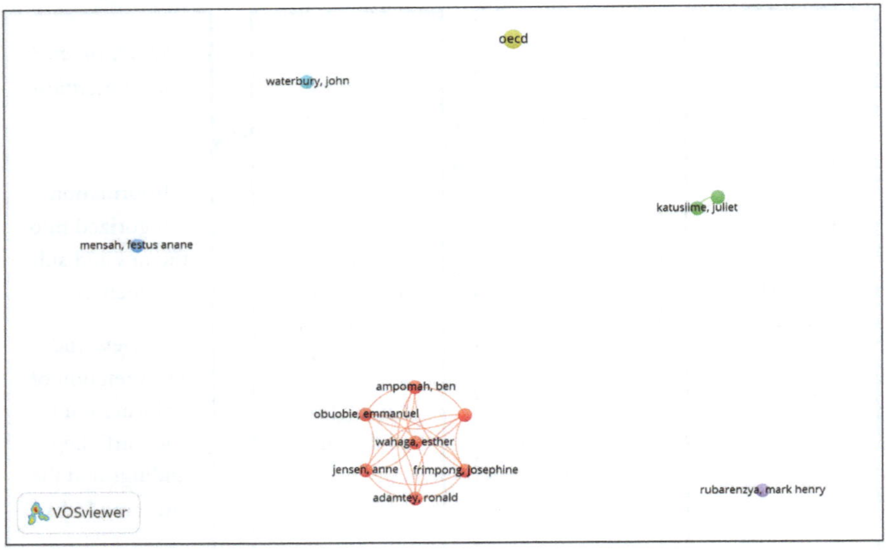

Fig. 3 Network analysis of the author trends

and intricate, suggesting that various nations frequently engage in cooperative relationships. The United States is positioned near the leading edge of a few nodes in Figs. 3 and 4, which also suggests that the node holds a significant position within the network structure.

4.3 Analysis of Sources of Journal Publication

Figure 6 shows that the Science of the total environment was the leading in publication, followed by integrated environmental issues, and the least being water research. This suggests that the node holds a significant position within the network structure among publishing houses.

4.4 Analysis by the Organizations

According to the analysis, the leading organization was Wageningen University and research, followed by the University of Chines Academy (Fig. 7). This was because these Universities publish in indexed journals than other Organizations across the world.

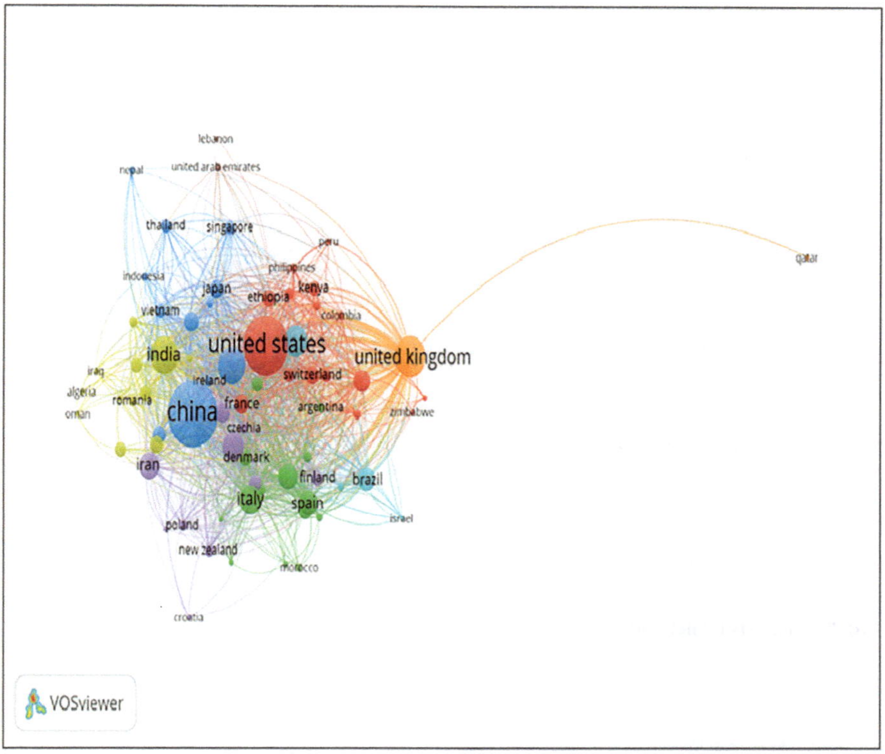

Fig. 4 Analysis by network

5 Conclusion

Despite sector changes and the use of integrated water resources management methodologies, Uganda still has problems with the effectiveness of its water resource management. The obstacles to effective water management include institutional funding and capacity, sector cooperation, management techniques, policy execution, and legal compliance. According to findings, the United States, the United Kingdom, China, and India were the leading in publications by the country networks. On the analysis of sources of journal by publications, the Science of the total environment was the leading. According to the analysis by the organizations, the leading organization was Wageningen University and research, followed by the University of Chines Academy of Sciences.

The work was performed in accordance with the state budget theme "Danger and risk of natural processes and phenomena" (121051300175-4) and "Evolution of the cryosphere under climate change and anthropogenic impact" (121051100164-0).
Ethics Approval and Consent to ParticipateNot applicable.

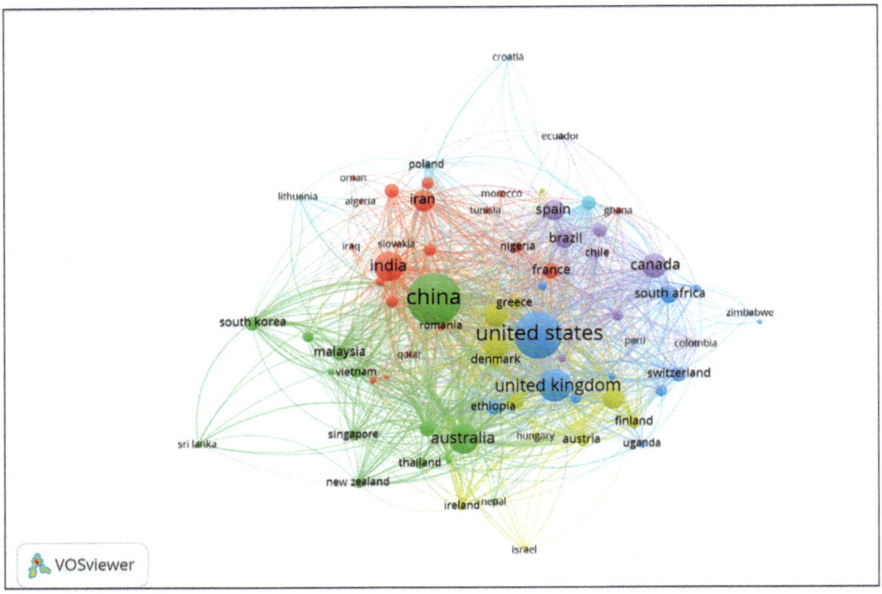

Fig. 5 Analysis by network

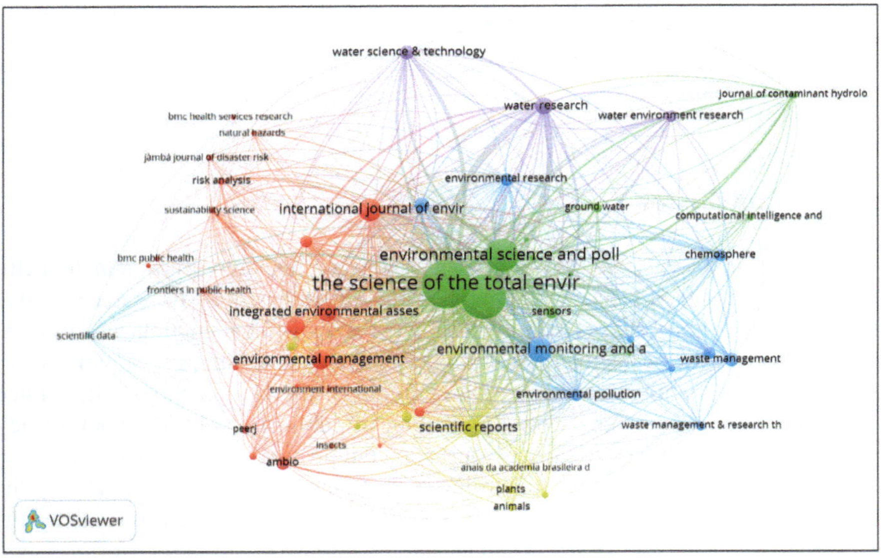

Fig. 6 Journals of publication

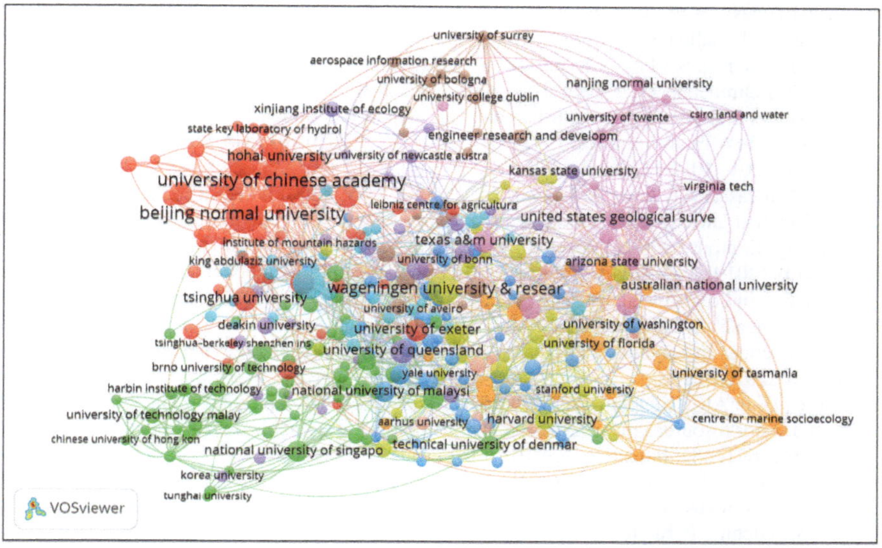

Fig. 7 Analysis by the organizations

Consent for Publication Not applicable.

Competing Interests The authors declare no competing interests.

References

Aayog N (2020) Niti Aayog annual report 2019-20 about policy and programme. Government of India, New Delhi

Act E (1999) Environment protection and biodiversity conservation act. Commonwealth of Australia, Canberra

Agarwal A, de los Angeles MS, Bhatia R, Chéret I, Davila-Poblete S, Falkenmark M, Villarreal FG, Jønch-Clausen T, Kadi MA, Kindler J (2000) Integrated water resources management. Global Water Partnership, Stockholm

Akello CE (2007) Environmental regulation in Uganda: successes and challenges. Law Env't Dev J 3:20

Alaerts GJ, Kaspersma JM (2022) Facing global transitions in water management: advances in knowledge and capacity development and towards adaptive approaches. Water Policy 24(5):685–707. https://doi.org/10.2166/wp.2022.301

Al-Jawad JY, Alsaffar HM, Bertram D, Kalin RM (2019) A comprehensive optimum integrated water resources management approach for multidisciplinary water resources management problems. J Environ Manag 239:211–224. https://doi.org/10.1016/j.jenvman.2019.03.045

Allen C, Metternicht G, Wiedmann T (2018) Initial progress in implementing the sustainable development goals (SDGs): a review of evidence from countries. Sustain Sci 13(5):1453–1467. https://doi.org/10.1007/s11625-018-0572-3

Alex S, Johnson, R (2019) Analysis of bacteriological quality of domestic water sources in Kabale municipality, Western Uganda. http://hdl.handle.net/20.500.12493/150

Ampaire EL, Jassogne L, Providence H, Acosta M, Twyman J, Winowiecki L, Van Asten P (2017) Institutional challenges to climate change adaptation: a case study on policy action gaps in Uganda. Environ Sci Policy 75:81–90. https://doi.org/10.1016/j.envsci.2017.05.013

Amugsi DA, Dimbuene ZT, Mberu B, Muthuri S, Ezeh AC (2017) Prevalence and time trends in overweight and obesity among urban women: an analysis of demographic and health surveys data from 24 African countries, 1991-2014. BMJ Open 7(10):e017344. https://doi.org/10.1136/bmjopen-2017-017344. PMid: 29079606. PMCid: PMC5665233

Attar H, Abu-Jassar AT, Amer A, Lyashenko V, Yevsieiev V, Khosravi MR (2022) Control system development and implementation of a CNC laser engraver for environmental use with remote imaging. Comput Intell Neurosci 2022. https://doi.org/10.1155/2022/9140156

Badmos OS (2019) An integrated remote sensing and urban growth model approach to curb slum formation in Lagos megacity. Universitäts-und Landesbibliothek Bonn, Bonn

Benson T, Ayiga N (2022) Classifying the involvement of men and women in climate smart agricultural practices in Kayonza sub-county, Kanungu District, Uganda. https://doi.org/10.11648/j.ijees.20220701.12

Benson D, Gain AK, Giupponi C (2020) Moving beyond water centricity? Conceptualizing integrated water resources management for implementing sustainable development goals. Sustain Sci 15(2):671–681. https://doi.org/10.1007/s11625-019-00733-5

Benson D, Gain AK, Rouillard JJ (2015) Water governance in a comparative perspective: from IWRM to a'nexus' approach? Water Altern 8(1):756–773

Bertule M, Glennie P, Koefoed Bjørnsen P, James Lloyd G, Kjellen M, Dalton J, Rieu-Clarke A, Romano O, Tropp H, Newton J (2018) Monitoring water resources governance progress globally: experiences from monitoring SDG indicator 6.5. 1 on integrated water resources management implementation. Water 10(12):1744. https://doi.org/10.3390/w10121744

Bolwig S (2002) Land use change and soil degradation in the southwestern highlands of Uganda. A contribution to the Strategic Criteria for Rural Investments in Productivity (SCRIP) program of the USAID Uganda Mission. The International Food Policy Research Institute, Washington, DC

Borchardt D, Bogardi JJ, Ibisch RB (2016) Integrated water resources management: concept, research and implementation. Springer, Cham. https://doi.org/10.1007/978-3-319-25071-7

Campbell LS, Masquillier C, Knight L, Delport A, Sematlane N, Dube LT, Wouters E (2022) Stay-at-home: the impact of the COVID-19 lockdown on household functioning and ART adherence for people living with HIV in three sub-districts of cape town, South Africa. AIDS Behav 26(6):1905–1922. https://doi.org/10.1007/s10461-021-03541-0. PMid: 34977957. PMCid: PMC8720535

Chapman DV, Sullivan T (2022) The role of water quality monitoring in the sustainable use of ambient waters. One Earth 5(2):132–137. https://doi.org/10.1016/j.oneear.2022.01.008

Clarke R (2013) Water: the international crisis, 1st edn. Routledge, New York, NY. https://doi.org/10.4324/9781315070261

Cohen D (2006) Trois leçons sur la société post-industrielle. Seuil, Paris, p 90

Conway MA (1996) Autobiographical knowledge and autobiographical memories. https://doi.org/10.1017/CBO9780511527913.003

Donkor FK, Chitakira M (2020) The nexus of water, sanitation, and hygiene (WASH) and sustainable development goals. In: Clean water and sanitation. Springer International Publishing, Cham, pp 1–10. https://doi.org/10.1007/978-3-319-70061-8_175-1

Ebeku KSA (2007) Constitutional right to a healthy environment and human rights approaches to environmental protection in Nigeria: Gbemre v. Shell revisited. Rev Eur Comp Int Environ Law 16(3):312–320. https://doi.org/10.1111/j.1467-9388.2007.00570.x

Fritsch O, Benson D (2013) Integrating the principles of integrated water resources management? River basin planning in England and Wales. Int J Water Gov 1(3-4):265–284. https://doi.org/10.7564/13-IJWG7

Gaudermann E (2022) Advancing private sector engagement in integrated water resources management

Goentzel J, Blair C, Russell T, Wiseman M, Gralla E, Steinberg S, Wetmore F (2022) Uganda farmer market engagement study final report

Griffiths J, Lambert R (2013) Free flow: reaching water security through cooperation. Unesco, Paris
Grover VI (2012) Impact of climate change on water and health. Crc Press, Boca Raton. https://doi.org/10.1201/b14323
Hernández JL, Nan D, Fernandez-Ayala M, García-Unzueta M, Hernández-Hernández MA, López-Hoyos M et al (2021) Vitamin D status in hospitalized patients with SARS-CoV-2 infection. J Clin Endocrinol Metabol 106(3):e1343–e1353. https://doi.org/10.1210/clinem/dgaa733
Huang HY, Lin YCD, Cui S, Huang Y, Tang Y, Xu J et al (2022) miRTarBase update 2022: an informative resource for experimentally validated miRNA–target interactions. Nucleic Acids Res 50(D1):D222–D230. https://doi.org/10.1093/nar/gkab1079
Katamba E (2018) The performance of Nile tilapia (Oreochromis niloticus) fingerlings at varying salinity levels. Makerere University, Kampala
Katusiime J, Schütt B (2020) Integrated water resources management approaches to improve water resources governance. Water 12(12):3424. https://doi.org/10.3390/w12123424
Koudstaal R, Rijsberman FR, Savenije H (1992) Water and sustainable development. Nat Res Forum 16(4):277–290
Kwesiga G, Greese J, Kelling A, Sperlich E, Schmidt B (2023) The Suzuki-Miyaura cross-coupling-Claisen rearrangement-cross-metathesis approach to Prenylated Isoflavones. J Org Chem 88(3):1649–1664. https://doi.org/10.1021/acs.joc.2c02698
Leal Filho W (2019) Encyclopedia of sustainability in higher education. Springer, Cham. https://doi.org/10.1007/978-3-030-11352-0
Maniple EB (2015) "I found myself staying"-A case study of the job embeddedness and retention of qualified health workers in rural and remote areas of Uganda. Doctoral dissertation, Royal College of Surgeons in Ireland
Marks SJ, Clair-Caliot G, Taing L, Bamwenda JT, Kanyesigye C, Rwendeire NE, Kemerink-Seyoum JS, Kansiime F, Batega DW, Ferrero G (2020) Water supply and sanitation services in small towns in rural-urban transition zones: the case of Bushenyi-Ishaka municipality, Uganda. NPJ Clean Water 3(1):1–9. https://doi.org/10.1038/s41545-020-0068-4
Merems JL, Shipley LA, Levi T, Ruprecht J, Clark DA, Wisdom MJ et al (2020) Nutritional-landscape models link habitat use to condition of mule deer (Odocoileus hemionus). Front Ecol Evol 8:98. https://doi.org/10.3389/fevo.2020.00098
Mugume I, Semyalo R, Wasswa P, Ngailo T, Odongo RI, Lunyolo J, Tao S (2021) Community views on water demands under a changing climate: the case of river Mpanga water catchment, Western Uganda. Afr J Environ Sci Technol 15(9):371–378. https://doi.org/10.5897/AJEST2021.3036
Nadi-Ravandi S, Batooli Z (2022) Gamification in education: a scientometric, content and co-occurrence analysis of systematic review and meta-analysis articles. Educ Inf Technol 27(7):10207–10238. https://doi.org/10.1007/s10639-022-11048-x
Nandiyanto ABD, Al Husaeni DF (2021) A bibliometric analysis of materials research in Indonesian journal using VOSviewer. J Eng Res. https://doi.org/10.36909/jer.ASSEEE.16037
Ngene BU, Nwafor CO, Bamigboye GO, Ogbiye AS, Ogundare JO, Akpan VE (2021) Assessment of water resources development and exploitation in Nigeria: a review of integrated water resources management approach. Heliyon 7(1):e05955. https://doi.org/10.1016/j.heliyon.2021.e05955. PMid: 33521352. PMCid: PMC7820563
Nyenje PM, Ocoromac D, Tumwesige S, Ascott MJ, Sorensen JPR, Newell AJ, Macdonald DMJ, Gooddy DC, Tindimugaya C, Kulabako RN (2022) Hydrogeology of an urban weathered basement aquifer in Kampala, Uganda. Hydrgeol J 30:1–19. https://doi.org/10.1007/s10040-022-02474-9
O'Hanlon S, Inouye SK (2020) Delirium: a missing piece in the COVID-19 pandemic puzzle. Age Ageing 49(4):497–498. https://doi.org/10.1093/ageing/afaa094
Ofumbi, M. (2020). Water and environment sector performance report 2020, Ministry of Water and Environment,. September 2
Onohuean H, Okoh AI, Nwodo UU (2021) Epidemiologic potentials and correlational analysis of vibrio species and virulence toxins from water sources in greater Bushenyi districts, Uganda.

Sci Rep 11(1):1–16. https://doi.org/10.1038/s41598-021-01375-3. PMid:34789791. PMCid: PMC8599681

Pandey VP, Shivakoti BR, Shrestha S, Wiberg D (2021) Localizing and mainstreaming global initiatives on water, climate change and sustainable development. In: Pandey VP et al (eds) Water, climate change, and sustainability. John Wiley & Sons, Hoboken, NJ, pp 1–19. https://doi.org/10.1002/9781119564522.ch1

Pourghasemi HR, Pouyan S, Heidari B, Farajzadeh Z, Shamsi SRF, Babaei S et al (2020) Spatial modeling, risk mapping, change detection, and outbreak trend analysis of coronavirus (COVID-19) in Iran (days between February 19 and June 14, 2020). Int J Infect Dis 98:90–108. https://doi.org/10.1016/j.ijid.2020.06.058

Qadir M, Smakhtin V, Koo-Oshima S, Guenther E (2022) Global water scarcity and unconventional water resources. In: Unconventional water resources. Springer, Cham, pp 3–17. https://doi.org/10.1007/978-3-030-90146-2_1

Sadeghi Pasvisheh R, Eurie Forio MA, Ho LT, Goethals PLM (2021) Evidence-based management of the Anzali wetland system (northern Iran) based on innovative monitoring and modeling methods. Sustainability 13(10):5503. https://doi.org/10.3390/su13105503

Shah A, Haq S, Rehman W, Waseem M, Shoukat S, Rehman MU (2019) Photocatalytic and antibacterial activities of paeonia emodi mediated silver oxide nanoparticles. Mater Res Express 6(4):045045. https://doi.org/10.1088/2053-1591/aafd42

Swamy L, Drazen E, Johnson WR, Bukoski JJ (2018) The future of tropical forests under the United Nations sustainable development goals. J Sustain For 37(2):221–256. https://doi.org/10.1080/10549811.2017.1416477

Tamale BN, Bulafu D, Isunju JB, Jamu AV, Baguma JN, Tigaiza A et al (2022) Pregnancy-related complications and associated factors among women attending antenatal care at a specialised maternal and child health national referral hospital, in Uganda. medRxiv, 2022-07. https://doi.org/10.1101/2022.07.29.22278187

Tang Y, Feng W, Chen Z, Nong Y, Guan S, Sun J (2021) Fracture behavior of a sustainable material: recycled concrete with waste crumb rubber subjected to elevated temperatures. J Clean Prod 318:128553. https://doi.org/10.1016/j.jclepro.2021.128553

Tränckner J, Helm B, Blumensaat F, Terekhanova T (2012) Integrated water resources management: approach to improve river water quality in the western bug river basin. In: Nalecz T (ed) Transboundary aquifers in the eastern borders of the European Union. Springer, Dordrecht, pp 61–78. https://doi.org/10.1007/978-94-007-3949-9_6

Turyasingura B, Ayiga N, Benzougagh B (2022a) Re-thinking on land degradation and its impacts on livelihoods of the farmers in Kanungu District, Uganda. https://doi.org/10.21203/rs.3.rs-1966742/v1

Turyasingura B, Mwanjalolo M, Ayiga N (2022b) Diversity at landscape level to increase resilience. A review. East Afr J Environ Nat Res 5(1):174–181. https://doi.org/10.37284/eajenr.5.1.723

Turyasingura B, Atuhaire A, Jennifer T, Rogers A (2023) A literature review of climate-smart landscapes as a tool in soil-water management in Sub-Saharan Africa. Int Res J Multidiscip Tech 5(2):18. https://doi.org/10.54392/irjmt2322

Turyasingura B, Chavula P (2022) Climate-smart agricultural extension service innovation approaches in Uganda. Int J Food Sci Agric. https://doi.org/10.26855/ijfsa.2022.03.006

UNESCO (2006) UNESCO institute for lifelong learning (UIL): annual report, 2006, 64p

Vörösmarty CJ, Osuna VR, Cak AD, Bhaduri A, Bunn SE, Corsi F, Gastelumendi J, Green P, Harrison I, Lawford R (2018) Ecosystem-based water security and the sustainable development goals (SDGs). Ecohydrol Hydrobiol 18(4):317–333. https://doi.org/10.1016/j.ecohyd.2018.07.004

Wang X, Tan L, Wang X, Liu W, Lu Y, Cheng L, Sun Z (2020) Comparison of nasopharyngeal and oropharyngeal swabs for SARS-CoV-2 detection in 353 patients received tests with both specimens simultaneously. Int J Infect Dis 94:107–109. https://doi.org/10.1016/j.ijid.2020.04.023

MIX
Papier aus verantwortungsvollen Quellen
Paper from responsible sources
FSC® C105338

If you have any concerns about our products,
you can contact us on
ProductSafety@springernature.com

In case Publisher is established outside the EU,
the EU authorized representative is:
Springer Nature Customer Service Center GmbH
Europaplatz 3, 69115 Heidelberg, Germany

Printed by Libri Plureos GmbH
in Hamburg, Germany